LIBRARY
NSCC, AKERLEY CAMPUS
21 WOODLAWN RD.
DARTMOUTH, NS B2W 2R7 CANADA

D1610980

HANDBOOK OF RECYCLING

HANDBOOK OF RECYCLING

STATE-OF-THE-ART FOR PRACTITIONERS, ANALYSTS, AND SCIENTISTS

Edited by

ERNST WORRELL AND MARKUS A. REUTER

AMSTERDAM • BOSTON • HEIDELBERG • LONDON
NEW YORK • OXFORD • PARIS • SAN DIEGO
SAN FRANCISCO • SYDNEY • TOKYO

ELSEVIER

Elsevier
225 Wyman Street, Waltham, MA 02451, USA
The Boulevard, Langford Lane, Kidlington, Oxford OX5 1GB, UK
Radarweg 29, PO Box 211, 1000 AE Amsterdam, The Netherlands

Copyright © 2014 Elsevier Inc. All rights reserved.

No part of this publication may be reproduced, stored in a retrieval system or transmitted in any form or by any means electronic, mechanical, photocopying, recording or otherwise without the prior written permission of the publisher

Permissions may be sought directly from Elsevier's Science & Technology Rights Department in Oxford, UK: phone (+44) (0) 1865 843830; fax (+44) (0) 1865 853333; email: permissions@elsevier.com. Alternatively you can submit your request online by visiting the Elsevier web site at http://elsevier.com/locate/permissions, and selecting *Obtaining permission to use Elsevier material*

Library of Congress Cataloging-in-Publication Data
Worrell, Ernst.
  Handbook of recycling : state-of-the-art for practitioners, analysts, and scientists/Ernst Worrell and Markus A. Reuter.
    pages cm
  Includes bibliographical references.
1. Recycling (Waste, etc.)–Handbooks, manuals, etc. I. Reuter, M. A. II. Title.
TD794.5.W69 2014
628.4'458–dc23
                              2014001188

British Library Cataloguing-in-Publication Data
A catalogue record for this book is available from the British Library.

ISBN: 978-0-12-396459-5

For information on all Elsevier publications visit our website at store.elsevier.com

Printed in the United States of America

14 15 16 17   10 9 8 7 6 5 4 3 2 1

  Working together to grow libraries in developing countries

www.elsevier.com • www.bookaid.org

Contents

List of Contributors xi

I
RECYCLING IN CONTEXT

1 Recycling: A Key Factor for Resource Efficiency
ERNST WORRELL, MARKUS A. REUTER

2 Definitions and Terminology
ERNST WORRELL, MARKUS A. REUTER

2.1 Introduction 9
2.2 Defining Recycling 10
2.3 Materials and Products 12
2.4 Applying the Product-Centric Approach—Metals 13
References 15

3 Recycling in Context
T.E. GRAEDEL, BARBARA K. RECK

3.1 Introduction 17
3.2 Metal Recycling Considerations and Technologies 17
3.3 Defining Recycling Statistics 19
3.4 Process Efficiencies and Recycling Rate Constraints 22
3.5 Perspectives on Current Recycling Statistics 23
3.6 Summary 25
References 25

4 Recycling Rare Metals
ROBERT U. AYRES, GARA VILLALBA MÉNDEZ, LAURA TALENS PEIRÓ

4.1 Introduction 27
4.2 Indium 28
4.3 Other Examples of Rare Metals 31
4.4 The Distant Future: Georgescu's Last Laugh? 36
References 37

5 Theory and Tools of Physical Separation/Recycling
KARI HEISKANEN

5.1 Recycling Process 40
5.2 Particle Size 40
5.3 Pulp Rheology 44
5.4 Properties and Property Spaces 45
5.5 Sampling 47
5.6 Mass Balances and Process Dynamics 48
5.7 Material Balancing 51
5.8 Liberation 55
5.9 Grade-Recovery Curves 57
References 61

II
RECYCLING - APPLICATION & TECHNOLOGY

6 Recycling of Steel
BO BJÖRKMAN, CAISA SAMUELSSON

6.1 Introduction 65
6.2 Scrap Processing and Material Streams from Scrap Processing 68
6.3 The Processes Used for Smelting Steel Scrap 70
6.4 Trends in Quality of the Scrap Available for Steel Production 72
6.5 Hindrances for Recycling—Tramp Elements 74
6.6 Purification of Scrap 76
6.7 To Live with Impurities 79
6.8 Measures to Secure Sustainable Recycling of Steel 79
References 81

v

7 Copper Recycling
CAISA SAMUELSSON, BO BJÖRKMAN

7.1 Introduction   85
7.2 Raw Material for Copper Recycling   86
7.3 Processes for Recycling   87
7.4 Challenges in Copper Recycling   91
7.5 Conclusions   93
References   94

8 Lead Recycling
BART BLANPAIN, SANDER ARNOUT, MATHIAS CHINTINNE, DOUGLAS R. SWINBOURNE

8.1 Introduction   95
8.2 The Lead-Acid Battery   96
8.3 Battery Preprocessing   97
8.4 Smelting   100
8.5 Alternative Approaches   107
8.6 Refining   107
8.7 Conclusions and Outlook   109
References   110

9 Zinc and Residue Recycling
JÜRGEN ANTREKOWITSCH, STEFAN STEINLECHNER, ALOIS UNGER, GERNOT RÖSLER, CHRISTOPH PICHLER, RENE RUMPOLD

9.1 Introduction   113
9.2 Zinc Oxide Production from Drosses   114
9.3 Electric Arc Furnace Dust and Other Pb, Zn, Cu-containing Residues   116
9.4 Zinc Recycling from Copper Industry Dusts   119
9.5 Fuming of Slags from Lead Metallurgy   121
References   123

10 Recycling of Rare Metals
ELINOR ROMBACH, BERND FRIEDRICH

10.1 Precious Metals   126
10.2 Rare Earth Metals   132
10.3 Electronic Metals   137
10.4 Refractory Metals (Ferro-alloys Metals, Specialty Metals)   140
10.5 Other Metals   145
References   148

11 Recycling of Lumber
AMBROSE DODOO, LEIF GUSTAVSSON, ROGER SATHRE

11.1 Introduction   151
11.2 Background   151
11.3 Key Issues in Post-use Management of Wood   154
11.4 Case Study Scenarios   157
11.5 Summary   161
References   162

12 Paper Recycling
HARALD GROSSMANN, TONI HANDKE, TOBIAS BRENNER

12.1 Important Facts about Paper Recycling   165
12.2 Stock Preparation for Paper Recycling   169
References   178

13 Plastic Recycling
LI SHEN, ERNST WORRELL

13.1 Introduction   179
13.2 Use of Plastics   180
13.3 Plastic Recycling   182
13.4 Mechanical Recycling   184
13.5 Impact of Recycling   186
13.6 Conclusions and Outlook   189
References   189
Further Reading   190

14 Glass Recycling
THOMAS D. DYER

14.1 Introduction   191
14.2 Types of Glass   191
14.3 Glass Manufacture   192
14.4 Glass Recovery for Reuse and Recycling   192
14.5 Reuse of Glass   194
14.6 Closed-Loop Recycling of Glass   194
14.7 Environmental Benefits of Closed-Loop Recycling of Glass   195
14.8 The Growth of Glass Recycling   196
14.9 Open-Loop Glass Recycling   198
14.10 Conclusions   206
References   206

15 Textile Recycling
JANA M. HAWLEY

15.1 Introduction 211
15.2 The Recycling Effort 212
15.3 Export of Secondhand Clothing 213
15.4 Conversion to New Products 213
15.5 Conversion of Mattresses 214
15.6 Conversion of Carpet 215
15.7 Wipers 215
15.8 Landfill and Incineration 216
15.9 Diamonds 216
15.10 Summary 217
References 217

16 Cementitious Binders Incorporating Residues
YIANNIS PONTIKES, RUBEN SNELLINGS

16.1 Introduction 219
16.2 Clinker Production: Process Flow, Alternative Fuels and Alternative Raw Materials 220
16.3 From Clinker to Cement: Residues in Blended Cements 222
16.4 Alternative Cements for the Future: Reducing the CO_2 Footprint while Incorporating Residues 225
16.5 Conclusions 227
References 227

17 Industrial By-products
JAANA SORVARI, MARGARETA WAHLSTRÖM

17.1 What is a By-product? 231
17.2 Major By-products and Their Generic Properties 232
17.3 Where and How to Use By-products 232
17.4 Technical and Environmental Requirements 247
17.5 Concluding Remarks 251
References 252

18 Recovery of Metals from Different Secondary Resources (Waste)
STEFAN LUIDOLD, HELMUT ANTREKOWITSCH

18.1 Introduction 255
18.2 Production of Ferroalloys from Waste 257
18.3 Recycling Concepts for Rare Earth Containing Magnets 265
References 267

19 Recycling of Carbon Fibers
SORAIA PIMENTA, SILVESTRE T. PINHO

19.1 Introduction 269
19.2 Carbon Fiber Recycling Processes 270
19.3 Composites Remanufacturing 274
19.4 Applications for Recycled Carbon Fibers and Composites 277
19.5 Life-Cycle Analysis of Carbon Fiber Reinforced Polymers 278
19.6 Further Challenges 278
19.7 Conclusions 281
References 281

20 Recycling of Construction and Demolition Wastes
VIVIAN W.Y. TAM

20.1 Introduction 285
20.2 The Existing Low-Cost Housing Technologies 286
20.3 Earth/Mud Building 287
20.4 Prefabrication Method 287
20.5 Lightweight Foamed or Cellular Concrete Technology 288
20.6 Stabilized Earth Brick Technology 288
20.7 Case Study 289
20.8 Cost-Effectiveness of Using Low-Cost Housing Technologies 291
20.9 Recycling Technologies and Practice 294
20.10 Conclusion 294
References 294

21 Recycling of Packaging
ERNST WORRELL

21.1 Introduction 297
21.2 Packaging Waste 297
21.3 Composition 300
21.4 Recovery and Recycling 301
21.5 Recovery and Collection Schemes 302
21.6 Concluding Remarks 305
References 305
Further Reading 306

22 Material-Centric (Aluminum and Copper) and Product-Centric (Cars, WEEE, TV, Lamps, Batteries, Catalysts) Recycling and DfR Rules

ANTOINETTE VAN SCHAIK, MARKUS A. REUTER

22.1 Introduction 307
22.2 Material-Centric Recycling: Aluminum and Copper 307
22.3 Product-Centric Recycling: Complex Sustainability Enabling and Consumer Products 313
22.4 Recycling Complex Multimaterial Consumer Goods: A Product-Centric Approach 329
22.5 Automotive Recycling/Recycling of ELVs Including Automotive Battery Recycling 344
22.6 Recycling of Waste Electrical and Electronic Equipment 347
22.7 Recycling of Lighting 353
22.8 Technology for Recycling of Batteries and Catalysts 359
22.9 Design for Recycling and Resource Efficiency 360
References 376

23 Separation of Large Municipal Solid Waste

JAN THEWISSEN, SANDOR KARREMAN, JORRIAN DORLANDT

23.1 Introduction 379
23.2 The Circular Process for Large Municipal Solid Waste 379
23.3 The Preconditions for Sorting Large Municipal Solid Waste 380
23.4 Collection System of Large Municipal Solid Waste 380
23.5 Sorting of Large Municipal Solid Waste 381
23.6 Sorting Installation 381
23.7 Sorting Process 382
23.8 Recycling Efficiency 382
23.9 The Future 383
Reference 383

24 Recovery of Construction and Demolition Wastes

VIVIAN W.Y. TAM

24.1 Introduction 385
24.2 Existing Recycled Aggregate Concrete Applications 385
24.3 Existing Concrete Recycling Methods 386
24.4 Cost and Benefit Analysis 388
24.5 Conclusion 395
References 396

25 Waste Electrical and Electronic Equipment Management

MATHIAS SCHLUEP

25.1 Introduction 397
25.2 Objectives of WEEE Management 398
25.3 WEEE Take-Back Schemes 398
25.4 Long-term Trends 401
References 402

26 Developments in Collection of Municipal Solid Waste

MAARTEN GOORHUIS

26.1 Introduction 405
26.2 Definition of Municipal Solid Waste 405
26.3 Quantities of Municipal Solid Waste 406
26.4 Quality of Municipal Solid Waste 408
26.5 Management of Municipal Solid Waste 408
References 417

III
STRATEGY AND POLICY

27 From Recycling to Eco-design

ELISABETH MARIS, DANIEL FROELICH, AMÉZIANE AOUSSAT, EMMANUEL NAFFRECHOUX

27.1 Introduction 421
27.2 Principle of Material Design for Recycling 421
27.3 Eco-design Strategies for Recycling 422
27.4 Is Recycling Really Less Impactful on the Environment? 422

27.5 Current Limits for Eco-design for Recycling
     Strategies 423
27.6 Market Demand 425
27.7 Conclusion 426
References 426

28 Recycling and Labeling
ELISABETH MARIS, AMÉZIANE AOUSSAT,
EMMANUEL NAFFRECHOUX, DANIEL FROELICH

28.1 Introduction 429
28.2 Functional Needs Analysis 429
28.3 Bibliographical Research on the Polymer Labeling
     Processes 431
28.4 First Results of Detection Tests with Polypropylene
     Samples 433
28.5 Conclusion 436
References 437

29 Informal Waste Recycling in Developing Countries
MATHIAS SCHLUEP

29.1 Introduction 439
29.2 Defining the Informal Sector 440
29.3 Informal Solid Waste Management 441
29.4 Informal e-Waste Recycling 442
References 443

30 Squaring the Circular Economy: The Role of Recycling within a Hierarchy of Material Management Strategies
JULIAN M. ALLWOOD

30.1 Is a Circular Economy Possible
     or Desirable? 445
30.2 Hierarchies of Material Conservation 449
30.3 When is Recycling Not the Answer? 465
30.4 Discussion 474
References 476

31 The Economics of Recycling
PIETER VAN BEUKERING, ONNO KUIK,
FRANS OOSTERHUIS

31.1 Introduction 479
31.2 Economic Trends and Drivers 479

31.3 Environmental and Social Costs
     and Benefits 481
31.4 Economic Instruments 485
31.5 Conclusions and Discussion 487
References 488

32 Geopolitics of Resources and Recycling
ERNST WORRELL

32.1 Introduction 491
32.2 Resources, Scarcity and Geopolitics 493
32.3 Recycling in the Geopolitical
     Context 494
References 495

33 Recycling in Waste Management Policy
ERNST WORRELL

33.1 Introduction 497
33.2 A Brief History of Waste Management 498
33.3 Integrating Recycling in Waste Management
     Policy Design 499
References 501

34 Voluntary and Negotiated Agreements
ERNST WORRELL

34.1 Introduction 503
34.2 Experiences in Recycling Policy 504
34.3 Lessons Learned 507
References 509

35 Economic Instruments
MAARTEN DUBOIS, JOHAN EYCKMANS

35.1 Introduction 511
35.2 Criteria to Compare Policy Instruments 511
35.3 Basic Environmental Policy Instruments Aimed at
     Stimulating Recycling 512
35.4 Incentives for Upstream Green
     Product Design 513
35.5 Multiproduct and Mixed Waste
     Streams 514
35.6 EPR and Recycling Certificates 515
35.7 Durable Goods 515
35.8 Imperfect Competition in Product and Recycling
     Markets 516

35.9 Policy Instruments in an International Market for Waste and Materials 517
35.10 Recycling and Nonrenewable Resources in a Macro Economic Perspective 517
35.11 Conclusion 517
References 518

36 Information Instruments
ERNST WORRELL

36.1 Introduction 521
36.2 Target Groups/Audience 522
36.3 Communication Tools 523
36.4 Messaging: Information and Communication 524
36.5 Conclusion 524
References 525

37 Regulatory Instruments: Sustainable Materials Management, Recycling, and the Law
GEERT VAN CALSTER

37.1 Introduction 527
37.2 Resource Efficiency and Waste Strategy—The Blurb 528
37.3 The EU Framework Directive on Waste, and Its View on Recovery and Recycling 529

Appendix 1: Physical Separation 101 537
KARI HEISKANEN

Appendix 2: Thermodynamics 101 545
PATRICK WOLLANTS

Appendix 3: Life-Cycle Assessment 555
JOHANNES GEDIGA

Index 563

List of Contributors

Helmut Antrekowitsch Chair of Nonferrous Metallurgy, Montanuniversitaet Leoben, Leoben, Austria

Jürgen Antrekowitsch University of Leoben, Austria

Julian M. Allwood Department of Engineering, University of Cambridge, Cambridge, UK

Améziane Aoussat Laboratoire Conception Produit Innovation, Chambéry (LCPI), Arts et Métiers ParisTech, Paris, France

Sander Arnout InsPyro, Leuven, Belgium

Robert U. Ayres INSEAD, Fontainebleau, France

Bo Björkman MiMeR—Minerals and Metallurgical Research Laboratory, Luleå University of Technology, Luleå, Sweden

Bart Blanpain High Temperature Processes and Industrial Ecology, Department of Metallurgy and Materials Engineering, KU Leuven, Leuven, Belgium

Tobias Brenner Papiertechnische Stiftung (PTS), Germany

Mathias Chintinne Metallo-Chimique, Beerse, Belgium

Ambrose Dodoo Sustainable Built Environment Group, Department of Building and Energy Technology, Linnaeus University, Växjö, Sweden

Jorrian Dorlandt Shanks/Van Vliet Groep, Nieuwegein, The Netherlands

Maarten Dubois Policy Research Center for Sustainable Materials, KU Leuven – University of Leuven, Leuven, Belgium

Thomas D. Dyer Division of Civil Engineering, University of Dundee, Dundee, UK

Johan Eyckmans Center for Economics and Corporate Sustainability (CEDON), KU Leuven – University of Leuven, Campus Brussels, Brussels, Belgium

Bernd Friedrich IME Process Metallurgy and Metal Recycling, RWTH Aachen University, Intzestraße 3, Aachen, Germany

Daniel Froelich Laboratoire Conception Produit Innovation, Chambéry (LCPI), Institut Arts et Métiers ParisTech, Chambéry, Savoie Technolac, Le Bourget du Lac, France

Johannes Gediga PE International AG, Leinfelden-Echterdingen, Germany

Maarten Goorhuis Senior Policy Advisor, Dutch Solid Waste Association, NVRD, Arnhem, The Netherlands

T.E. Graedel Center for Industrial Ecology, Yale University, New Haven, CT, USA

Harald Grossmann Technische Universität Dresden, Germany

Leif Gustavsson Sustainable Built Environment Group, Department of Building and Energy Technology, Linnaeus University, Växjö, Sweden

Toni Handke Technische Universität Dresden, Germany

Jana M. Hawley Textile and Apparel Management, University of Missouri, Columbia, MO, USA

Kari Heiskanen Aalto University, Espoo, Finland

Sandor Karreman Shanks/Van Vliet Groep, Nieuwegein, The Netherlands

Onno Kuik Institute for Environmental Studies (IVM), VU University, Amsterdam, The Netherlands

Stefan Luidold Chair of Nonferrous Metallurgy, Montanuniversitaet Leoben, Leoben, Austria

Elisabeth Maris Laboratoire Conception Produit Innovation, Chambéry (LCPI), Institut Arts et Métiers ParisTech, Chambéry, Savoie Technolac, Le Bourget du Lac, France

Gara Villalba Méndez Universitat Autònoma de Barcelona, Spain

Emmanuel Naffrechoux Laboratoire Chimie moléculaire Environnement (LCME), Université de Savoie, Le Bourget du Lac, France

Frans Oosterhuis Institute for Environmental Studies (IVM), VU University, Amsterdam, The Netherlands

Laura Talens Peiró Institute for Environmental Protection and Research, Ispra, Italy

Christoph Pichler University of Leoben, Austria

Soraia Pimenta Department of Mechanical Engineering, Imperial College London, South Kensington Campus, London, UK

Yiannis Pontikes KU Leuven, High Temperature Processes and Industrial Ecology Research Group, Department of Metallurgy and Materials Engineering, Kasteelpark Arenberg, Heverlee, Belgium

Silvestre T. Pinho Department of Aeronautics, Imperial College London, South Kensington Campus, London, UK

Barbara K. Reck Center for Industrial Ecology, Yale University, New Haven, CT, USA

Markus A. Reuter Outotec Oyj, Espoo, Finland and Aalto University, Espoo, Finland

Gernot Rösler University of Leoben, Austria

Elinor Rombach IME Process Metallurgy and Metal Recycling, RWTH Aachen University, Intzestraße 3, Aachen, Germany

Rene Rumpold University of Leoben, Austria

Caisa Samuelsson MiMeR—Minerals and Metallurgical Research Laboratory, Luleå University of Technology, Luleå, Sweden

Roger Sathre Sustainable Built Environment Group, Department of Building and Energy Technology, Linnaeus University, Växjö, Sweden

Mathias Schluep World Resources Forum Association (WRFA), St.Gallen, Switzerland

Li Shen Copernicus Institute of Sustainable Development, Utrecht University, Utrecht, The Netherlands

Ruben Snellings Ecole Polytechnique Fédérale de Lausanne (EPFL), Laboratory of Construction Materials, Institute of Materials, Ecublens, Switzerland

Jaana Sorvari Finnish Environment Institute, Helsinki, Finland

Stefan Steinlechner University of Leoben, Austria

Douglas R. Swinbourne School of Civil, Environmental and Chemical Engineering, Melbourne, VIC, Australia

Vivian W.Y. Tam School of Computing, Engineering and Mathematics, University of Western Sydney, Penrith, NSW, Australia

Jan Thewissen Shanks/Van Vliet Groep, Nieuwegein, The Netherlands

Alois Unger University of Leoben, Austria

Pieter van Beukering Institute for Environmental Studies (IVM), VU University, Amsterdam, The Netherlands

Geert van Calster Department of International and European Law, University of Leuven, Leuven, Belgium and Brussels Bar, Brussels, Belgium

Antoinette van Schaik MARAS—Material Recycling and Sustainability, The Hague, The Netherlands

Margareta Wahlström VTT Technical Research Centre of Finland, VTT, Finland

Patrick Wollants KU Leuven - Faculty of Engineering, Department Metallurgy and Materials Engineering (MTM), Kasteelpark Arenberg, Leuven, Belgium

Ernst Worrell Copernicus Institute of Sustainable Development, Utrecht University, Utrecht, The Netherlands

PART I

RECYCLING IN CONTEXT

Recycling: A Key Factor for Resource Efficiency

Ernst Worrell[1], Markus A. Reuter[2,3]

[1]Copernicus Institute of Sustainable Development, Utrecht University, Utrecht, The Netherlands,
[2]Outotec Oyj, Espoo, Finland; [3]Aalto University, Espoo, Finland

Materials form the fabric of our present society; materials are everywhere in our lives, and life as we know it would be impossible without them. In fact, terms as the "Bronze Age" and "Iron Age" demonstrate that materials have really defined our society in history. Moreover, materials will play a key role in the transition of our society toward sustainability. The challenge of sustainability is rooted in the way that we now process resources to make materials and products, which are often discarded at the end of life. This *linear* economy is now running into its limits given the large demand for materials and resources of an increasing (and increasingly affluent) global population.

Industrial society has become extremely dependent on resources, as it produces more, builds an increasingly complex society and accumulates an incredible volume of resources. Figure 1.1 depicts the global production of the key materials used in society and shows an extremely rapid growth in the past few decades, as emerging economies like China develop. Today, China produces half of all the cement, steel and other commodities in the world. Mankind now dominates the global flows of many elements of the periodic table (Howard and Klee, 2004). The materials are drawn from natural resources. However, the Earth's resources are not infinite, but until recently, they have seemed to be: the demands made on them by manufacturing throughout the industrialization of society appeared infinitesimal, the rate of new discoveries outpacing the rate of consumption. Increasingly we realize that our society may be approaching certain fundamental limits. This has made access to materials an issue of national security of many nations, especially also to ensure that emerging new "sustainable" technologies can be supplied with metals and materials.

Historically, industry has operated as an open system, transforming resources to products that are eventually discarded to the environment, as depicted in Figure 1.2. This, coupled to the massive increase in the use of resources, has led to growing impacts on the environment, as large amounts of energy, greenhouse gas emissions and other emissions to the environment are directly tied to the production and use of the resources, while also affecting land use change, the use of water and

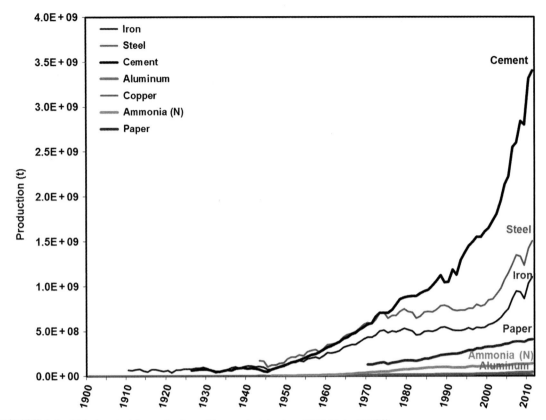

FIGURE 1.1 Global materials production of key materials since 1900 (Data: USGS).

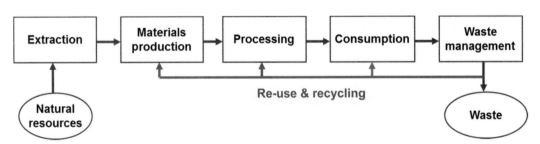

FIGURE 1.2 Closing the loops. Shifting from a linear economy (in blue) to a circular economy requires the closing of loops to ensure that products and materials are reused or recycled (in red). (For interpretation of the references to color in this figure legend, the reader is referred to the online version of this book.)

other environmental resources. Moreover, our resources use results in increasing amounts of solid wastes, which are discarded or incinerated. Waste is becoming a large problem, as we are running out of land for landfilling, and end-of-life waste treatment has negative environmental and health impacts. This is especially a problem for emerging economies, where material use (and hence discarding of it) is growing very rapidly, while limited waste management infrastructure exists.

Materials consumption in the United States now exceeds 10 t/person/year, while the global average consumption has grown to about 5 t/annum. The global average is growing rapidly, given expected population growth and developing patterns for the majority of the population living in developing countries. This makes a reevaluation of the way that we use resources necessary. To maintain our level of welfare, services by resources should be provided more efficiently using less (environmental) resources per unit of activity; i.e. we must improve the resource efficiency of our society. There are several ways that we can improve the resource efficiency of society:

- Use resources more efficiently in the provision of an activity or product (including lifetime lengthening).
- Use less resource-related services.
- Reuse product and services.
- Recycle the resources and materials in products.

Waste is only waste if it cannot be used again or if its economic value, including dumping costs, is not sufficient to make its exploitation economically feasible. Economic recycling enables waste to become a resource; however, various aspects hinder it becoming totally reusable. Recycling is the reprocessing of recovered materials at the end of product life, returning them into the supply chain.

Potentially, it could be done at a rate comparable with the rate with which we discard resources, but then the system must be carefully designed to minimize inevitable losses. Also, the energy needed for recycling is generally substantially less than the energy needed to produce the material from ores, which in the process also creates large amounts of valueless byproducts or wastes with little or no economic value, and that sometimes contains harmful compounds at the mining or processing site. While some (bulk) materials are well recycled, others can (currently) not be recycled, especially due to their complex connections because of functionality reasons in consumer products. Figure 1.3 shows that a level of sophistication in metal recovery from recyclates could exceed that of geological resources due to this depicted complexity.

For this reason, for example, despite all the policy attention to the strategic supply problems surrounding rare earth metals, less than 1% of the rare earth metals in waste are currently being recycled as these go lost owing to the aforementioned complexity. The fraction of a material that can reenter the life cycle will depend both on the material itself and on the mineralogy of the product from which it is being recovered, as (still) the quality and purity of the recovered material determine its future applicability.

The economics of recycling will depend on the degree to which the material has become dispersed, as well as the matrix (e.g. complex consumer product, building, transport, packaging) within which it is recovered. Although recycling has far-reaching environmental and social benefits, market forces determine if a material or complex products can ultimately be recycled and their contained metals, materials and compounds recovered. And market forces often fail to value externalities from environmental pollution or future scarcity, making for an "uneven" playing field. In various countries, these externalities and (other) market failures have provided the incentive to design policies to support resource efficiency in general, and recycling, specifically.

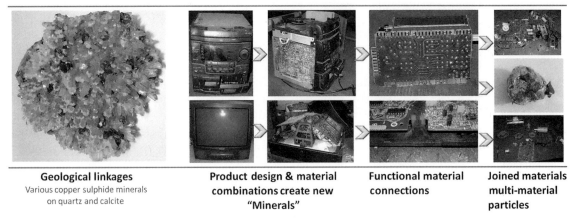

FIGURE 1.3 Conventional extraction processes can recovery various elements from geological ores economically, while much work has to be done to recover all metals from complex designer copper "minerals".

Recyclable wastes are often collected by cities and municipalities, selling them into a market of traders and secondary processors who reprocess the materials to eventually sell them to manufacturers. In the recycling market, prices fluctuate according to the balance of supply and demand, the prices of materials made from primary resources, as well as the behavior and organization of markets and its stakeholders (e.g. the role of increased market power concentration, and speculation of e.g. silver and copper). This couples the price of the recycled material to that of the primary or virgin material. This becomes more complex when minor elements associated with the ores are priced, as supply and demand are geologically linked to the extraction of the bulk base metals such as aluminum, copper, nickel, lead and zinc. The same holds for critical elements that "hitchhike" with the mining of other more common elements, as no separate mines exist for these critical or strategic elements. The markets are also affected by economic or policy interventions. Legislation setting a required level of recycling for vehicles, electronic products and packaging is now in force in most European countries, while other nations have similar programs and plans for more. A lively international trade in recycled resources has emerged, due to the local costs of separating materials from products, and the increasing resources appetite from rapidly developing nations like China and India. This has made recycling a strategic and geopolitical issue as well.

In short, recent economic and global developments have put recycling in the spotlight again, necessitating a critical assessment of the role of recycling in the context of a resource efficient society.

It is the objective to show in this book how material- and product-centric recycling can be harmonized to maximize resource efficiency (UNEP 2013). Figure 1.4 depicts this complex interplay among materials, products and different stakeholders to help maximize the recovery of all materials.

This book will provide the basis, in terms of fundamental theories of recycling, take-back

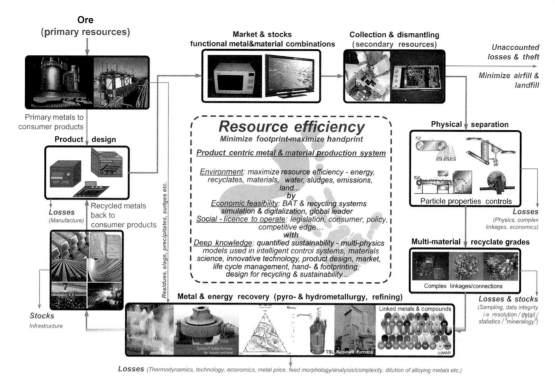

FIGURE 1.4 Recycling helps enable resource efficiency. This book will cover the various aspects detailed in this figure and figures in chapter 1.

systems and collection, used technology, economics, design-for-recycling, consumer behavior, material efficiency and policy, to understand the state of the art and the challenges of recycling in the larger context. It is intended to provide professionals, analysts and decision makers with a solid background in order to show how recycling can be used to improve resource efficiency.

The book is organized in several parts:

I. General aspects of recycling
II. Recycling technologies and applications
III. Recovery and collection
IV. Material efficiency
V. Economics of recycling
VI. Recycling and policy

Each part consists of a number of chapters. Each of the chapters is authored by key experts in the field. Experts come from academia, industry and the policymaking community, providing for a strong basis from theory to practice today, describing the lessons learned and the state of the-art in recycling of a wide variety of resources, products and waste flows.

Part I provides the background and context for recycling by discussing the basics of recycling in science and society, and putting recycling in the perspective of resource efficiency and the development toward a resource-efficient and sustainable *circular* economy.

Part II is the technical "body" of the handbook, describing the state of the art of recycling material, product and waste streams. Each chapter describes a specific material or waste stream, covering the current situation as well as technological state of the art. Starting with specific materials, this part of the book develops toward more complex waste streams that consist of a typical type of products (e.g. vehicles and electronics, as shown in Figures 1.3 and 1.4) or a difficult and varying mix of products (e.g. construction and demolition wastes, bulky household wastes).

More detailed discussions of relevant theory are also included in the Appendices.

Subsequent chapters discuss the different types of recovery schemes for key waste flows, ranging from curbside collection systems to postconsumer separation technologies.

Recovery and collection technology is an integral part of recycling. While conventional economics of a few stakeholders often determine the collection system, it is important to understand that the system will affect the material quality of the recycled materials, and hence should be considered in the full context and not stand alone.

In *Part III* the handbook will focus on the position of recycling within the total life-cycle of resources and the need to increase the efficiency of our total resources system. Hence, the role, effectiveness and efficiency of recycling should be evaluated within the context of and relative to all opportunities to improve the resource efficiency of our society. Redesigning products and reusing products are a key part of this overall strategy. Design for recycling and the fundamental link to the physics of the recycling system are key issues included in the discussions.

In the next parts of the Handbook of Recycling, attention is shifted to the economics and policy aspects of recycling. Furthermore, the economics of recycling are discussed, addressing the costs, benefits and externalities of recycling in more detail. This part also focuses on policies to stimulate reuse and recycling, bringing together knowledge on the type of policy instruments used, the experiences, effectiveness and efficiency.

A number of appendices to the Handbook of Recycling provide background data on some of the tools and data needed in evaluating recycling technology and opportunities. The appendices provide a brief introduction into topics such as physical separation and sorting, thermodynamics and extractive metallurgy, process simulation, life-cycle assessment, and simulation to guide the readers to more advanced texts on these topics. These are crucial in understanding the physics underlying resource efficiency. This knowledge provides the basis for innovation in the system, creating innovative new solutions and technologies, or also determining which systems and processes should be made redundant to ensure resource efficiency is maximized.

Combined, the parts of the handbook bring together a unique collection of material on recycling, from technology to policy, providing state-of-the-art discussions and information from a wide variety of backgrounds and experiences. This will help the reader understand the dynamics, context and opportunities for recycling, within the larger picture of shifting our economic production system to a circular system. We hope that this will contribute to the realization of a circular economy and efficient use of resources in our society.

Finally, to show the difference between material and product centric recycling, aluminium recycling was integrated into a chapter that shows the link between product design, product complexity and metal recovery. This permits the rigorous simulation based analysis of the limits of recycling by considering the effect of for example the dissolution of minor elements in (less noble) metals and other losses from the system on resource efficiency.

References

Klee, R.J., Graedel, T.E., 2004. Elemental Cycles: A Status Report on Human or Natural Dominance. Annu. Rev. Environ. Resour. 29, 69–107.

UNEP, 2013. Metal Recycling: Opportunities, Limits, Infrastructure, a Report of the Working Group on the Global Metal Flows to the International Resource Panel. In: Reuter, M.A., Hudson, C., Schaik, A. van, Heiskanen, K., Meskers, C., Hageluken, C. (Eds.), 316 p.

CHAPTER 2

Definitions and Terminology

Ernst Worrell[1], Markus A. Reuter[2,3]

[1]Copernicus Institute of Sustainable Development, Utrecht University, Utrecht, The Netherlands,
[2]Outotec Oyj, Espoo, Finland; [3]Aalto University, Espoo, Finland

2.1 INTRODUCTION

Recycling is not a goal in itself, but rather an essential tool out of a whole toolbox to better manage natural resources. Materials consumption in the United States now exceeds 10 t/person/year, while the global average consumption has grown to about 5 t/annum. The global average is growing rapidly, given expected population growth and developing patterns for the majority of the population living in developing countries. This rapidly increasing demand for resources has initiated various initiatives, such as "Factor 4, 5 or 10" to reduce the total amount of resources needed, while still fulfilling the needed services provided by materials and resources in today's society (e.g. Von Weizsäcker et al., 2009).

Historically, industry producing the materials has operated as an open system, transforming resources to products that are eventually discarded to the environment. This, coupled with the massive increase in the use of resources, has led to growing impacts on the environment. The massive use of materials results in increasing amounts of solid wastes, which are discarded or incinerated. This results not only in a loss of valuable materials, but also in negative environmental and health effects.

Our massive use of resources also contributes to a potential depletion of economically (or environmentally) recoverable reserves of materials. In fact, one could say that natural stocks (i.e. geologic reserves) are transferred to anthropogenic stocks (i.e. capital goods in our society). This makes recycling an important source of material, and this importance will only increase in the future. This is reflected in terms as the "urban mine" (i.e. minerals and metals contained in the urban infrastructure and buildings) and "urban forest" (i.e. wood fibers and paper).

A distinction needs to be made between non-renewable materials, such as minerals (including oil) and metals, and renewable materials (e.g. wood and biomass). Note that the two are interrelated because of the need for nutrients (e.g. nitrogen, phosphorus and potassium) and micronutrients (e.g. selenium and many others). Recycling of renewable materials contributes to a more efficient supply of resources, since primary resources (e.g. forest, land, water, energy) are saved and emissions such as greenhouse gases are reduced (Laurijssen et al., 2010). The success of paper recycling in many countries illustrates the need for recycling of renewable resources.

To maintain our level of welfare, services by resources should be provided more efficiently using less (environmental) resources per unit of activity, i.e. improve the resource efficiency of our society. This means that we need to move from a linear economy, which extracts resources from the environment and discharges the wastes to the environment, to a circular economy, one that reuses and recycles products and materials in a material-efficient system, extracting and wasting as little as possible (see Chapter 1). There are several ways that we can improve the resource efficiency of society, of which recycling is one. Waste is only waste if it cannot be used again or if its economic value including dumping costs is not sufficient to make its exploitation economically feasible. In historical societies, recycling was much more prevalent. The Industrial Revolution allowed for a massive reduction in the cost of materials, resulting in a reduced emphasis on reuse and recycling in today's society. However, with increasing costs for primary materials, recycling has enabled waste to become a resource. While some materials are well recycled, others can (currently) not be recycled. In this chapter, recycling is put in context of a resource-efficient economy, and critical issues are defined that will contribute to understanding and positioning recycling. This is illustrated by applying the concepts to the case of metals.

2.2 DEFINING RECYCLING

Recycling is the reprocessing of recovered materials at the end of product life, returning them into the supply chain. Recycled material may also be called "secondary" in contrast to "primary" material that is extracted from the environment. Hence, primary and secondary in the context of recycling do not express a difference in quality.

Recycling has a key role to play in a resource-efficient economy. In past decades, recycling was mainly considered a waste management issue, whereas today the vision is slowly moving toward resource efficiency as a driver for recycling. This places recycling in a wider context. In various countries a variation on the "waste management hierarchy" (for example, also known as the 3 Rs in the United States, which stands for reduce, reuse and recycle) was introduced, which still forms the basis for waste management in most countries:

1. Reduce or avoid waste
2. Reuse the product
3. Recycle
4. Energy recovery
5. Treatment and landfilling

Figure 2.1 provides an overview of the typical waste management chain, including the treatment options of the hierarchy.

Reduction (or avoidance) describes the impact of material efficiency and demand reduction to minimize the amount of material that is needed

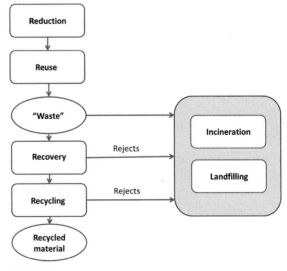

FIGURE 2.1 Simplified depiction of the typical waste management chain.

to satisfy a material service (Worrell et al., 1995). This may also include reducing the need for the service in the first place. Other options are lengthening the service lifetime of a product (either by design or through repair), or increasing the yield in the supply chain of a product (i.e. reducing material or off-spec product losses in the different production steps of a product) (Allwood et al., 2011). Note that this may reduce the amount of material available for recycling, as e.g. less "home" or "new" scrap may be generated (see below). Material efficiency is increasingly getting attention again, as resource efficiency is slowly gaining traction.

Reuse allows for the reuse of the product in which the material is contained, by (re-)designing a product for multiple uses (e.g. refillable bottle versus single-use bottles) or setting up a market for reusable goods (of which many can be found, both in industry and households). Exchange systems can be very effective means to reuse products, as, for example, evidenced by the success of online auction systems such as eBay and others in many countries.

Recycling aims at recycling the materials contained in the products that are recovered from the waste stream. Potentially, recycling can be done at a rate comparable with the rate with which we discard resources, but then the system must be carefully designed to minimize inevitable losses. The fraction of a material that can reenter the life cycle will depend both on the material itself and on the product it was part of, as (still) the quality and purity of the recovered material determine its future applicability. Recycling rates are defined in various ways, affecting the figures dramatically (UNEP, 2011). First of all, the recycling rate is determined by the volume of material that is recovered or actually recycled. This volume can also include material that is generated during the production of the material, manufacturing of products, or at end-of-life. For example, in metals, the following categories of recovered material are discerned (Graedel et al., 2011):

- *Home scrap*: scrap material arising internally in production sites or mills as rejects, e.g. from melting, casting, rolling or other processing steps.
- *New (pre-consumer) scrap*: scrap from fabrication of the material into finished products.
- *Old (post-consumer) scrap*: scrap from obsolete products that is recovered, traded and sold to plants for recycling.

Furthermore, the rate is affected by the basis on which it is calculated, e.g. the volume of material sold in the market, produced in a country or region, or the total amount of the material available in the waste.

Energy Recovery generally applies to the recovery of (part of the) embodied energy in the materials in the products, using a number of techniques, including the production of refuse-derived fuel for industrial processes (e.g. in cement making) or specialized boilers, incineration with energy recovery in waste-to-energy (WTE) facilities, or through anaerobic digestion of biologic/organic materials in the waste. Note that the latter process may also take place in a landfill, and the landfill gas may be recovered for energy production. The efficiency of energy recovery of these systems may vary largely, and could be very low (e.g. 12–15% for older WTE facilities).

Treatment and landfilling are waste management techniques to reduce the environmental and health impacts (if properly controlled) of waste, and do generally not result in recycling or recovery from resource. In many developing countries, but also the United States, landfilling is still the main waste management option, while in developing countries this is often done in uncontrolled and non-sanitary landfills, resulting in negative impacts on the local environment, water and air quality, as well as human health. In some specific cases, old landfills have been mined to recover some of the materials contained in the landfill. In practice, this has only been economically interesting for selected metals and is determined by local economic conditions.

Figure 2.1 also distinguishes various indicators to measure the success of a recycling program, i.e. recovery and recycling rates. These terms are often used, but also often not clearly defined. Hence, caution is needed when interpreting reported rates.

Recovery rate refers to the volume of material recovered from a waste stream. However, different definitions are found in the literature. Typically, it can be defined as the volume of material recovered from a waste stream divided by the amount of material in the generated waste.

Recycling rate often refers to the volume of material collected for recycling—generally including any material rejected during processing—divided by the volume (weight) of waste generated. However, more correctly, the rejected material should be subtracted, and only material marketed for recycling after processing should be included. Differences may hence be found in where the volume of recycled material is counted, and how the volume of material in the waste is estimated. The most rigorous way would be to dynamically simulate the material cycles in all its complexity, but data are often lacking.

Recycling efficiency. The total amount of material available for reuse will be affected by any material losses (due to e.g. quality, color or processing) during the recycling process itself. This can be defined as the recycling efficiency, or output of the recycling process divided by the input. For metal recycling, the recycling efficiency would be defined as the amount of scrap melted (output)/amount of scrap recovered (input).

The relation between the three indicators can be given as:

$$\text{Recycling Rate} = \text{Recovery Rate} * \text{Recycling Efficiency} \quad (2.1)$$

Recycled content. This is the fraction of recycled or secondary material in the total input into a production process.

2.3 MATERIALS AND PRODUCTS

Typically, recycling focuses on materials, while the first two steps in the waste management hierarchy focus on products. An integrated view on recycling in the waste management hierarchy also puts a central focus on the product (Allwood et al., 2011), which will enable reduction, reuse and also the efficient recovery of recyclable materials from a product. In other words, an efficient waste management system should be centered on the product, and less on the material, or what is called a product-centric approach. A *mineral-centric approach* or in other words a *product (mineralogy) centric* view (UNEP, 2013) is required to maximize resource efficiency rather than a simpler material-centric view that considers things material by material. It is this depth that lies at the heart of the recycling, recycling simulation models for optimization of resource efficiency and design for recycling/resource efficiency. It is the application of this depth that will enhance closing of the loop because it will permit a much deeper understanding between all actors than the current understanding.

This will help us to better understand, sample, quantify products and recycling on element/compound/alloy level, and simulate the performance of recycling systems, also in relation to product design. This rigor in the recycling will also help to increase the general level of sophistication in the field and bring it to a similar level of detail and sophistication as common in the producing industry, something which is very important when discussing such initiatives as design for recycling, resource efficiency and eco-design, and labeling for recycling/environmentally optimized products.

A *product-centric* approach considers how to increase the recycling of a product (e.g. an LCD screen, mobile phone, car, solar panel) in its entirety and therefore considers the complex thermodynamic and physics aspects and interactions that affect their recovery. This necessarily involves consideration of what will happen to the many different materials within the product, and enables decision makers to more easily look at how the products are collected and how design affects outcomes. However, it is to be noted that design for recycling is not the golden bullet it is made to be, as functionality often determines the material connections, overriding their incompatibility for recycling. However, some companies have used similar approaches to design equipment (or components thereof) for reuse, allowing the use of remanufactured parts in e.g. new copiers (e.g. Xerox) or machinery (e.g. Caterpillar).

Also, note that the order of the steps goes (generally) hand in hand with a decreasing amount of energy recovered in the processing of the material. Reduction will obviously save the largest amount of energy, while reuse recovers more from the energy embodied in the product/materials than recycling. However, after a certain degree of reuse, inevitably the product will land in the recycling chain and its materials will be recovered. The energy needed for recycling is generally substantially less than the energy needed to produce the material from ores. Landfilling will recover only the smallest part of the embodied energy, in fact, only from the organic fraction if landfill gas would be recovered.

While the hierarchy forms the theoretical basis for a waste management strategy, in practice in many countries some of the steps in the hierarchy may be lacking for specific waste or product streams. This failure is one of the reasons why there is still a very large potential for improving resource efficiency in today's economy.

2.4 APPLYING THE PRODUCT-CENTRIC APPROACH—METALS

Metals, their compounds and alloys have unique properties that enable sustainability in innovative modern infrastructures and through modern products. Through mindful product design and high (end-of-life) collection rates, metals and their compounds can enabling sustainability, and other products can be recovered as well; thus recovery and therefore recycling of metals can be high. However, limitations on the recycling rate can be imposed by the (functionality driven) linkages and combinations of metals and materials in products (UNEP, 2013).

Figure 1.3, Figure 1.4 and Figure 2.2 shows various factors that can affect the resource efficiency of metal processing and recycling. The interaction therefore of primary and secondary recovery of metals not only drives the sustainable recovery of elements from minerals but also provides the recycling loop that recovers metals from complex products and therefore enables the maximum recovery of all elements from designer minerals. It is self-evident that "classical" minerals processing and metallurgy play a key role in maximizing resource efficiency and ensuring that metals are true enablers of sustainability. Thus key to recycling of complex consumer goods is:

- Mineral processing and metallurgy — foundation
  - The link of minerals to metal has been optimized through the years by economics and a deep physics understanding.
  - There is a good understanding between all actors from rock to metal.
- Product-centric vis-à-vis metal-centric recycling
  - Designer minerals (e.g. cars) as shown in Figure 1.3, Figure 1.4 and Figure 2.2 are far more complex than geological minerals, complicating recovery.

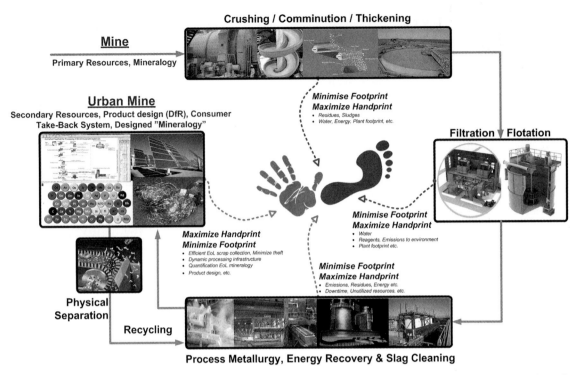

FIGURE 2.2 Design for resource efficiency. Optimally linking mining, minerals processing, (BAT) Best Available Technique(s) for primary and secondary extractive metallurgy, energy recovery, original equipment manufacturers and product design, end-of-life products, recyclates, residues and wastes, while minimizing resource losses.

- To "close" the loop requires a much deeper understanding between all actors of the system than is the case currently. Resource efficiency will improve if this is achieved.
- Metal/material-centric recycling is a subset of product-centric recycling
  — Various definitions exist for material-centric recycling of metals as documented in a report by UNEP (2011), for example, how much of the end-of-life (EoL) metal is collected and enters the recycling chain (old scrap collection rate), the recycling process efficiency rate and the EoL-recycling rate (EoL-RR).
- This deeper understanding of recycling will help to develop sensible, physics-based policies.

The use of available minerals processing and process metallurgical theoretical depth to describe the system shown in Figure 2.2 is required to understand the resource efficiency of the complete system. A fundamental description of the system also shows what theory and methods still have to be developed to innovate the primary and recycling fields further. It is evident from Figure 2.2 that the rigorous theory developed in the classical minerals and metallurgical processing industry over the years and more recently adapted for recycling is very useful to quantify the various losses shown in Figure 2.2. "Classical" minerals processing and process metallurgy therefore both have a significant role to play in a modern resource-constrained society. Identifying the detailed metal, compound, and other contents in all flows will help in optimizing the recycling system, as is already the case for the maximum recovery of metals in concentrates from known ore and product streams, giving a rather precise mass balance for all total, compound and elemental flows (see Figure 1.3).

The recycling and waste processing industry has much to gain to implement and adapt techniques and thinking of our industry rather than following the conventional bulk flow approaches of a material-centric mindset of waste management and derived legislation, which are often colored by this thinking.

Three major factors determine the outcomes of recycling expressed as a recyclability index (RI): (1) the way waste streams are mixed or pre-sorted during collection; (2) the physical properties and (3) design of the end-of-life products in those waste streams. These factors all affect the final recovery and subsequent production of high quality metal, material and alloy products. These factors interrelate in ways that make it impossible to optimize one without taking into account the others. To get the best results out of recycling, the stakeholders of the recycling system (e.g. in design, collection, processing) need to take into account what is happening in the other parts of the system. They also need to consider how to optimize along the chain the recycling of several metals found within one product, rather than only focusing on one or two major metals (and their alloys and alloying elements) and ignoring the rest of the periodic table.

Figure 1.3, Figure 1.4 and Figure 2.2 provides an overview of all the actors and aspects that have to be understood in a *product-centric* systemic and physics-based manner in order to optimize resource efficiency. Also a clear understanding of the various losses that occur is imperative (many governed by physics, chosen technology and linked economics), which also requires a deep compositional understanding of all residues, but also the understanding of unaccounted flows (poor statistics, data as well as collection) and the economics of the complete system are critical. Especially also understanding and controlling the dubious and illegal flows as well as theft, etc. will help much to maximize recovery, but this is a relatively simple task organized by leveling the playing field by suitable policy. Maximizing resource efficiency, and therefore design for resource efficiency, considers and embraces Figure 2.2 in its totality. This requires rigorous modeling techniques to pin-point, understand and minimize all losses. It also requires a detailed understanding of the technology of recycling, both physical and metallurgical, as discussed in detail by Reuter and Van Schaik (2012) and Van Schaik and Reuter (2012).

In summary, it is extremely important for resource efficiency to step away from the material perspective to the product perspective. A particular focus will be the recycling of the high-value, lower-volume metals that are essential elements of today's and tomorrow's high tech products, as applied in complex multimaterial design such as electronics and vehicles (also aircraft) or generation and storage of renewable energy. These metals, such as gallium, rare earth elements, platinum group elements and indium, are often scarce, essential to sustainable growth and yet typically lost in current recycling processes.

References

Allwood, J., Ashby, M.F., Gutowski, T.G., Worrell, E., 2011. Material efficiency: a white paper. Resources, Conservation & Recycling 55, 362–381.

Laurijssen, J., Marsidi, M., Westenbroek, A., Worrell, E., Faaij, A., 2010. Paper and biomass for energy? The impact of paper recycling on biomass availability, energy and CO_2 emissions. Resources, Conservation & Recycling 54, 1208–1218.

Reuter, M.A., Van Schaik, A., 2012. Opportunities and limits of recycling: a dynamic-model-based analysis. MRS Bulletin 37, 339–347.

UNEP, 2011. Recycling Rates of Metals – a Status Report of the Working Group on the Global Metal Flows to the

International Resource Panel. Graedel, T.E., Allwood, J., Birat, J.-P., Reck, B.K., Sibley, S.F., Sonneman, G., Bucher, M., Hagelüken, C. 44 p.

UNEP, 2013. Metal Recycling: Opportunities, Limits, Infrastructure, a Report of the Working Group on the Global Metal Flows to the International Resource Panel. Reuter, M.A., Hudson, C., van Schaik, A., Heiskanen, K., Meskers, C., Hagelüken, C. 316 p.

Van Schaik, A., Reuter, M.A., 2012. Shredding, sorting and recovery of metals from WEEE: linking design to resource efficiency. In: Goodship, V. (Ed.), Waste Electrical and Electronic Equipment (WEEE) Handbook. University of Warwick, UK and A Stevels, Delft University of Technology, The Netherlands, pp. 163–211.

Von Weizsäcker, E., Hargroves, K.C., Smith, M.H., Desha, C., Stasinopoulos, P., 2009. Factor Five: Transforming the Global Economy through 80% Improvements in Resource Productivity. Earthscan, London, UK.

Worrell, E., Faaij, A.P.C., Phylipsen, G.J.M., Blok, K., 1995. An approach for analysing the potential for material efficiency improvement. Resources, Conservation & Recycling 13 (3/4), 215–232.

# CHAPTER 3

Recycling in Context

T.E. Graedel, Barbara K. Reck
Center for Industrial Ecology, Yale University, New Haven, CT, USA

3.1 INTRODUCTION

Of the different resources seeing wide use in modern technology, metals are unusual in that they are inherently recyclable. This means that, in principle, they can be used over and over again, thus minimizing the need to mine and process virgin materials while saving substantial amounts of energy and water. These activities also avoid the often significant environmental impacts connected to virgin materials extraction.

Recycling indicators have the potential to demonstrate how efficiently metals are being reused, and can thereby serve the following purposes:

- Determine the influence of recycling on resource sustainability by providing information on meeting metal demand from secondary sources
- Provide guidance for research needs on improving recycling efficiency
- Provide information for life-cycle assessment analyses
- Stimulate informed and improved recycling policies

Notwithstanding these promising attributes of recycling, the quantitative efficiencies with which metal recycling occurs are not very well characterized, largely because data acquisition and dissemination are not vigorously pursued. It is worthwhile, however, to review what we do know, and to consider how that information might best be improved.

3.2 METAL RECYCLING CONSIDERATIONS AND TECHNOLOGIES

Figure 3.1 illustrates a simplified metal and product life cycle. The cycle is initiated by choices in product design: which materials are going to be used, how they will be joined and which processes are used for manufacturing. Choices made during design have a lasting effect on material and product life cycles. They drive the demand for specific metals and influence the effectiveness of the recycling chain during end-of-life (EOL).

When a product is discarded, it enters the EOL phase. It is separated into different metal streams (recyclates), which have to be suitable for raw materials production in order to ensure that the metals can be successfully reused. The cycle is closed if scrap metal, in the form of

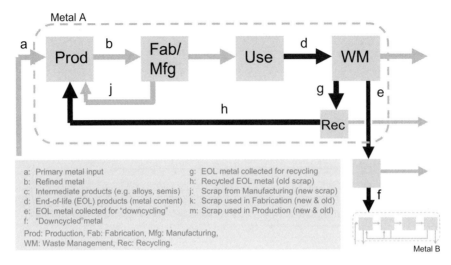

FIGURE 3.1 Flows related to a simplified life cycle of metals and the recycling of production scrap and end-of-life products. Boxes indicate the main processes (life stages): Prod, production; Fab, fabrication; Mfg, manufacturing; WM&R, waste management and recycling; Coll, collection; Rec, recycling. Yield losses at all life stages are indicated by dashed lines (in WM referring to landfills). When material is discarded to WM, it may be recycled (e), lost into the cycle of another metal (f, as with copper wire mixed into steel scrap), or landfilled. The boundary indicates the global industrial system, not a geographical entity. *Reproduced with permission from Graedel et al., 2011a.*

recyclates, returns as input material to raw materials production. The cycle is open if scrap metal is lost to landfills and other repositories (e.g. tailings, slag).

The different types of recycling are related to the type of scrap and its treatment:

- The *home scrap* portion of new scrap recycling, in which metal is essentially recovered in its pure or alloy form within the facility of the metal supplier. This type of recycling tends to be economically beneficial and easy to accomplish. It is generally absent from recycling statistics, however, because it takes place within a single facility or industrial conglomeration.
- The *prompt scrap* portion of new scrap recycling, in which metal is essentially recovered in its pure or alloy form from a fabrication or manufacturing process and returned to the metal supplier for reprocessing and reuse. This type of recycling is generally economically beneficial and easy to accomplish. It may not be identified in recycling statistics, but can sometimes be estimated from process efficiency data.
- EOL, metal-specific ("functional") recycling, in which the metal in a discarded product is separated and sorted to obtain recyclates that are returned to metal suppliers for reprocessing and reuse. This type of recycling is generally accomplished for high-value metals, especially if they are easily accessible (Streicher-Porte et al., 2005; Dahmus and Gutowski, 2007). The processes are straightforward if the metal is in pure form, but are often challenging and expensive if the metal is a small part of a very complex product (Chancerel and Rotter, 2009; Oguchi et al., 2011).
- EOL, alloy-specific recycling, in which an alloy in a discarded product is separated and returned to raw materials production for

recovery as an alloy. Often it is not the specific alloy that is remelted to make the same alloy, but any alloy within a certain class of alloys that is remelted to make one or more specific alloys by adding small amounts of other alloying elements to achieve the desired elemental composition. For example, a mixture of austenitic stainless steel alloys might be remelted and the resulting composition adjusted by addition of reagents or virgin metal to make a specific austenitic alloy. A similar approach is followed in aluminum recycling.

- EOL metal-unspecific reuse (nonfunctional recycling, or "downcycling"), in which the metals or alloys in a discarded product are downgraded or downcycled by incorporation into a large-magnitude material stream in which its properties are not required. This prevents the metal or alloy from being dissipated into the environment, but represents the loss of its function, as it is generally impossible to recover it from the large-magnitude stream. The recycled metal does not replace primary metal in metal production, so that the energy benefits of recycling cannot be taken advantage of. Equally discouraging is the fact that the recycled metal potentially lowers the quality of the produced metal alloy by becoming an impurity or tramp element. Examples are small amounts of copper in iron recyclates that are incorporated in recycled carbon steel, or alloying elements being incorporated in slag during final recovery of the major alloy element.

3.3 DEFINING RECYCLING STATISTICS

There are four approaches to measuring the efficiency of EOL metal recycling (Graedel et al., 2011a):

1. How much of the EOL metal (in products) is collected and enters the recycling chain (as opposed to metal that is landfilled)? (*Old scrap collection rate*)
2. What is the *EOL recycling rate* (metal-specific)? (*EOL-RR*)
3. What is the efficiency in any given recycling process (i.e. the yield)? (*Recycling process efficiency rate*)
4. What is the nonfunctional EOL recycling rate (downcycling)? (*nonfunctional EOL-RR*)

Figure 3.1 provides an annotated waste management and recycling system from which the EOL metrics can be calculated:

1. Old scrap collection rate = e/d
2. EOL-recycling rate = g/d
3. Recycling process efficiency rate, example EOL = g/e
4. Nonfunctional EOL recycling rate = f/d

The *recycled content* (*RC, sometimes termed the "recycling input ratio"*) describes the fraction of recycled metal contained within the total metal flow metal production. In the simplified diagram of Figure 3.1, it is defined as $(j+m)/(a+j+m)$. The calculation of the recycled content is straightforward at the global level, but difficult if not impossible at the country level. The reason is that information on the recycled content of imported produced metals is typically not available (flow b, i.e. the share of $m/(a+m)$ in other countries is unknown), in turn making a precise calculation of the recycled content of flow c impossible.

A final metric, the *old scrap ratio* (*OSR*), provides information on the composition of the scrap used in metal production. It is the fraction of old scrap g in the recycling flow (g + h). Recycled content and OSR are closely linked in that the OSR reveals the share of old versus new scrap used in metal production, thus providing information on the efficiencies at different life cycle stages.

In terms of its significance, the most important metric is the EOL recycling rate, which indicates how effectively discarded products are recovered and recycled. Of limited relevance for metals, however, is the widely used metric "recycled content", for two reasons. First, calculations can usually be carried out only at the global level, leaving little room for incentives at the national level. Second, the share of available old scrap depends on the level of usage a lifetime ago. As this use rate was typically much less than today, there is not enough old scrap available to allow for a recycled content close to 100%. Note also that a high share of new scrap may be the result of an inefficient manufacturing process, and is therefore of limited relevance as a measure of recycling merit.

In order to arrive at and present global estimates of metal recycling statistics that are as comprehensive as possible, a detailed review of the recycling literature was conducted by the United Nations Environment Programme (Graedel et al., 2011b). The three periodic table displays in Figures 3.2–3.4 illustrate the consensus results in compact visual display formats.

The EOL-RR results in Figure 3.2 relate to whatever form (pure, alloy, etc.) recycling occurs. To reflect the level of certainty of the data and the estimates, data are divided into five bins: >50%, 25–50%, 10–25%, 1–10% and <1%. It is noteworthy that for only 18 of the 60 metals are the EOL-RR values above 50%. Another three metals are in the 25–50% group, and three more in the 10–25% group. For a very large number, little or no EOL recycling is occurring.

Similarly, Figure 3.3 presents the recycled content data. Lead, ruthenium, and niobium are the only metals for which RC >50%, but 16 metals have RC in the >25–50% range. This

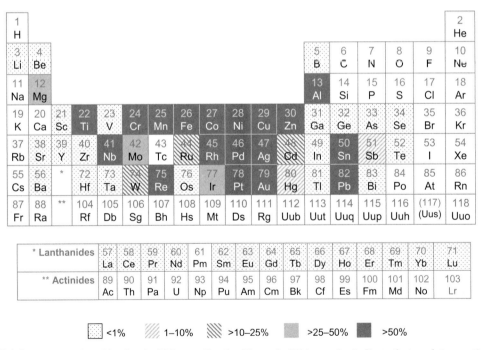

FIGURE 3.2 The periodic table of end-of-life recycling for 60 metals. White entries indicate that no data or estimates are available.

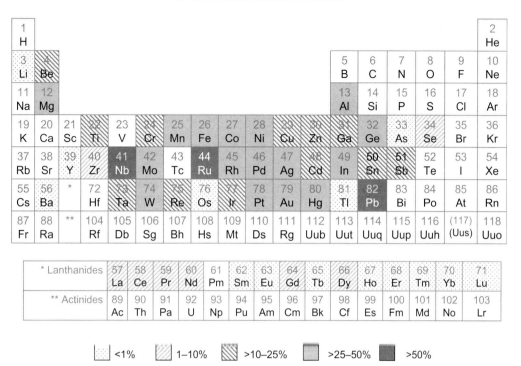

FIGURE 3.3 The periodic table of recycled content for 60 metals. White entries indicate that no data or estimates are available.

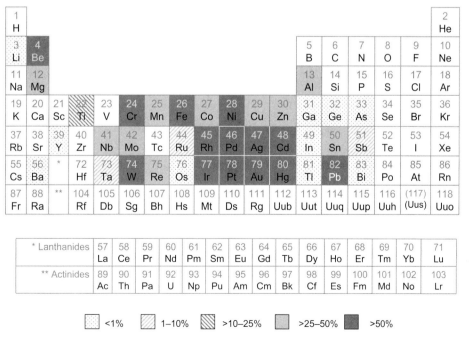

FIGURE 3.4 The periodic table of old scrap ratios for 60 metals. White entries indicate that no data or estimates are available.

reflects a combination in several cases of efficient reuse of new scrap as well as better than average EOL recycling.

The OSR results (Figure 3.4) tend to be high for valuable materials, because they are utilized with minimal losses in manufacturing processes and collected at EOL with relatively high efficiency. Collection and recycling at EOL are high as well for the hazardous metals cadmium, mercury and lead. Overall, 13 metals have OSR >50%, and another 10 have OSR in the range >25—50%.

For cases in which relatively high EOL-RR are derived, the impression might be given that the metals in question are being used more responsibly than those with lower rates. In reality, rates tend to reflect the degree to which materials are used in large amounts in easily recoverable applications (e.g. lead in batteries, steel in automobiles). In contrast, where materials are used in small quantities in complex products (e.g. tantalum in electronics), recycling is much less likely, and the rates will reflect this challenge.

3.4 PROCESS EFFICIENCIES AND RECYCLING RATE CONSTRAINTS

A common perception of the recycling situation is that if a product is properly sorted into a discard bin it will be properly recycled. This turns out never to be even approximately correct, because the recycling system comprises a number of stages (Figure 3.5): collection, preprocessing (including separation and sorting), and end processing (usually in a smelter). Losses occur at every stage, and generally the stage with the lowest recycling efficiency is the very first: collection. Higher efficiencies at the subsequent stages cannot make up for low first-stage performance, as suggested by the efficiencies shown in Figure 3.5.

Even if efficient collection occurs, efficiencies lower than 100% (which is always the case) combine to generate low EOL-RRs over time. Figure 3.6 shows the situation. Each stage has an imperfect process efficiency; if those efficiencies are multiplied together over several metal use lifetimes, even well-run recycling processes eventually dissipate the metal. Studies have shown that a unit of the common metals iron, copper and nickel is only reused two or three times before being lost (Matsuno et al., 2007; Eckelman and Daigo, 2008; Eckelman et al., 2012), because no process is completely efficient, and losses occur at every step (Figure 3.6).

Finally, product design plays an important role in the recycling efficiency of EOL products. First, does the product design allow for easy accessibility and disassembly of the relevant components? For example, precious metals contained in personal computer motherboards are easily accessible for dismantling and will be recycled, while circuit boards used in car electronics are typically not accessible for recycling (Hagelüken, 2012). Second, can the diverse mix of materials used in complex products be technologically separated at EOL? This challenge of material liberation goes back to thermodynamic

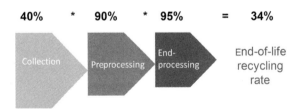

FIGURE 3.5 The steps involved in the recycling sequence. *Adapted from Hagelüken, 2012.*

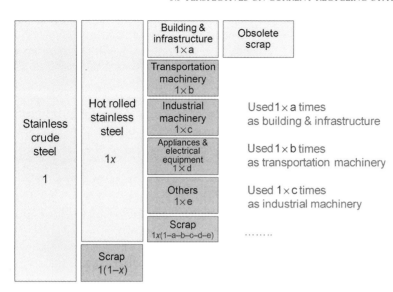

FIGURE 3.6 The efficiencies in the initial step of Markov recycling of a metal cycle. *(Reproduced with permission from Eckelman et al. (2012)).* The efficiency of conversion of stainless crude steel to hot-rolled stainless steel is x, and the resulting stainless steel is divided among five uses, each with its own conversion efficiency. If the metal is later recovered as obsolete scrap when the products are discarded, the process chain and its inevitable losses must be revisited.

principles, but also to the fact that material combinations in products are often very different from material combinations found in ores (Reuter et al., 2013; van Schaik and Reuter, 2004). Well-established technologies from the mining industry can thus be utilized only if elements not found in the respective ores can be removed beforehand (Nakajima et al., 2010).

3.5 PERSPECTIVES ON CURRENT RECYCLING STATISTICS

As can be seen from the figures, there are large differences in recycling rates among the specialty metals, but differences also exist between the different applications of the metals. Some insights into the causes of the relatively low recycling rates in Figures 3.2–3.4 are discussed below:

1. *Hardly any recycling*. This designation is applicable to specialty metals such as antimony, arsenic and barium. These metals are mainly used in oxide or sulfate form, and many of the applications are highly dispersive. Collection is thus very difficult (drilling fluid remains in the hole, preservative remains in wood) or dependent on collection of the product (flame retardants in electronics), hence the old scrap in the recycling flow is very low, as is the recycling flow itself.

2. *Mainly new scrap recycling*. This designation is applicable to specialty metals such as indium and germanium (e.g. Yoshimura et al., 2011). The recycled content is above 25%, but other recycling statistics are very low. These metals are largely used in such applications as (opto)-electronics and photovoltaics. During manufacturing, a large amount of new scrap, such as spent sputtering targets, sawdust or broken wafers, is created. All of this material is recycled, and contributes to a high recycled content in the material supply to the manufacturing stage. Old scrap in the recycling flow is currently low due to the difficulty in collecting the products. Furthermore, the metal content in the products can be low, and recycling technology for these metals in EOL products is often lacking. Yet, research into the

recycling of a variety of specialty metals from discarded products is increasing, and promising first results exist (Rollat, 2012; Yoshida and Monozukuri, 2012).

3. *Old scrap recycling—metal specific*, rhenium, for example. Rhenium is used in superalloys and as a catalyst in industrial applications, which together make up most of its total use. This closed industrial cycle, as well as the high value of Re, ensures very good collection (Duclos et al., 2010). Furthermore, good recycling technologies are in place to recover the metal. New scrap is recycled as well. Because the rhenium demand is growing, and this is met largely by primary production, the share of recycled Re in the overall supply is low.

4. Old scrap recycling—metal unspecific, beryllium, for example. Beryllium–copper alloys are used in electronic and electric applications. The collection of these devices is generally good, and the Be follows the same route as the copper and ends up at copper smelters/recyclers. During the recycling process, the Be is usually not recovered but is diluted in the copper alloy or, most often, transferred to the slag in copper smelters. Hence the old scrap collection rate is quite high but the EOL recycling rate is low.

The way in which a product is designed is also a strong factor in whether recycling occurs. Dahmus and Gutowski (2007) have shown that

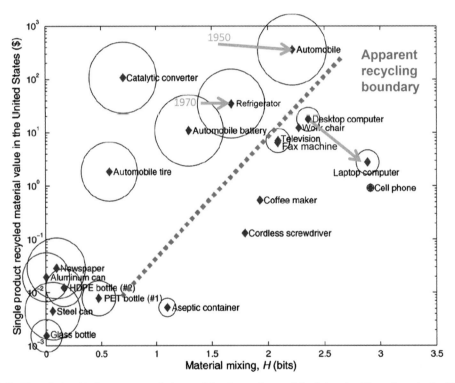

FIGURE 3.7 The relationship between recycled material value and material mixing, for 20 products in the United States, c. 2005. The area of the circle around each data point is proportional to the product recycling rate; products with no circle are generally not recycled. The arrows indicate a trend to increased material mixing, both at the product level (in the case of automobiles and refrigerators) and through substitution (in the case of computers). *Adapted from Dahmus and Gutowski, 2007.*

the greater the degree of material complexity in a product, the smaller the probability that recycling will occur (Figure 3.7). In fact, Figure 3.7 indicates that some products have actually increased their required level of material complexity in the past decade, so the product design-recycling situation is trending in the wrong direction.

3.6 SUMMARY

Many different approaches have been taken to quantify the rates at which metals are recycled. Inevitably, recycling rates have been defined in different ways, and this has made it difficult to determine how effectively recycling is occurring. Adopting the recycling rate definitions specified in this chapter will deal with this challenge.

An important realization regarding metal recycling is that it is a sequence of steps. If any one step is done poorly, the efficiency of the entire sequence suffers. Attention needs to be paid to each of the steps, because one step may be the most inefficient for some types of products, other steps for others.

The key questions, of course, are whether overall recycling efficiencies can be improved and, if so, by how much? That is, can materials cycles be transformed from open (i.e. without comprehensive recycling) to closed (i.e. completely reusable and reused), or at least to less open than they are at present? These are issues that turn out to be quite complex, to involve everything from product designers to policies for pickup of discarded electronics. The full range of this detail has seldom been presented to those who are most interested, but much of it will be explored in detail in subsequent chapters of this book.

References

Chancerel, P., Rotter, S., 2009. Recycling-oriented characterization of small waste electrical and electronic equipment. Waste Management 29 (8), 2336–2352.

Dahmus, J.B., Gutowski, T.G., 2007. What gets recycled: an information theory based model for product recycling. Environmental Science & Technology 41 (21), 7543–7550.

Duclos, S.J., Otto, J.P., Konitzer, D.G., 2010. Design in an era of constrained resources. Mechanical Engineering 132 (9), 36–40.

Eckelman, M.J., Daigo, I., 2008. Markov chain modeling of the global technological lifetime of copper. Ecological Economics 67 (2), 265–273.

Eckelman, M.J., Reck, B.K., Graedel, T.E., 2012. Exploring the global journey of nickel with Markov models. Journal of Industrial Ecology 16 (3), 334–342.

Graedel, T.E., Allwood, J., Birat, J.-P., Buchert, M., Hagelüken, C., Reck, B.K., Sibley, S.F., Sonnemann, G., 2011a. What do we know about metal recycling rates? Journal of Industrial Ecology 15 (3), 355–366.

Graedel, T.E., Allwood, J., Birat, J.-P., Buchert, M., Hagelüken, C., Reck, B.K., Sibley, S.F., Sonnemann, G., 2011b. Recycling Rates of Metals – A Status Report, a Report of the Working Group on the Global Metal Flows to UNEP's International Resource Panel.

Hagelüken, C., 2012. Recycling the platinum group metals: a European perspective. Platinum Metals Review 56 (1) 29–35.

Matsuno, Y., Daigo, I., Adachi, Y., 2007. Application of Markov chain model to calculate the average number of times of use of a material in society – an allocation methodology for open-loop recycling – part 2: case study for steel. International Journal of Life Cycle Assessment 12 (1), 34–39.

Nakajima, K., Takeda, O., Miki, T., Matsubae, K., Nakamura, S., Nagasaka, T., 2010. Thermodynamic analysis of contamination by alloying elements in aluminum recycling. Environmental Science & Technology 44 (14), 5594–5600.

Oguchi, M., Murakami, S., Sakanakura, H., Kida, A., Kameya, T., 2011. A preliminary categorization of end-of-life electrical and electronic equipment as secondary metal resources. Waste Management 31 (9–10), 2150–2160.

Reuter, M.A., Hudson, C., van Schaik, A., Heiskanen, K., Meskers, C., Hagelüken, C., 2013. Metal Recycling. Opportunities, Limits, Infrastructure. A Report of the Working Group on the Global Metal Flows to UNEP's International Resource Panel.

Rollat, A., 2012. How to satisfy the rare earths demand. Rhodia Rare Earth systems initiatives. In: Déjeuner-conférence sur "Les défis de l'approvisionnement en terres rares". European Society for Engineers and Industrialists, Brussels.

Streicher-Porte, M., Widmer, R., Jain, A., Bader, H.P., Scheidegger, R., Kytzia, S., 2005. Key drivers of the e-waste recycling system: assessing and modelling e-waste processing in the informal sector in Delhi. Environmental Impact Assessment Review 25 (5), 472–491.

van Schaik, A., Reuter, M.A., 2004. The time-varying factors influencing the recycling rate of products. Resources Conservation and Recycling 40 (4), 301–328.

Yoshida, M., Monozukuri, N., 2012. Mitsubishi develops eco-friendly, low-cost Ga recycling technology. In: Tech-On! Nikkei Business Publications, Inc. (NikkeiBP). http://techon.nikkeibp.co.jp/english/NEWS_EN/20121203/254051/.

Yoshimura, A., Daigo, I., Matsuno, Y., 2011. Construction of global scale substance flow of indium from mining. Journal of the Japan Institute of Metals 75 (9), 493–501.

CHAPTER 4

Recycling Rare Metals

Robert U. Ayres[1], Gara Villalba Méndez[3], Laura Talens Peiró[2]

[1]INSEAD, Fontainebleau, France; [2]Institute for Environmental Protection and Research, Ispra, Italy; [3]Universitat Autònoma de Barcelona, Spain

4.1 INTRODUCTION

The application of material flow analysis (MFA) to trace metals throughout their life cycle is an exercise for quantifying the magnitude of the uses and losses and identifying where they occur. Furthermore, by distinguishing between losses that are dissipative versus losses that are recoverable, the MFA can help us assess the potential for recycling of the material. This is a very practical analysis for geologically scarce metals used in new technology products such as smartphones, wind turbines, and solar panels. These metals e.g. gallium, germanium, indium, tellurium, tantalum, and platinum group metals (PGMs), are of special concern in terms of geological scarcity. They also pose a problem for recycling due to the trend toward miniaturization in their uses. Many of these metals are not found anywhere in high concentrations but are distributed as contaminants of other "attractor" metals to which they are chemically similar. For this reason, we call these metals "hitchhikers" when they accompany "attractors". For example, molybdenum, rhenium, selenium, silver, and tellurium are hitchhiker metals the production of which depends to a large extent on the mining and smelting of copper (Talens Peiró et al., 2013). Losses occur initially in the production process that, for the case of the hitchhiker metals, includes the mining of the "attractor" ore, mineral processing, and further extraction, separation, and refining processes. A very simple calculation can be performed to estimate the theoretical amount of the hitchhiker metal that could be extracted based on the quantity of the metal present in the ore. Even though the potential metal production estimated this way does not consider technological limitations and unavoidable losses, it serves to identify which metals offer opportunities for improved resource management and which ones are already being managed efficiently. We have performed these calculations for hitchhiker metals for which one might expect revolutionary demand due to their use in new technologies. We present the results in Table 4.1.

Metals for which potential mine production is comparable to current (2010) output do not have much margin for increasing recovery during the production phase. For example, cobalt from nickel ores is extracted efficiently (99%), whereas recovery from copper sulfide ores such as carrolite is less efficient (8%) resulting

TABLE 4.1 Potential and Actual Production of Hitchhiker Metals in 2010

"Attractor" Metals or Mineral Ores			"Hitchhiker" Metals				
Name	Reserves (10^6 t)	Production (10^6 t)	Name	Reserves (10^3 t)	Current Mine Production (t)	Potential Mine Production (t)	Recovery Efficiency (%)
Iron ore	87,000	2400	Rare earth oxides[a]	110,000	54,000	4,114,280	1
			Niobium[a]	2900	63,000	89,140	71
Bauxite	28,000	211.00	Gallium[b]	n.a.	106	10,550	1
Copper	630	16.20	Cobalt[c]	7300	31,000	408,800	8
			Rhenium[d]	2.5	46	9370	<1
			Molybdenum[d]	9800	133,000	281,050	47
			Tellurium[e]	22	475	1050	45
			Selenium[e]	88	3250	4210	77
Zinc	250	12.00	Germanium[e]	0.45	84	597	14
			Indium[f]	n.a.	574	1454	39
			Gallium[e]	n.a.	—	420	—
Nickel	76	1.55	Cobalt[g]	7300	44,000	44,600	99
			PGMs[d]	66	11	17	63
Tin	5.2	0.26	Niobium[d]	2900	2	373	1
			Tantalum[d]	110	102	746	14

Sources: (a) Drew et al. (1991), (b) Jaskula (2011), (c) Shedd (2011), (d) Habashi (1997), (e) USGS (1973), (f) Tolcin (2011), (g) Berger et al. (2011).

in losses of up to 378,000 t of cobalt. The table also serves to point out a potential source for gallium from zinc (420 t) that is presently not being exploited (Berger et al., 2011). On the other hand, the PGMs selenium and molybdenum are currently being produced near their mine potential from current sources. These metals are correspondingly important to recover and recycle to meet future demand. For the rest of the metals listed in Table 4.1, a significant fraction of the potential is ending up as waste.

Similarly, the subsequent life cycle stages have unnecessary losses and consequently opportunities for recovery. In the next few pages, we illustrate how material flow analysis is used to quantify the losses and recycling potential during the whole life cycle of several critical metals: indium, europium, gallium, and platinum group metals (PGMs).

4.2 INDIUM

In our economy, we use the life cycle of indium to illustrate how metal losses can be quantified from mining to end-of-life (EOL) based on consumption data for 2010. Based on knowledge of the processes and end-uses of indium, we differentiate between dissipative and recoverable fractions. The analysis is illustrated in Figure 4.1, which indicates the losses of indium during its life cycle in the present situation

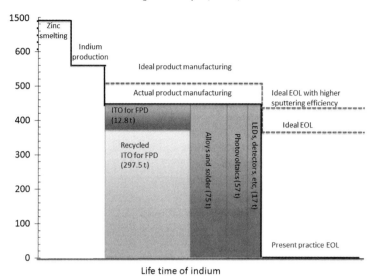

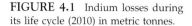
FIGURE 4.1 Indium losses during its life cycle (2010) in metric tonnes.

(2010) and the potentially recoverable fraction marked by a more sustainable practice represented by the dotted line.

Indium is of special interest for several reasons. It is classified as a "critical metal" by numerous reports because it is a scarce metal that plays an important role in the solar energy industry, in flat screens, and in other new technology applications (US National Research Council and Committee on Critical Mineral Impacts of the US Economy, 2008; Buchert et al., 2009; European Commission, 2010a; Erdmann and Graedel, 2011). For many of these uses, there are no substitutes that offer the same performance. Furthermore, given the miniaturization trend required by new technologies, indium is used in very small quantities that impede its separation, recovery, and recycling. For example, a 2005 cell phone screen contained as little as 5.5 mg of indium per phone (the metals presently recovered from cell phones such as gold and silver represent 350 mg and 3.7 g, respectively) (Yoshioka et al., 1994; Hagelüken, 2007b). PC monitors contain between 79 and 82 mg of indium, whereas bigger liquid crystal display (LCD) television screens use 260 mg (Buchert et al., 2009). However, there are so many of these products that these small amounts of indium add up to a potentially important source for future supply. Without recycling or implementing substitutes, the sustainability of these technologies is questionable.

Indium is most commonly recovered as a by-product of the zinc smelting process from zinc-sulfide ores (sphalerite). Other minor sources are from tin and lead refining processes, which at the present time are not exploited. The content of indium in zinc concentrate varies depending on the geographic location of the mine. For example, Rodier gives 0.027% (by weight) of indium content in zinc concentrate for Kidd Creek, Ontario, Canada; 0.010% in Polaris, Northwest Territories, Canada; and 0.004% in Balmat, New York (Rodier, 1980). Based on the intermediate concentration of 0.010%, an average zinc content of 55% in zinc concentrate, and a global production of eight million tonnes of zinc in 2010, we estimate that the amount of indium that could have potentially been recovered from zinc-bearing ores in 2010 is 1454 t. The USGS reports that 574 t of indium were produced globally for the same year (Tolcin, 2011). In other words, a third of the total indium currently available from zinc refineries is not recovered. This coincides with the amount Indium Corporation estimates is lost to tailings or non-indium-capable refineries (about two-thirds of all indium in zinc ores) (US Department of Energy, 2011). There is a margin for improving recovery at this initial stage, but it is still not economically attractive to do so.

High purity indium is used to make compounds with other metals for the manufacture of products. The most important is indium-tin-oxide (ITO) (78% indium, 17.5% oxygen, and 4.5% tin). Thanks to its electrical conductivity and optical transparency, ITO is used for transparent electrodes in products such as LCDs, TVs, and touch screens. Based on a previous study, we estimate that 425 t of indium were used for the production of ITO in 2011 (Talens Peiró et al., 2013). ITO comes as a powder and is deposited as a thin, film coating on a substrate via a process called "sputtering" that is reportedly only 3% efficient.

Sputtering deposition techniques vary, but the most common ones are electron-beam evaporation and physical vapor evaporation. Approximately 3% of the indium ends up successfully on the substrate, although 70% is target residual material (recycled by reducing to indium metal), 20% is deposited onto the surface of tools and chamber walls, 5% is etched from the substrate, and 2% ends up in faulty panels (Yoshioka et al., 1994). These percentages are based on a study from 2007, and it is possible that sputtering has become more efficient since then (Goonan, 2012). However, for lack of updated process data, based on these percentages we calculate that 12.8 t of indium are embodied in flat-panel display (FPD) products, 297.5 t are recycled to make more ITO, and 114.7 t are dissipated during the FPD production process. These quantities are shown in the product manufacture stage of Figure 4.1. The upper dotted line represents a greater amount of indium available if the sputtering process resulted in half the indium losses (58 t instead of 114.7 t). The figure also shows the fraction of the indium in ITO that is not successfully placed on the substrate and is recycled to make ITO.

Presently there is no indium recycling from post-consumer products that employ LCDs such as TVs, personal computers (PCs), mobile phones, and car navigation systems. The indium is diluted and dissipated in waste management schemes (recycling, landfill, etc.) at the EOL of the products. Sharp Corporation in Japan has invented a process for recovering ITO from LCDs, which involves crushing the screens into small chips and treating them in acid solution. A large-scale implementation in the future could result in an important source of indium (Kawaguchi, 2006). Given the growth in consumption of touch screens and flat-panel TVs, as well as the increase in average screen size, this recoverable indium is crucial for future supply. In Figure 4.1, the ideal EOL stage includes the indium embodied in LCDs, whereas the present situation shows no recovery.

Indium is also used to make alloys with base metals to lower their melting point. The indium-based alloys are used as fusible alloys (needed for devices such as fire sprinkler systems), holding agents, and solders. Indium solders are preferred in printed circuit board (PCB) assembly because indium is lead-free and has excellent wetting properties. Alloys of indium with precious metals such as gold and palladium are used in dental work due to their ductile nature. We estimate that 75 t of indium were destined for alloys and solders in 2010, of which 6 t were destined for PCBs (Talens Peiró et al., 2013). Some of these applications render the indium unrecoverable because it is used in such small quantities. PCBs are presently being recycled to recover copper, gold, silver, and palladium, but other critical metals such as indium are not being recovered (Buchert et al., 2009). The use of indium in solders and alloys can be considered dissipative at the EOL. The "ideal" scenario represented in Figure 4.1 does not include the 75 t of indium destined for alloys and solders as recoverable.

The solar energy industry is also an important consumer of indium. According to industry figures, 57 t of indium were used in the form of copper-indium-gallium-diselenide (CIGS) employed as a semiconductor in photovoltaic cells in the form of thin-film photovoltaic material (Talens Peiró et al., 2013). ITO is used as an anode in the photovoltaic cells to increase the light conversion efficiency because of its light trapping properties and high transparency. Other indium compounds, such as indium antimonide (InSb), indium nitride (InN), and indium phosphide, are also used as semiconductors in photovoltaic cells and in other applications such as infrared detectors and thermal imaging cameras. Photovoltaic cells have a lifetime of 20—30 years, after which no recovery of indium is presently taking place. Potentially the indium present in the cells could be recovered in a similar fashion as the FPDs, and this is represented in the "ideal" scenario represented by the dotted line in Figure 4.1.

Other semiconductor applications such as laser diodes, fiber optic telecommunications, detectors, and light-emitting diodes (LED) were responsible for 17 t of indium. Indium in the form of indium gallium nitride (InGaN) is used as a light-emitting layer in blue and green LEDs. Indium gallium arsenide (InGaAs) is a popular material in infrared detectors. We are not aware of any recovery of the indium used in these products, but potentially, this is a recoverable fraction because the indium is not dissipated during use.

As is illustrated by Figure 4.1, less than 1% of indium is presently recovered at EOL as has been shown by other authors (Graedel et al., 2011). However, what the figure also shows are the fractions of the indium produced that are not dissipative and that could potentially be recovered. Recovering indium from LCDs and photovoltaic cells could result in 384 t of indium at EOL (minus unavoidable losses). If we assume a 50% improvement in the efficiency of the sputtering process, the recoverable indium at EOL is further increased to 442 t. The use phase is not included in the diagram because we know of no losses of indium that occur during the use of the products described previously.

4.3 OTHER EXAMPLES OF RARE METALS

Other metal life cycles can be analyzed in a similar fashion. We next explain three examples of scarce metals: europium, gallium, and platinum group metals (PGMs), all used in low quantities, yet crucial for different technologies.

Europium is one of the 17 so called rare earth metals (REMs). Its major mineral source is bastnäsite, a carbonate fluoride mineral found in association with (some) iron mineral ores (such as magnetite, hematite, goethite, and limonite). The content of bastnäsite in iron ores varies from 4—17%. Bastnäsite contains about 0.1—0.5% of europium, a very small concentration compared to lanthanum, cerium, and neodymium, which are found in 42—50%, 23—33%, and 12—20%, respectively (Gupta and Krishnamurthy, 1992). The commercial application of europium is based on its phosphorescence. It is added to semiconductors in trace amounts, either in the +2 or +3 oxidation state, to improve the emission of light in various wavelengths. Divalent europium (Eu^{+2}) tends to give blue phosphors, whereas trivalent europium (Eu^{+3}) gives red phosphors. Both of them combined with terbium-based phosphors yield white light (Gupta and Krishnamurthy, 2005).

In 2010, the entire global production of europium (404 t) was used in phosphors for cold cathode fluorescent lamps (CCFLs), in LEDs fitted in LCDs for background illumination, and in LED for light bulbs. Although most of the market is now dominated by LCDs using LEDs, to estimate the amount of europium stock in EOL LCDs, we need to know the amounts contained by each type of background lighting. The amount of europium in LCDs using CCFLs varies from 8.10 mg for TV to 0.13 mg for notebook computers, whereas LCDs using LEDs contain an even lower amount: 0.09 mg for TVs compared to 0.03 mg for notebooks. Europium is also used to give a reddish color in warm white LEDs that have a correlated color temperature lower than 3000 K. The average weight of europium in warm white LEDs varies from 0.4 to 0.9 μg (Buchert et al., 2012). Considering that losses of europium from mining to the manufacturing of phosphors are estimated to be 25%, we estimate that the remaining 75% is all contained in LCDs and LEDs.

At present, europium is rarely recycled (Du and Graedel, 2011; Graedel et al., 2011). The most important obstacle is collection due to its use in very tiny amounts. As an example, recovering one ton of europium would require the collection of at least 1.3 trillion units of white light LEDs. From a technological perspective, the recovery of europium also requires detailed knowledge about the composition of the phosphors. For instance, phosphors generally contain other rare earth elements such as yttrium, cerium, and lanthanum in the support matrix, not to mention terbium and gadolinium as dopants. The combined use of rare earths in phosphors inhibits their recovery because they all have very similar properties. Hence, the separation of any one from the others requires many processing steps.

Although gallium is a relatively common element, it mainly occurs as a trace amount in bauxite and zinc ores such as sphalerite. At present, almost all gallium is obtained as a by-product of alumina production from bauxite, which contains an average of 60 ppm (Gray and Kramer, 2005). In 2010, the world production of refined gallium was 161 t. Of that, 106 t were obtained as crude gallium and the remaining 55 t from preconsumer recycling (Jaskula, 2010). In 2010, 65% of gallium output was obtained as a "hitchhiker" of aluminum smelting. Table 4.1 estimates that the current production of gallium represents only 3% of the total amount contained in bauxite. It also identifies zinc ore as a potential source of gallium that is not being exploited at present.

About 66% of the gallium consumed in 2010 was used in integrated circuits (ICs). Phosphors accounted for 18%, thin films 2%, and the remaining 14% of gallium consumption was for research and development, specialty alloys, and other applications (European Commission, 2010b; Talens Peiró et al., 2013). The most important gallium compounds are arsenides, nitrides, and phosphides. Gallium arsenide (GaAs) and aluminum gallium arsenide (AsGaAl) can convert electrical signals into optical signals at high speed with low power consumption, and better resistance to radiation compared to other compounds (Habashi, 1997). They are widely used as substrates in the manufacture of semiconductor components as transistors and ICs for the electronics and telecommunications industry. For example, a mobile phone contains from 0.3 to 1.5 mg of gallium, and fourth generation smartphones use up to 10 times that amount (Talens Peiro and Villalba Méndez, 2011; Jaskula, 2013). Gallium nitrides (GaN) are primarily applied in light-emitting diodes (LEDs) for the backlighting of LCDs for TVs and notebooks. Buchert et al. estimates that an LCD TV contains 4.90 mg and a notebook 1.60 mg of gallium (Buchert et al., 2012). Gallium phosphides and phosphides complex of gallium, aluminum and indium are also used for the production of optoelectronic components and ICs. Gallium is also used in thin-film solar cells as CIGS and in the triple-junction cells: indium gallium phosphide (GaInP$_2$) (Green and Emery, 2011).

More than 85% of gallium ends up in theoretically nondissipative uses: electronics, phosphors, and thin film photovoltaic panels. The numerous steps during manufacturing and the high quality requirements for most of the electronic products using gallium lead to substantial processing losses. Typically only 20–30% of the gallium is finally embodied in the end-products (Koslov et al., 2003). For instance, in the production of GaAs substrates, 30% of the initial gallium is embodied in the final product; the remaining 70% is lost during such processes as etching and polishing. Eichler estimates that nearly 35% of the losses can still be recovered in the different steps (Eichler, 2012).

In CIGS manufacturing, gallium can be deposited in the absorber cell layer by various technologies: electron-beam (EB), electrochemical deposition, and co-evaporation. The amount of gallium deposited for electron-beam and co-evaporation is only 20% of the total gallium input. For electrochemical deposition, the amount deposited is 30% of the total input (Kamada et al., 2010). Based on the fact that a 1 MW CIGS panel contains about 3.5–4 kg of gallium, we estimate that 7.5–16.5 t of gallium are wasted during the manufacturing stage (Kalejs, 2009; Christmann et al., 2011). New developments to minimize the loss of gallium during CIGS manufacturing are based on using inkjet printing for deposition. This technology could reduce the amount of gallium wasted by 90% and thus reduce the cost of thin-film solar cells (Quick, 2011).

At present, gallium is only recovered from new scrap (preconsumer scrap) generated during the manufacturing of semiconductors, mainly from GaAs wafer waste, and from liquid phase deposition. There are several options for the recovery of gallium from semiconductor waste: thermal dissociation, oxidation with oxygen, nitriding in ammonia, and chlorination with chlorine gas (Koslov et al., 2003). Most of the difficulty in recycling gallium is due to the problem of separation from the other components of contact alloys such as indium, tin, germanium, lead, silver, copper, and gold. In 2010, about 55 t of gallium were produced from preconsumer scrap (Jaskula, 2013).

Many efforts to recycle gallium focus on increasing the collection of EOL products. In Europe, the PV Cycle association, which represents more than 90% of the EU photovoltaic market, recently set up a voluntary collection and recycling scheme. In 2010, PV Cycle collected about 80 t of modules. In 2011, this grew to 1400 t and, in the first two quarters of 2012, 2250 t (Neidlein, 2012). In the EU, photovoltaic module take back and recycling became mandatory in February 2014. Even though thin films represent only 5% of the current photovoltaic panels market, they may eventually dominate the future market because they have more potential for cost reduction. Moss et al. estimates that thin films will have a market share of 18% by 2020; thus, gallium demand would increase to 14 t (Moss et al., 2011).

PGM is the term used for a group of six metals: platinum, palladium, rhodium, ruthenium, iridium, and osmium. They tend to occur together in primary and secondary mineral deposits. In primary deposits, they are found in platiniferrous ores in association with iron sulfides and sulfides of nickel, cobalt, and copper. Economical deposits mainly contain sperrylite ($PtAs_2$), cooperite (PtS), stibiopalladinite (Pd_3Sb), laurite (RuS_2), ferroplatinum (Fe–Pt), polyxene (Fe–Pt–other platinum metals), osmiridium (Os–Ir), and iridium platinum (Ir–Pt). PGMs can also be found in secondary deposits formed by weathering and washing of primary deposits. The content of PGM varies depending on the type of mineral deposit. In South Africa, the average platinum grade is about 8 g/t of ore, but it can reach 27g/t in deposits associated with nickel ore. In Russia, where platinum is obtained as a by-product of nickel and copper, the average grade is about 5 g/t. The average grade can reach 15 g/t in sulfide deposits (Renner, 1997). In 2010, the world production of PGM was 689 t; 456 t were obtained from platinum sulfides and arsenides, 222 t from recycling, and 11 t were obtained as hitchhikers of nickel (Talens Peiró et al., 2013). According to Table 4.1, the current production of PGM as a hitchhiker of nickel is near its potential level.

Among the PGM group, we can differentiate subgroups based on production quantities. Platinum and palladium are a first group of metals produced in multiples of 100 t/year. Rhodium and ruthenium are a second group produced in multiples of 10 t/year; iridium and osmium can be regarded as a third group, produced in very small quantities. In brief, rhodium, ruthenium, iridium, and osmium can be regarded as by-products (or "hitchhikers") of platinum and palladium. In 2010, the global production of PGMs was: 249 t of platinum, 360 t of palladium, 36 t of rhodium, 35 t of ruthenium, 7 t of iridium, and 2 t of osmium.

PGMs are exceptional oxidation catalysts. They are electrically and thermally conductive and offer a high oxidation resistance combined with extraordinary catalytic activity, chemical inertness, high melting point, temperature stability, and corrosion resistance (Lofersky, 2010). Almost 50% of the demand for PGMs is for use in automobile catalytic converters, 14% is used for jewelry, 9% for investment, 4% for other electrical uses, 1% for catalysts in the chemical industry, 1% for alloys with electrochemical applications, less than 1% for the glass industry, and the remaining 22% to other unspecified uses (Lofersky, 2011). See Figure 4.2. Some PGMs are lost from their extraction in the manufacturing of end-products. Dissipative uses of PGM include dental restorative materials and in the glass industry for the development of flat-panel displays as LCDs, plasma (PDP), and as LEDs (Buchert et al., 2009). The PGMs used as catalysts in the chemical and petrochemical sectors are mostly close-loop recycled, and the amount dissipated during use is negligible. Platinum used in jewelry accumulates over time, but some is recycled. Recoverable fractions of PGM include automobile catalytic converters and from electrical uses. Figure 4.2 shows platinum's dissipative (regarded as lost), stock, and recycling fractions.

As illustrated in Figure 4.2, almost 40% of platinum is lost in nonrecoverable uses, 18% is added to the stock of end-products, and 23% is recycled. Platinum is lost when it is used in medical applications, in the glass industry, and for other uses such as automotive sensors, the coating of aircraft turbine blades, and spark

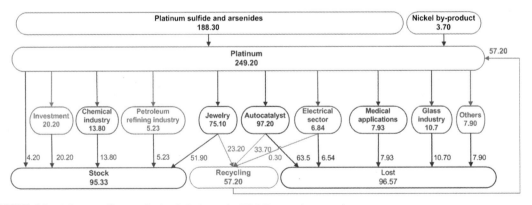

FIGURE 4.2 Substance flow analysis of platinum in 2010 (in metric tonnes).

plugs. It is also lost from the fraction of automobile exhaust converters and electrical equipment not collected and adequately recycled.

In 2010, automobile catalytic converters used 99 t of platinum, 204 t of palladium, and 30 t of rhodium (Butler, 2011). Catalytic converters are part of the exhaust system designed to accelerate the oxidation of carbon monoxide (CO), nitrogen oxides (NO_x), and unburnt hydrocarbons. They are composed by a ceramic or metal honeycomb structure of channels coated with so-called "wash-coating" catalysts that are composed of a combination of PGMs and rare earth oxides (Borgwardt, 2001; Lifton, 2007). The best known combinations of PGMs are platinum—rhodium (Pt—Rh), platinum—palladium (Pt—Pd), and a three-way combination of all three metals (Pt—Pd—Rh). The first two combinations accelerate the complete oxidation of carbon monoxide and hydrocarbons. The three-way catalyst also reduces nitrogen oxides to pure nitrogen and oxygen. PGMs are fixed in the wash-coat surface usually by impregnation or by coating with a solution of hexachloroplatinic (IV) acid ($H_2PtCl_6 \cdot 6H_2O$), palladium chloride ($PdCl_2$), and rhodium chloride ($RhCl_3$) salts. As the solvent evaporates, a dry layer of the PGM salts results in the surface of the honeycomb structure (Ravindra et al., 2004).

PGM losses during the wash-coat formation range from about 2—6% as maximum (Patchett and Hunnekes, 2012). Taking into account that current catalytic converters contain an average amount of about 2 g of PGM, about 4—12 mg of PGM are lost during manufacturing (Yentekakis et al., 2007). Based on the data shown in Figure 4.2, we estimate that between 2 and 6 t of platinum were lost during wash-coat formation in 2010. PGMs are also lost during automobile engine operation due to the rapid change of oxidative/reductive conditions, high temperature, and due to mechanical abrasion. The average quantity of PGMs released into the environment is in the range of 65—180 ng/km (Ravindra et al., 2004).

A possible way to estimate the amount of PGM lost during vehicle operations is to consider the current stock of vehicles in service and the average distance traveled by them.

A more accurate estimate of the amount of PGMs potentially recoverable can be compiled by counting deregistered vehicles. A case study in Germany estimated that only 41% of catalytic converters from deregistered vehicles reach dismantler facilities, and about 70% of the PGMs contained in those converters are recovered. The remaining 59% of catalytic converters are not separated from the vehicle and are not recycled. Many (perhaps most) are exported to areas without adequate recycling infrastructure. In some cases, convertors are not removed from car bodies before being shredded, or they are mistakenly sorted into the wrong fractions, from which separation is not feasible (Hagelüken, 2007a). In 2010, there were 33.7 t of platinum, 41.2 t of palladium, and 7.3 t of rhodium recycled from automobile catalytic converters (Lofersky, 2011).

In the electrical sector, PGMs are used for active components such as transistors, ICs, thick-film hybrid circuits and semiconductor memories, and also in passive components, which include multilayer capacitors, thick-film resistors, and conductors. The most important PGM used in electronics is ruthenium in combination with platinum, palladium, and iridium in alloys. Ruthenium is used in small quantities to increase wear resistance for electric contacts, and in microelectronics in computer hard disks, multilayer ceramic capacitors, and in hybridized ICs. At present, neither ruthenium nor iridium is recycled (Lofersky, 2011).

In 2010, only platinum and palladium were recycled from electrical and electronic waste. The amount of platinum recovered represented 5% of the amount consumed, as shown in Figure 4.2. In 2010, there were 113.7 t of palladium recycled from electronic waste, whereas the global sales of palladium for electronics were 43.9 t (Lofersky, 2011). These figures show that the recycling system and the technology for recovering palladium are functioning. In fact, in 2010, recycled palladium supplied 44% of the world's production. Platinum and palladium were also recycled from jewelry: 23.2 t of platinum and 2.5 t of palladium.

4.4 THE DISTANT FUTURE: GEORGESCU'S LAST LAUGH?

Georgescu-Roegen proclaimed that "matter matters" and asserted that the "entropy law" is the "taproot" of economics (Georgescu-Roegen, 1971). His point was that dissipation of matter (i.e. chemical elements) due to the impossibility of perfect recycling constitutes a "fourth law" of thermodynamics and an unavoidable limit to economic growth. He was wrong about the fourth law and wrong about the technical impossibility of recycling, given an unlimited supply of useful energy (exergy) (Ayres, 1999). This technical error has somewhat discredited his entire thesis. However, the long-term prospects for technologies that utilize geologically scarce "hitchhiker" metals are very poor from an economic perspective.

As noted earlier, the short- to medium-term situation is that most rare metals are currently underexploited in the sense that the potential supply from mining of the important primary ("carrier") metals (iron, aluminum, copper, zinc, nickel, etc.) is larger than the actual output. In some cases, the potential supply is hundreds of times larger than the current consumption; therefore, there is no immediate need for recycling. But in other cases (e.g. cobalt and molybdenum), current output is reasonably close to maximum potential output, given the current demand for the associated carrier metal (e.g. copper). Of course, the potential supply of copper is also limited, although not (it seems) for many decades to come.

With regard to standard industrial metals such as iron and aluminum, the ores or possible ores are plentiful, and most of the uses are not inherently dissipative. There are dissipative uses of iron such as nails and small hardware items, as well as rust from structural beams and iron oxide pigments. There are dissipative uses of aluminum, especially aluminum foil and aluminum bottle caps. Hence iron and aluminum will be mined to some extent even in the very distant future unless those "wasteful" uses are also eliminated. Nickel, cobalt, and molybdenum will be recycled to a high degree because their metallurgical uses, such as in stainless steel or refractory (heat resistant) alloys, are such that collection and separation are not too difficult. Copper, zinc, and lead are very easily recycled. In the case of copper, it is easy to recycle copper roofing, copper tubing for water, and copper from household wiring, motor and generator windings, and so forth. But copper for computer chips, brass cartridges for bullets, and copper chemicals are very difficult or impossible to recycle. In the case of lead, old water pipes, radiation shielding, and lead-acid batteries are easily recycled, but lead-based solder is not, and lead bullets and lead shot for shotguns are quite dissipative. Zinc has few metallic uses except for some kinds of hardware. Zinc coating for iron ("galvanizing") is one of that metal's biggest uses. Recovery of zinc from galvanized scrap iron is possible but not very profitable. Similarly, PGMs are easily recycled from industrial catalysts, or fuel cells, but it is getting very hard to recover PGMs from automobile exhaust systems unless the EOL vehicles are carefully dismantled prior to scrapping operations. However, in the very long run, automotive vehicles with internal combustion engines that burn gasoline or diesel may be largely replaced by electric vehicles.

For the major metals, we may never eliminate the need for some mining activity, but the resource base is large enough that no supply crisis is likely for centuries to come. Recycling will never be 100% effective, thanks to the Second Law, but fairly high recovery rates from demolition scrap and scrap cars can be foreseen. Also, the value of secondary metals (especially gold and silver) is already keeping some mines open. Similarly, the increasing value of some very scarce metals, such as indium or tellurium, may eventually justify the investment in large-scale recovery and in the recycling of certain electronic items such as touch screens or PV panels.

Unfortunately, new IT products, such as iPhones and iPads, use more and more of the scarce metals in trace amounts. It is important to ensure that this trend toward microminiaturization does not prevent us from recovering these metals. In the future, this will require much higher prices—such as those that we now see for gold and platinum. From a systems perspective, this may have to be accomplished by innovative means such as "renting" the scarce materials. The implications of such schemes will require extensive study and analysis.

References

Ayres, R.U., 1999. The second law, the fourth law, recycling and limits to growth. Ecological Economics 29 (3), 473–483.

Berger, V.I., Singer, D.A., et al., 2011. Ni-Co Laterite Deposits of the World. Database and Grade and Tonnage Models. United States Geological Survey.

Borgwardt, R., 2001. Platinum, fuel cells and future US road transport. Transportation Research, Part D: Transport and Environment 6 (3), 199–207.

Buchert, M., Manhart, A., et al., 2012. Recycling Critical Raw Materials from Waste Electronic Equipment. Sustainable Innovation and Technology Transfer Industrial Sector Studies. Oeko-Institut e.V.

Buchert, M., Schueler, D., et al., 2009. Critical Metals for Future Sustainable Technologies and Their Recycling Potential. Sustainable Innovation and Technology Transfer Industrial Sector Studies. UNEP.

Butler, J., 2011. Platinum 2011. PGM Market Review. Johnson Matthey.

Christmann, P., Angel, J.M., et al., 2011. Panorama 2010 du marché du gallium. Compagnie Européenne d'Intelligence Stratégique (CEIS), pp. 1–53.

Drew, L.J., Qingrun, M., et al., 1991. The geology of the Bayan Obo iron-rare earth-niobium deposits, Inner Mongolia, China. Materials Science Forum 13 (70–72), 13–32.

Du, X., Graedel, T.E., 2011. Uncovering the global life cycles of the rare earth elements. Scientific Reports 1, 145.

Eichler, S., 2012. Green gallium arsenide (GaAs) substrate manufacturing. In: The International Conference on Compound Semiconductor Manufacturing Technology, Boston, Massachusetts, USA.

Erdmann, L., Graedel, T.E., 2011. Criticality of non-fuel minerals: a review of major approaches and analyses. Environmental Science & Technology 45 (18), 7620–7630.

European Commission, 2010a. Annex V to the Report of the Ad-hoc Working Group on Defining Critical Raw Materials. European Commission, Brussels, pp. 220.

European Commission, 2010b. Critical Raw Materials for the European Union. European Commission.

Georgescu-Roegen, N., 1971. The Entropy Law and the Economic Process. Harvard University Press, Cambridge, MA.

Goonan, T.G., 2012. Materials Flow of Indium in the United States in 2008 and 2009. US Geological Survey, Denver.

Graedel, T.E., Allwood, J., et al., 2011. What do we know about metal recycling rates? Journal of Industrial Ecology 15 (3), 355–366.

Gray, F., Kramer, D.A., 2005. Gallium and gallium compounds. Kirk-Othmer Encyclopedia of Chemical Technology. Kirk-Othmer, John-Wiley & Sons.

Green, M.A., Emery, K., 2011. Solar cell efficiency tables (version 37). Progress in Photovoltaic Research and Applications 19 (1), 84–89.

Gupta, C.K., Krishnamurthy, N., 1992. "Extractive metallurgy of rare earth. International Materials Reviews 37 (5), 197–248.

Gupta, C.K., Krishnamurthy, N., 2005. Extractive Metallurgy of Rare Earths. CRC Press.

Habashi, F. (Ed.), 1997. Handbook of Extractive Metallurgy. Wiley-VCH.

Hagelüken, C., 2007a. Closing the loop — recycling automotive catalysts. Metall 61, 24–39.

Hagelüken, C., 2007b. RO7 Metals Recovery from E-scrap in a Global Environment. OEWG, Basel. Umicore.

Jaskula, B.W., 2010. Mineral Commodity Summaries: Gallium. US Department of the Interior, Washington, DC. United States Geological Survey, pp. 28–59.

Jaskula, B.W., 2011. 2009 Minerals Yearbook: Gallium. US Department of the Interior, Washington, DC. United States Geological Survey I, pp. 1–21.

Jaskula, B.W., 2013. Minerals Commodity Summaries: Gallium. US Department of the Interior, Washington, DC. United States Geological Survey I, pp. 58–59.

Kalejs, J., 2009. NSF Workshop. Manufacturing Challenges for PV in the 21st Century, Arlington, VA, USA.

Kamada, R., Shafarman, W.N., et al., 2010. Cu(In,Ga)Se$_2$ film formation from selenization of mixed metal/metal-selenide precursors. Solar Energy Materials and Solar Cells 94 (3), 451–456.

Kawaguchi, Y., 2006. Coming full circle. Recycling Magazine.

Koslov, S.A., Potolokov, N.A., et al., 2003. Preparation of high-purity gallium from semiconductor fabrication waste. Inorganic Materials 39 (12), 1257–1266.

Lifton, J., 2007. By-products III: Rhodium. Resource Investor News.

Lofersky, P.J., 2010. Mineral Commodity Summaries: Platinum-Group Metals. US Department of the Interior, Washington, DC. United States Geological Survey, pp. 120–121.

Lofersky, P.J., 2011. 2010 Minerals Yearbook: Platinum-Group Metals. US Department of the Interior, Washington, DC. United States Geological Survey, pp. 1–12.

Moss, R.L., Tzimas, E., et al., 2011. Critical Metals in Strategic Energy Technologies. Assessing Rare Metals as Supply-Chain Bottlenecks in Low-Carbon Energy Technologies. JRC – Institute for Energy and Transport, pp. 1–162.

Neidlein, H.C., 2012. PV module recycling mandatory for all EU members by Q1 2014. PV Magazine.

Patchett, J., Hunnekes, E.V., 2012. Gas Catalysts Comprising Porous Wall Honeycombs. http://www.patentgenius.com. B. Corporation. US, Sample, D, pp. 1–19.

Quick, D., 2011. Researchers Cut Waste and Lower Cost of 'CIGS' Solar Cells Using Inkjet Printing Technology. Gizmag.

Ravindra, K., Bencs, L., et al., 2004. Review: platinum group elements in the environment and their health risk. Science of the Total Environment 318 (1), 1–43.

Renner, H., 1997. Platinum group metals. In: Habashi, F. (Ed.), Handbook of Extractive Metallurgy. Wiley-VCH (Section 25).

Rodier, D.D., 1980. Lead-zinc-tin '80. In: Cigan, J.M., Mackay, T.S., O'keefe, T.J. (Eds.), Proceedings of World Symposium on metallurgy and environmental control. The Metallurgical Society of AIME, Las Vegas.

Shedd, K.B., 2011. Mineral Commodity Summaries: Cobalt. US Department of Interior, Washington, DC. United States Geological Survey, pp. 46–47.

Talens Peiro, L., Villalba Méndez G., 2011. Results from the Workshop "Preliminary Assessment of Multifunctional Mobile Phones". Brussels.

Talens Peiró, L., Villalba Méndez, G., et al., 2013. Material flow analysis of scarce metals: sources, functions, end-uses and aspects for future supply. Environmental Science & Technology 47 (6), 2939–2947.

Tolcin, A.C., 2011. Mineral Commodity Summaries: Indium. US Department of Interior, Washington, DC. United States Geological Survey, pp. 74–75.

US Department of Energy, 2011. Critical Materials Strategy. US Department of Energy, pp. 1–196.

US National Research Council and Committee on Critical Mineral Impacts of the US Economy, 2008. Minerals, Critical Minerals and the US Economy. National Academy Press, Washington, DC.

USGS, 1973. United States Mineral Resources. US Gov't Printing Office, Washington, DC.

Yentekakis, I.V., Konsolakis, M., et al., 2007. Novel electropositively promoted monometallic (Pt only) catalytic converters for automotive pollution control. Topics in Catalysis 42–43, 393–397.

Yoshioka, K., Nakajima, T., et al., 1994. Sources of Total Factor Productivity for Japanese Manufacturing Industry: 1962–1988. Keio University, Tokyo, Japan.

Theory and Tools of Physical Separation/Recycling

Kari Heiskanen

Aalto University, Espoo, Finland

Materials for recycling may consist of end-of-life (EOL) product streams, byproducts and waste streams from original equipment manufacturing and the production of components, and finally also rejects, byproducts and waste streams from raw-material producers. A common feature is that all consist of compounds. The elements of the compounds can be recycled only by chemical or metallurgical means.

Recycled products can be characterized by the properties they have as a function of size. Properties can be physical or chemical in nature. They arise from the mass/area distribution of the compounds of a stream. These compounds can either be dominant in the product or form varying parts of them. As an example, a freight railroad car consists mostly of different carbon steels made by alloying three or four elements, but a mobile phone consists of a multitude of compounds made out of approximately 60 elements.

The chemical complexity of a compound as well as a metal alloy is not a function of particle size, and thus cannot be reduced by physical means. In practice, mixtures of very fine matter in a continuous matrix, such as pigments or flame retardants in plastic, also show a similar behavior, in that the complexity is not a function of particle size. Some of the most miniaturized electronic devices also belong to this group, because the required particle size is too small for practical purposes.

Most of the products manufactured are characterized by a variable scale showing a degree of particle size dependency. Some of them are complex and often contain components that fall into the first category in which complexity cannot be reduced. Good examples of this are EOL electronics or cars. At the other end of the spectrum are simple products that consist of few materials, and in which the joints between compounds (alloys) are easily breakable: for example, the freight railroad car.

Many waste streams, like metal production slag, have voluminous matrix components that cannot be reduced in complexity. These can be treated by the removal of the matrix into streams where the valuable components are concentrated.

Compounds in a recyclable stream are distributed and connected in different ways, affecting the size dependency. This is termed *liberation*. If the particle size is made finer, the liberation will change in a way that is typical to the recyclable material and method of particle reduction.

This characteristic is termed a *liberation curve*, named for the similar phenomenon in natural minerals. Particles in a recycled set will contain different mass fractions of different compounds. This is discussed later more in detail. These particles will also exhibit different physical and chemical properties that react to physical forces and the chemical environment in different ways. The chemical interactions may become complex and lead to unwanted reactive results, affecting recycling rates negatively.

This essential feature can be captured in a product-based approach to recycling, in contrast to a material-based approach. The former implicitly takes into account the whole cycle from the EOL phase to the production of renewed raw material. It also allows the interactions caused by the stream complexities to be modeling, to achieve improved recycling rates by technology and systems development. The latter is adequate for bookkeeping but does not address the effects of complexity. As an example, an error often made in material-based recycling discussions is to mix steel recycling with iron recycling. Steel forms a wide family of iron-based alloys. Many of the over 6000 alloys can be recycled together. However, for the production of recycled steel, there are limits to the content of several other metals included in the scrap, which either need to be diluted with primary material, separated away from the feed scrap stream or lost into production slag, fumes and dusts. Examples are copper and tin.

In addition to the complexity of a single product, practical recycling streams have another level of complexity arising from the collection phase where different products are collected into a combined stream. This may dilute the economic value of a stream substantially by changing the mass ratio of high-value compounds relative to low-value compounds. Because different products have different liberation curve characteristics, combining different kinds of streams will also increase the complexity of the liberation characteristics.

A complex set of properties makes recycling difficult. As an example, freight railroad car recycling does not pose difficulties, whereas mobile phone recycling does.

5.1 RECYCLING PROCESS

The collection of recyclable materials should be designed so that an unnecessary increase in stream complexity is avoided. Much can be done at the origin of the recycling process. In industry cuttings, turnings, rejects, etc., should be sorted carefully. The same applies to EOL goods from households, a much more difficult task. The optimal degree of presorting is dictated by the collection system costs and structures, location and process capabilities of treatment facilities, and economic incentives available for different actors.

After collection, the streams often tend to be too complex and unsuitable for final processing. Many valuable elements and compounds may be present at too low a value to merit the high cost of final processing. Thus, the normal step following collection is first to reduce the particle size to a more suitable finer size, and then to perform a mechanical separation step or several steps using the physical property differences between the particles in the feed stream. These streams are either further purified physically or sent for further processing by chemical or metallurgical means. Some material streams may already have reached a saleable commercial quality after physical treatment. Physical treatments range from manual sorting to sophisticated automated systems.

In all of these steps, the particle size, shape and density of individual pieces is important and affects the recycling process outcome.

5.2 PARTICLE SIZE

Any mass of material to be recycled consists of particles. The particles can be characterized by their size. Usually the shape of particles differs from a sphere, which is the only geometric form that has a well-defined unique size, its diameter. All other geometric and irregular forms have different sizes, depending on the technique used for measurement. The most common method for scrap sizing is sieving. The particle size x_A is characteristic of an aperture through which the particle passes. The sieve surfaces are usually woven wire cloths with square apertures. In this case, the size is the side length of the square. The surface can also be made with a punched square or round hole. Another common method for measuring particle size is to measure its settling velocity in liquid or air. Then, the size is given as the size of a sphere with the same settling velocity as the particle. This is called Stokes diameter, x_s, for small particles settling at laminar velocities (Reynolds number < 0.2), or more generally, drag diameter, x_d. Particles can be illuminated by light and their projections are measured to obtain the projected area diameter, d_p. Other image-processing diameters are the Feret diameter, d_F, and the Martin diameter, d_M. They are respectively the distance between parallel tangents to the image and the cord length of the particle image measured in a defined direction. Particle volume and particle surface can also be measured to yield the volume diameter, d_v, and surface diameter, d_s, which are the diameters of a sphere with the same volume or area as the particle. For fine fractions, laser diffraction is currently the most common method. The size is the equivalent diameter of a sphere, with the same optical properties, that produces a diffraction pattern similar to the particle. It is, however, more complex than that, because the diffraction pattern measured is the total light energy falling on the sensor array and the particle size distribution is inversely computed from those data using known material optical models and diffraction theories. As the particles are suspended in a fluid and pass the measurement cell at some velocity, the measured light energy is an average over some time, and thus the obtained particle size is also a time average of the mass passed through the measurement cell.

Several average particle sizes can be used to estimate recycling system behavior. The most common is the average volume/surface diameter, the Sauter mean diameter (x_{32}). It is the diameter of a sphere that has the same volume-to-area ratio as all particles in the whole sample.

Particle sizes and their distributions are most often given in discrete classes for historical reasons, because sieving naturally creates classes. The class divisions are given usually in geometric series. The traditional series is the Tyler series, in which the subsequent size of an aperture between sieves has a ratio $\sqrt{2}$. The base size is a 74-μm sieve. This is often called the 200 mesh (200#) sieve, because there are 200 wires per inch in the woven sieve cloth. The international ISO 565/ISO 3310 standard is based on a base sieve of 1 mm and has a geometric ratio of R20/3, i.e., every third in a series of $\sqrt[20]{10}$ from 125 mm to 32 μm. Finer sizes follow R10 series ($\sqrt[10]{10}$) down to 20 μm.

Any set of particles (the sample) will have a particle size distribution. It can be expressed either as the frequency distribution $q(x)$ or as a cumulative distribution finer than size x, $Q(x)$.

$$Q_r(x) = \int_{x_{min}}^{x_{max}} q_r(x)dx \approx \sum_{i=1}^{n} q_{ir}(x). \quad (5.1)$$

There are four types of particle size distributions distinguished by the subscript r in the equation above: number, length, area and volume. The most common are number and volume (mass) distribution. The convention is that the subscript r is given a value from 0 to 3 subsequently from number to volume distribution. So, $Q_3(x)$ means a cumulative volume distribution and $q_0(x)$ is a number frequency distribution.

Number distributions are important to planning sampling campaigns because statistical errors and analytical confidence limits depend on the number of particles sampled. Often, reaction rates are related to the reactive surface and the mass balances of such reactions correlate with the mass distributions.

Real distributions can be estimated with simple mathematical distributions. All are suitable only for mono-modal distributions. The simplest representation is an exponential function called the Gates–Gaudin–Schuhmann (GGS) equation

$$Q_r(x) = \left(\frac{x}{x_0}\right)^\alpha, \quad (5.2)$$

where x_0 is the size at which all particles are finer and α is the slope (width) of the distribution. A second widely used form of distribution is the Rosin–Rammler–Sperling–Bennett equation

$$Q_r(x) = 1 - \text{Exp}-\left\{-\left(\frac{x}{x_n}\right)^m\right\}, \quad (5.3)$$

which is capable of describing the ends of the distribution better than GGS. In the equation, x_n is the size at which 62.3% of particles are finer. Parameter m is the slope (width) of the distribution.

The third equation given here is the log-normal distribution

$$q_r(x) = \frac{1}{\sigma_{\ln}\sqrt{2\pi}} Exp\left\{-\frac{1}{2}\left(\frac{\ln(x_r/x_{\mu,r})}{\ln s_g}\right)^2\right\}, \quad (5.4)$$

where $x_{\mu,r}$ is the median and $\ln(s_g)$ is $\ln\left(\frac{x_{84,r}}{x_{50,r}}\right)$ (x_{50} = size at which 50% is finer).

For multi-modal particle distributions, one has to combine two or more distributions. Multi-modality is often observed in recycled feeds because the material properties (for example, brittleness) in a product can vary a lot.

5.2.1 Translational Velocity of Particles

A single particle moves in a fluid medium (liquid or gas) obeying classical mechanics. Two dimensionless numbers are useful for evaluating the behavior of particles settling a fluid.

Reynolds number Re is the ratio of inertial forces to viscous forces. For particles, we can write it as

$$\text{Re} = \frac{\rho v d}{\mu}, \quad (5.5)$$

where d is the characteristic length, i.e., particle diameter and μ the dynamic viscosity (Pa s). At low Reynolds numbers, the viscous forces dominate and the flow around the particle is smooth. A limit is considered typically to be Re $\leq$ 0.2. When the Reynolds number exceeds Re $\geq$ 1000, the inertial forces dominate and the fluid forms a distinct turbulent wake at the aft of the particle.

The other important dimensionless variable is the drag coefficient C_d. When a particle moves through a fluid, it must displace fluid elements from its path. This consumes energy, which can be understood as a force F_d slowing particle velocity. This drag force has two components that are important to velocities used in recycling. The first: skin friction, is caused by the fluid viscosity; and the second, form drag, is caused by the pressure difference between the fore and aft of the particle

$$C_d = F_d \frac{2}{\rho_f v^2 A}. \quad (5.6)$$

Drag coefficient varies as a function of velocity, particle size and shape, fluid density and viscosity. Drag coefficient is a function of the Reynolds number Re.

The drag force, buoyancy and gravitational force are the main forces controlling the settling of a particle in a quiescent fluid. The acceleration of a particle is

$$m\frac{dv}{dt} = mg + mg\frac{\rho_f}{\rho_s} + \rho_f v^2 C_d A/2, \quad (5.7)$$

where m is the mass; v, the velocity of the particles; ρ density (subscript f for fluid and s for solids); C_d, drag coefficient; and A, the area of the particle perpendicular to the direction of movement. As the velocity increases, the drag force will also increase until the acceleration comes to zero and the particle has obtained terminal velocity.

For fine particles (typically below 60 μm for solid particles), the drag coefficient can be estimated to be $C_d = 24/Re$ (for spherical particles). For terminal velocity, this leads to the well-known Stokes equation

$$v_{St} = \frac{d^2 g(\rho_s - \rho_f)}{18\mu}. \quad (5.8)$$

For very high Reynolds numbers, the drag coefficient is essentially a constant $C_d \approx 0.44$. This leads to a terminal settling velocity equation for large particles (also known as Newton's equation)

$$v_N \approx \frac{3 d_p (\rho_s - \rho_f)}{\rho_f}, \quad (5.9)$$

For intermediate sizes, no closed solutions exist. Turton and Clark (Turton and Clark, 1987) presented a useful approximation using dimensionless numbers. For dimensionless velocity v^*, they give as a function of dimensionless size d^*

$$v^* = \left[\left(\frac{18}{d^{*2}}\right)^{0.824} + \left(\frac{0.321}{d^*}\right)^{0.412}\right]^{-1.214}. \quad (5.10)$$

The dimensionless size d^* can be obtained from

$$d^* = \left(\frac{3}{4} C_d Re^2\right)^{(1/3)} = d_p \left(\frac{g(\rho_s - \rho_f)\rho_f}{\eta^2}\right)^{(1/3)} \quad (5.11)$$

and dimensionless velocity v^* from

$$v^* = \left(\frac{4Re}{3C_d}\right)^{(1/3)} = v_0 \left(\frac{\rho_f^2}{\eta g(\rho_s - \rho_f)}\right)^{(1/3)}. \quad (5.12)$$

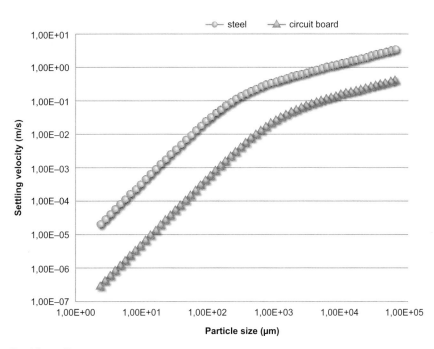

FIGURE 5.1 Particle settling.

For a known size, first calculate the dimensionless size (Eqn (5.11)) and use it to estimate dimensionless velocity (Eqn (5.10)). Then, solve the real velocity from Eqn (5.12).

The differences in settling velocity are large when comparing a piece of steel or a piece of circuit board resin, as can be seen in Figure 5.1. A resin piece with a diameter of 4.3 mm obtains a terminal velocity of 0.1 m/s. A steel particle is only 0.23 mm in diameter.

Non-spherical particles often behave in an erratic way, depending on their Reynolds number and the shape itself. In laminar conditions, the particles tend to become oriented so that the total drag force is minimized. Platy particles tend to wobble and flow in an erratic way. The drag coefficient tends to be a decade higher than for spheres of the same density and mass. Needle-shaped particles translate in a laminar flow with the longest dimension aligned with the flow.

In turbulent flow conditions, the particles tend to become oriented so that skin friction is minimized. Platy particles are moving wobbling in a position where the highest surface area is perpendicular to the flow. This causes the drag coefficient to be up to 2 decades higher than for a respective sphere. Needle-shaped particles wobble in a turbulent flow.

As a general rule of thumb, platy particles may be an order or even two orders of magnitude larger (largest dimension) to obtain the same terminal velocity.

5.3 PULP RHEOLOGY

Pulp rheology substantially affects the flow behavior of a separator. There, the most important variable is the volume concentration of solids ϕ in the suspension given by

$$\phi = \frac{\theta_p - 1}{\theta_s - 1}, \quad (5.13)$$

where specific weight $\theta_s = \rho_s / \rho_{water}$.

5.3.1 Apparent Density

In separators, where solid particles are dispersed in a fluid, the apparent specific density of the dispersion increases as

$$\theta_p = \frac{100}{\left(\frac{P}{\theta_s}\right) + 100 - P}, \quad (5.14)$$

where P is the suspension solid content percentage by mass.

5.3.2 Apparent Viscosity

Most pure fluids are Newtonian in their behavior. Any small stress will cause a shear and the fluid moves. The ratio is called viscosity. When the solids content increases in a fluid, the behavior of the fluid resembles increasing viscosity effects. Thomas (Thomas, 1965) proposed the following equation for the viscosity effects of suspended solids

$$\frac{\mu_a}{\mu_0} = 1 + 2{,}5\phi + A\phi^2 + B \exp(C\phi), \quad (5.15)$$

where ϕ is the solids volume concentration. Heiskanen and Laapas (Heiskanen and Laapas, 1979) proposed slightly different parameters to the equation. (Table 5.1).

The difference between the predictions is typically below 4% when ϕ is below 30% but then increases quickly, because the Thomas equation predicts substantially higher apparent viscosities at higher volume concentrations.

5.3.3 Hindered Settling

When the solids content of a fluid, notably water, increases, the translational velocity of

TABLE 5.1 Parameters for the Apparent Viscosity Equation

	A	B	C
Thomas (Thomas, 1965)	10.05	0.00273	16.6
Heiskanen and Laapas (Heiskanen and Laapas, 1979)	14.1	0.0274	−16.6

TABLE 5.2 Parameters for the Richardson and Zaki Equation

Re < 0.2	$b = 4.65$
0.2 < Re < 1.0	$b = 4.36/\text{Re}^{-0.03}$
1.0 < Re < 500	$b = 4.4/\text{Re}^{0.1}$
Re > 500	$b = 2.39$

the particles will decrease. For spherical particles, we get as an experimental equation for the ratio for hindered settling (Richardson and Zaki (Richardson and Zaki, 1954))

$$\frac{v_h}{v_{St}} = (1 - \phi)^b, \quad (5.16)$$

where b is a function of Re. (Table 5.2).

A small particle with a density of 3000 kg/m³ in a 30% by weight (12.5% by volume) slurry has only about 55% of the free settling velocity. Increasing the solids fraction by weight to 50% reduces the settling velocity further to about 28%. For very large particles, the ratios are 72 and 50%, respectively.

To evaluate the effect of particle density in a solid suspension, the following approximate equation can be used for small particles

$$\frac{x_1}{x_2} \approx \sqrt{\frac{\theta_2 - \theta_p}{\theta_1 - \theta_p}} \quad (5.17)$$

and

$$\frac{x_1}{x_2} \approx \frac{\theta_2 - \theta_p}{\theta_1 - \theta_p} \quad (5.18)$$

for large particles.

The size ratio between a steel particle and an aluminum one increases by 12% for fine particles and 24% for coarse particles, when the specific gravity increases from 1.0 to 1.4.

5.4 PROPERTIES AND PROPERTY SPACES

Any particle in a size class can be characterized by its properties averaged over its volume or surface. Of course, this property can be simply its chemical composition, but the composition of compounds and the set of consequent physical properties are more useful. Any set of particles will have a distribution of property values (Figure 5.2). This distribution can be treated as any statistical distribution with a mean and a variance. Let us denote the particle size as i and the properties as $j, k, \ldots$. The size classes can be single ISO-565 fractions or any combination deemed applicable, or any other sizes. For example, it can be too fine for processing at -4 mm, optimal for processing at $4-64$ mm, and too coarse and unliberated at $+64$ mm.

Properties must be considered by their utility. Only those properties that are important to the separation stage at hand and the requirements of further processing need to be considered. It is advisable that there be as few property classes as possible.

For example, if low purity steel scrap is to be treated, the simplest technology is to perform magnetic separation. Because most carbon steels exhibit ferromagnetism, magnetic susceptibility can consist of two classes: with or without this property. Because steel scrap also contains elements detrimental to the steel-making process, for high-quality steels we need to add composition properties such as a fractional content of copper in, say, 10%-unit steps (in any single particle). However, this approach quickly leads to a high number of combinations of properties.

The property space shows a way to model recycling (Figure 5.3). Models that track the changes in numbers of particles between the different volumes of property space are called population models. The simplest space is a one-dimensional binomial space of similar-size white and red beads; a description of a mobile phone after complete shredding will need an N-dimensional space with a large number of classes for every property, which is impossible to model. However, these

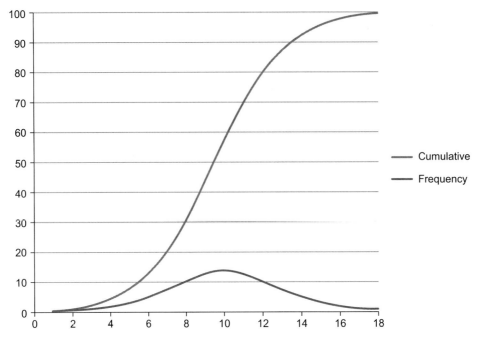

FIGURE 5.2 Property distribution.

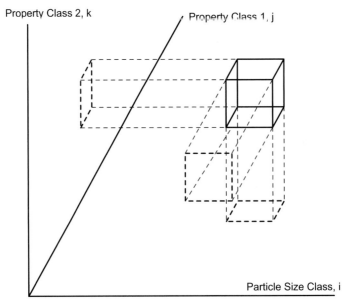

FIGURE 5.3 A three-dimensional property space R with a volume of particles belonging a class $R_{i,j,k}$ (with a surface area of S_c).

population models can still be useful in modeling size reduction and separation in recycling.

Particles can move to and from the property class volume by several mechanisms. First, they can move through the boundary by a convective motion. This can be by mechanical abrasion, by chemical reactions starting to take place as the temperature or chemical environment changes, by melting, or by any such process. They can also move by finite steps as a result of breakage or agglomeration. They can be destroyed in one volume for the progeny particles to arrive into several others. There can also be physical additions and withdrawals of particles into and from a given property volume.

$$\frac{\partial}{\partial t}\int_{R_c} N\Phi(x)dx = -\int_{S_c} N\Phi \mathbf{u} \cdot \mathbf{n}\, d\theta - D + B - Q + A, \quad (5.19)$$

where N is the number of particles, $\Phi(x)$ is the frequency size distribution of particles, $\mathbf{u}$ is a vector describing the rate of particles passing the boundary S_c, D is the disappearance and B is the birth of particles resulting from some physical or chemical action, and Q and A are the removal and addition, respectively, of the particles' byproduct and feed streams.

For many practical applications, the number distribution can be substituted with a mass distribution.

One way to use the property space for mass balancing and data reconciliation is to define for each stream f a flow rate Q_{fj} of a phase j (property class 1). Each phase consists of components k (property class 2). We can define $\mathbf{P}_{kjf}$ to be the fraction of component k in phase j in stream f. We can define X_{ijf} to be the fraction weight of particle size class 1 of phase j in stream f. As the last definition, we have T_{kijf} as the fraction of component k in particle size class i of phase j in stream f.

For many technical purposes, a one-dimensional cut from the mass-based property space, i.e., a frequency distribution or histogram of a single property mass, is interesting, as will be discussed next.

5.5 SAMPLING

Properties described earlier can be treated as property distributions with a mean and standard deviation. The variability is always a function of particle numbers sampled, not mass. There are properties that are integrative, such as the chemical composition. They are independent of particle size and can therefore be shredded and comminuted to finer sizes to increase the number of particles sampled. There are also properties that are size dependent, such as specific surface area of the material.

Even if the unknown real distribution is skewed, the sampled distribution tends to be closer to a normal distribution. It is usually assumed that the sampled distribution of the mean is normally distributed. Then, the limit clause

$$\text{var}(\bar{x}) = \frac{\text{var}(x)}{n} \quad (5.20)$$

stipulates that the variance of the mean is an nth fraction of the variance of a single measurement. It also says that doubling the sample size will reduce the variance to half. For a small number of samples (<30), the distribution follows Student t distribution, which is a wider distribution than the normal distribution because of the uncertainty in estimating the standard deviation.

Gy (Gy, 1979) developed a sampling theory that is in general use. The variance that is caused by the inhomogeneity of the material itself is called the fundamental variance. This error will remain even when the sampling is performed in an ideal way. The fundamental variance is related to the third power of the largest particles present in the sampled material

$$\mathrm{var}(x_{\mathrm{fundam}}) = \frac{Cx_{95}^3}{n\Delta m}, \quad (5.21)$$

where x_{95} is the size at which 95% are finer, Δm is the increment size, n is the number of increments, and C is a constant depending on the property distribution, liberation, particle shape and width of the size distribution. As can be seen in Eqn (5.21), a reduction in size of the largest pieces reduces the fundamental variance rapidly.

There are several sources for error in performing the sampling. The total variance of sampling consists of the fundamental variance and variances of error taking place in assaying and sample selection owing to wrong delimitation of the sample (for example, loss of material from increment) and owing to integration errors caused by the discrete sampling of a continuous variability.

$$\mathrm{var}(z) = \mathrm{var}(\mathrm{fundamental}) + \mathrm{var}(\mathrm{assaying}) + \sum \mathrm{var}(\mathrm{sample\,selection}) \quad (5.22)$$

The dimension of sampling can be defined as the spatial directions of a sampler to obtain a representative sample. A one-dimensional example is a material stream falling from a transport belt by a sampler that traverses it. A three-dimensional example is a heap of material, in which sampling points need to be distributed in three dimensions, a practical impossibility.

Materials for recycling can be sampled by

- Random sampling
- Systematic sampling
- Stratified sampling.

These can be also performed as a two-stage process or as a sequential process. Random sampling is discouraged in most cases because true randomness is difficult to obtain. This is especially the case for three-dimensional examples. A random grab sample will have a high variance and will often be prone to errors caused by nonideal sampling. The most accurate sampling method is systematic sampling from a one-dimensional case, i.e., an automatic sampler sampling a moving stream of material. This also applies to all secondary sampling before assaying.

If the material has a tendency to segregate, one can try stratified sampling, in which different strata of the material are sampled separately and the result is obtained by weighing the results by stratum masses.

For cheap materials, sampling can be performed using a two-stage process, in which the first sampling may be, for example, first selecting randomly the wagons (m) from a train of M wagons to be sampled in more detail

$$s_{\mathrm{ts}}^2 = \left(\frac{M-m}{M-1}\right)\frac{\sigma_b^2}{m} + \frac{\sigma_w^2}{mn}, \quad (5.23)$$

where σ_b is the standard deviation between wagons, and σ_w, within wagons.

A sequential process is often used for quality standards. If the first set is clearly within or outside defined limits, the sampling is discontinued and the lot is either accepted or rejected. If the result falls between the defined limits, sampling is continued with a second set.

For a variable (y) computed from a set of measured variables as $y=f(x_1,x_2,\ldots,x_n)$, the variance is computed as follows

$$\mathrm{var}(y) = \left(\frac{\partial y}{\partial x_1}\right)^2 \mathrm{var}(x_1) + \left(\frac{\partial y}{\partial x_2}\right)^2 \mathrm{var}(x_2) + \ldots + \left(\frac{\partial y}{\partial x_n}\right)^2 \mathrm{var}(x_n).$$

$$(5.24)$$

5.6 MASS BALANCES AND PROCESS DYNAMICS

Recycling can also be understood as a materials handling operation, where material is transported, concentrated and purified during the treatment. The process always consists of the units handling the material with temporal holdups, units of moving the material, and storage units. They all are combined by an average flow of material.

5.6.1 Mass Balances

Mass balances can be written either over a single unit in the process or over larger parts of the process. In mechanical recycling, one often starts with steady-state mass balances, where the recycled mass and its constituents are assumed to be constant in any flow.

$$F = C + T$$
$$Fc_{af} = Cc_{ac} + Tc_{at}$$
$$Fb_{bf} = Cc_{bc} + Tc_{bt} \quad (5.25)$$
$$\ldots$$
$$Fm_{mf} = Cc_{mc} + Tc_{mt},$$

where capital letters denote the total mass flow of feed (F), product (C), and tails (T), and c_{af}, c_{ac}, c_{at}, c_{bf}, c_{bc}, c_{bt},...,c_{mf}, c_{mc}, and c_{mt} are the concentrations of the property of interest ($a,b,\ldots m$) in streams f, c, and t, respectively.

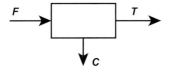

FIGURE 5.4 A simple separator.

As will be discussed later, these equations never hold completely, but contain random sampling and assay errors, and therefore do not close completely and will require data reconciliation.

For a separation as in Figure 5.4, recovery of material with the property of interest can be calculated from the mass balance

$$R_a = \frac{Cc_{ac}}{Fc_{af}} = \frac{a_{af}(a_{ac} - c_{at})}{c_{ac}(c_{af} - c_{at})}. \quad (5.26)$$

For a two-product case (Figure 5.5), in which properties a and b are divided to both C and S product streams, the recoveries become

$$R_a = \frac{c_{ac}[(c_{af} - c_{as})(c_{bs} - c_{bt}) - (c_{bf} - c_{bs})(c_{as} - c_{at})]}{c_{af}[(c_{ac} - c_{as})(c_{bs} - c_{bt}) - (c_{ac} - c_{bs})(c_{as} - c_{at})]}$$
$$R_b = \frac{c_{bc}[(c_{ac} - c_{af})(c_{af} - c_{bt}) - (c_{bc} - c_{bf})(c_{as} - c_{at})]}{c_{bf}[(c_{ac} - c_{as})(c_{bs} - c_{bt}) - (c_{ac} - c_{bs})(c_{as} - c_{at})]}. \quad (5.27)$$

For dynamic situations, where a property is changing in time, we have to use dynamic mass balance equations. Examples are shredding, smelting, and leaching.

$$\frac{d}{dt}(Vc_a) = Fc_{af} - Pc_{ap} - Vr_a$$
$$\frac{d}{dt}(Vc_g) = Fc_{gf} - Pc_{gp} - Vr_g, \quad (5.28)$$

where V is the volume of reacting space, F is the feed, P is the product, and r is the reaction rate. These equations can be written in state-space

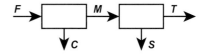

FIGURE 5.5 A two-stream separation case.

notation. If the reaction between a and g is of first order $r = kc_a$, we get

$$\dot{X}(t) = \mathbf{A}(t)\mathbf{x}(t) + \mathbf{B}(t)\mathbf{u}(t)$$

$$\mathbf{x} = \begin{bmatrix} c_a \\ c_g \end{bmatrix}, \mathbf{A} = \begin{bmatrix} -\left(k + \dfrac{1}{\tau}\right) & 0 \\ k & -\dfrac{1}{\tau} \end{bmatrix},$$

$$\mathbf{u} = \frac{1}{\tau}\begin{bmatrix} c_{af} \\ c_{gf} \end{bmatrix}, \mathbf{B} = \mathbf{I}. \qquad (5.29)$$

Equation (5.28) can be used as a starting point for population balance modeling of recycling.

5.6.2 Process Dynamics

The holdup or storage variation (mass stored W) can be expressed as the difference between incoming and outgoing flows in the ideal case.

$$\frac{dW(t)}{dt} = Q_i(t) - Q_o(t). \qquad (5.30)$$

The change in buildup is mathematically an integrating process, but for our needs the important point is that it takes time to change the inventory.

If material is transported a given distance L at a velocity v, it will show a transportation lag $\tau = L/v$. For a property $p(t)$ entering the transport system at time $t = 0$, we can write

$$p_{out}(t) = p_{in}(t - \tau). \qquad (5.31)$$

In Laplace domain[1] the transfer function of this is

$$G(s) = e^{-\Delta ts}. \qquad (5.32)$$

This is a transfer function of a pure time delay.

In recycling, we can use the well-known limiting cases. In the first case, the reactor is not mixed (plug flow reactor) (Eqn (5.32)); in the second case, it is instantaneously fully mixed (an ideal [fast] reactor). For the fully mixed case, we have

$$G(s) = \frac{1}{\tau s + 1}. \qquad (5.33)$$

If such a mixer is disturbed by a change in the feed composition, we can multiply the transfer function with the Laplace transform of the disturbance to obtain the response. By performing the inverse Laplace transform back to time space, we get the response in time. If the property entering the fully mixed reactor is a step change with a Laplace transform of $(1/s)$, we get for the response

$$C(s) = \frac{1}{s}G(s) = \frac{1}{s}\frac{1}{\tau s + 1} => C(t) = 1 - e^{-t/\tau}. \qquad (5.34)$$

For n reactors in series with a constant residence time in all of them

$$G(s) = \frac{1}{(\tau s + 1)^n}. \qquad (5.35)$$

By using these simple components, we can predict the dynamic response of a flowsheet (Figure 5.6) by first constructing a signal flow diagram (Figure 5.7).

By summing all of the transport lags and taking into consideration only the two largest time constants (the third has only a minor effect on the dynamic response), we get the following Figure 5.8. Using the combined notation of the figure, the variation in the product $p(s)$ related to the required quality set point $q(s)$ is

$$p(s) = \frac{G_X(s)G_Y(s)}{1 + G_X(s)G_Y(s)G_M(s)}q(s). \qquad (5.36)$$

In designing the control circuit in Figure 5.8, one has to ensure that the system is stable and that the control result is adequate. The overall stability of the control circuit can be answered by solving the function $1 + G_X(s)G_Y(s)G_M(s) = 0$. The roots must be negative or have negative real parts. For a constant set point, the dynamic response of

---

[1] Laplace transform $F(s) = \int_0^\infty f(t)e^{-st}dt$.

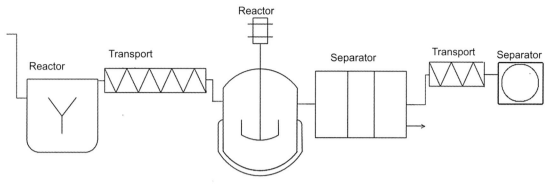

FIGURE 5.6 A recycling flowsheet.

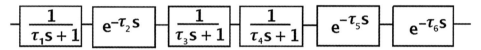

FIGURE 5.7 Flowsheet as a signal flow diagram.

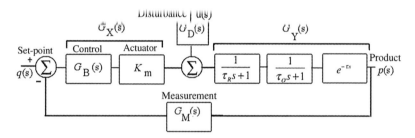

FIGURE 5.8 Signal flow diagram for feedback control for product quality.

(product/disturbance) is defined by the open-loop transfer function $G_Y(s)G_X(s)G_M(s)$.

5.7 MATERIAL BALANCING

5.7.1 Linear Data Reconciliation

A prerequisite for meaningful recycling computations is to perform data reconciliation, which allows consistent and closed process balances to be obtained. This allows one to generate a good understanding of the operation and its trends for further process improvement. Closed balances are also needed for process accounting and performance estimates. The aim is obtain consistent estimates for recycling process variables subject to model constraints.

A recycling process can be thought as a network of nodes connected to perform the recycling task at hand.

The node behavior can be expressed by constraining state equations, which need to be fulfilled when performing data reconciliation. For the balance of the process, we can write in matrix form

$$\mathbf{Cx} = 0, \qquad (5.37)$$

where $\mathbf{C}$ is the constraint equation matrix (with elements c) and $\mathbf{x}$ is the vector of flows connecting the units.

For this, we need a set of measurements of process variables such as particle size, material and chemical composition, and so forth, to write a constraining model for any single node. These are typically conservation equations. Nodes can consist of units combining or separating streams (physical separators), and reactors (chemical and metallurgical reactors and furnaces). The two first kinds are characterized by the conservation of all variables; the reactors always conserve the total mass and the masses of elements but may not conserve other variables.

We also need an estimate of the uncertainty of the measured variables. An assumption made in the reconciliation process and in formulating the previous equation is that the errors involved are not gross errors (bias) but are always randomly distributed.

It is typical that variables are measured only from some of the streams. Thus, we have measured and unmeasured variables. Some can also be calculated from information obtained from other streams. If a variable is not measured but can be calculated, it is called observable. Of course, if we have no way to obtain the value of a variable if it is unobservable.

called topological. It considers measurements from a stream. If the size distribution of all three streams of a size separator is measured, all measurements are redundant because we can compute the values of one stream from the two others and also have a direct measurement. If one of the measurements is not performed, the remaining stream variables become nonredundant. One of the streams is then nonmeasured but observable. If a second stream measurement is omitted, the two streams not measured become unobservable. A nonredundant variable becomes unobservable if its measurement fails. The second kind of redundancy arises from repeated measurements. This is important because it gives information about the standard deviation of the variable.

Estimability is a slightly broader definition than observability, which is reserved only for nonmeasured variables. A variable is estimable if it is measured or nonmeasured but observed.

A typical flowsheet in recycling is depicted in Figure 5.9.

For simple material flow constraints, we get for matrix $\mathbf{C}$:

$$\mathbf{C} = \begin{bmatrix} \overset{S_1}{1} & \overset{S_2}{-1} & \overset{S_3}{-1} & \overset{S_4}{} & \overset{S_5}{} & \overset{S_6}{} & \overset{S_7}{} & \overset{S_8}{} & \overset{S_9}{} & \overset{S_{10}}{} & \overset{S_{11}}{} & \overset{S_{12}}{} & \overset{S_{13}}{} & \overset{S_{14}}{} \\ & & 1 & -1 & -1 & & & & & & & & & \\ & & & & 1 & -1 & -1 & & & & & & & \\ & & & & & & 1 & & -1 & -1 & & & & \\ & & & & & 1 & & & 1 & -1 & & & & \\ & & & & & & & & & 1 & -1 & -1 & & \\ & & & & & & & & & & & 1 & -1 & -1 \end{bmatrix}. \quad (5.38)$$

If the variable is measured and can also be calculated, it is redundant; if it can be obtained only by the measurement itself, it is nonredundant. The first kind of redundancy can be

By first arranging the streams as unmeasured and measured and using the basic matrix operators, we can have the unmeasured part arranged into observable and unobservable parts

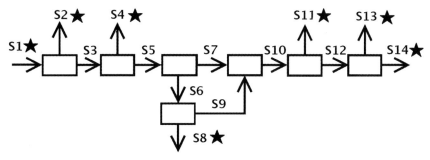

FIGURE 5.9 A typical flowsheet for recycling (H denotes sampled stream).

by developing the matrix canonical form. The canonical form is

$$C = \begin{bmatrix} S_3 & S_5 & S_{12} & S_{10} & S_6 & S_7 & S_9 & S_1 & S_2 & S_4 & S_8 & S_{11} & S_{13} & S_{14} \\ 1 & & & & & & & -1 & 1 & & & & & \\ & 1 & & & & & & -1 & 1 & 1 & & & & \\ & & 1 & & & & & & & & & 1 & & -1 \\ & & & 1 & & & & & & & & 1 & 1 & 1 \\ & & & & -1 & -1 & & -1 & 1 & 1 & & & & \\ & & & & 1 & & -1 & 1 & -1 & -1 & -1 & & & \\ & & & & & 1 & 1 & & & & & -1 & -1 & -1 \end{bmatrix} \quad (5.39)$$

In the example, streams S_3, S_5, S_{12}, and S_{10} are observable because they only have one nonzero element in the column. Streams S_6, S_7, and S_9 are unobservable. All the measured streams are nonredundant. A material reconciliation cannot be performed for this sampling scheme.

Equation (5.37) can be written with the help of Eqn (5.39) to show the measured (M) and unmeasured (U) variables (Rao and Narasimhan (Rao and Narasimhan, 1996))

$$[C_U \ C_M] \begin{bmatrix} x_U \\ x_M \end{bmatrix} = 0. \quad (5.40)$$

We get from the system[2]

$$c_R x_R = 0. \quad (5.41)$$
$$x_O = c_{RO} x_R + c_{NRO} x_{NR} \quad (5.42)$$
$$c_{UO} x_{UO} = -c_{RUO} x_R - c_{NRUO} x_{NR}. \quad (5.43)$$

The first equation cannot be satisfied by measurements and forms the basis for data reconciliation. The second equation allows us to calculate the observable variables. The third equation cannot be solved.

The redundant variables will never satisfy Eqn (5.41). We have

$$c_R x_R = r, \quad (5.44)$$

where **r** is a vector for residuals.

For linear systems (i.e., only simple mass flow conservation equations for nodes $x_i = Q_i$) reconciliation procedure is a minimization problem where the objective function is minimized subject to a set of constraints.

In matrix for the equation can be written as

$$Min\left\{ [\overline{Q}_R - Q_R]^T s_R^{-1} [\overline{Q}_R - Q_R] \right\} \quad (5.45)$$

with the constraint $c_R \overline{Q}_r = 0$. This is a least-square minimization problem.

---

[2] R, redundant; NR, nonredundant; O, observable; UO, unobservable.

Standard Lagrange multipliers can solve this minimization.

$$L = [\overline{Q}_R - Q_R]^T s_R^{-1} [\overline{Q}_R - Q_R] - \lambda^T (c_R \overline{Q}_R). \quad (5.46)$$

The condition for optimality is that $\frac{\partial L}{\partial \overline{Q}_R} = 0$ and $\frac{\partial L}{\partial \lambda} = 0$.

The answer becomes

$$\overline{Q}_R = \left[I - s_R c_R^T (c_R s_R c_R^T)^{-1} c_R \right] Q_R. \quad (5.47)$$

This is the minimum level of data reconciliation to be performed.

5.7.2 Nonlinear Data Reconciliation

If the sampling has also given information about components, they can be included into data reconciliation. In addition to Eqn (5.37), we have

$$c_R M_j = 0, \quad (5.48)$$

where M is the component flow of P components $m_{jk} = Q_j c_{jk}$; $k = 1...p$. Components can be liberation classes, size classes, elements, etc.

The equation for data reconciliation now becomes (like Eqn (5.45))

$$\begin{aligned} Min \big\{ & [\overline{Q}_R - Q_R]^T s_{QR}^{-1} [\overline{Q}_R - Q_R] \\ & + [\overline{m}_R - m_R]^T s_{PR}^{-1} [\overline{m}_R - m_R] \big\} \end{aligned} \quad (5.49)$$

subject to the constraints $c_R \overline{Q}_R = 0; c_R \overline{M}_j = 0;$ $\sum_{j=1}^{K} \overline{y}_{R,j} = 0$. The procedure with Lagrange multipliers is analogous.

As it is written, Eqn (5.49) contains redundant flow rates and concentrations from the same redundant flows. However, concentrations can be measured without a subsequent flow rate measurement. These concentration measurements can sometimes be used to obtain flow rate values for flows that otherwise would be unobservable. This makes the computations complex, because the redundancy is affected. In programs designed for data reconciliation, this is often solved by taking into consideration all variables. The unmeasured variables are given a large variance. This approach can lead to large optimization systems and numerical problems prohibiting convergence.

For a single separator, where all streams are measured, we can either minimize the sum of squares of the closure residuals (to be shown here) or minimize the sum of squares of the component adjustments. For a steady state, we can start by dividing the balance equation by Q_f (Figure 5.4) to get

$$c_{fk} - Cc_{ck} - (1 - C)c_{tk} = r_k. \quad (5.50)$$

The sum subject to minimization is now

$$S = \sum_{k=1}^{p} (r_2)^2. \quad (5.51)$$

We get as a result

$$\overline{C} = \frac{\sum_{k=1}^{p} (c_{fk} - c_{tk})(c_{ck} - c_{tk})}{\sum_{k=1}^{p} (c_{ck} - c_{tk})^2}. \quad (5.52)$$

To distribute the closure errors, we can write

$$\Delta_{fk} - \overline{C}\Delta_{ck} - (1 - \overline{C})\Delta_{tk} = r_k. \quad (5.53)$$

This can again be solved by using Lagrangian multipliers. The Lagrangian becomes

$$\begin{aligned} L = & \sum_{k=1}^{p} \left(\Delta_{fk}^2 + \Delta_{ck}^2 + \Delta_{tk}^2 \right) \\ & + 2 \sum_{k=1}^{p} \lambda_k \left(r_k - \Delta_{fk} + \overline{C}\Delta_{ck} + (1 - \overline{C})\Delta_{tk} \right). \end{aligned} \quad (5.54)$$

Finding by derivation the minima for the variables (including λ_k) we get, after manipulation,

$$\begin{aligned} \Delta_{fk} &= \frac{r_k}{1 + \overline{C}^2 + (1 - \overline{C})^2} \\ \Delta_{ck} &= \frac{-\overline{C}r_k}{1 + \overline{C}^2 + (1 - \overline{C})^2} \quad (5.55) \\ \Delta_{tk} &= \frac{-(1 - \overline{C})r_k}{1 + \overline{C}^2 + (1 - \overline{C})^2} \end{aligned}$$

5.8 LIBERATION

The mix of compounds in particles of various origins, from complete devices or parts of them, or any type of byproducts such as slag, may vary from a single compound to a mix of several compounds. A particle consisting of a single compound is called liberated. A mix of two compounds is called a binary, and with the same logic, ternary particles have three compounds.

For the optimal processing of compounds, the aim is to maximize the mass of liberated particles in a set of recycled material. As mentioned before, it poses an optimization problem, because breaking the material into too-fine particles causes the processes to operate under nonoptimal conditions.

All recycled materials have a specific way of breaking when exposed to an impacting, compressing, or shearing force large enough. By studying the progeny particles, one can form a model of liberation breaking. This kernel function can be determined by either textural modeling or probabilistic methods (van Schaik et al. (van Schaik et al., 2004), Gay (Gay, 2004)).

As pointed out by Gay (Gay, 2004), the approach of direct liberation classes soon becomes numerically expensive for multiphase particles. It can be avoided by using parent particle to progeny particle relationships by using a liberation kernel function K.

$$g_j = \sum_i f_i K_{ij}, \quad (5.56)$$

where f_i and g_j are the composition frequencies of parent particle type i and progeny particle type j, respectively, and K_{ij} is the kernel function. When the kernel is known, it can be used to calculate the frequency distribution of type j progeny particles originating from type i parent particles

$$p_{ij} = f_i K_{ij}. \quad (5.57)$$

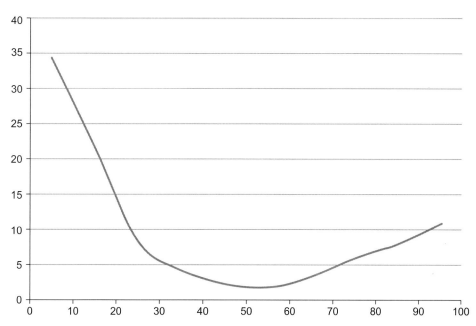

FIGURE 5.10 Frequency distribution of a single property.

Gay (Gay, 2004) proposed using the probability entropy method to solve Eqn (5.57) under the constraints that the original composition distribution of parent particles is satisfied, the original composition distribution of progeny particles is satisfied, and the average mineral composition of the resulting progeny particles is the same as the parent particles. The constraints can be formulated as follows

$$\sum_j p_{ij} = f_i \text{(for each } i\text{)}$$
$$\sum_i p_{ij} = g_j \text{(for each } j\text{)} \quad (5.58)$$
$$\sum_j p_{ij}(c_{jm} - C_{im}) \text{(for each } i \text{ and } m\text{)}$$

where c_{jm} is the composition of the mth mineral for the jth progeny particle and C_{im} is the composition of the mth mineral for the ith parent particle.

He obtains as a solution

$$\ln(p_{ij}) - 1 = \lambda_{1i} + \lambda_{2j} + \lambda_{3im}(c_{jm} - C_{im}), \quad (5.59)$$

where the Lagrange multipliers λ_1, λ_2, and λ_3 are for the respective constraints.

After shredding, the frequency distribution of a property might look like the curve in Figure 5.10. One can see that there is a large mass of particles that does not possess the property, and a good mass of particles with a high degree of the defined property.

The distribution gives the possibility of separating the stream into a stream low in the property and a stream rich in the property. As an example, say, the property is aluminum (Al)

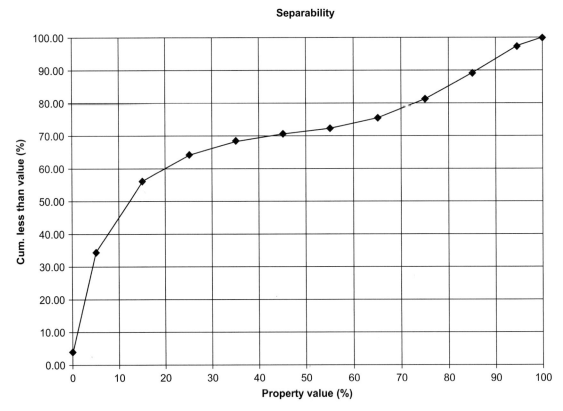

FIGURE 5.11 Separability curve.

mass content. We can separate an Al-rich and an Al-poor product from the feed.

By calculating the total mass content of, say, Al cumulatively, starting from the poorest fraction, we get a curve called the separability curve, which shows how much Al is present in the fraction below a property limit (Figure 5.11). The example reveals that 70% of Al values are particles, which carry more than 50% non-AL matter.

5.9 GRADE-RECOVERY CURVES

A relation always exists between the grade of a separated product and its recovery (recyclability). There are two basic reasons for that relationship. First, the liberation of particles subjected to separation is not complete. Second, the related response to a physical force or chemical potential gradient will cause different particles to react in different ways.

A liberation-based grade-recovery curve (Figure 5.12) can be constructed directly from the separability curve with easy computations. Taking the entire feed stream as a product, the recovery is 100% and the product grade is the same as the feed: in our Al case, 33.7% Al. Leaving the fractions with less than 10% Al away reduces the mass almost to 50% according to Figure 5.10, but only loses 5% of the Al. It also increases the product quality to 49% Al. Continuing in this way yields a full grade-recovery curve owing to the lack of liberation. As can be seen, with this material a 90% Al purity product can only be obtained with a 55% recovery.

Breaking the material into smaller particles will improve the separability curve (Figure 5.13). This experimental value is from the same scrap sample but is substantially finer in particle size.

Compared with Figure 5.11, the central part of the curve is much flatter, indicating that no significant mass of middling particles exists. This leads to the following grade recovery curve (Figure 5.14).

The recovery for a 90% pure product has increased to 85%, a substantial improvement in the theoretically obtainable result.

The importance of the liberation-based grade-recovery curve depends on the property interactions and the nature of the following processes.

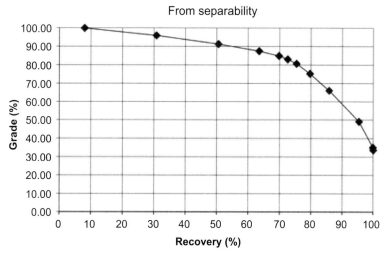

FIGURE 5.12 Grade-recovery curve.

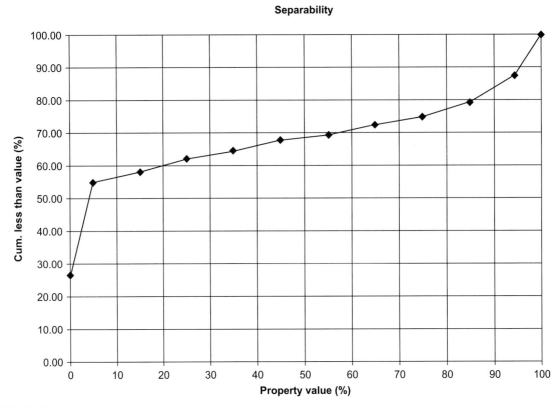

FIGURE 5.13 Separability curve after further breakage.

5.9.1 Mechanical Separations

As will be discussed in more detail in the Appendix, mechanical separation is based on a balance of forces. The force for separation is chosen according to the properties of the particles to be separated. It can be a body force such as gravity, centripetal force or magnetic force, or a surface force induced by surface property modifications. This active separation force is directed by the separator design so that the trajectory of particles affected by the force becomes different from the trajectory of particles not affected by the force. Mass forces can be used to enhance the difference in particle trajectories.

Figure 5.15 depicts a separator with a flow from one feed point to two streams. All particles start from the same point, but the acting separation force (lower picture) takes the particles to the upper discharge, whereas a similar particle without the affecting force will report to the lower discharge.

As stated, separation is a particulate process in which particle size, shape, and density affect the outcome in addition to the active separation force. For large particles, the mass forces are the most important forces. As the mass decreases to the third power of diminishing particle size and area only to the second power, at some point the surface forces will become dominant. The

5.9 GRADE-RECOVERY CURVES

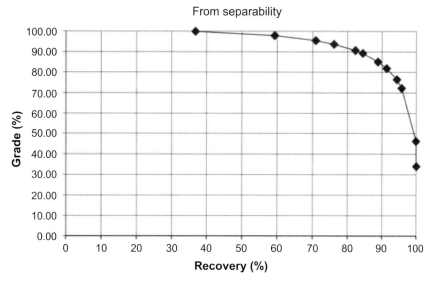

FIGURE 5.14 Grade-recovery curve after further breakage.

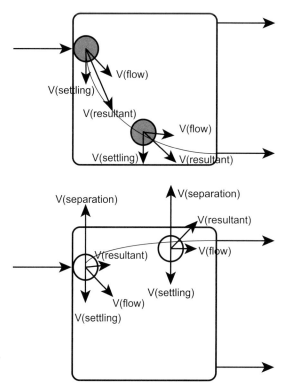

FIGURE 5.15 Particle trajectories in a separator, when a separating force acts at some particle classes.

I. RECYCLING IN CONTEXT

smaller the particle, the more will surface forces such as drag and viscosity and even electrostatic and van der Waals forces affect the total force balance. The task is to optimize particle size for liberation and for efficient separation.

For mechanical separations, some operational deficiencies always exist owing to property and size distributions, apparent viscosity effects, turbulence, and boundary flows.

Any particle entering a separator will have a probability of entering one of the product streams. The Separation cut point is the value of the property in which particles have an equal probability of entering either of the two product streams. It is often denoted as ψ_{50}. In Figure 5.16, the cut point density is 2705 kg/m (Heiskanen and Laapas, 1979).

The separation efficiency curve can be calculated from reconciled data as the percentage of the mass in each property class to report to the chosen product. The shaded areas are misplaced particles. Vertical shading represents light particles reporting in heavies or sinks, and the horizontal shading is heavy particles found in the light fraction. The steepness of the curve can be used as a measure of quality, imperfection.

$$I = \frac{\psi_{75} - \psi_{25}}{2\psi_{50}}. \quad (5.60)$$

A separation efficiency curve can also be computed for secondary effects. In Figure 5.16, the substance F probability curve has a similar slope with an offset. This indicates a slight concentration to the heavy fraction, but homogeneously.

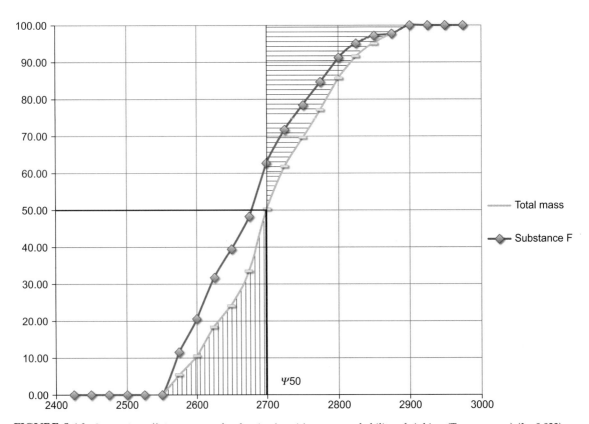

FIGURE 5.16 Separation efficiency curve for density (x-axis) versus probability of sinking (Tromp curve) ($I = 0.023$).

Chemical methods for recycling i.e. hydro- and pyrometallurgical recycling technology, theory and systems are discussed in various chapters of Part II.

References

Gay, S., 2004. A liberation model for comminution based on probability theory. Minerals Engineering 17, 525–534.

Gy, P., 1979. Sampling of Particulate Material: Theory and Practice. Elsevier, Amsterdam, 431 pp.

Heiskanen, K., Laapas, H. On the effects of the fluid rheological and flow properties in the wet gravitational classification. Proceedings of XIII International Mineral Processing Conference, 1979, Warsaw, pp. 183–204.

Rao, R.R., Narasimhan, S., 1996. Comparison of techniques for data reconciliation of multicomponent processes. Industrial & Engineering Chemistry Research 35, 1362–1368.

Richardson, J.F., Zaki, W.N., 1954. Sedimentation and fluidization: part I. Transactions of the Institution of Chemical Engineers 32, 35–53.

Thomas, D.G., 1965. Transport characteristics of suspension: VIII. A note on the viscosity of Newtonian suspensions of uniform spherical particles. Journal of Colloid Science 20 (3), 267.

Turton, R., Clark, N., 1987. An explicit relationship to predict spherical particle terminal velocity. Powder Technology 53, 127–129.

van Schaik, A., Reuter, M.A., Heiskanen, K., 2004. The influence of particle size reduction and liberation on the recycling rate of end-of-life vehicles. Minerals Engineering 17, 331–347.

PART II

RECYCLING – APPLICATION & TECHNOLOGY

CHAPTER 6

Recycling of Steel

Bo Björkman, Caisa Samuelsson

MiMeR – Minerals and Metallurgical Research Laboratory, Luleå University of Technology, Luleå, Sweden

6.1 INTRODUCTION

Recycling of steel is a very old business, and it should be pointed out that in most developed parts of the world we have a very well-developed system of scrap collectors, scrap-processing companies and steel plants utilizing the scrap. For the scrap-treating companies and the steel plants, steel scrap is a very valuable resource. From the scrap that is collected, losses occur to residues in the scrap processing and in material streams that are recycled in other metal cycles, e.g. iron units present in scrap sorted as scrap for copper making.

Steel scrap in steelmaking is usually recycled to either an electric arc furnace (EAF), in what usually is labeled scrap-based steelmaking, or to the basic oxygen furnace (BOF), in ore-based steelmaking, where there exists a considerable need for cooling to avoid too high temperatures as a result of exothermic reactions. It should be pointed out that in the steel produced in ore-based steelmaking, about 10–20% is from scrap used for cooling in the BOF. The world production of steel was 1547 Mt in 2012 (World Steel Association, 2013), a record high production. Of these 29.3% was produced in the EAF process. Almost all of the steel production based solely on scrap is produced in the EAF process. Assuming then 100% of scrap use in an EAF and about 15% of the steel produced in ore-based steelmaking originating from scrap used as coolant gives about 40% of steel production in the world based on scrap. Some references indicate a lower value, thus indicating that in developing countries ore is more often used as coolant in BOF steelmaking than scrap. The share of steel produced from scrap in Western Europe in 2012 was about 50–55%, calculated based on the same assumptions (Eurofer, 2013). This figure is in good agreement with statistics on the scrap consumption in Western Europe, indicating about 55% of steel production is based on scrap.

In addition to much lower emissions to the environment in general, steel production in an EAF process based on scrap contributes much less to the emission of CO_2 and has a much lower total energy consumption. The focus in recent years has been on research toward decreased emissions of greenhouse gases from steelmaking in among others the European ULCOS project (ULCOS, 2013), and in this context a comprehensive evaluation of CO_2

emissions and energy consumption in benchmark BOF steelmaking and EAF steelmaking as well as for different alternative processing routes has been presented. The benchmark emissions of CO_2 from the BOF route and the EAF route are about 1770 and 380 kg CO_2 per ton liquid steel, respectively, and the benchmark total energy consumption is about 4900 and 1100 kWh, respectively, per ton of liquid steel (e.g. Birat et al., 2004).

Figure 6.1 illustrates the increase in world steel production by region in recent years, showing a drastic increase in steel consumption, mainly driven by economic development in China. China's steel production has increased from 182 Mt in 2002 to 716 Mt in 2012. About 46% of today's world production of crude steel is produced in China. It is important when considering the share of steel produced from scrap that as long as the world consumption of steel increases in the way illustrated in Figure 6.1, and as a majority of the steel is in final products that have a quite long life time, a considerable amount of the steel produced has to come from ore-based steel production, despite the high recycling rate of collected scrap, cf. Table 6.1.

Figure 6.2 illustrates the use of steel split into different product groups. The by far largest use of steel is in construction, an application which

TABLE 6.1 Steel Industry Recycling Rates

Application Area	Estimate for 2007 (%)	Target for 2050 (%)
Construction	85	90
Automotive	85	95
Machinery	90	95
Appliances	50	75
Containers	69	75
Total	83	90

From World Steel Association (2013).

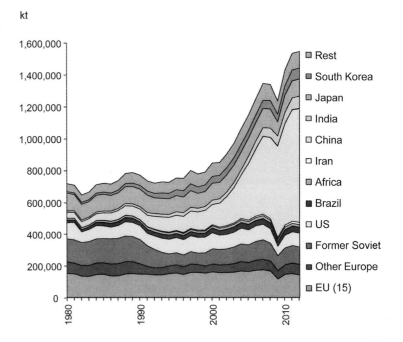

FIGURE 6.1 Crude steel production by country and region, 1980–2012 (Swedish Steel Producers Association, 2013).

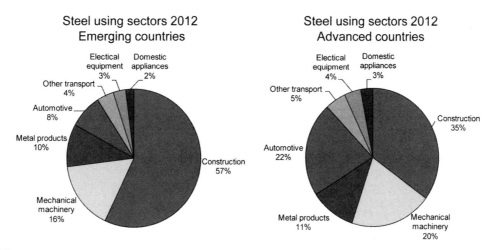

FIGURE 6.2 The distribution of steel use in emerging and advanced countries across sectors. *From World Steel Association.*

has a very long time in use, cf. Table 6.2, but most of the steel applications have a quite long time in use.

Steel stocks in use have reached saturation in several developed countries and are not increasing anymore, even declining in some countries, whereas some developed countries and most of the rest of the world still have increasing stocks of steel in use. The saturation level for steel stock in use is at 13 ± 2 t/capita (Pauliuk et al., 2013). In China, the country that showed the fastest-growing steel consumption so far in this twenty-first century, consumption are expected to peak between 2015 and 2020 (Pauliuk et al., 2012). As a saturation of steel stocks in use occurs, the fraction of steel production based on scrap can and will increase. In 2050, the share of steel produced from scrap has been estimated as about 80% (Pauliuk et al., 2012). Other estimates give a figure of 50% of the world steel production based on scrap in 2050 (Swedish Steel Producers Association, 2012).

Steel is an alloy and is produced into a large variety of different alloys depending on the application intended for the steel, usually divided into what is labeled low-alloyed steel and high-alloyed steel. Depending on the alloy, steel contain alloying elements like manganese, nickel, chromium, tungsten, titanium, niobium etc., in varying amounts, for high-alloyed steels is from 1% and upward. The alloys are often added as ferroalloys, by which is meant an alloy comprising iron and the alloying element. Ferroalloy production almost always consumes more resources than iron production, and greenhouse gas emissions are, for example, 2–20 times higher for the ferroalloys than for iron.

TABLE 6.2 Steel Product Lifespan (Brooks and Pan, 2004)

Product	Years
Buildings	20–60
Major industrial	40
Heavy industrial machinery	30
Rails	25
Consumer durables	7–15
Vehicles	5–15
Steel cans	<1

One type of high-alloyed steel is stainless steel, based on the noncorrosive properties that chromium and nickel introduces to the steel. The simplest stainless steel grade is composed of 18 and 8 wt% of chromium and nickel, respectively, but the chromium content is in some grades much higher and stainless steel is also usually alloyed with Mo, W, Mn etc.

6.2 SCRAP PROCESSING AND MATERIAL STREAMS FROM SCRAP PROCESSING

6.2.1 Scrap Classification

Scrap used can be of several different origins, as illustrated in Figure 6.3. Home scrap is the scrap that emanates from recycling within the plant, e.g. reverts from the furnace surrounding, from ladles, from separation of metals contained in slag, melts that has to be recycled due to some specifications that are not fulfilled as well as pieces cut off during heat treatment, rolling and finishing of the products. This type of scrap is usually well sorted into different qualities and with low content of impurity elements, raising the possibility of recycling the scrap back to the production of the same steel quality.

Manufacturing scrap is the scrap coming from the manufacturing of the consumer goods; it can be very well sorted and of high purity and definitely has the potential to be so. Old scrap or obsolete scrap is the scrap coming from spent consumer products or other origins, as collected by the scrap dealers. The scrap is generally sorted into different scrap categories depending on composition and size of the scrap as illustrated in Table 6.3.

As mentioned earlier, steel has a high recovery of collected scrap. However, as also is obvious from the sorting specifications given in Table 6.3, the steel scrap is very seldom sorted into different alloys, thus the recovery of alloying elements into the same type of steel is much lower.

6.2.2 Scrap Processing

Larger scrap from, e.g. cars, white wares and WEEE, is usually treated in a shredder, cf. Figure 6.4, where the scrap is fragmented into smaller pieces with the aim of obtaining different materials in separate pieces to facilitate sorting. A typical process scheme for the upgrading and sorting of different materials from the shredder is given in Figure 6.5.

The steel scrap is loaded into the shredder and fragmented into about hand-sized pieces. Very small and light particles, mainly with organic content, is pneumatically conveyed and collected as a dust. The material then passes a number of separation steps, at a minimum consisting of magnetic separation, sieving and air separation, whereby the majority of the steel is collected in the magnetic fraction, nonferrous metals are collected in the nonmagnetic fraction and very small and light particles are separated from the scrap fractions. These process steps can then be complemented with a number of different separation steps to improve the sorting: density separation, weak magnetic separation, eddy-current separation, hand sorting, automatic sorting based on color or physical properties, etc.

Typically the material coming from the shredder is not completely separated. Pieces

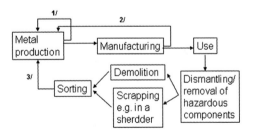

FIGURE 6.3 Scrap recycling. Labels indicates 1/home scrap, 2/manufacturing scrap and 3/old or obsolete scrap.

TABLE 6.3 European Steel Scrap Specifications (Eurofer, 2013)

			Impurity Content		
Category	Specification	Density	Cu	Sn	Cr, Ni, Mo
Old scrap	E3	≥0.6	≤0.250	≤0.010	∑ ≤ 0.250
	E1	≥0.5	≤0.40	≤0.020	∑ ≤ 0.30
New scrap	E2	≥0.6		≤0.30	
Low residuals	E8	≥0.4		≤0.30	
Uncoated	E6	≥1		≤0.30	
Shredded	E40	≥0.9	≤0.250	≤0.020	
Steel	E5H				
Tunings	E5M		≤0.40	≤0.030	∑ ≤ 1
High residual	EHRB	≥0.5	≤0.450	≤0.030	∑ ≤ 0.350
Scrap	EHRM	≥0.6	≤0.40	≤0.030	∑ ≤ 1.0
Fragmented, from incineration	E46	≥0.8	≤0.50	≤0.070	

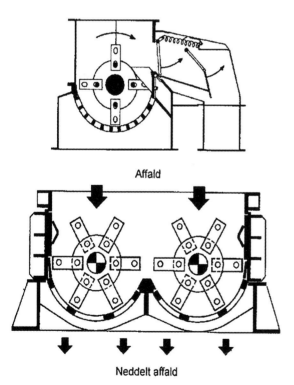

FIGURE 6.4 Schematic illustration of a shredder.

with steel together with copper, aluminum or other metals are quite common (e.g. van Schaik, 2004; Reuter et al., 2005, van Schaik et al. 2002), and depending on the sorting process a piece with, e.g. steel together with copper, might either end up in the steel or copper material after sorting due to inappropriate liberation in the shredding or due to inappropriate particle size for the separation process used. Copper content in steel scrap is one thing that limits the possible steel qualities that can be produced, an issue that will be further discussed in Section 6.5. For the moment we can simply just note that the requirements on impurity content in the final steel is very much dependent on the steel quality produced, from above 0.3% Cu to below 0.1% Cu for different steel grades. Therefore, the tradition has been that steel qualities with very low limits on copper content, which usually also have a higher price, are produced in ore-based steelmaking, the ore usually having low copper content. Copper content in steel scrap from a shredder plant has been about 0.25% to 0.3% and rising. An interesting observation is that

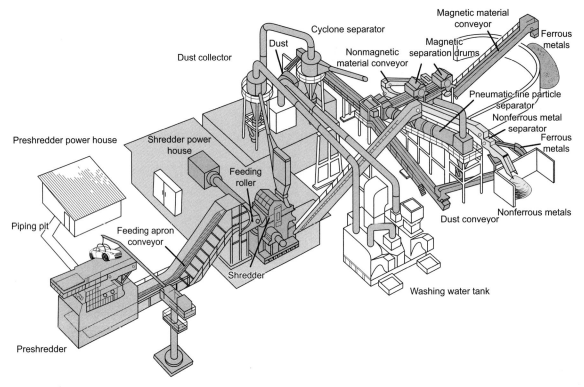

FIGURE 6.5 Typical process scheme for sorting of material from a shredder into different material categories. *From www.morita119.com.*

shredded steel scrap in Sweden has decreasing copper content nowadays, about 0.22% Cu, because the copper has become very valuable, therefore giving an incentive for handpicking of larger copper pieces present in the steel scrap.

Equipment for sorting of scrap based on properties such as color and magnetic properties is already on the market and is to some extent also used by the scrap-treating companies. An alternative is sorting based on the chemical composition as determined, e.g. through spectroscopic techniques. Scrap sorting of aluminum using laser induced breakdown spectroscopy (LIBS) has already proven to work very well (Gesing et al., 2003; Aydin et al., 2004). The use of the same technique for sorting of steel scrap has been tested recently in a research project within the Steel Eco-Cycle program in Sweden (Gurell et al., 2012). The technique has proven to be feasible to rapidly determine different type of steel alloys and works well as an aid at the scrap yard to determine the exact use of different scrap deliveries to optimize the use of alloying elements and not risk contamination by unwanted impurities. The technique is still too slow for use as an online sorting aid at a conveyor belt in a shredder plant.

6.3 THE PROCESSES USED FOR SMELTING STEEL SCRAP

Figure 6.6 illustrates the process scheme for production of steel from both ore and scrap. Scrap-based steelmaking is almost 100% based

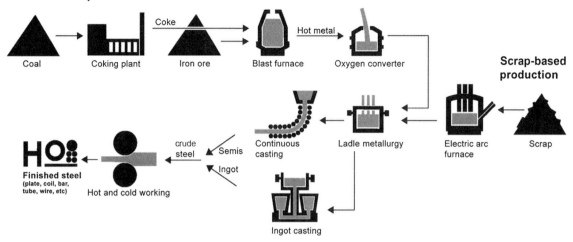

FIGURE 6.6 Schematic process scheme of steelmaking based on ores or scrap (Swedish Steel Producers Association, 2011).

on smelting the scrap in an EAF, cf. Figure 6.7, where electricity is used as energy source for the smelting. Development of the process with slag foaming through injection of a carbon source and oxygen, oxygen lancing, preheating of the scrap, bottom tapping, cocombustion of fuel in burners, etc. has led to a substantial decrease in energy consumption and also has added some flexibility in the use of energy source. Depending on the actual processing technology, electricity consumption for smelting steel scrap in a modern EAF is in the range

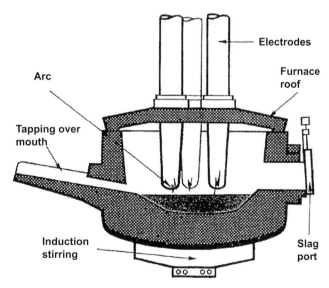

FIGURE 6.7 Principals of an Electric Arc Furnace (EAF) for smelting steel scrap.

of 400–450 kWh/t steel. Slag formers are added to take up some of the impurities coming with the steel. The steel is then refined in ladle furnace processes, casted, rolled and finished. A recent trend is that also within scrap-based steelmaking in an EAF, more and more virgin iron units are used, in the form of direct reduced iron (DRI) or hot briquetted iron (HBI), which will be discussed further in Sections 6.7 and 6.8. The furnace can either be operated on AC current with three electrodes, which is the dominating alternative and also the version illustrated in Figure 6.7, or operated on DC current with one electrode.

The EAF is a very efficient smelting unit but not a very good reactor type for carrying out refining reactions or for adding large amounts of nonmetallic material. Scrap smelting in an EAF is thus dependent on subsequent refining to give a high-quality steel.

Ore-based steelmaking, cf. Figure 6.6, is usually based on a sintered ore and coke charged to a blast furnace (BF), where the iron oxides in the ore are reduced to metal. As the material in the blast furnace is always in contact with carbon in coke, the hot metal tapped from the blast furnace will be almost saturated with carbon, typically about 4.5 wt% carbon in hot metal. Some tenths of a percent of silicon will also be dissolved in the tapped hot metal.

After tapping from the blast furnace, the hot metal is usually refined at least from sulfur, sometimes also from phosphorus. The refined hot metal is decarburized with pure oxygen in the converter, usually a BOF. Reactions taking place are:

$$[Si]_{Fe} + O_2 = SiO_2 (slag) \quad (6.1)$$

$$[C]_{Fe} + 1/2\,O_2 = CO\,(g) \quad (6.2)$$

$$[Mn]_{Fe} + 1/2\,O_2 = MnO\,(slag) \quad (6.3)$$

$$Fe + 1/2\,O_2 = FeO\,(slag) \quad (6.4)$$

All these reactions are exothermic, generating a lot of heat. Silicon content in the metal will very rapidly decrease to almost zero, whereas carbon content decreases during the whole blow, until the blow is stopped at about 0.05% C left in the steel. Oxidation of Mn and Fe is of course only partial. Hot metal from the blast furnace usually has a temperature of about 1300–1400 °C when it arrives at the converter. As a consequence of the heat evolution from the exothermic reactions, large quantities of coolant have to be added to ensure that the temperature is not raised above approximately 1700 °C. Coolant is either ore or scrap. If scrap is used as coolant the amount added is usually in the range 15–20% of the amount of steel produced. Since the volumes of steel produced in the BF–BOF route are large, the amounts of scrap consumed within what is usually labeled ore-based steelmaking are quite large.

Stainless steel is produced in a process flow similar to the process route illustrated in Figure 6.6 for producing scrap-based ordinary steel except for one important difference. Most stainless steel produced in the Western world is based on scrap melted in an EAF, but because the amount of stainless steel scrap not is enough, also ordinary steel scrap has to be used. The lack of alloying elements such as chromium and nickel in the latter makes it necessary to alloy with ferrochromium and ferronickel or pure nickel. Carbon is then introduced with the ferroalloys and a refining step using oxygen to convert carbon into CO(g) has to be introduced, often carried out in an Argon Oxygen Decarburization (AOD)-converter. Stainless steel can be either ferritic or austenitic, a difference that can be of large importance for recycling, as discussed further in Section 6.4.

6.4 TRENDS IN QUALITY OF THE SCRAP AVAILABLE FOR STEEL PRODUCTION

Steel is produced in a large number of different alloy compositions containing different amount of alloying elements like Cr, Mn, Nb, B, V, etc. In the final product the steel may be provided with a coating of zinc, pigments etc. The share of coated steel produced is globally increasing, resulting in an increasing amount of recirculated scrap with different types of coatings. The scrap used in steel making tends therefore to be a more complex material. In conventional scrap melting processes (steel converters or EAF's), some of the impurities evaporate and leave the process with the off-gas and with particulate matter and are collected in the gas cleaning. Others either are oxidized and report to the slag or are dissolved in the steel, or both.

The average North American car today contains not only more of metals other than steel (Al, Mg, Cu) and more of other materials but also more of speciality alloyed steels, as shown by the difference in composition when data for 1975 and 2007 are compared, cf. Figure 6.8 (Bevans et al., 2013). As can be seen, the share of steel present as mild, low-alloyed steel has drastically decreased. Projections from the automotive industry indicate that the advanced high-strength steels will account for up to 65% of the vehicle's body and structure by 2020 (Bevans et al., 2013). Some of these steels can contain as much as up to 24 wt% Mn and 6 wt% Al. To be able to also recover the alloying elements with high efficiency will then be of utmost importance.

From 1992 to 2001, the generation of steel packaging increased by 22% in the EU, and an increasing part is recycled (Baeyens et al., 2010). In 2007, the amount of steel used for packaging in EU was slightly less than 5 Mt. The steel used for packaging is mainly of two types: steel coated with a very thin layer of tin or steel coated with chromium and chromium oxide. Both types can also be combined with a polymer film.

When stainless steel is recycled through a shredder plant, the ferritic part of the stainless steel will be collected with the ordinary carbon steel scrap because ferritic stainless steel is magnetic. Whereas the recovery rate of austenitic stainless steel can be expected to be high, much of the ferritic stainless steel will be recovered as carbon steel scrap. Oda et al. carried out an analysis of the substance flow of chromium in steel in Japan and predicted that the mean chromium content in EAF carbon steel will gradually increase and reach 0.24% in 2030 (Oda et al., 2010).

Zinc present as a coating on steel will during melting at reducing conditions evaporate and is collected in the gas cleaning and is not any problem for steel quality. Provided that the zinc content in the dust collected is high enough, the dust is a raw material for extraction of zinc in, e.g. Waelz-kilns and fuming furnaces. In the

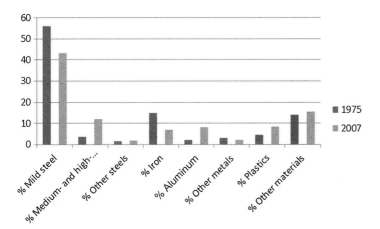

FIGURE 6.8 Material composition (wt%) of an average North American vehicle from 1975 to 2007. *Adapted from Bevans et al. (2013).*

ore-based steel industry, on the other hand, the share of scrap used is lower, and the zinc content is not at a level where extraction of zinc is economically feasible. In this case, the zinc content becomes an obstacle for internal recycling of dust and sludge to the blast furnace, which cannot tolerate high zinc loads.

6.5 HINDRANCES FOR RECYCLING—TRAMP ELEMENTS

6.5.1 Definition of Tramp Elements

Tramp elements are those elements that are present in steel, but usually not intentionally added, and are difficult to refine from the steel, and therefore will be kept in the steel cycle once entered, coming back over and over again when steel is recycled. Some of these have detrimental effects on the steel properties, others being alloying elements for some steel grades, but not for all. As iron is a quite un-noble metal, not as e.g. aluminum, but still, the iron will react with a refining agent before the tramp element reacts, thereby making refining from these elements difficult. For the elements defined as tramp elements no commercially feasible refining processes exist today. Figure 6.9 illustrates common elements in steel, how they are used and controlled and which ones that usually are referred to as tramp elements.

Sources for tramp elements can be many, ore, scrap, reductants used, fluxing agents or coming with alloying additions. Usually the ore used has much less impurity content than the scrap used, although depending of the ore, some impurities may come with the ore, like Cr, As and Cd. The behavior of several impurity and tramp elements during oxidation in, e.g. the BOF process, is illustrated in Table 6.4 (Vallomy, 1985).

As Figure 6.9 shows, some of these elements are added intentionally as alloying elements in certain steel grades, like Mo, Cr, Ni and Cu, but are detrimental for the quality of the steel in other steel grades or are lost to the slag in the processing of another type of steel grade. Low-alloyed steel is often produced by melting the scrap in quite oxidizing conditions, resulting in a substantial loss of chromium to the slag even when chromium-containing low-alloyed steel is produced. Stainless steel can either be ferritic or austenitic stainless steel or both, and as ferritic stainless steel is magnetic it will report to the ordinary carbon steel fraction during separation after shredding. In the upgrading from a shredder plant, steel scrap is not sorted to any larger extent than into an ordinary steel fraction and a stainless steel fraction. Also, in the

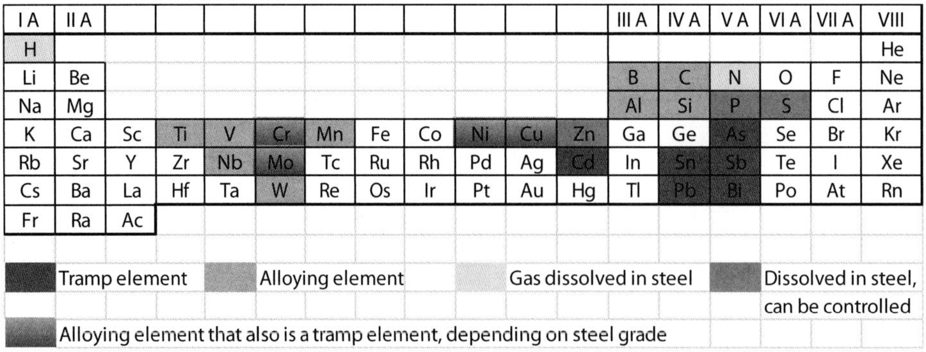

I A	II A											III A	IV A	V A	VI A	VII A	VIII
H																	He
Li	Be											B	C	N	O	F	Ne
Na	Mg											Al	Si	P	S	Cl	Ar
K	Ca	Sc	Ti	V	Cr	Mn	Fe	Co	Ni	Cu	Zn	Ga	Ge	As	Se	Br	Kr
Rb	Sr	Y	Zr	Nb	Mo	Tc	Ru	Rh	Pd	Ag	Cd	In	Sn	Sb	Te	I	Xe
Cs	Ba	La	Hf	Ta	W	Re	Os	Ir	Pt	Au	Hg	Tl	Pb	Bi	Po	At	Rn
Fr	Ra	Ac															

Tramp element	Alloying element	Gas dissolved in steel	Dissolved in steel, can be controlled
Alloying element that also is a tramp element, depending on steel grade			

FIGURE 6.9 Common elements in steel, how they are used and which ones that are referred to as tramp elements.

6.5 HINDRANCES FOR RECYCLING—TRAMP ELEMENTS

TABLE 6.4 Distribution of Some Common Alloying, Impurity and Tramp Elements to Different Phases During Decarburization of Steel

To Bath			To Slag		To Gases	
Sb	Cr	B	Al	B	Zn	Pb
As	Pb	Cd	Be	Cr		
Bi		P	Ca	Cd		
Co		Se	Hf	P		
Cu		S	Mg	Se		
Mo		Te	Si	S		
Ni		V	Ti	Te		
Ta		Zn	Zr	V		
Sn						
W						

Legend: ■ Totally ▨ Mostly ▧ Partially

manufacturing industry, it is very seldom that the scrap is sorted into various types of alloys and depending on the content of alloying elements. This means that in addition to containing tramp elements that have detrimental influence on the steel quality, alloying elements valuable in the production of some steel grades might get lost.

6.5.2 Influence on Steel Quality by Tramp Elements

Copper is known to induce surface cracks during hot rolling if reheating is carried out in oxidizing atmosphere (Vallomy, 1985), at contents above 0.15% (Marique, 1996). Presence of Sn has the same effect at a content higher than 0.05 wt% and a combination of Cu + Sn enhances the effect. The presence of these elements is also responsible for an increased probability for transverse cracking during continuous casting (Mintz, 1999). The mechanism of crack formation due to copper is due to a preferential oxidation of iron at the surface and liquid copper precipitating at the steel surface when the copper content in remaining austenite exceeds the solubility limit. The liquid copper is wetting the iron and thus easily penetrates into the austenite grains, giving rise to the cracks (Kajitani et al., 1996). By increasing the Sn content, the domain of liquid phase is widened, which is why Sn increases the probability of crack formation (Yamamoto et al., 2006). Marique (1996) showed that the rejection rate due to crack formation during wire rod rolling was drastically increased at copper content above 0.15%. The crack depth was shown to increase with increasing reheating temperature above 1000 °C and increasing copper content (Kajitani et al., 1996). Decreased silicon content in the steel also increases the upper temperature limit for crack initiation (Kajitani et al., 1996). Calvo et al. (2007) have shown that the width and depth of the cracks depend on both the reheating cycle and the as-received condition, i.e. the condition from the caster.

Several tramp elements have a negative effect on cold forming and cold drawing, e.g. Sn, Cu, As and Cr. The upper limit for copper and tin for deep drawing grades is 0.1 and 0.02%, respectively (Vallomy, 1985). Ductility deteriorated when testing was performed in oxidising atmosphere (Mintz et al., 1995) and it was proposed that observed precipitation of Cu_2S was responsible and that an equal amount of nickel could prevent this from happening. In inert atmosphere the influence of Cu on hot ductility is confusing, with no clear conclusions. Another phenomenon reported is that increased nickel content has been shown to reduce the scale removability from rolled steel (Asai et al., 1997).

These are only a few examples, and excellent reviews of the existing knowledge on influence of tramp elements on steel properties can be found in the literature (e.g. Vallomy, 1985; Marique, 1996; Herman and Leroy, 1996; Mintz and Crowther, 2010).

6.5.3 Importance of Tramp Elements for a Sustainable Recycling of Steel

There was much focus on tramp elements in steel, especially during 1980s and 1990s, but not as much later on. This period was characterized by stagnant level of world steel production and a rapidly increasing share of steel produced from scrap. Scrap sorting and knowledge about the rate for accumulation of tramp elements was not very well developed and hence steel producers looked on increasing levels of e.g. copper in the steel scrap with fear. In this century steel production has increased rapidly and hence a larger potion of steel is produced from ores. Scrap sorting has become more efficient, and the increased prices for metals like copper has even made hand-picking of larger copper pieces from the steel scrap on the conveyor belt in the shredder plant economically feasible. Thus, e.g. copper content in the shredded scrap has not continued to increase but instead decreased on average. Nevertheless, scrap-based steel producers sometimes need to dilute the steel scrap with virgin iron units from HBI or DRI, if they are producing steel grades with very low tolerance for tramp elements. If and when the world once again comes into a situation with small or no increase in the consumption of steel and the share of scrap-based steelmaking is increasing, the question of how to handle the tramp elements will once again be of fundamental importance for sustainable recycling of steel.

6.6 PURIFICATION OF SCRAP

6.6.1 Dezincing of Galvanized Scrap

Two main principles are used, leaching and thermal treatment. Leaching in both acid solutions (HCl, H_2SO_4) and in basic solution ($NaOH$) is discussed in the literature (Sen and Roy, 1975; Niedringshaus et al., 1992; Koros, 1992; Rome, 1992; Ijomah and Ijomah, 2003). When removing zinc at high pH, Zn is transferred to Zn^+ and recovered either in the form of sodium zincates (Na_2ZnO_2 or $NaHZnO_2$) precipitated from a saturated solution or zinc recovered by electrolysis. In acid leaching, zinc is recovered via a saline solution of $Zn(NO_3)_2$, $ZnSO_4$, or $ZnCl_2$. Leaching at high pH is a selective process which not affect the iron, but is a time-consuming process, while acid leaching is less selective but much faster. Since leaching is a surface reaction, the surface area available becomes important and the leaching proceeds much faster for shredded scrap than for bundles. A leaching process with a treatment time of 6 h at high pH with high, >90%, recovery of the zinc through electrolysis was described by Niedringshaus et al. (1992). Formation of cyanide was indicated as one possible problem. One commercialized process, the Meretec process, is based on this concept. The concept is said to be competitive and fully functional.

Several studies describe thermal treatment of scrap. The possibility of evaporating Zn, Ni, Cr and Sn from coated materials using chlorine is one alternative (Dapper et al., 1978). Because $ZnCl_2$ is more stable than $FeCl_2$ and the $ZnCl_2$ is much more volatile, a selective evaporation of Zn can be obtained as long as there is metallic Zn present. Rinsing of nonlron oxides and metals using HCl from the combustion of PVC has been described by Fray (1999). Tee and Fray have shown that to selectively evaporate compounds in the coatings oxychlorination with controlled ratio of $O_2:Cl_2$ in the gas mixture is necessary (Tee and Fray, 1999a,b; Tee and Fray, 2005). Zn boils at a temperature of 906 °C, and it has been shown that vaporization of zinc in CO/CO_2 and N_2 gas mixtures is very rapid and almost complete at temperatures higher than 850 °C (Ozturk and Fruehan, 1996).

Another alternative for removing highly volatile compounds present in coatings on steel is vacuum treatment. A method to clean scrap by heating and vacuum treatment was suggested by Okada et al. (1995). High removal efficiency was obtained already at 700 °C as compared to preheating with CO/CO_2 gas mixtures. Almost 97% zinc recovery was achieved.

Steel scrap is often preheated before charging to the EAF. This has several purposes: (1) to decrease the meltdown time in the EAF, (2) to remove water and especially in cold climates to remove ice, which otherwise can result in violent gas explosions and (3) if waste heat can be used, this will result in an overall saving of the energy need. However, the presence of halogens and organic matter together with the scrap may result in the formation of dioxins and furans and therefore preheating temperature is according to regulations in e.g. Sweden limited to 300 °C in conventional scrap preheating equipment. Possibilities to increase the scrap preheating temperature to temperatures where a simultaneous evaporation of e.g. zinc can be obtained would have many benefits for the steel industry. This concept has been studied by Larsson et al. (2012) in a project in the Swedish research program "The Steel Eco-cycle" (Steel Eco-cycle, 2013) and is also part of an ongoing European research project financed within RFCS (Research Fund Coal and Steel).

The idea is to separately combust energy-containing waste as auto shredder residue (ASR) with oxygen. The hot gas is used for preheating scrap. Off-gases from this preheating are partially recirculated and mixed with gases from the combustion to control the temperature of the hot gases entering the scrap preheating, but also giving enough gas volumes to ensure rapid enough preheating. Preheating of the scrap is controlled at <900 °C in a separate shaft reactor. Exhaust gases are cleaned and chlorides are captured.

Results from the small-scale tests showed that chlorine-rich gases, e.g. exhaust gases, from ASR combustion can be a suitable gas for cleaning and preheating scrap (Larsson et al., 2012). Pilot tests have shown that it is possible to achieve a uniform scrap preheating at 650 °C, where the zinc removal efficiency is high. Preheating to 650 °C generates a metal loss of less than 1%. The process concept, including oxyfuel combustion, dedicated gas scrubbing and exhaust gas recirculation, has been proved. The low gas volumes and low amount of gas vented to the surrounding will allow for treatment of the gas and the removal of harmful elements in a cost-efficient manner.

A number of different methods to remove zinc from surface coatings of steel have been proposed during the years but very few have yet reached a commercial stage. For new methods to be commercially introduced, they have to be cost-competitive with the traditional way of dealing with zinc in steel scrap, namely the processing of zinc rich dust from the EAF in Waelz kilns, the dominant process route, or, e.g. in a zinc fuming furnace. The low zinc content in the dust from ore-based steelmaking makes it very expensive to send the dust for zinc recovery. The strategy for many ore-based steel producers has therefore been to purchase only comparably clean scrap as coolant. Market shortness of clean scrap in the future would perhaps make the ore-based steel industry interested in scrap rinsed from zinc, if the prices can be comparable with the cost for use of ores as coolant.

6.6.2 Detinning of Steel Scrap

Electrolytic detinning of tinplate scrap has been in commercial use for a long time. Scrap pressed into bundles serve as anodes immersed in a caustic soda bath at about 85 °C. Tin is deposited on a steel cathode as a sponge material, which is scraped off (Savov and Janke, 1998). The tin content in the steel is below 0.02 wt% after this processing. The process requires rather large volumes to be economical and is claimed to be suitable only for prompt scrap. Because the process is a surface process, it will be applicable only where tin is present as a surface coating. A proposed alternative method for detinning tinplate scrap has been to react the surface with a sulfur-containing gas at moderate temperature to form a tin sulfide layer. The tin content in the steel could be decreased to levels of about 0.1 wt% from above 0.3% originally. At higher temperatures the diffusion of tin from the surface layer and into the steel becomes fast and limits the possibilities for detinning based on surface reactions. Other proposed methods include selective evaporation or by oxidizing (Savov et al., 2000).

6.6.3 Others

It is technically and sometimes also economically quite possible to refine steel from metals or impurities present as a surface coating using leaching or evaporation techniques because these methods are acting on the surface. For impurities or tramp elements present in the steel alloy, it is a different question. Although sometimes detrimental for the properties of the steel, the content in the steel is low, a few hundredths up to a few tenths of weight percentage, and thus they have a low thermodynamic activity. Furthermore, iron is sometimes the least noble metal. Both factors make it difficult to find a suitable reagent. In addition, steel has a comparably low value per tonne and the volumes of steel scrap are huge in comparison with other metals, necessitating a refining concept to be very cost effective.

Many different processes for refining steel from copper have nevertheless been proposed, including dissolution in a solvent like Al, Pb, sodium sulfide—iron sulfide, soda-iron sulfide or aluminum sulfide—iron sulfide, reacting with a chlorinating agent or vacuum treatment of the steel. The thermodynamics of the removal of tramp elements from steel scrap was evaluated by Tsukihashi (2003). The conclusion was that calcium—metal halide fluxes have an excellent refining ability. No figures were given on the amount of different tramp elements that possibly could be removed. None of these suggestions have, however, reached a commercial stage. Several aspects of tramp element control in steel have been studied by a group of steel producers and research institutes in the European Union (Boom and Steffen, 2001).

To give one example, the high distribution of some impurity elements to a lead phase in contact with a carbon-containing iron phase could be utilized (Yamaguchi and Takeda, 2003a,b). A miscibility gap exists between an iron phase with 95.4%Fe—4.5%C—0.1%Pb and 99.9% Pb—0.1%Fe. Copper, tin, zinc, gold, silver and palladium are primarily distributed to the lead phase. An equal amount of iron and lead makes 70% of the copper and tin contained in the iron to report to the lead phase. The temperature for the study was 1453 K. The quite high amount of lead needed to dissolve a considerable amount of copper and tin makes such a process dependent on the value of recoverable noniron elements, not only for the purpose of cleaning the steel scrap.

6.7 TO LIVE WITH IMPURITIES

In a world with an increased share of steel produced from scrap, the issue of the impurity content in the scrap will be of high importance. The question is illustrated in Table 6.5 showing the typical content of impurity elements (given as the sum of content in wt% of the elements Cu, Sn, Ni, Cr, Mo) in different types of raw materials, as well as the limits for the content of these impurity elements for the production of different type of steels.

Although these data are somewhat old and the separation processes at the shredder plants and the sorting of scrap have improved, they illustrate that it is difficult to produce all steel grades based on 100% of scrap, if the scrap is not well sorted before it enters a shredder plant.

That is also the reality for scrap-based steel producers in the business sectors for high-quality sheet products, speciality tools and machinery as well as scrap-based production of iron powder. They are all dependent on having access to a very well sorted scrap, usually home or manufacturing scrap, and to a much lesser extent on obsolete scrap. If well-sorted scrap not is available to affordable prices, ore-based iron units suitable for charging to an EAF is their only alternative today.

TABLE 6.5 Typical Content of the Impurity Elements Cu, Sn, Ni, Cr, Mo (Given as the Sum of These Elements in wt%) in Different Type of Raw Materials for Production of Steel and the Requirements for Production of Different Steel Grades

Raw Material	Total Impurity Content in wt% (Sum of $Cu + Sn + Ni + Cr + Mo$)
Direct reduced iron	0.02
Pig iron	0.06
No. 1 factory bundles	0.13
Bushelling	0.13
No. 1 heavy melting	0.20
Shredded auto scrap	0.51
No. 2 heavy melting	0.73
STEEL GRADE	
Tin plate for draw and iron cans	0.12
Extra deep drawing quality sheet	0.14
Drawing quality and enameling steels	0.16
Commercial quality sheet	0.22
Fine wire grades	0.25
Special bar quality	0.35
Merchant bar quality	0.50

Adapted from Vallomy (1985).

6.8 MEASURES TO SECURE SUSTAINABLE RECYCLING OF STEEL

Steel is, based on volume, the most recycled metal, and there exists a well-functioning business structure for the recycling of steel. Losses are inevitable in the processing and handling of any metal. Losses occur to byproduct streams and waste in the processing (slag, dust, sludge) and into other material cycles, material fractions aimed for energy recovery or into waste that is landfilled. These losses can and will be decreased as technology develops and as awareness of the issue grows among all the stakeholders involved in the production of products, from designers to manufacturers. The amount of collected steel that is lost from the recycling cycle is nevertheless probably small in comparison with the total production. Not at least if metal prices increase in comparison with the living standard, what is worth recycling will be recycled. If and when the increase in world consumption of steel decreases, there will be greater possibilities of producing a large amount of the steel from recycled scrap.

The largest question for the steel and scrap processing industry in order to obtain long-term sustainable steel recycling is perhaps the question of scrap quality and avoiding quality losses when recycling steel. As the share of steel produced from ore has increased in the last decade, accumulation of tramp elements has not been an issue of high importance recently, but it is an issue that will have to be tackled in the future.

Several actions can be taken to minimize the detrimental effects of tramp elements:

- Design for recycling
- Improve sorting at the shredder plant
- Improve processing at the steel plant
- Live with impurities based on development of new alloys
- Dilute scrap with ore-based iron units
- Improve the understanding of steel flow in the society

Design for recycling. By proper choice of materials placed together in different parts of a construction and by the choice of fittings between different materials, it is possible to ensure a good separation of crucial parts, by manual dismantling or in the following shredding and processing of the scrap (Reuter et al., 2005). Although design for recycling has been on the table within the research communities dealing with sustainability and recycling for long time, the impression is that the issue has not yet really reached the designers and construction engineers. Issues on appearance, functionality and manufacturing are still considered more important.

Improved sorting at the shredder plant. Techniques for sorting of scrap where physical properties or composition is utilized have been developed and new techniques will surely emerge. Hitherto, more advanced sorting techniques have not been implemented at larger scale, possibly because the steel producers are not willing to pay for the extra cost through an increased scrap price and also because the need has not been there, as "pure" ore-based iron units are cheaper. Rem et al. (2012) has shown by the introduction of a shape-sensitive magnetic separator that it is possible to pre-sort scrap into two products. One product is a bulky, thin-walled steel fraction of high purity, the other a volumetrically small flow of heavy parts containing the contaminants. The latter is very well suited for cost-effective hand sorting or sensor sorting.

A better knowledge of the composition of the scrap will increase possibilities to utilize the scrap for production of the same type of alloy, thereby not only securing the recycling of iron units but also securing a better utilization of the alloying elements and decreasing the losses of the same.

Improved processing at the steel plant. Direct casting of thin strips or in-line strip production (ISP) is a technique that will give higher tolerances for impurity elements such as copper due to a faster cooling and making a reheating procedure unnecessary (Arvedi, 2010; Spitzer et al., 2003).

However, as long as not all steel would be produced in these new thin-strip casting processes, steels produced with higher content of e.g. copper would have to be kept in a separate recycling loop.

To live with impurities through development of new alloys that can accept a higher content of impurity and tramp elements. Although introducing detrimental properties in many steel grades with the present processing of steel, some of the tramp elements are also used as alloying elements for some steel grades to improve corrosion resistance and enhance the mechanical properties.

It is possible to improve the properties of steel by precipitating nanoscale copper sulfides (e.g. Liu et al., 2006, 2007; Yamamoto et al., 2006) through a rapid solidification and/or controlling the balance between Cu, S and Mn content or by adding phosphorus, which promotes precipitation of copper sulfide from ferrite instead of from austenite and suppresses precipitation of MnS at higher temperature. Precipitation of copper sulfide is preferable compared to MnS. From a steel recycling point of view, it must, however, be noted that the copper content in the alloys studied is low, below 0.1 wt%. In extra-low-carbon titanium added interstitial free steel sheet, Cu is the useful alloying element for increasing hardness. Addition of Cu or Cu + Ni at a level of 0.5 and 0.4 wt%, respectively, has been shown to increase the volume fraction of retained austenite and to improve elongation and the strength ductility balance (Kim et al., 2002, 2003).

As pointed out regarding ISP technology, if the new alloys developed contain substantial amounts of tramp elements and as long as many steel grades require a very low content of tramp elements, it has to be possible to keep these steels in a separate recycling loop by well-functioning collection and sorting.

Dilution of scrap with ore-based iron units. The last choice will, as it already is, be to dilute impure scrap with ore-based iron units produced in the form of DRI or HBI.

Finally, an improved understanding of steel flows in the society, on a steel grade level, would give us a tool to model the possible accumulation of tramp and impurity elements depending on reasonable scenarios for future economic and technological development. Many Material Flow Analysis (MFA) and Substance Flow Analysis (SFA) studies of the steel flow have been carried out, but very few on steel grade level. Some exceptions exist (e.g. Oda et al., 2010) where the recycling of different types of stainless steel was analyzed.

References

Arvedi, G., 2010. Achievements of ISP steelmaking technology. Ironmaking and Steelmaking 37, 251–256.

Asai, T., Soshiroda, T., Miyahara, M., 1997. Influence of Ni impurity in steel on the removability of primary scale in hydraulic descaling. ISIJ International 37, 272–277.

Aydin, Ü., Makowe, J., Noll, R., 2004. Fast identification of light metal alloys by laser-induced breakdown spectroscopy for material recycling: development of a demonstrator for automatic sorting of aluminium scrap. VDI Berichte, 101–112, 171.

Bauyens, J., Brems, A., Dewil, R., 2010. Recovery and recycling of post-consumer waste materials. Part 2. Target wastes (glass beverage bottles, plastics, scrap metal and steel cans, end-of-life tyres, batteries and household hazardous waste). International Journal of Sustainable Engineering 3, 232–245.

Bevans, K., et al., 2013. Importance of advanced high-strength steels and electronic units on the recycling of automobiles: a review. Iron and Steel Technology 10, 266–277.

Birat, J.-P., Hanrot, F., Danloy, G., 2004. CO_2 mitigation technologies in the steel industry: a benchmark study based on process calculations. In: Scanmet II, 2nd International Conference on Process Development in Iron and Steelmaking, June 2004, Luleå, Sweden, pp. 73–80.

Boom, R., Steffen, R., 2001. Recycling of scrap for high quality steel products. Steel Research 72, 91–96.

Brooks, G., Pan, Y., 2004. Developments in steel recycling technology. In: Green Processing 2004, 2nd International Conference on the Sustainable Processing of Minerals. Australasian Institute of Mining and Metallurgy, pp. 65–72. Publication Series No 2.

Calvo, J., Cabrera, J.M., Rezaeian, A., Yue, S., 2007. Evaluation of the hot ductility of a C-Mn steel produced from scrap recycling. ISIJ International 47, 1518–1526.

Dapper, G., Sloferdijk, W., Verbach, C.A., 1978. Removal of surface layers from plated materials: upgrading of scrap. Conservation Recycling 2, 117–121.

Eurofer, The European Steel Assiociation, 2013. European Steel Scrap Specification. www.eurofer.eu/docs/.

Fray, D.J., 1999. Use of poly(vinyl chloride) as chlorinating agent in recycling of metals. Plastics, Rubber and Composites 28, 327–329.

Gesing, A., et al., 2003. Assuring continued recyclability of automotive aluminium alloys: chemical composition based sorting of wrought and cast Al shred. In: Das, S.K. (Ed.), Aluminium 2003. TMS, pp. 3–14.

Gurell, J., Bengtson, A., Falkenström, M., Hansson, B.A.M., 2012. Laser induced breakdown spectroscopy for fast elemental analysis and sorting of metallic scrap pieces using certified reference materials. Spectrochimica Acta B: Atomic Spectroscopy 74–75, 46–50.

Herman, J.C., Leroy, V., 1996. Influence of residual elements on steel processing and mechanical properties. Iron and Steelmaker 23, 35–43.

Ijomah, M.N.C., Ijomah, A.I., 2003. Chemical recycling of galvanized steel scrap. Indian Journal of Chemical Technology 10.

Kajitani, T., Wakoh, M., Tokumitsu, N., Ogibayashi, S., Mizoguchi, S., 1996. Influence of temperature and strain on surface crack to residual copper in carbon steel. In: 79th Steelmaking Conf., Pittsburgh, pp. 621–626.

Kim, S.-J., Lee, C.G., Lee, T.-H., Oh, C.-S., 2003. Effect of Cu, Cr and Ni on mechanical properties of 0.15 wt.% C TRIP-aided cold rolled steels. Scripta Materialia 48, 539–544.

Kim, S.-J., Lee, C.G., Lee, T.-H., Oh, C.-S., 2002. Effects of copper addition on mechanical properties of 0.15C-1.5Mn-1.5Si TRIP-aided multiphase cold-rolled steel sheets. ISIJ International 42, 1452–1456.

Koros, P.J., 1992. Recycling of galvanized steel scrap – issues and solutions. In: 1992 Steel Making Conference Proceedings, p. 687.

Larsson, M., Hahlin, P., Ångström, S., 2012. The future of scrap preheating and surface cleaning. In: Scanmet IV – 4th International Conference on Process Development in Iron and Steelmaking, 10–13 June 2012, Luleå, Sweden, pp. 83–92.

Liu, Z., Kobayashi, Y., Kuwabara, M., Nagai, K., 2007. Interaction between phosphorus micro-segregation and sulfide precipitation in rapidly solidified steel-utilization of impurity elements in scrap steel. Materials Transactions 48, 3079–3087.

Liu, Z., Kobayashi, Y., Yang, J., Nagai, K., Kuwabara, M., 2006. Effect of nano-scale copper sulfide precipitation on mechanical properties and microstructure of rapidly solidified steel with tramp copper element. Materials Transactions 47, 2312–2320.

Marique, C., 1996. Scrap recycling and production of high quality steel grades in Europe. In: 79th Steelmaking Conf., Pittsburgh, pp. 613–619.

Mintz, B., Crowther, D.N., 2010. Hot ductility of steels and its relationship to the problem of transverse cracking in continuous casting. International Materials Reviews 55, 168–196.

Mintz, B., 1999. Influence of composition on the hot ductility of steels and to the problem of transverse cracking. ISIJ International 39, 833–855.

Mintz, B., Abushosha, R., Crowther, D.N., 1995. The influence of small additions of Cu and Ni on the hot ductility of steels. Materials Science and Technology 11, 474–481.

Niedringshaus, J.C., Rodabaugh, R.D., Leeker, J.W., Steribick, A.E., 1992. A technical evaluation of dezincification of galvanized steel scrap. In: 1992 Steelmaking Conference Proceedings, p. 725.

Oda, T., Daigo, I., Matsuno, Y., Adachi, Y., 2010. Substance flow and stock of chromium associated with cyclic use of steel in Japan. ISIJ International 50, 314–323.

Okada, Y., Raheachi, Y., Fujio, S., 1995. Development of method for removal of zinc from automobile body scraps. In: Galvatech '95 Conference Proceedings, pp. 549–554.

Ozturk, B., Fruehan, R.J., 1996. Vaporization of zinc from steel scrap. ISIJ International 36, 239–242.

Pauliuk, S., Wang, T., Müller, D.B., 2013. Steel all over the world: estimating in-use stocks of iron for 200 countries. Resources, Conservation and Recycling 71, 22–30.

Pauliuk, S., Wang, T., Müller, D.B., 2012. Moving toward the circular economy: the role of stocks in the Chinese steel cycle. Environmental Science and Technology 46, 148–154.

Rem, P.C., van den Broeck, F., Bakker, M.C.M., 2012. Purification of post-consumer steel scrap. Ironmaking and Steelmaking 39, 504–507.

Reuter, M., et al., 2005. The Metrics of Material and Metal Ecology; Harmonizing the Resource, Technology and Environmental Cycles. Elsevier, Amsterdam, Netherlands.

Rome, C.L., 1992. A contribution to the recycling of coated steels – the results of a consultation of experts. In: 1992 Steelmaking Conference Proceedings, p. 687.

Savov, L., Janke, D., 1998. Recycling of scrap in steelmaking in view of the tramp element problem. Metall 52, 374–383.

Savov, L., Tu, S., Ja, R., Janke, D., 2000. Methods of increasing the rate of tin evaporation from iron based melts. ISIJ International 40, 654–663.

Sen, P.K., Roy, S., 1975. Recovery of zinc from galvanized steel scrap. Transactions of the Indian Institute of Metals 28 (5).

Schaik, A., van, Reuter, M.A., Boin, U., Dalmijn, W.L., 2002. Dynamic modelling and optimisation of the resource cycle of passenger vehicles. Minerals Engineering 15, 1001–1016.

van Schaik, A., 2004. Theory of Recycling Systems – Applied to Car Recycling (thesis). Delft Univ. Techn., ISBN: 90-9018839.

Spitzer, K.-H., Rüppel, F., Viščorová, R., Scholz, R., Kroos, J., Flaxa, V., 2003. Direct strip casting (DSC) — an option for the production of new steel grades. Steel Research International 74, 724—731.

Steel Eco-cycle, 2013. www.stalkretsloppet.se.

Swedish Steel Producers Association, 2011. http://www.jernkontoret.se/.

Swedish Steel Producers Association, 2012. http://www.jernkontoret.se/.

Swedish Steel Producers Association, 2013. http://www.jernkontoret.se/.

Tee, J.K.S., Fray, D.J., 1999a. Recycling of galvanised steel scrap using chlorination. In: EDP Congress 1999. The Minerals, Metals & Materials Society, pp. 883—891.

Tee, J.K.S., Fray, D.J., August 1999b. Removing impurities from steel scrap using air and chlorine mixtures. JOM, 24—27.

Tee, J.K.S., Fray, D.J., 2005. Recycling of galvanised steel scrap using chlorination. Ironmaking and Steelmaking 32, 509—514.

Tsukihashi, F., 2003. Thermodynamics of removal of tramp elements from steel scrap. In: Yazawa International Symposium: Metallurgical and Materials Processing: Principles and Techologies; Materials Processing Fundamentals and New Technologies, vol. 1, pp. 927—936.

ULCOS, 2013. www.ulcos.org.

Vallomy, J.A., 1985. Adverse effects of tramp elements on steel processing and product. Industrial Heating 52, 34—35, 37.

World Steel Association. World Steel in Figures 2013. www.worldsteel.org.

Yamaguchi, K., Takeda, Y., 2003a. Impurity removal from carbon saturated liquid iron using lead solvent. Materials Transactions 44, 2452—2455.

Yamaguchi, K., Takeda, Y., 2003b. Impurity removal from carbon saturated liquid iron using lead solvent. In: Yazawa International Symposium: Metallurgical and Materials Processing: Principles and Techologies; Materials Processing Fundamentals and New Technologies, vol. 1, pp. 485—492.

Yamamoto, K.-I., Shibata, H., Mizoguchi, S., 2006. Precipitation behavior of copper, tin and manganese sulfide at high temperature in Fe-10%Cu-0.5%Sn alloys. ISIJ International 46, 82—88.

CHAPTER 7

Copper Recycling

Caisa Samuelsson, Bo Björkman

MiMeR - Minerals and Metallurgical Research Laboratory, Luleå University of Technology, Luleå, Sweden

7.1 INTRODUCTION

Copper is a metal that is used in many applications such as electric and electronic equipment, automobiles, and plumbing because of its properties. Copper has a high thermal and electric conductivity and is relatively corrosion resistant. It can be alloyed with, e.g. Zn, Sn or Ni, forming brasses and bronzes. The use of copper is, however, not exclusive to our modern society; traces of copper usage can be traced as far back as 7000 BC It is also claimed that copper was the first metal to be extracted through a metallurgical process (in Asia around 4000 BC). In principle copper can be recycled endlessly without loss of quality. However, some elements integrated in products with copper may cause problems for both processes and products, for example, antimony (Sb) and aluminum (Al).

The increasing amount of more complex scrap as well as the fast changes in material composition in some products are a challenge for recycling. In addition, copper ores tend to have an increased complexity. Understanding metallurgical phenomena is one of the key issues in meeting the demand for an increased recycling and sustainable copper extraction.

Recycling is an important part of the supply of raw materials to the copper refining and manufacturing facilities. However, there are large variations in recycling between different countries, which depends on such factors as efficiency in collection/recovery, historical usage and application; for example, copper grade is less critical when used in pipes and roofing. Also, the market and customers play an important role; copper scrap is mainly produced in cities whereas copper smelters often are located in more remote areas. Recycling of nonferrous metals is discussed in an extensive work by Henstock (1996), in which among many other issues, aspects such as benefits and limitations of recycling are considered. The collection and treatment of secondary material have to be economically favorable or deduced by governmental regulations. Benefits of recycling are also lost if the collection and processing consume more resources than extraction from primary sources.

There is a large difference in scrap consumption between different parts of the world. The Middle East and Asia accounted for about 64% of world copper recovered from scrap in 2011, the Western European countries accounted for 22% and the countries of North and South

America accounted for 11% (Jolly, 2013). The use of copper scrap also reflects the copper demand and prices; for example, during the period of higher prices, 2004 to 2008, world production of refined copper from scrap also increased.

In 2007–2012, roughly between 15% and 18% of the refined copper produced originated from copper scrap (International Copper Study Group, 2013). This figure takes into account only the copper produced through refining; direct-melted or reused copper is excluded. Counting direct-melted copper scrap, the figure is considerably higher; for example, in 2009 more than 45% of copper used in Europe came from recycled material and the global recycling rate was 34% (European Copper Institute, 2013). Recovery of copper from end uses varies greatly from product to product. For example, studies have shown that in Japan nearly 100% of copper wire from electric power, telecommunications and railways was recovered. Recovery from small electrical and electronic appliances was only 20%, for automobiles the recovery fraction was 48%, for industrial machinery a little over 81% and for buildings about 68% (Ayres et al., 2002).

7.2 RAW MATERIAL FOR COPPER RECYCLING

7.2.1 Copper Scrap

Recycling of copper is based on a large variety of raw materials, ranging from low-grade copper scrap containing only a few percent Cu to very high-grade copper as well as pure copper close to 100% Cu (Table 7.1). Thus there are several options for recycling processes, within both primary plants and secondary plants treating only scrap material.

Copper scrap that is adequately clean can be directly recovered through remelting without further refining, whereas scrap of lower grade has to be refined in similar processes as primary copper.

Copper scrap is often classified according to its source: (1) direct or "home scrap", which is the scrap generated at the smelter/refinery and has the highest purity, usually recycled internally at the plant; "new scrap" is generated at downstream metal fabricators, e.g. trimmings, boring and croppings, which also usually is recycled at the smelter/refinery.

Post-consumer scrap or old copper scrap can be divided into a number of different grades, which also may vary between different countries. The Institute for Scrap Recycling Industry Inc. has in their guidelines for nonferrous scrap NF-2013, for about 50 different copper and copper alloy specifications (Institute of Scrap Recycling, 2013). There are three main categories of unalloyed scrap, each containing several grades (Jolly, 2013). No. 1 copper scrap has a minimum of 99% copper and is often just remelted. No. 2 copper consists of unalloyed scrap, with a copper content of at least 94% with a nominal content of 96% Cu. Light copper scrap contains between 88% and 92% copper. These types of copper scrap usually have to be refined to obtain the quality required. Refinery brass has a

TABLE 7.1 Examples of Different Types of Scrap and Secondary Copper-Containing Material

Material	Typical Cu Content (wt%)	Source
Pure no. 1 copper scrap	99%	Semi-finished products, wire, strip, cuttings
No. 2 copper scrap	94–98%	Miscellaneous unalloyed wire
Light copper scrap	88–92%	Sheet, gutters, boilers, wires
Red brass scrap	75–85%	Valves, machinery parts
Shredder material	60–65%	Cars
Electronic scrap	5–30%	Electronics
Copper slag	1–8%	Copper smelter slag
Copper dust	1–50%	Copper smelter dust

minimum of 61.3% copper and a maximum of 5% iron; it consists of brass and bronze solids as well as alloyed and contaminated copper scrap. Copper alloy scrap can be classified by type of alloy or by its end use. Brass and bronze scrap is generally remelted without refining, provided they can be sorted into composition categories.

Other types of copper-containing secondary material are, for example, shredder material, which has a copper content of 60–65%, and drosses, ashes and slags from foundries containing 20–25% copper (Lossin, 2005). This type of material is usually smelted under reducing conditions, converted and refined. In addition, there are low-grade materials such as sludges and combustion ashes that have potential for recycling and recovery of copper.

Another important type of copper bearing scrap is waste electric and electronic equipment (WEEE). This type of scrap is often very complex, with low metal content and organic material, etc., which renders it special challenges. It is usually not only the copper that makes it economical to recycle but more the content of precious metals.

7.2.2 Smelter Residues

Copper recycling may also include copper recovery from process residues such as slag, dust and sludges. Often these types of copper-containing residues are internally recycled within the plant. However, there is a need for bleeding and thus an outlet from the plant.

Slag from copper-smelting operations usually contain between 1.5% and 8% copper depending on process used. After internal recycling, reduction and/or settling, the slag contains between 0.7% and 2% Cu and is often discarded. It is estimated that about 2.2 t of slag is generated per ton of copper produced.

Dust generated at smelters is usually recycled to a large extent at the plant; as such it is difficult to find figures about the amount produced.

The amount and composition also vary, depending largely on process, process systems used and raw materials processed. Both hydrometallurgical and pyrometallurgical methods to treat dust for recovery of valuable metals have been proposed.

7.3 PROCESSES FOR RECYCLING

Copper scrap is smelted in primary and secondary smelters. Type of furnace and process steps depend on copper content of the secondary raw material, other constituent, size, etc. In the case of oxide scrap material, reducing conditions are required, which can be achieved through carbon and iron along with fluxing agents. Depending on the raw material and process, other metals such as Zn, Pb and Sn can be recovered from the fume. Depending on the quality/grade of the scrap, refining is required, which is usually done through electrorefining.

A considerable amount of copper scrap is used in primary copper smelters, and the amount and type depend on which process routes exist for primary concentrate. In primary smelters, heat is generated in smelting operations, especially in the Peirce-Smith converting of matte into blister copper, where the heat evolution is so great that coolants have to be used. Thus copper-containing scrap is to large extent used as a coolant in ore-based copper production. The matte-converting process is normally divided into two main stages: the slag-making stage and the copper-making stage. In the slag-making stage, impure scrap with a lower copper content can be used, whereas in the copper-smelting stage, pure copper scrap is used, especially spent anodes from the electrolysis process. Copper-containing scrap can also be added to the smelting stage, depending on type of smelting furnace used and type of scrap. For example, smelters using an electric smelting furnace can process a large amount of secondary oxidic copper-containing material.

Copper scrap can be smelted in a number of different furnaces: blast furnaces, reverberatory, rotary, bath smelting or electric furnaces. Smelters dedicated to treating copper scrap use different concepts for smelting depending on the kind of scrap material. Small Peirce-Smith converters are commonly used for impure copper alloy scrap like red brass scrap. The Contimelt process is a continuous two-stage process to melt and treat, e.g. high-grade copper scrap, black copper and blister copper. Several of the more recently built plants are using bath smelting with top-submerged lance Isasmelt/Ausmelt technology or top blown rotary converters (TBRC)/Kaldo or a combination of submerged lance and TBRC for smelting and refining. Processing and recycling of copper is extensively described in a recent publication by Langner (Langner, 2011). In the book by Davenport et al. (Davenport et al., 2002), the chemistry of copper recycling in primary and secondary smelters is discussed.

The Contimelt process is based on two interconnected furnaces, including a heart shaft furnace where high-grade scrap, blister copper and black copper are smelted. Oxy-gas burners supply heat for melting. The molted copper flows to a cylindrical furnace, where it is deoxidized using natural gas and cast in anodes (Figure 7.1).

Isasmelt/Ausmelt is based on cylindrical bath furnaces using a steel lance for injection of, e.g. natural gas, oil, oxygen or air into the melt. The lance is submerged into the bath, creating a well-stirred melt. Formation of a slag coating protects the lance (Figure 7.2).

The TBRC or Kaldo furnace is a rotating and tilting furnace that uses lances for heating and blowing purposes. The furnace is a compact and energy-efficient reactor. Good mixing is achieved through rotation; however, this may also lead to abrasion of the refractory lining. Oxygen and fuel are added via lances, which blow onto the surface of the melt. The operation is normally a batch process. The furnace is used for smelting, converting and slag treatment in the copper, lead and precious metals processing industries (Figure 7.3).

The combination of bath smelting and TBRC is used in the Kayser recycling system at Aurubis to produce copper from secondary raw materials. The system is based on a two-step process consisting of a smelting and conversion step. In the first step a submerged lance furnace is used to produce a liquid metallic phase

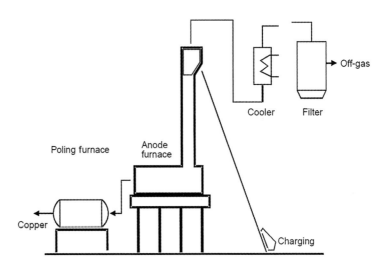

FIGURE 7.1 The Contimelt process. *From Integrated Pollution Prevention and Control (2001), p. 105.*

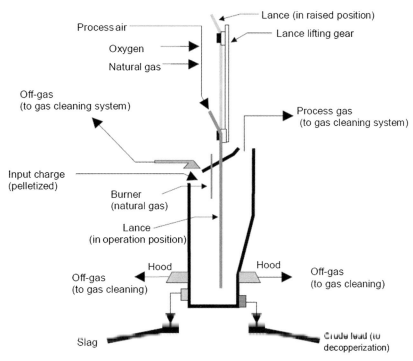

FIGURE 7.2 Isasmelt/Ausmelt process. *From Integrated Pollution Prevention and Control (2001), p. 94.*

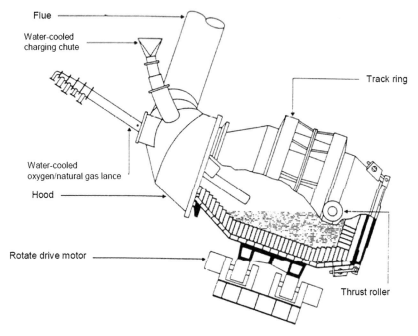

FIGURE 7.3 Top blown rotary converter. *From Integrated Pollution Prevention and Control (2001), p. 95.*

containing approximately 80% Cu (black copper). The black copper flows to the TBRC where it is enriched to 95% Cu and then further refined in an anode furnace and through electrolysis. Copper alloy scrap is added to the TBRC and to the anode furnace (www.Aurubis.com).

7.3.1 Recycling of WEEE

The metal composition of different electronic scraps varies widely depending on age, origin and manufacturing. The copper content may vary between 3 and 27 wt% (Cui and Zhang, 2008). The dominating method to treat WEEE is through the pyrometallurgical route. Recycling of WEEE is discussed in a separate chapter and will thus be mentioned only briefly here. After collection, sorting and mechanical upgrading, the electronic scrap containing nonferrous metals is treated in a primary or secondary smelter. Typically the scrap is smelted under reducing conditions producing a so-called black copper, which is further treated in oxidizing atmosphere, removing impurities through slagging and vaporization. At Boliden Rönnskär smelter, the scrap is smelted in a Kaldo unit, and the black copper produced is then charged to the Peirce-Smith converter.

7.3.2 Slag Treatment for Recovery of Copper

There are several options to treat slag, depending on the metal content in the slag. Flotation can be used for slag that has been slow-cooled, crushed and grinded. Mere settling of the slag to allow mechanically entrained matte droplets to settle and thus be separated from the slag can be obtained in electric furnaces without any addition of reductants. Carbothermic slag reduction is done to enhance the separation of metals and subsequently metal recovery. Production of high-grade mattes results in a highly oxidized slag containing considerable amounts of copper oxides, which requires reduction of the slag to reduce copper and other metals into other phases, e.g. matte, speiss and gas phase. Carbothermic reduction of copper slag or slag settling is usually carried out in an electric arc furnace, rotary holding furnace or Teniente slag-cleaning furnace. In the Teniente slag-cleaning furnace, the slag is reduced in a batch process by injection of an air/oil mixture followed by settling of matte/metallic particles. An increased interest in developing methods for slag treatment to reduce the environmental impact of landfill and to recover valuable metals otherwise lost can be seen in Chile (Palacios and Sánchez, 2011; Acuna and Sherrington, 2005; Demetrio et al., 2000).

Slag fuming is a well-established process that has been in industrial operation for more than 80 years. Slag fuming is traditionally used to vaporize zinc from zinc-containing slags, mainly lead blast furnace slag, but also is applied in a few plants for copper-smelting slag. Reduction is achieved by using pulverized coal, lump coal or natural gas. At Rönnskär Smelter slag from the electric smelting furnace is treated in batches in a fuming furnace to recover zinc (Hansson et al., 2010). The slag is reduced, generating a cleaned slag suitable for use in, for example, construction work.

7.3.3 Dust Handling

One example of a plant that treats a variety of materials is Kosaka Smelter and Refinery in Japan (Watanabe and Nakagawara, 2003). The company has developed a technique to treat complex concentrates containing many impurities. The copper-smelting unit also processes secondary materials such as printed circuit board scraps and sludge. It was considered necessary to keep track of impurities owing to the complexity of raw materials. One way to overcome the build-up of impurities in the plant due to recycling was to install a hydrometallurgical plant to treat dust from smelting units. The combination of several smelting process units and a hydrometallurgical plant was shown to be a feasible way to remove impurities from the plant and recover valuable metals.

An overview of smelting processes is shown in Figure 7.4. As shown in the figure, extraction of copper from primary and secondary sources is often integrated. For example, up to 30% copper scrap can be used in converting copper matte (Langner, 2011). On the other side is the 20—25% of global output of copper from direct-melt scrap (Moskalyk and Alfantazi, 2003).

7.4 CHALLENGES IN COPPER RECYCLING

It can be foreseen that the future raw materials for extraction of base metals will be of lower grade. Complex primary ores and secondary materials with a complex composition will need to be treated for recovery owing to the high demand for metals and the depletion of rich ore reserves.

The dominating source for extraction of copper is sulfide ores, with the main mineral chalcopyrite, $CuFeS_2$. The other sources of raw

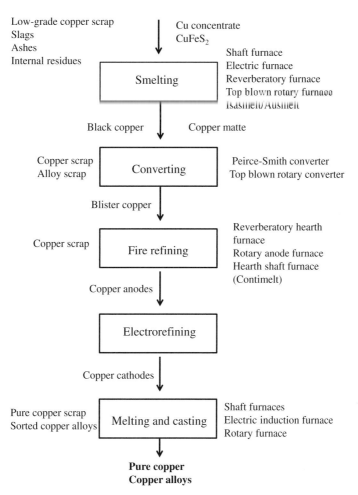

FIGURE 7.4 Overview of material flows and processes.

material for extraction of copper and precious metals are scrap, including electronic scrap. The composition of secondary material is often very complex, containing a variety of metals along with plastics and ceramic material. A further challenge is that the composition changes continuously.

Extraction of copper can be done via various process routes, including hydro- and pyrometallurgical methods and the combination of both. At the metallurgical part of the process chain in the extraction of copper, similar challenges occur whether the copper originates from primary or secondary sources, and primary and secondary raw materials are also often mixed in the processes. Smelting and refining are carried out in a number of process units, and intermediate products are recycled within the plant, which may cause accumulation of impurity elements. There are a number of different process routes considering both primary and secondary raw materials that have the potential to efficiently treat complex raw materials and extract the metal values. It is important to look at the extraction chain as a whole to consider different process routes and combinations of methods for efficient extraction of metals.

Elements not removed prior to the smelting unit, whether they comes from scrap or concentrates, need to be separated and bled from the smelter through slag, dust or other intermediate materials. Very high-purity copper, typically 99.99% Cu, can be obtained through electrorefining. The efficiency of the electrolysis is, however, sensitive to impurities such as antimony, bismuth and arsenic. The absolute limits of acceptable levels in the copper anodes are debated, but a typical anode composition is 50—1100 ppm Sb, 10—600 ppm Bi and 240—4000 ppm As (Larouche, 2001). Antimony is one of the minor elements together with arsenic and bismuth that is often discussed as the impurity element that is most problematic in the smelting route of copper extraction, owing to its affinity to liquid copper. Arsenic is relatively easily removed to the gas phase, but bismuth is not as easily vaporized as arsenic, but it is reported to vaporize when converting matte to white metal. Antimony has a lower partial pressure and is thus not as readily vaporized in roasting and smelting operations. Nowadays it is rather common to use antimony as flame retardants and as such it may enter copper smelters along with for example complex sulfide concentrates.

Removal of antimony prior to smelting operation would be beneficial and even more so if the antimony could be turned into a product. Recent research has been shown that pretreatment of complex copper sulfide concentrates with alkaline sulfide solution has the potential to selectively dissolve antimony from the concentrate; by electrowinning antimony can be recovered from the leach liquors (Awe, 2013). This is one example where hydro- and pyrometallurgical methods can be combined, facilitating the possibility of extracting copper from a complex material and at the same time extracting valuable by-products. Although in this example it is applied on a primary raw material, it would have consequences for recycling capacities of smelters treating both primary and secondary materials. Removing antimony from primary concentrates may allow for higher antimony intake from secondary material. This exemplifies the importance of considering the whole process chain from different raw materials. The interaction between different process routes and raw materials is schematically shown in Figure 7.5.

Slag is a molten oxide phase with the main purpose to transfer elements such as iron from the metal phase, e.g. copper. With increased recycling of different types of impure scrap and residues, an increased load of slag-forming elements can be expected at the smelter. One such element that has a large impact on slag properties is aluminum. The content of alumina in slag will affect properties such as viscosity, density and melting properties, which are important to control during processing.

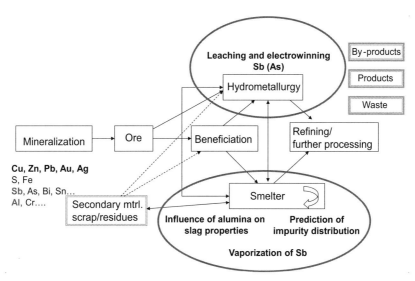

FIGURE 7.5 Simplified schematic flow of possible routes for different types of materials for copper extraction.

Aluminum will also influence phases formed upon cooling of the slag, which subsequently affect the postprocessing properties of the slag, such as leaching (Mostaghel, 2012).

Challenges in the metallurgy of recycling complex feed include the formation of various phases, properties of the phases such as slag and distribution of minor elements between different phases. Tin and indium are two elements of interest in recycling of WEEE. A study has shown the importance of understanding distribution behavior in smelting conditions to evaluate the feasibility for recovery of such elements through a certain smelting route (Anindya, 2012).

Thermodynamic modeling and simulation can be used to understand and optimize reactions occurring in the different process units and the distribution of elements. There are two great challenges when using thermodynamic for process simulation: data used must be valid and evaluated for the actual composition range, and the reactions should not be controlled by kinetics.

Although modeling will not solve all problems connected with metallurgical processes, it can be a good tool when used in combination with industrial experience, and experimental studies in lab scale, to describe metallurgical processes such as slag chemistry, phase separation and element distribution.

7.5 CONCLUSIONS

It can be assumed that there will continue to be a demand for copper also for future generations, as it has properties that are difficult to compete with in certain applications. The recycling of copper is, for example, in Europe, well-functioning with established business for collection, treatment and processing of high-grade scrap. This is shown in the rather large proportion of the copper produced originating from secondary sources. Although the recycling rates for copper are good, some challenges can be foreseen, such as a scarcity of pure and high-grade scrap and an increased amount of products containing a mixture of materials and with low copper concentrations, which means that the processing industry must deal with more impurities.

References

Acuna, C.M., Sherrington, M., 2005. Slag cleaning processes: a growing concern. Materials Science Forum 475–479, 2745–2752.

Anindya, A., 2012. Minor Elements Distribution during the Smelting of WEEE with Copper Scrap (Doctoral thesis). RMIT University.

Awe, S.A., 2013. Antimony Recovery from Complex Copper Concentrates through Hydro- and Electrometallurgical Processes (Doctoral thesis). Luleå University of Technology.

Ayres, R.U., Ayres, L.W., Råde, I., January 2002. The Life Cycle of Copper, its Co-products and By-products. Mining, Minerals and Sustainable Development (MMSD), No. 24. International Institute for environment and Development IIED.

Cui, J., Zhang, L., 2008. Metallurgical recovery of metals from electronic waste: a review. Journal of Hazardous Materials 158, 228–256.

Davenport, W.G., King, M., Schlesinger, M., Biswas, A.K., 2002. Extractive Metallurgy of Copper, fourth ed.

Demetrio, S., Ahumada, J., Angel Duran, M., Mast, E., Rojas, U., Sanhueza, J., Reyes, P., Morales, E., August 2000. Slag cleaning: the Chilean copper smelter experience. JOM 52 (8), 20–25.

European Copper Institute, 2013. http://www.eurocopper.org.

Hansson, R., Holmgren, H., Lehner, T., 2010. Recovery of recycled zinc by slag fuming at the Rönnskär smelter. In: Siegmund, A., et al. (Eds.), Lead-Zinc 2010. TMS.

Henstock, M.E., 1996. The Recycling of Non-ferrous Metals.

Institute of Scrap Recycling Industries, Inc., 2013. www.isri.org.

Reference Document on Best Available Techniques in the Non Ferrous Metals Industries Integrated Pollution Prevention and Control (IPPC), December 2001.

International Copper Study Group, 2013. http://www.icsg.org.

Jolly, J.J., 2013. The U.S. Copper-base Scrap Industry and its By-products A1309-12/13. www.copper.org.

Langner, B.E., 2011. Understanding Copper-Technologies, Markets, Business, first ed.

Larouche, P., 2001. Minor Elements in Copper Smelting and Electrorefining (M. Eng. thesis). McGill University, Montreal, Canada.

Lossin, A., 2005. Ullmann's Encyclopedia of Industrial Chemistry (Copper).

Moskalyk, R.R., Alfantazi, A.M., 2003. Review of copper pyrometallurgical practice: today and tomorrow. Minerals Engineering 16, 893–919.

Mostaghel, S., 2012. Influence of Alumina on the Zinc Slag Fuming Processes. An Experimental Study on Physical Properties and Leaching Behaviour of the generated Fayalite-type Slag (Doctoral thesis). Luleå University of Technology.

Palacios, J., Sánchez, M., 2011. Wastes as resources: update on recovery of valuable metals from copper slags. Transactions of the Institution of Mining and Metallurgy C 120 (4).

Watanabe, K., Nakagawara, S., 2003. The behavior of impurities at Kosaka Smelter. In: Kongoli, F., Itagaki, K., Yamauchi, C., Sohn, H.Y. (Eds.), Yazawa International Symposium: Metallurgical and Materials Processing: Principles and Technologies, High temperature metal production, vol. 2, pp. 521–531.

CHAPTER 8

Lead Recycling

Bart Blanpain[1], Sander Arnout[2], Mathias Chintinne[3], Douglas R. Swinbourne[4]

[1]High Temperature Processes and Industrial Ecology, Department of Metallurgy and Materials Engineering, KU Leuven, Leuven, Belgium; [2]InsPyro, Leuven, Belgium; [3]Metallo-Chimique, Beerse, Belgium; [4]School of Civil, Environmental and Chemical Engineering, Melbourne, VIC, Australia

8.1 INTRODUCTION

Lead is a soft, malleable, ductile, bluish-white, dense metallic element, extracted chiefly from galena (PbS) and found in ores together with zinc, silver and copper (ILA, 2013a).

Today about 80% of lead is used in lead-acid batteries. A further 6% of lead is used in the form of lead sheet by the building industry. There are a number of other smaller volume applications for metallic lead such as radiation shielding, cable sheathing and various specialized applications, such as earthquake dampers (ILA, 2013b).

Lead is one of the first metals produced by man—with beads of it dating from 6500 BCE having been found in Anatolia.

The mine and metal production, and the metal use of the last five years, are given in Table 8.1. It is notable that in spite of the gradually more stringent ban of lead from a number of its traditional applications (such as pigment, antiknocking agent, solder alloy, plumbing, gun shot), the level of lead usage has doubled in the last 30 years (ILA, 2013a). About 90% of all lead is used in readily recyclable products, conserving precious ore reserves for future generations. Lead enjoys one of the highest recycling rates of all materials in common use today. The ILZSG has estimated that secondary lead accounts for 56% of total refined lead metal output globally and that this figure rises to 73.7% on an ex China basis (White, 2013). Because of the relatively unreactive nature of lead and the resulting relative ease of refining lead (see later under lead refining), the purity of the recycled lead is identical to that of primary metal from mining (ILA, 2013b).

TABLE 8.1 World Refined Lead Production and Usage 2008–2013 (in kt) (ILZSG, 2013)

Year	2008	2009	2010	2011	2012
Mine production	3812	3810	4291	4683	5236
Metal production	9198	9197	9804	10,545	10,525
Metal use	9190	9206	9776	10,396	10,469

In general three major sources of secondary lead are distinguished: (1) batteries and battery-derived streams (lead alloy, antimonial (hard) lead, lead sulfate, sulfuric acid, polypropylene, lead oxide), (2) metallic lead (alloy) from plumbing, weights or cable sheathing and (3) intermediate products from nonferrous smelting/refining such as drosses and slags. The ILZSG, without having detailed information, estimates that about 80% of the secondary lead originates from used lead-acid batteries (White, 2013). Most recycling processes combine the first two sources, although there are some that are capable of handling all three, leading up to more complex bullions and refining treatments afterward. Today there is a large range of smelting technologies in use for secondary smelting. We can distinguish between dedicated technologies such as rotary furnaces and the Varta blast furnace and processes such as QSL, Ausmelt, Isasmelt, SKS, Kivcet and the traditional lead blast furnace that principally treat primary lead concentrates. The use of a particular technology is as much a result of the expectations on the complexity of the feed at the time of plant design as it is of the inertia of capital investments in changing primary and secondary resource markets.

Lead production is one of the pivotal metallurgical flowsheets in primary and also secondary metal production (UNEP, 2013), as through lead production associated metals such as Cu, Sn, Sb, Bi, and Ag are also co-produced. It should also be noted that non-lead plants generate substantial amounts of lead. Interesting examples of such secondary smelting operations can be found for precious metals recycling and Cu/Sn recycling (Campforts et al., 2013; Goris, 2013).

8.2 THE LEAD-ACID BATTERY

Representing about 80% of the use of lead, batteries are the dominant application of lead as well as the dominant secondary resource for lead metal production. The lead-acid battery was invented in 1859 by Gaston Planté and is now widely used as a stationary battery and the Starting-Lighting-Ignition (SLI)-type battery (see for instance Crompton, 2000; Figure 8.1).

The negative pole, or anode, is composed of a lead grid; the positive pole, or cathode, is a grid pasted PbO_2. The electrolyte solution is an aqueous sulfuric acid solution. Due to the discharge reaction

$$PbO_2 + Pb + 2H_2SO_4 \rightarrow 2PbSO_4 + 2H_2O \quad (8.1)$$

the two electrodes are converted into $PbSO_4$. The half-reaction during discharge at the anode is:

$$Pb + HSO_4^- \rightarrow PbSO_4 + H^+ + 2e^- \quad (8.2)$$

and the half-reaction during discharge at the cathode is:

$$PbO_2 + 3H^+ + HSO_4^- + 2e^- \rightarrow PbSO_4 + 2H_2O \quad (8.3)$$

During recharging these reactions proceed from right to left. The cell voltage is 2.0 V and the power capacity of these batteries can now be up to 35–40 Wh/kg. The weight continues to be the most important disadvantage of these batteries, notwithstanding the remarkable electrical performance achieved. The grid plates are produced from lead alloys to improve the mechanical strength, creep resistance and castability, while minimizing the hydrogen evolution reaction. Commonly used materials are an Sb alloy containing Se, Sn and As and a Ca alloy containing Sn, Al and often Ag (Siegmund and Prengaman, 2001).

The lead-acid battery is now a complex consumer product made of several materials. The composition of a lead-acid battery is shown in Table 8.2. The main components are lead, either as a metal, oxide or sulfate, and sulfuric acid is another important fraction. Also the polypropylene is valuable and can be recycled (Jolly and Rhin, 1994).

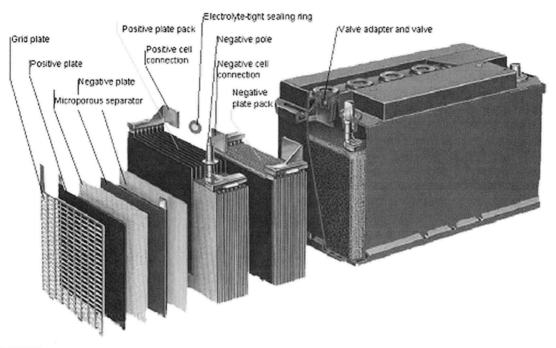

FIGURE 8.1 Schematic of an automotive lead-acid battery (Darling, 2013).

More than 97% of all lead-acid batteries are recycled today, the highest recycle rate of all consumer products. According to research by Battery Council International, 55% of aluminum soft drink and beer cans are recycled, 45% of newspapers, 26% of glass bottles and 26% of tyres (Battery Council International, 2013).

8.3 BATTERY PREPROCESSING

Lead-acid batteries, the major source of secondary lead, are composed of several materials as indicated above. An important step in their recycling is therefore to process the batteries first by separating the distinct material fractions.

8.3.1 Acid Drainage

Upon reception at the recycling plant, a first step common to all Western recycling approaches is the removal of the sulfuric acid. Various simple but sufficient strategies are in place in order to crack the battery casings and

TABLE 8.2 Composition of Typical Lead-Acid Battery Scrap (Gravita, 2013b; Badanoiu and Buzatu, 2012)

Component	(wt%)
Lead (alloy) components (grid, poles, …)	25–30
Electrode paste (fine particles of lead oxide and lead sulfate)	35–50
Sulfuric acid (10–20% H_2SO_4)	10–20
Polypropylene	5–8
Other plastics (PVC, PE, etc.)	4–7
Ebonite	1–3
Others materials (glass, …)	<0.5

drain the acid from them. Dropping the batteries from a height of a few meters or driving over them with a front loader are the most common. The acid is collected into a basin for settling. In some cases, a press filter is used to recover additional solids and deliver cleaner acid. Depending on the local situation, it can be either transported for use in other applications (e.g. alkali waste neutralization; Stevenson, 2004) or neutralized using sodium or calcium hydroxides. In the case of neutralization, the resulting gypsum needs to be disposed of or may be mixed with the battery paste for smelting.

8.3.2 Breaking and Sorting

Breaking or dismantling batteries further, after drainage, separates the lead-bearing materials from the plastics or other fractions. Automated breakers are used in the most advanced plants, but manual or semiautomated dismantling lines are also in use. Large stationary batteries with a metal casing cannot be treated in automated breakers. They have to be dismantled in less automated ways, often with custom-built auxiliary equipment.

Modern automated breaking and sorting lines are available from dedicated engineering companies and have capacities up to 50 t of batteries per hour (Wirtz, 2013). However, more often plants are smaller (Engitec, 2013a) and source local materials because of the high transport costs for heavy battery scrap. The unit consists of a hammer mill or roller crusher, followed by hydrodynamic or sink/float separators to remove and sort plastics, and a further separation of the metals from the paste, based on their size (Stevenson, 2004). A flowsheet of a typical plant with a breaker and separators is given in Figure 8.2.

Separating the different materials has the advantage that the polypropylene casing can be recovered and sold for recycling into new plastic products. Further, besides the lead-rich fractions, a mixed heavy plastic fraction is also formed that contains combined plastic-metal particles and plastics other than

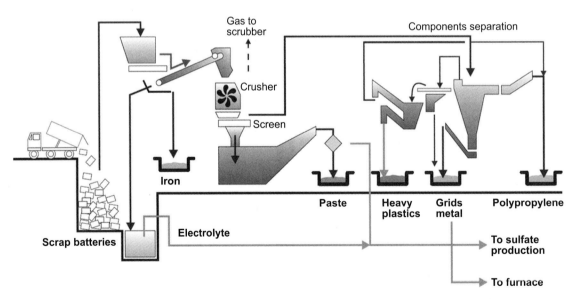

FIGURE 8.2 Typical flowsheet for a preprocessing plant: acid drainage, crusher and sorting (Engitec, 2013b).

polypropylene. This fraction contains the polyethylene, as well as decreasing amounts of PVC, from separators, with other materials such as ebonite or glass. It can be added to a lead smelting furnace or used for energy recovery in a hazardous waste incinerator (Behrendt and Fisher, 2013).

The fractions used for lead recycling are the metallic fraction and the battery paste. The metallic fraction, containing the grids and connectors, is generally sent to rotary furnaces in the same plant, without the need of an extensive smelting cycle as the amount of oxides/sulphates is minimal. The paste is mainly also locally smelted, possibly after desulfurization, but can also be shipped to a primary smelter.

8.3.3 Desulfurization of the Paste

In order to facilitate local smelting, sulfur is often removed from the batteries' paste fraction. Paste desulfurization relies on the reaction of lead sulfate to lead carbonates, by the addition of a reagent to the slurry. The most commonly used reagent is sodium carbonate, with the reaction being:

$$PbSO_4(s) + Na_2CO_3(aq) \rightarrow PbCO_3(s) + Na_2SO_4(aq)$$

An excess of carbonate is needed, as it is also consumed to neutralize any remaining acid. The desulfurized paste contains about 0.5% insoluble sulfur and 1.5–2.5% sodium, which Engitec claims can be further reduced to <0.2% Na and <0.2% S using more advanced superdesulfurization (Olper and Maccagni, 2009). This process uses CO_2 to fully convert Na–Pb carbonate-hydrates to $PbCO_3$ in the paste.

Another reagent that can be used is ammonium carbonate, which leads to ammonium sulfate. The reactor design is more complicated, but in some countries the market for ammonium sulfate, used for fertilizer production, is healthier than for sodium sulfate, used for the production of detergents and glass (Olper and Maccagni, 2009). Instead of sodium carbonate, sodium hydroxide may also be used, leading to lead oxide instead of carbonate (Stevenson, 2004).

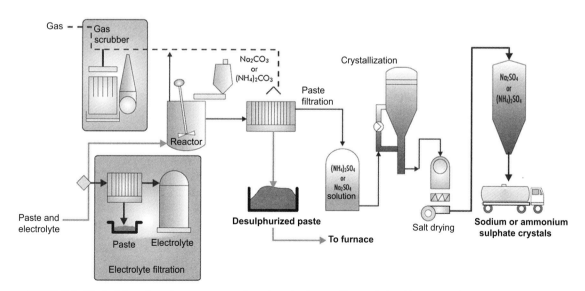

FIGURE 8.3 Typical flowsheet for a preprocessing plant: hydrometallurgical desulfurization (Engitec, 2013b).

The resulting sodium or ammonium sulfate needs to be removed from the solution, and this is achieved by evaporation and crystallization. The water is then recycled to the breaker. A flowsheet of a typical plant for desulfurization is given in Figure 8.3.

8.4 SMELTING

8.4.1 Primary Smelting

Dedicated recycling processes will be discussed in the next section, but as a substantial fraction of secondaries or of their side streams are treated at primary smelters and as they are trying to shift their focus to recycling, we commence our discussion with primary smelting processes. Primary smelting processes treat a lead-bearing feed that is derived from mined ores. Those ores, almost all of which contain lead as "galena" (PbS), are upgraded by mineral processing methods to a high-grade concentrate. The mineral particles in the concentrate are very fine grained, with typically 80% of the particles smaller than 40–50 μm.

The traditional approach to extracting lead from galena concentrates utilizes the blast furnace. However, blast furnaces require the feed to be coarse enough to allow gas to be blown through the charge in the shaft without excessive pressure drop. It also requires the charge to be in the form of oxides because the blast furnace uses carbon as the reductant. The fine-grained sulfide concentrates are treated by a process known as "sintering" that both oxidizes the galena and partially fuses it into large, strong and porous lumps. During sintering the concentrates are mixed with an excess of crushed return sinter to form a porous bed. A thin layer of charge is laid down on a continuous belt of steel boxes with slatted bottoms. This thin layer is ignited and then the main layer of charge is laid down. Air is blown up through the bed as it travels toward the end of the sinter machine, and a high temperature band moves upwards through the bed. No additional fuel is required in the sinter mix because the oxidation of PbS is very exothermic. Sulfur is removed as sulfur dioxide gas (SO_2) and the high temperature causes partial fusion of the bed into large lumps (Sinclair, 2009). Sinter lumps, together with coke, are charged to the top of the blast furnace and air is blown in through tuyeres near the bottom. Carbon combusts to carbon monoxide (CO), which travels upwards through the packed bed, and this reduces the PbO in sinter to lead bullion. Other oxides in the charge, such as silica (SiO_2), iron oxide (FeO) and added lime (CaO) flux form slag.

The sinter/blast furnace route is restricted in its ability to accept significant amounts of battery paste as part of the feed. The oxidic part of the paste provides no heat during sintering, while the $PbSO_4$ in the paste incompletely decomposes very endothermically. This results in a peak temperature in the sinter bed which is insufficient for successful partial fusion (Sinclair, 2009) and carries sulfur into the blast furnace where it complicates smelting by forming a molten sulfidic matte phase. Nevertheless, traditional smelters have moved to accept from 10% to 25% of their feed as secondaries (Siegmund, 2000)—especially in Japan.

Since the 1980s lead smelters have been under pressure to improve their practices. Most significant of these was the imposition of increasingly strict environmental legislation for both SO_2 and fugitive lead emissions. Other pressures were the rising costs of energy and manpower. The result was the introduction of novel "direct smelting processes" that feed concentrates, fluxes and oxygen into one or two closed reactors. Such reactors achieve much greater levels of SO_2 capture, much more efficient use of the exothermic heat of oxidation of sulfur and reduced manpower through extensive automation than the sinter/blast furnace route. They utilize oxygen or oxygen-enriched air and so greatly lower the volume of off-gases produced and therefore the quantity of dangerous emissions released into the workplace and atmosphere (Siegmund, 2000).

The significance of the development of direct smelting processes to lead recycling is that, by eliminating the sinter machine, they open up the possibility of feeding battery paste and metallics, alongside concentrates, into the reactors. This had the important benefit of creating a new supply of lead-bearing materials at a time when lead mine closures were occurring and there were difficulties in sourcing adequate quantities of concentrates to maintain the financial viability of smelters. It also meant that the use of secondary materials as feed was largely restricted to the developed countries where a sufficient quantity of such materials could be sourced (Siegmund, 2000). Feeding secondary materials to primary smelters also has the advantage that expensive desulfurization of battery paste is unnecessary because the SO_2 produced from lead sulfate ($PbSO_4$) simply joins the SO_2 stream from PbS oxidation. However, there are also problems associated with the feeding of lead secondary materials into direct smelting reactors. The endothermic oxidation of lead sulfate causes difficulties with the heat balance in the reactor, the possible presence of chloride in the feed from the PVC separators in batteries affects the recovery of minor elements and the very low concentrations of silver in battery scrap versus galena concentrates robs the smelters of a significant revenue stream (Stevenson, 2004).

Siegmund (2003), Stephens (2005) and Hayes et al. (2010) have reviewed the direct smelting technologies and the four more commercially important of these will be discussed here, with emphasis on their capacity to accept secondary materials in conjunction with primary feeds. They fall into two distinct groups according to their process chemistry: three that produce most lead directly from oxidation of PbS, which are all bath smelting processes and which differ only in reactor design—QSL, SKS and Ausmelt/Outotec TSL; and the single process that produces all lead from reduction of slag produced by the complete oxidation of PbS to PbO in a flash reactor—Kivcet.

The chemical differences in these processes can best be visualized using Figure 8.4 below. It was produced using computational thermodynamics software and illustrates the outcome of reacting 1000 kmol of PbS, contained within a typical lead concentrate, with increasing amounts of oxygen and appropriate quantities of silica flux at 1100 °C. It can be seen that the amount of lead formed increases to a maximum at 2000 kmol of oxygen for this feed, then rapidly decreases. There is a small amount of PbS dissolved in the lead, so the lead is not pure but contains some sulfur in solution. The amount of PbO reporting to the slag initially increases slowly but then increases rapidly as lead is oxidized. The amount of lead reporting to the gas phase is considerable and is mostly PbS(g), but the amount drops to effectively zero when all lead is present as PbO in slag. If $PbSO_4$ is added to the input to represent battery paste, then almost all of it reports to the metal phase as lead via decomposition of $PbSO_4$ to PbO, then through the reaction with PbS to lead.

The three bath smelting processes operate at the maximum point on the lead production curve. Their aim is to maximize the amount of lead formed by oxidation so that the amount of carbon needed for reduction of PbO from slag is minimized. The amount of PbS volatilizing to the gas phase during oxidation is quite large, and this causes hygiene challenges and a large recycle of condensed fume. The amount of lead reporting to the slag is also large, its concentration typically being from 35 to 45 wt%. A consequence of so much PbO being in the slag is that it lowers the liquidus temperature to approximately 1000–1050 °C. As PbO is reduced from slag, the liquidus temperature

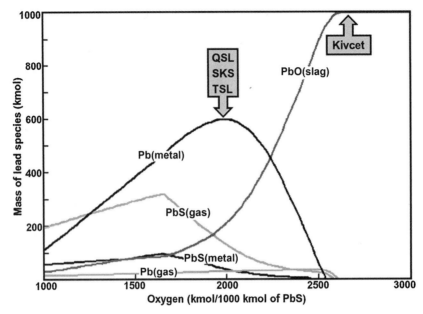

FIGURE 8.4 The amount of lead species formed from the reaction of 1000 kmol of PbS as a function of the amount of oxygen provided, at 1100 °C.

rises to approximately 1250 °C, so the reduction stage must be much hotter than the oxidation stage. This is a problem because the reduction reactions are endothermic, so heat must be supplied to the reduction stage.

The single flash smelting process operates at the point where the oxidation of PbS to PbO in slag is complete. The aim is to minimize the amount of lead reporting to the gas and so to minimize hygiene problems and fume recycle. All lead comes from the subsequent reduction of PbO from slag with carbon, so the carbon consumption is high.

8.4.1.1 QSL

The Queneau-Schumann-Lurgi (QSL) furnace is a refractory-lined horizontal cylinder, approximately 30 m long and 3 m in diameter, with separate oxidation and reduction zones. They are separated by a weir to minimize back-mixing of slag and lead bullion, and the reactor has a slight slope from the oxidation zone toward the reduction zone. Slag flows toward the reduction end, from where it is tapped, while lead bullion flows in the opposite direction toward a siphon in the oxidation zone. Lead-bearing feed and fluxes drop into the bath in the oxidation zone, oxygen is blown in through injectors in the bottom of the reactor and primary lead, lead-rich slag and SO_2 gas are formed. Coal dust and oxygen-enriched air are supplied through bottom injectors in the reduction zone to reduce the lead content of the slag to about 2 wt%. Oxygen is blown into the reduction zone gas space to postcombust CO gas as a heat supply. The QSL process was designed to accept a wide range of secondary materials that are moist and vary in size from fine materials to larger lumps. It is reported that the lead-bearing feed to the QSL reactor at Berzelius Stolberg GmbH comprises over 50% secondary materials, and that it has operated with 100% secondaries feed during trials (Meurer et al., 2005).

8.4.1.2 SKS

Li and Suo (2010) have reviewed the evolution of the Chinese Shui Kou Shan (SKS) oxygen bottom blowing process that carries out oxidation and reduction reactions in two separate reactors. The oxidation stage has very similar reactor geometry to the QSL reactor—but the first-generation process cast the high lead slag and fed it as lumps to a blast furnace for reduction. The subsequent second-generation process has a reduction stage that is similar to the QSL reduction stage but uses side-blowing tuyeres. The third-generation reduction reactors have graphite electrodes whose function is to raise the slag temperature and so decrease the final lead content of the discharge slag to less than 3 wt% Pb. It is reported that up to 50% of the feed to the oxidation stage can be spent acid battery paste. The SKS process is now said to dominate the Chinese lead smelting industry.

8.4.1.3 TSL

Top submerged lance (TSL) smelting was developed in Australia and differs from the QSL and SKS processes in that it uses a stationary vertical refractory-lined cylinder as the reactor and injects oxidizing gas, and fuel, if necessary, through a large single vertical lance. The lance dips into the slag bath and is protected by a frozen slag layer. Sofra and Hughes (2005) have described the application of the Ausmelt TSL process to primary lead smelting. Typically two reactors are used, one for oxidation to produce primary lead and the second for carbon reduction of the slag. However, some operations dispense with the reduction stage and handle the lead-rich slag in other ways. Weser-Metall GmbH at Nordenham in Germany have a TSL reactor that accepts a feed having about 70% as lead secondary materials, and their operations have been described by Kerney (2010). The most stable furnace operation occurs when a slag containing approximately 50 wt% Pb is produced. Secondary materials have a low content of slag-forming oxides so little primary oxidation slag is produced. For this reason Weser-Metall decided not to use a slag reduction step but to sell the slag to other smelters.

8.4.1.4 Kivcet

The Kivcet process and its operating characteristics have been described in detail by Siegmund (2003). The furnace consists of two sections: a reaction shaft followed by an electric furnace, separated by a partition wall that ensures the gases formed in both sections cannot mix. Dried feed is injected at the top of the reaction shaft and is both oxidized and melted to a PbO-rich slag as it falls. Coarse coke (5–15 mm) is also added to the feed and forms a floating layer on the slag at the bottom of the shaft. Slag droplets are partially reduced to lead as they percolate through this coke layer. Slag flows to the electric furnace where three in-line electrodes heat the slag and final reduction of PbO takes place. The Kivcet process can treat a variety of secondary materials; the furnace at TeckCominco in Canada has a lead-bearing feed that is typically about 65% secondaries, only part of which arises from battery recycling. Coarse materials such as battery grids and poles must be screened out of the feed.

8.4.2 Secondary Smelting

Here we discuss dedicated secondary smelting technologies. The choice of technology is dependent on a number of items including the battery pretreatment method, the size of the operation, the source, market and availability of scrap feed and the technological capability of the region. There are three major pretreatment options: (1) no separation, (2) separation without desulfurization and (3) separation with desulfurization. In all processes, reduction of the lead compounds (sulfate, oxide or carbonate) is the critical chemical reaction. The most-used process without separation is the shaft furnace, whereas with separation, it is the short rotary furnace.

8.4.2.1 Shaft Furnace

Batteries are treated in a one-step continuous process, using a shaft furnace (Varta process) with afterburner. The process uses whole batteries with casing; only drainage is required (Figure 8.5).

The smelting energy is provided by (possibly preheated or oxygen enriched) air firing up the coke in the hot burden. The burden of the furnace is composed of (petroleum) cokes, iron scrap, drained batteries and other recyclables such as cable lead or refinery drosses. In the shaft, as the burden moves down, it is gradually heated by the hot gases formed at the tuyere level. The plastics pyrolyze or disintegrate with slowly rising temperatures and a large fraction evaporates in the form of volatile and complex organic compounds; therefore, afterburning is required to eliminate dioxins and furans. Next, the metallic fractions melt and collect in the hearth. At higher temperatures, deeper down in the shaft, decomposition and reduction of $PbSO_4$ and PbO takes place, by Fe and CO gas. Metallic iron from scrap reacts with $PbSO_4$, according to the reaction:

$$Fe + PbSO_4 + CO \rightarrow FeS + Pb + 2CO_2$$

This reaction produces a matte, which is sold for metallurgical treatment. Lead losses are limited, as with increasing iron content, the solubility for lead in the matte decreases to a few percent. Ninety-five percent of the total sulfur input is captured in the matte. The remaining sulfur input leaves the furnace as SO_2 gas and is scrubbed in the gas cleaning section.

In the hearth the slag melts at temperatures around 1100 °C. The slag and matte are tapped at regular intervals at slightly higher temperatures. The slag is of the calcium ferrosilicate type, and a large fraction of it is recycled to the furnace feed for process stability. The lead is collected through a syphon.

The process gases leaving the furnace are afterburnt, preferably with heat recovery. Injection of lime at high temperature reduces the SO_2 emissions. After cooling and dedusting, the gases are released through a stack.

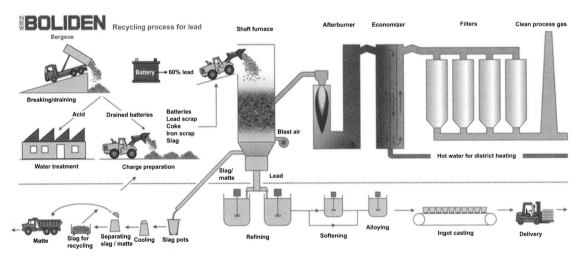

FIGURE 8.5 Shaft furnace secondary lead production (Martensson, 2013).

Matte and filter dusts need further treatment. The matte is treated in other pyrometallurgical flowsheets. The lead fraction in the filter dust justifies recycling to the furnace, but due to the chlorine content it is first hydrometallurgically treated (Kunicky, 2013).

8.4.2.2 Short Rotary Furnace

By far the largest smelting group is the numerous variants of the rotary furnace. The ease and simplistic nature of their operation, although hides a metallurgical complexity, has allowed many new entrants into the market, with turnkey plants available and "up and running within six months. Coupled with the soda-iron slag, the system has a wide tolerance for error, allowing for the operator to smelt lead without having the plant fully synchronized in both metallurgical and production terms (Stevenson, 2013). If the various battery fractions have been well separated (plastics—metals—oxides/sulfates), they can receive specific treatment. The oxide/sulfate fraction needs reductive smelting to produce a lead bullion. Desulfurized paste, although low in sulfur, also needs reduction. The metal fraction may move straight to the refinery, but as they are contaminated with sulfate due to corrosion (Olper and Maccagni, 2009), and may contain some plastics, they are often treated in the same reductive smelting process to avoid emissions.

The sorting of battery fractions allows full use of the batch nature of the short rotary process. If operations allow for different batch cycles, paste can be smelted separately from metal fractions. The paste leads to very pure soft lead, whereas the metallics contain the antimonial lead parts and other alloying elements. The antimony content, although limited, can then be concentrated into bullion for use at the refinery when producing antimonial lead.

The short rotary furnace is operated as a batch process. The units are relatively small, and it is common to operate two to four furnaces for one breaker. Heating is provided by a fuel or gas burner, with air or increasingly with oxygen, at the front or back of the furnace. The furnace is fed through a front opening, generally with a dedicated feeder. This feeder consists of a hopper, wide enough to be filled using a front loader, and an Archimedes screw on the bottom pushing the material through a tube into the furnace (e.g. Gravita, 2013b). Conveyor belts may also be used (Forrest and Wilson, 1990). The feed consists of lead containing fractions (paste, metal, and recirculated drosses), together with iron, and one or more fluxes (Figure 8.6).

The purpose of the iron, similar to the shaft furnace process, is to capture the sulfur as FeS. As FeS melts at relatively high temperatures for the rotary furnace (around 1200 °C), fluxes are needed to decrease the melting point of the matte. Two strategies are employed: creating a sodium rich slag/matte commonly termed the soda-iron slag, or creating a calcium ferrosilicate slag. The soda-iron method is more widely used throughout the world (Stevenson, 2013).

The first strategy consists of the addition of soda ash (Na_2CO_3), which leads to a sulfur capture in sulfate as well as in sulfide form. At less reduction, the following reaction can take place:

$$PbSO_4 + Na_2CO_3 + CO \rightarrow Pb + Na_2SO_4 + 2CO_2$$

Both Na_2CO_3 and Na_2SO_4 have melting points below 900 °C, which facilitates the reaction. At higher reduction, sodium sulfide is formed:

$$PbS + Na_2CO_3 + CO \rightarrow Pb + Na_2S + 2CO_2$$

Sodium sulfide also has a relatively high melting point (1180 °C), but when FeS and Na_2S are combined in comparable quantities (25–75% FeS), they form a sulfide matte with liquidus below 800 °C (Figure 8.7). The solubility of other compounds such as metallic iron

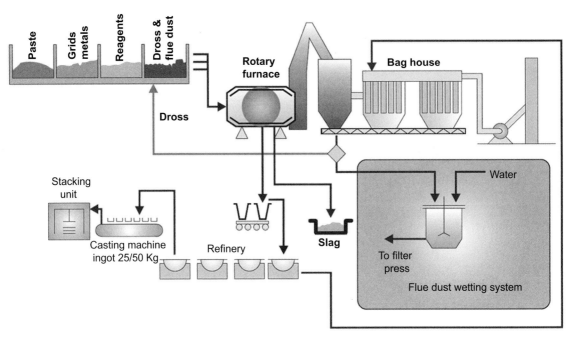

FIGURE 8.6 Rotary furnace secondary lead production flowsheet (Engitec, 2013b).

or iron oxide in this matte is expected to be considerable, as it is for FeS itself (Arnout et al., 2011), but is not well understood. The low melting points lead to a simple and robust process that can operate below 1000 °C, but the environmental compatibility of the slag is a source of concern. The slag is hygroscopic and pulverizes easily on reacting to hydrates such as "erdite" and further dissociation to compounds such as Na_2SO_4, NaOH. Further, the solubility of sodium compounds in water may allow for some release of heavy metals to the environment, although many are bound to the slag (Stevenson 2009).

If soda-iron slag production is not allowed or leads to too high landfill or treatment costs, a second strategy is used. As in all smelting processes, iron is added to capture sulfur as the sulfide. In this approach, lime and silica fluxes are then added to create a calcium ferrosilicate slag (Stevenson, 2004). Depending on the scrap, it may already contain considerable amounts of iron oxide or silica from separators. The lead alloys may lead to calcium and aluminum oxides in the slag. The silicate slag has considerable solubility for calcium and iron sulfide. This approach may require operating at higher temperatures. Foam depressants or slag modifiers may be used to prevent slag foaming or to decrease slag viscosity (Stevenson, 2004). The silicate structure provides a more stable matrix for the heavy metals, resulting in considerably lower leaching rates (Knežević et al., 2010), which can be further reduced by transforming the slag into a geopolymer material (Onisei et al., 2012).

In the case of prior paste desulfurization, the smelting operation in the Short Rotary furnace is simplified because of lower quantities of sulfur in the charge. Similar mattes or slags are created, but in much smaller quantities.

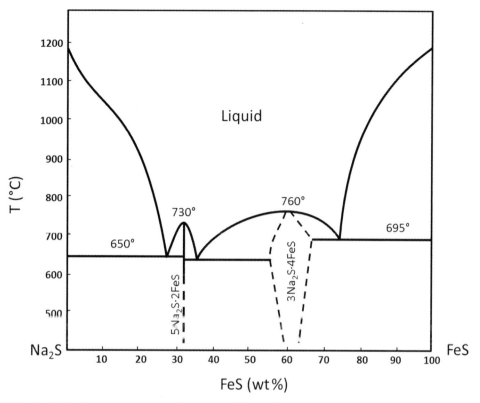

FIGURE 8.7 Na$_2$S-FeS phase diagram (after Kopylov and Novoselov, 1964).

8.5 ALTERNATIVE APPROACHES

Recent developments on further hydrometallurgical treatment of the desulfurized paste have led to a chloride-based flow cell technology for electrolytic reduction to lead metal (Pizzocri and Maccagni, 2013). The system consists of ammonia chloride and hydrogen peroxide leaching. Pilot and demonstration plants have been built. An important advantage would be the elimination of reductive smelting and the associated difficulty in controlling diffuse lead dust and SO$_2$ emissions.

Besides smelters specializing in battery scrap, some relatively straightforward remelting facilities are in place. For example, a lead sheet producer may collect nearly uncontaminated lead sheet and combine it with lead bullion from other plants to produce new alloys.

8.6 REFINING

The lead bullion may contain many impurities, depending on the original lead alloys in the battery. Liquid lead is an excellent solvent for a wide range of elements, and the bullion can contain a large number of elements such as Cu, Fe, Zn, S, As, Sn, Sb, Bi, the precious metals, and sometimes also Ni, Co, Te. Secondary lead, especially from lead-acid batteries, is typically less complex in composition than primary bullions from lead concentrate smelting. It tends to contain higher amounts of Sb and considerably lower amounts of precious metals.

The refining of lead can be done both pyrometallurgically and electrometallurgically. In refining, we can distinguish between primary and complex secondary bullions, secondary bullion and oxidized bullion.

1. For primary and complex secondary bullions, the full range of refining is required, starting with decoppering followed by Sn, Sb, As oxidation, precious metals refining and Bi removal. Electrolytic refining is also a possibility.
2. For secondary bullion, copper, tin and antimony are the main impurities to be removed. As the input of arsenic, silver and bismuth is limited, no specific refining is carried out for these elements.
3. Some smelters operate with highly oxidizing conditions. Impurities such as Sn, As and Sb report to the Pb-rich slag, and the bullion only contains Cu and Ag. The subsequent refining process is then limited to the removal of copper and silver.

8.6.1 Pyrorefining

Pyrorefining is the most widely used refining method (Hayes et al., 2010). It consists of a series of treatments, all of which aim at the elimination of one or a family of similar impurities. It typically occurs in a sequence of batch processes.

8.6.1.1 Decoppering

In primary metal refining the copper is removed by liquation: as the temperature is lowered the solubility of copper in liquid lead is exceeded and on top of the molten bath a removable copper rich dross is formed. This is partially done at the furnace before the bullion enters the refinery. The Cu-rich particles float to the surface where they assemble in a lead-rich copper dross. If nickel or cobalt are present, they follow the copper. The remaining copper or the initial copper for secondary metal processing, where copper bullion levels are much lower, can be removed by selective sulfidization (i.e. fine decoppering). This is done by stirring sulfur or pyrite into the lead bath at around 330 °C. The sulfur reacts to form CuS and Cu_2S. Sulfur also reacts with lead, but this reaction is much slower in the presence of Ag and Sn. Because of these slow kinetics, the lead can be decoppered to 0.001 wt% Cu. Fine decoppering is necessary for the efficiency of the desilvering process, during desilvering copper will interact with the added zinc and so lower the removal of silver. It is also makes the composition of the silver crust more complex and so crust treatment is made more difficult.

8.6.1.2 Removal of Sn, As, Sb

After decoppering, the liquid lead is heated and oxidized. Oxidation can be done by air-/oxygen injection or the Harris process. Air/oxygen injection is typical for secondary bullion. Tin is oxidized at 450–600 °C by air injection or simple stirring. The resulting product is a Pb/Sn oxidic dross, containing mainly PbO. The antimony removal ("softening") is carried out at higher temperatures (600–750 °C) and with oxygen. Often, a separate reactor is used, to increase the antimony content of the slag and produce a lead-antimony slag.

In the Harris process, typically used for more complex bullion, the molten lead is contacted with a molten mixture of caustic soda (NaOH) and sodium nitrate ($NaNO_3$) at 450 °C. Arsenic, tin and antimony are removed as arsenates, stannates and antimonates. In practice, arsenic and tin are first removed together, followed by antimony. The arsenic, tin and antimony salts are then treated hydrometallurgically, and eventually calcium arsenate, calcium stannate and sodium antimonate are produced.

8.6.1.3 Separation of the Precious Metals

Subsequently and mostly relevant for primary bullions, zinc is added to the molten lead in the Parkes process. This creates a ternary Zn−Ag−Pb alloy (triple alloy) containing Zn−Ag intermetallics that can be skimmed off. The silver content in the lead bullion is reduced to around 2−8 ppm. The excess zinc (0.5−0.6%) is removed from the lead by vacuum distillation and treatment with molten caustic soda to levels below 1 ppm. The triple alloy (30% Ag, 7% Zn and 60% Pb) is vacuum distilled in order to extract the zinc, which is reused in the process, and then further processed in the precious metals refinery.

Finally, bismuth is removed by addition of calcium and magnesium in the Kroll−Betterton process, through which a foam is formed that contains Ca_3Bi_2 and Mg_3Bi_2. $NaNO_3$ is added as a final treatment in order to remove excess Ca and Mg, after which refined lead is obtained with a purity of 99.97−99.99%.

8.6.2 Electrolytic Refining

Lead with an even higher purity can be obtained by electrolytic refining of impure lead anodes in fluorosilicic acid H_2SiF_6 (Betts process). Although less used than pyrometallurgical refining, both primary and secondary lead producers produce refined lead using the Betts process, especially when the Bi level of the bullion is high (Gonzalez-Dominguez et al., 1991). The lead is cast into anodes after decoppering. While dissolving the anodes during the electro-refining, the elements that are electrochemically more noble than lead are collected in the anode slimes from which Cu, As, Sb, Bi and precious metals may be recovered. Fe and Zn, which are less noble than Pb, go into solution, but are not deposited as impurities on the cathode. Sn and Sb, however, can be reduced at the cathode. Due to the difference in current efficiency at the anode and cathode, a slight increase in the lead content in the electrolyte takes place. The lead content is kept constant by removing some lead with a number of separate electrowinning cells using graphite anodes. The purity of the lead obtained is above 99.99%.

8.7 CONCLUSIONS AND OUTLOOK

Lead production and use has been growing steadily over the last decades, mainly due to the fact that lead-acid batteries remain the battery of choice for standard SLI applications and stationary systems. It is expected that this will not change in the near future (Prengaman, 2009; Wilson, 2011). Recently small electric motorbikes have adopted the lead-acid battery, although the outlook for this application may change. At the same time carbon-enhanced lead-acid battery designs are being developed specifically for hybrid electric vehicles (Battery Council International, 2013).

Due to lead's physico-chemical properties as well as its use in products that are conducive to closing its materials life cycle, lead recycling has been widely practiced and may be considered exemplary. Even then, it is to be noted that the fraction of recycled lead in the total global production of the metal is estimated to be only slightly above 50%.

There are a wide range of lead production technologies in use. Each technology has its advantages and drawbacks, and choices are made based on feed types as well as local situation and preference. The adaptability of most technologies to at least partially treat secondary materials is remarkable, so much so that it is now difficult to make a clear distinction between primary and secondary lead producers. It may be expected that the drive toward the development and use of flexible, clean and energy-efficient technologies will continue for some time.

References

Arnout, S., Nagels, E., Blanpain, B., 2011. Thermodynamics of lead recycling. In: Proceedings of EMC 2011, p. 363.

Battery Council International, 2013. The Advanced Lead-Acid Battery, the Greatest Unknown Recycling Story. The Advanced Lead-Acid Battery Consortium (02.11.13.). http://www.alabc.org/.

Badanoiu, G., Buzatu, T., 2012. Structural and physico-chemical analysis of waste from used lead-acid batteries. U.P.B. Sci. Bull., Series B 74 (1), 246−254.

Behrendt, H.-P., Fisher, M., 2013. Lead recycling and hazardous waste incinerator with tradition and future. In: Proceedings of EMC 2009. GDMB, p. 411.

Campforts, M., Foerier, S., Vasseur, K., Meskers, C., 2013. Valorisation of critical metals from high temperature residues at Umicore. In: 3rd Slag Valorisation Symposium, pp. 235−247.

Crompton, T.R., 2000. Battery Reference Book, third ed. Butterworth-Heinemann, Oxford, UK.

Darling, 2013. http://www.daviddarling.info/encyclopedia/L/AE_lead-acid_battery.html (02.11.13.).

Engitec, 2013a. http://www.engitec.com/PDF/LEADREF.pdf (Nov 2013).

Engitec, 2013b. http://www.engitec.com/PDF/CXSYSTEM.pdf (Nov 2013).

Forrest, H., Wilson, J.D., 1990. Lead recycling utilising short rotary furnaces. In: Lead-Zinc '90. TMS.

Gonzalez-Dominguez, J.A., Peters, E., Dreisinger, D.B., 1991. The refining of lead by the Betts process. Journal of Applied Electrochemistry 21, 189−202.

Goris, D., June 2013. Do you know Metallo Chimique? In: Recycling Metals from Industrial Waste, Workshop, Golden, CO.

Gravita, 2013a. http://www.gravitatechnomech.com/Lead-scrap/Lead-battery-scrap.html# (02.11.13.).

Gravita, 2013b. http://www.gravitatechnomech.com/equipments/rotary-furnace.html (Nov 2013).

Hayes, P.C., Schlesinger, M.E., Steil, H.-U., Siegmund, A., 2010. Lead smelter survey. In: Siegmund, A., Centomo, L., Geenen, C., Piret, N., Richards, G., Stephens, R.I. (Eds.), Lead Zinc 2010, 5th Decennial Symposium. TMS, pp. 343−414.

ILA, 2013a. http://www.ila-lead.org/lead-facts/statistics (02.11.13.).

ILA, 2013b. http://www.ila-lead.org/lead-facts/lead-recycling (02.11.13.).

Jolly, R., Rhin, C., 1994. he recycling of lead-acid batteries: production of lead and polypropylene. Resources, Conservation and Recycling 10, 137−143.

Kerney, U., 2010. Recyclex PB production in Nordenham with bath smelting technology − an update. In: Siegmund, A., Centomo, L., Geenen, C., Piret, N., Richards, G., Stephens, R.I. (Eds.), Lead Zinc 2010, 5th Decennial Symposium. TMS, pp. 415−427.

Knežević, M., Korać, M., Kamberović, Z., Ristić, M., 2010. Possibility of secondary lead slag stabilization in concrete with presence of selected additives. Metallurgical & Materials Engineering (MJoM) 16 (3), 195−204.

Kopylov, N.I., Novoselov, S.S., 1964. The Cu_2S-FeS-Na_2S system. Russian Journal of Inorganic Chemistry 9 (8), 1038.

Kunicky, Z., 2013. Hydrometallurgical Treatment of Flue Dust from Lead Blast Furnace, Especially for Chlorine and Thallium Removal. Presented at the Meeting of the GDMB Lead Experts Committee, Aachen, Germany, May 15−17.

Li, D., Suo, Y., 2010. Oxygen bottom-blowing smelting lead smelting technology by bottom blowing electro-thermal reduction. In: Siegmund, A., Centomo, L., Geenen, C., Piret, N., Richards, G., Stephens, R.I. (Eds.), Lead Zinc 2010, 5th Decennial Symposium. TMS, pp. 483−486.

Martensson, C., 2013. "New Boliden Bergsöe, Landskrona, Sweden: Safety and Environmental Work at Bergsöe", in GDMB Lead Experts Meeting.

Meurer, U., Püllenberg, R., Griesel, 2005. Developing QSL technology to an economically superior process. In: Fujisawa, T., Dutrizac, J.E., Fuwa, A., Piret, N.L., Siegmund, A.H.-J. (Eds.), Proc. International Symposium on Lead and Zinc Processing; Lead & Zinc '05. MMIJ (Mining and Materials Processing Institute of Japan), pp. 547−558.

Olper, M., Maccagni, M., 2009. The secondary Pb challenge − the Engitec approach. Proceedings of EMC, 441.

Onisei, S., Pontikes, Y., Van Gerven, T., Angelopoulos, G.N., Velea, T., Predica, V., Moldovan, P., 2012. Synthesis of inorganic polymers using fly ash and primary lead slag. Journal of Hazardous Materials 205−206, 101−110.

Pizzocri, M., Maccagni, M., 2013. F.A.S.T Pb process: its impact on the new secondary lead facility. Proceedings of EMC 2013, p. 549.

Prengaman, R.D., 2009. Lead industry − transition, evolution, and growth (abstract). In: Proceedings of EMC 2009, p. XVII.

Siegmund, A.H.-J., 2000. Primary lead production − a survey of existing smelters and refineries. In: Dutrizac, J.E., Gonzalez, J.A., Henke, D.M., James, S.E., Siegmund, A.H.-J. (Eds.), Proc. Lead-Zinc 2000. TMS, pp. 55−116.

Siegmund, A., 2003. Modern applied technologies for primary lead smelting at the beginning of the 21st century.

In: Kongoli, F., Itagaki, K., Yamauchi, C., Sohn, H.Y. (Eds.), Proc. Yazawa International Symposium "Metallurgical and Materials Processing: Principles and Technologies", High-Temperature Metal Production, vol. 2. TMS, pp. 43–62.

Siegmund, R., Prengaman, R.D., 2001. Grid alloys for automobile batteries in the new millennium. JOM 53 (1), 38–39.

Sinclair, R., 2009. The Extractive Metallurgy of Lead (CD). Australasian Institute of Mining and Metallurgy. ISBN: 9781921522024.

Sofra, J., Hughes, R., 2005. Ausmelt technology operation at commercial lead smelters. In: Fujisawa, T., Dutrizac, J.E., Fuwa, A., Piret, N.L., Siegmund, A.H.-J. (Eds.), Proc. International Symposium on Lead and Zinc Processing; Lead & Zinc '05. MMIJ (Mining and Materials Processing Institute of Japan), pp. 511–528.

Stephens, R.L., 2005. Advances in primary lead smelting. In: Fujisawa, T., Dutrizac, J.E., Fuwa, A., Piret, N.L., Siegmund, A.H.-J. (Eds.), Proc. International Symposium on Lead and Zinc Processing; Lead & Zinc '05. MMIJ (Mining and Materials Processing Institute of Japan), pp. 45–72.

Stevenson, M.W., 2004. Recovery and recycling of lead-acid batteries (Chapter 15). In: Mosely, P.T., Garche, J., Parker, C.D., Rand, D.A.J. (Eds.), Valve-Regulated Lead-Acid Batteries. Elsevier, pp. 491–512.

Stevenson, M., 2009. Presentation at the International Secondary Lead Conference, Macau.

Stevenson, M., 2013. Private communication.

UNEP, 2013. Reuter, M.A., Hudson, C., van Schaik, A., Heiskanen, K., Meskers, C., Hagelüken, C. Metal Recycling: Opportunities, Limits, Infrastructure. A Report of the Working Group on the Global Metal Flows to the International Resource Panel.

White, P., private communication, 12 Nov 2013. ILZSG press release, 4 Oct 2013.

Wirtz, http://www.wirtzusa.com/battrec/product/44.htm Nov 2013.

Wilson, D., 2011. Lead – a bright future for the grey metal? World of Metallurgy Erzmetall 2011 (4), 196.

CHAPTER 9

Zinc and Residue Recycling

Jürgen Antrekowitsch, Stefan Steinlechner, Alois Unger, Gernot Rösler, Christoph Pichler, Rene Rumpold

University of Leoben, Austria

9.1 INTRODUCTION

Annual global production of zinc is more than 13 million tons. More than 50% of this amount is used for galvanizing while the rest is mainly split into brass production, zinc-based alloys, semi manufacturers and zinc compounds such as zinc oxide and zinc sulfate (World Bureau of Metal Statistics).

Zinc is recycled at all stages of production and use—for example, from scrap that arises during the production of galvanized steel sheet, from scrap generated during manufacturing and installation processes, and from end-of-life products. The presence of zinc coating on steel does not restrict steel's recyclability, and all types of zinc-coated products are recyclable. Zinc-coated steel is recycled along with other steel scrap during the steel production process—the zinc volatilizes and is then recovered. For the zinc and steel industries, recycling of zinc-coated steel provides an important new source of raw material. Historically, the generation of zinc-rich dusts from steel recycling was a source of loss from the life-cycle (landfill); however, technologies today provide incentive for steel recyclers to minimize waste. Thus, the recycling loop is endless—both zinc and steel can be recycled again and again without losing any of their physical or chemical properties (International Zinc Association, 2013; www.zinc.org/sustainability/resourceserve/zinc_recycling_closing_the_loop; http://www.zinc.org/applications/news/zinc_environmental_profiles_published).

Depending on the composition of the scrap being recycled, it can either be remelted or returned to the refining process. A classification of recycling related to different scrap types and treatment processes is given in Table 9.1.

The overall recycling rate is difficult to define because of the wide lifetime range of the various zinc products. Therefore, typical values are between 15% and 35%, depending on which types of scraps are considered as recycled materials. However, the recycling rate of zinc is increasing, underlining the considerable importance of this industry in the world's zinc production (International Zinc Association, 2013; www.zinc.org/sustainability/resourceserve/zinc_recycling_closing_the_loop; http://www.zinc.org/applications/news/zinc_environmental_profiles_published).

9. ZINC AND RESIDUE RECYCLING

TABLE 9.1 Sources and Type of Secondary Zinc (www.zinc.org/sustainability/resource-serve/zinc_recycling_closing_the_loop)

Scrap Source	Type of Scrap			Recovery Process
	Residue and Drosses (NS)	Whole Products (OS)	Steel Filter Dust (OS)	
Brass	✔	✔	—	Remelting
Die casting	—	✔	—	Remelting
Galvanizing	✔	—	✔	Remelting (NS) Refining (OS)
Rolled zinc	✔	✔	—	Remelting
Other	✔	✔	✔	Remelting (NS) Refining (OS)

OS, old scrap; NS, new scrap.

In the following sections, the main processes for zinc recycling from different scraps and residues are described. An overview of the various residues as well as the related recycling processes and products is provided in Figure 9.1.

The remelting of alloys is one major field in zinc recycling. However, because of its rather simple process technology, it is not described in detail.

9.2 ZINC OXIDE PRODUCTION FROM DROSSES

The way the production of zinc oxide is carried out today was developed by Le Claire in France in 1840, by burning metallic zinc in air—the so-called "French process". At the same time in America, a different production route was developed—the so-called "American" or

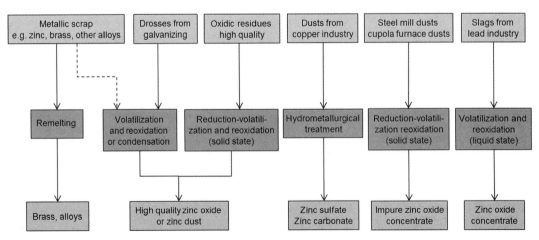

FIGURE 9.1 Main processes for zinc recycling from different residues.

"direct process". The name originated from the possibility to make direct use of oxidic ore and other oxidic secondary materials such as ashes, clinker and top dross from hot tip galvanizing (Zinc REACH Consortium; Yanlong and Longda, 2010; UCA Lanka Pvt Ltd).

Either the French or the American process makes use of the volatility of metallic zinc. So metallic zinc is melted or oxides are reduced to a metal and vaporized at temperatures above 907 °C—typically about 1000 °C. Due to an excess amount of air, the metallic zinc immediately reacts to ZnO in an exothermic reaction and a bright luminescence. Because of that, purification can be achieved within the production process. Critical elements are those with similarly high or higher values, like cadmium, bismuth, magnesium or manganese. In both the French (indirect) and American (direct) processes, air is used as a cooling and an oxidation agent for the vaporized zinc. Furthermore, the excess air works as a transportation medium for the zinc oxide particles to the product filter house (Zinc REACH Consortium; Yanlong and Longda, 2010; UCA Lanka Pvt Ltd).

The third method used for ZnO production is a hydrometallurgical one based on zinc hydroxide or zinc carbonate production as a semi-finished product. The advantage of this method is the possibility of several cleaning steps before the product is obtained. Because of that, a wide variety of possible raw materials can be used, but it goes hand in hand with increasing process complexity. The semi-finished product is then processed to ZnO using thermal treatment. Depending on the thermal conditions, the specific surface area is adjusted (Zinc REACH Consortium; Yanlong and Longda, 2010; UCA Lanka Pvt Ltd).

As shown in Figure 9.2, a distinction can be made for the main processes used for ZnO production, including their raw materials.

An alternative to the three processes in Figure 9.2 for high purity zinc oxide production is a fourth special option, namely the Larvik furnace. The Larvik furnace can be split into three zones—melting, vaporization and separation. The input material is similar to that of the French process, but one special difference is that this concept has a significantly higher acceptance of impurities, like lead, iron, etc. Possible materials are again bottom dross, crude skimming or grinded skimming. Additionally, the required melting heat is provided by the enthalpy of the off-gas in the vaporization zone. Nonmeltable components are removed from the liquid metal surface in the melting zone. Special attention is given to aluminum in

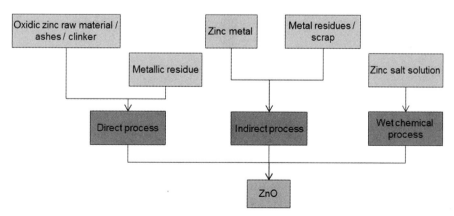

FIGURE 9.2 Different ways for ZnO production and utilized raw materials.

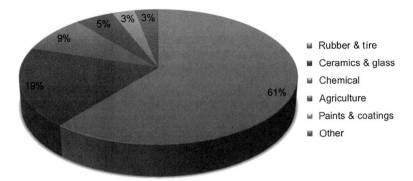

FIGURE 9.3 Breakdown of global zinc oxide usage into its utilization areas.

the input material, because a skin on the zinc metal surface can be formed by aluminum oxide, preventing continuous vaporization. A phosphorus addition is necessary to form a eutectic alloy with iron with the aim of a low liquidus temperature (lower than 1500 °C). This offers the opportunity for a continuous separation of the iron. The iron alloy sinks to the bottom as a result of the higher density and is collected in the casting area. The same occurs with the lead present, forming a third liquid metal layer in the furnace.

9.2.1 Quality, Market and Utilization Areas

Commercial grades of zinc oxide can be found—depending on their quality—in a wide range of utilization areas, starting from specialties such as the semi-conductor or the tire industry in case of high purity products (typically French process), its usage in everyday situations like sun creams or as a product with slightly lower quality in the ceramics industry or as a food additive for animals (typically American process). Aside from the product quality, properties such as specific surface area, coatings or individual customer requirements influence the decision for the chosen/required ZnO production process.

Compared by weight, most of the high purity ZnO produced is manufactured by the French process, having a market share of about 60%. American quality zinc oxide follows with slightly less than 40%, and only a minor amount, typically utilized in special fields, is manufactured by the wet route (Zinc REACH Consortium). The estimated global capacity of zinc oxide production was 1,887,000 metric tons per year in 2010. The real consumption was 1,326,000 tons worldwide in the fields of utilization as depicted in the pie chart in Figure 9.3 (Schlag, 2011; Hassall, 2010).

9.3 ELECTRIC ARC FURNACE DUST AND OTHER PB, ZN, CU-CONTAINING RESIDUES

When steel scrap is processed in an electric arc furnace (EAF), about 15—23 kg of dust are formed per tonne of steel, which, referring to steel production for 2012, translates to between 6.7 and 10.3 million tpa of EAF dust being produced. Table 9.2 provides a summary of the composition of EAF dust as well as other zinc- and lead-containing residues.

Because of economic reasons, low landfill costs, relatively low zinc and lead metal prices, etc., the Waelz kiln, in Figure 9.4, is still the preferred route for processing EAF dust to produce a Waelz oxide and other products, as shown in Table 9.3. The first step in this process is an agglomeration of the raw materials with

TABLE 9.2 Typical Zn & Pb-containing Raw Materials as well as Important Zinc Drosses and Ashes (Schneider et al., 2000; James, 2000)

	Zn (%)	Pb (%)	Fe (%)	FeO (%)	Cu (%)	Al (%)	Cl (%)	C (%)	SiO_2 (%)	H_2O (%)
Filter cake cupola furnace	31	3	—	10	0.2	—	0.4	11	15	45
Dust from copper and brass industry	43	20	—	0.6	3	—	5	0.6	1	0.3
Dust cupola furnace	31	0.1	—	23	—	—	0.5	5	1.3	0.4
EAF dust	23	0.1	—	35	0.1	—	0.6	2	1	10.6
Lead dust	2	68	—	5	—	—	0.2	2	—	10.4
Zinc-lead oxide	44	15	—	4	0.4	—	4	0.6	1	1.8
Neutral leach residue	18	7	—	33	1.6	—	—	0.2	3	30
Galvanization dross	92—94	1.0—1.6	1—3	—	—	—	—	—	—	—
HD galvanizing ash	60—75	0.5—2.0	0.2—0.8	—	—	—	2—5	—	—	—
Cont. galvanizing ash	65—75	0.1—0.5	0.2—0.8	—	—	0.1—0.5	0.5—2	—	—	—
Die-casting dross	90—94	0.1—0.2	Low	—	—	1—7	—	—	—	—
Sal skimmings	45—70	0.5—2.0	0.2—0.8	—	—	—	15—20	—	—	—
Brass fume	40—65	0.5—7.0	1—2	—	—	—	2—7	—	—	—
Die-casting ash	55—60	0.1—0.2	Low	—	—	3—10	Low	—	—	—
Ball mill ash	55—65	0.1—1.0	0.2—0.8	—	—	—	2—5	—	—	—
Zinc dust/overspray	92—95	1—2	0.1—0.5	—	—	—	—	—	—	—

EAF, electric arc furnace.

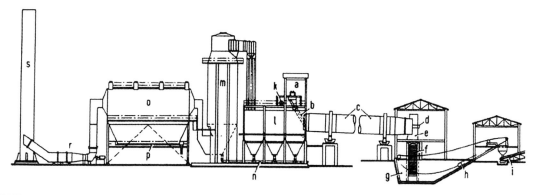

FIGURE 9.4 Waelz kiln for the processing of EAF dust (Pawlek, 1983). a: Material feeding; b: reel carriers; c: Waelz kiln; d: burner; e: discharge head; f: slag reel carriers; g: slag pit; h: scrapper; i: slag removal; k: sealed air fans; l: dust settling chamber; m: evaporation cooler; n: low-grade Waelz oxide transport; o: electrostatic precipitator; p: Waelz-oxide transport; r: flue gas blower; s: stack.

TABLE 9.3 Typical Analyses of the Products of a Waelz Kiln (Meurer, 2000)

	Zn (%)	Pb (%)	Cd (%)	F (%)	C (%)	FeO (%)	Fe(metallic)/Fe	Basicity
Waelz slag	0.2–2.0	0.5–2	<0.01	0.1–0.2	3–8	30–50	80–90	
Basic								1.5–4.0
Acidic								0.2–0.5
Waelz oxide	55–58	7–10	0.1–0.2	0.4–0.7	0.5–1	3–5	—	
Leached Waelz oxide	60–68	9–11	0.1–0.3	0.08–0.15	1–1.5	4–7	—	

reduction agents and slag-forming additives. As the process would not be economical, the minimum content of Zn in the feed material should not be less than 18%. A slight decline of the kiln leads the material to move through the furnace very slowly while it is heated up by the gas which is counter current flowing through the kiln (Kozlov, 2003; Rütten, 2009).

Minimum temperatures of 1000 °C are necessary for a total reduction of zinc- and lead-containing residues. Figure 9.5 illustrates the principal reactions that take place in the slowly rotating Waelz kiln, which can be longer than 100 m. The reduced Zn, because of its high vapor pressure, is volatilized as Zn gas and then combusted to form ZnO powder, which is collected in a filter, as shown in Figure 9.4. The combusting zinc liberates heat, which is an important part of the heat balance of the kiln. Similarly, other metals such as Pb, Cd and Ag (if present in other residues also fed to the furnace) are also volatilized and collected in the flue dust. Although the Waelz kiln process is the best available technology, improvements in the whole process were introduced in the past few years. An example is the reduction and subsequent reoxidation of the iron in the kiln which is within the feed material and later in the Waelz slag, to reduce the coke and natural gas consumption and increase the throughput and zinc output, respectively. Usually the fed rate of coke is overstoichiometric regarding the requirements but with this invention, the coke consumption is lower, which is also combined with lower CO_2 emissions (Kozlov, 2003; Rütten, 2009).

The process itself starts in the upper part of the furnace, where the feed material is charged or the Waelz oxide leaves the furnace, respectively. Here the charged material is dried at temperatures up to 150 °C. In the following section

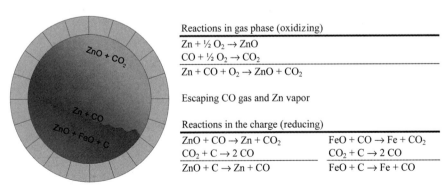

FIGURE 9.5 The main reactions in a Waelz kiln (Antrekowitsch and Griessacher, 2009).

volatile matter evolves from the coke since the temperature increases up to 500 °C. At 900 °C the sulfates and carbonates are decomposed and halides are volatilized. The main reaction of the zinc starts in the second half of the Waelz kiln. The final step in the kiln is the reoxidation of the iron at the discharge head to form the slag at temperatures up to 1120 °C. Then the Waelz slag is discharged and led to a slag granulation plant for cooling down. The oxygen supply for reoxidation of zinc and the iron is delivered by leakage air at the discharge head of the slag in the lower part of the kiln. Because of the optimization of the process (reduction of iron), there is almost no additional energy necessary for the endothermic chemical reactions in the Waelz kiln (Kozlov, 2003; Rütten, 2009).

The high zinc-containing Waelz oxide leaves the furnace in the upper part of the kiln with the off-gas (N_2, CO_2, H_2O, O_2 and C_2) and is collected in a dust settling chamber (Kozlov, 2003; Rütten, 2009). The Waelz oxide produced is processed in a zinc plant to generate London Metal Exchange grade zinc. However, unwanted compounds, especially halogens present in the EAF dust, are also introduced into the Waelz oxide. Therefore, most Waelz kilns need to have an additional soda-washing step to allow low enough fluorine and chlorine contaminations so that utilization in primary zinc metallurgy is possible without harming the electrolysis process. Even though the washing step removes a major amount of the halogens, it is generally not possible to substitute more than 15% of the primary concentrate by Waelz oxide as a result of the above-mentioned reasons.

Over the last decades, a significant number of alternative processes have arisen. To save energy by avoiding high temperatures, numerous hydrometallurgical concepts have been developed. Almost all of these concepts have failed because of their inability to dissolve zinc ferrite, a major component of the EAF dust, or because of complex process routes. Today, only the EZINEX process can be named as the hydrometallurgical alternative in operation on an industrial scale. Nevertheless, the share of hydrometallurgical techniques in EAF dust recycling is far below 5%.

Pyrometallurgical alternatives are mostly based on carbothermal reduction, as is the Waelz process. The main difference is the facility used, and for some of the developments, the attempt to also recover iron in a recyclable form. Typical examples are:

- Rotary hearth furnaces (zinc oxide and iron sponge)
- Multiple hearth furnace (zinc oxide and iron alloy)
- Direct melting, e.g. inductive (zinc oxide and iron alloy)

None of the concepts described were able to bring more than one or two units to industrial scale because of problems with excessively high energy consumption, refractory lining and remarkably low quality of the iron product.

Therefore, the Waelz process currently dominates EAF dust recycling with a share of more than 90%.

9.4 ZINC RECYCLING FROM COPPER INDUSTRY DUSTS

Nowadays, the processed primary and especially secondary raw materials in copper smelting contain significant amounts of zinc. Table 9.4 gives an overview of the copper sources used and their zinc content.

Basically, the primary production can be divided in smelting, converting, pyrometallurgical and finally hydrometallurgical refining. Secondary materials are processed similarly. Only the smelting step and the converting conditions vary, based on the raw material input.

As can be seen in Table 9.4, the secondary sources show quite high zinc concentrations, while primary concentrates are rather low in zinc. Nevertheless, the necessity of cooling scrap in the converting and refining step leads to zinc

TABLE 9.4 Raw Materials for Copper Production (Edelstein, 2013; Rentz et al., 1999; Muchova et al., 2011; Ayhan, 2000; Yamane et al., 2011; Barroso, 2010; Balladares, 2011; Rumpold and Antrekowitsch, 2013)

Type of Material	Cu (%)	Zn (%)
Copper concentrate	~30	0–2
Copper alloy scrap	36–98	0–43
Slags	10–50	2–10
Drosses	10–50	2–20
Sludges	0–40	0–20
Dusts	1–30	5–40
Nonferrous shredder	10–60	0–20
Electronic scrap	7–20	1–6

as well as compositions of their flue dusts (Barroso, 2010; Balladares, 2011; Rumpold and Antrekowitsch, 2013; European Commission, 2013; Hanusch and Bussmann, 1995; Litz, 1999; Piret, 1995).

Besides the given valuable metals, such materials contain harmful contaminants like halogens from secondary sources. Primary zinc production, in particular the electrowinning process, is very sensitive to chlorine and fluorine. Consequently, flue dusts from the copper industry are not suitable for application in the primary zinc industry, although high zinc contents are present (Antrekowitsch and Offenthaler, 2010).

Based on the complexity of these materials, their recycling requires specific procedures, including hydrometallurgical or pyrometallurgical processes.

The hydrometallurgical methods offer numerous options to recover the valuable metals from flue dust. In principle they are leached completely or partly, followed by a selective extraction. Very often a halogen removal is carried out first, since moderate to low values can be achieved by simple water or soda washing.

The conventional technology for the recycling of secondary copper dusts is leaching with ulfuric acid followed by cementation and finally crystallization of zinc sulfate. Lead and tin are not soluble at these conditions and can be separated by filtration. The lead- and tin-rich filter

inputs also in the primary production route. Because of the high vapor pressure and affinity to oxygen, impurities can be removed easily via the slag and dust phase. The slags from copper metallurgy are treated by a reduction step to recover entrained copper. Consequently, enclosed zinc oxide is also reduced and volatilized. That means the major portion of zinc is concentrated in the flue dusts, which represent potential zinc secondary raw materials. Table 9.5 shows the different primary and secondary process steps

TABLE 9.5 Flue Dusts from the Copper Industry (European Commission, 2013; Hanusch and Bussmann, 1995; Litz, 1999; Piret, 1995)

%	Smelting	Converting	Slag Treatment	Refining	Smelting	Converting
	Primary				Secondary	
Zn	0.1–10	5–70	25–60	5–40	20–60	25–70
Pb	0.1–5	2–25	2–15	2–20	5–50	5–30
Sn	0.1–1	0.1–4	—	—	0.2–5	1–20
Cu	5–30	10–25	0.5–2.5	15–25	2–12	2–15

cake can be processed in the lead industry. In the next step, copper is removed from the solution by cementation. Therefore, iron scrap or zinc represent potential cementation agents. The advantage of zinc is that no iron is added in the process. When iron is already present in the solution, scrap can be used as a cementation agent, since it is cheaper and has to be removed anyway. Finally, zinc is recovered by the crystallization of zinc sulfate (Piret, 1995; Jha et al., 2001).

Another relatively similar approach is to separate the zinc selectively via solvent extraction from the solution and to recover it by electrolysis. This technology is known as the ZINCEX process, but because of the risk of chlorine and organic contamination of the electrolyte, it is not widely applied as of yet (Piret, 1995; Jha et al., 2001; Martin et al., 2002).

Beside the typical sulfuric acid-based processes, ammonia can also be used to recover valuable metals. The EZINEX process represents the most common method using this medium. In so doing, ammonia chloride is used to dissolve the heavy metal fraction (except iron), followed by solution purification via cementation with zinc. In the end, the pure zinc is recovered by a novel electrolysis, which represents the core of the EZINEX technology. Instead of the sulfate system, which is usually used in the zinc industry, a chloride-based electrolysis is applied (Piret, 1995; Jha et al., 2001; Olper and Maccagni, 2008).

The leaching of metal oxides with NH_3-$(NH_4)_2CO_3$ is an alternative ammonium treatment, but because of the interference of chlorides, halogen removal in advance is necessary. After the solution purification by cementation, zinc carbonate can be precipitated (Jha et al., 2001; Schlumberger and Bühler, 2012).

A relatively new method is given by the so-called FLUREC process. It leaches the heavy metal fraction with a hydrochloric acid-containing solution. The subsequent cementation removes lead, copper, etc. and enables a selective solvent extraction. After reextracting the zinc into a sulfuric acid solution, high purity zinc can be recovered by conventional electrolysis (Jha et al., 2001; Schlumberger and Bühler, 2012).

The pyrometallurgical processes are used only rarely for copper dust recycling. The available process reduces the metal compounds contained. Furthermore, soda is added to fix sulfur in the slag. High temperatures lead to a zinc concentration in the flue dust. Remaining lead and tin form a marketable alloy. In fact, only high tin and lead contents in combination with low halogen concentrations justify such a processing. Therefore, this process is not suitable for modern secondary copper dusts (Piret, 1995; Steinlechner and Antrekowitsch, 2013).

9.5 FUMING OF SLAGS FROM LEAD METALLURGY

Lead concentrates are usually accompanied by zinc in the form of zinc sulfide. During roasting, the zinc sulfide turns into zinc oxide. Especially in the shaft furnace process for lead production, the reduction rate is set in such a way that lead is reduced while zinc stays as oxide and mainly ends up in the slag. Typical zinc values allowing an economical recovery can be found in the range of 8–12%, depending on the current zinc price (Koch and Janke, 1984; Krajewski and Krüger, 1984; Ullmann's Encyclopedia, 2007; Püllenberg and Höhn, 1999; Jak and Hayes, 2002).

The slag fuming is based on the reduction with carbon according to the following equation:

$$ZnO(s) + CO(g) \rightarrow Zn(g) + CO_2(g)$$

Zinc evaporates and is reoxidized in the off-gas system, generating a high zinc-containing dust that is collected in the baghouse.

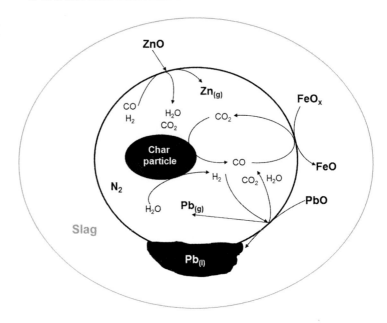

FIGURE 9.6 Reactions between gas, char and slag during the slag fuming process (Püllenberg and Höhn, 1999).

A typical furnace for this process is a rectangular-shaped hearth furnace. The processing temperature of the slag is above 1200 °C. The char particles mixed with air are blown into the molten slag, generating reducing gas bubbles. Figure 9.6 illustrates such a gas bubble with an enclosed char particle explaining the relevant reactions between gas, char and slag.

The reduced zinc accumulates as zinc vapor in the gas bubbles. Triggered by the Boudouard reaction, the generated $CO_2(g)$ reacts with carbon from the char to $2CO(g)$. This regeneration of $CO(g)$ leads to constant high reducing conditions in the gas bubble, which results in a high zinc recovery yield. If a $Zn(g)$-containing gas bubble reaches the bath level, the zinc vapor goes into the furnace atmosphere and reacts via postcombustion with $O_2(g)$ or $CO_2(g)$ to produce ZnO. Typical zinc recovery yields of about 87.5% are observed. Depending on the process conditions, lead oxide is also recovered from the slag. In Table 9.6, typical data for the slag fuming process are presented.

TABLE 9.6 Typical Data for Fuming of Lead Blast Furnace Slag (Koch and Janke, 1984; Krajewski and Krüger, 1984; Ullmann's Encyclopedia, 2007; Sundström, 1969; Jak and Hayes, 2002)

	Capacity (t/day)	Cu (%)	Pb (%)	Zn (%)	Fe (%)	CaO (%)	SiO$_2$ (%)
Charged slag	655	0.8	0.1–2.6	10.7–18.3	25.6–30.5	1.9–15.2	21.0–31.7
Treated slag	600	0.4	0.01–0.03	2.6–0.8	26.4–34.2	5.3–17.7	30.0–38.3
Dust	119	0.1	12.5–13.9	58.6–66.0	0.2	—	0.3
Matte	16	22.6	0.2	0.7	41.8	—	—

References

Antrekowitsch, J., Griessacher, T., 2009. The current status of recycling of electric arc furnace dusts and perspectives for the future. In: Proc.: MicroCad, Miskolc, pp. 27–38.

Antrekowitsch, J., Offenthaler, D., 2010. Die Halogenproblematik in der Aufarbeitung zinkhältiger Reststoffe. BHM 155, 31–39.

Ayhan, M., 2000. Das neue HK-Verfahren für die Verarbeitung von Kupfer-Sekundär-materialien, Heft 87 der Schriftenreihe. GDMB, pp. 197–207.

Balladares, E., 2011. Procesamiento Pirometalurgico De Minerals De Cobre En Chile: Technologias Actuales Y Emergentes, Presentation at XVI. Congreso Colombiano de Minera, Columbia.

Barroso, A.R.F., 2010. Copper Recycling, ENGR5187-Solid Waste Processing and Recycling. Laurentian University.

Edelstein, D.L., 2013. Copper, 2011 Minerals Yearbook. US Geological Survey, 20.1–20.25.

Hanusch, K., Bussmann, H., 1995. Behavior and removal of associated metals in the secondary metallurgy of copper. In: Third International Symposium Recycling of Metals and Engineered Materials, Point Clear, Alabama, pp. 171–788.

Hassall, C., 2010. Presentation: Zinc Oxide in China, IZA Conference, Cancun.

International Zinc Association: Zinc a Sustainable Material, www.zinc.org/applications/news/zinc...a_sustainable_material (30.10.13.).

Jak, E., Hayes, P., 2002. Phase equilibrium and thermodynamics of zinc fuming slag. Canadian Metallurgical Quarterly 41 (2), 163–174.

James, S.E., October 2000. Recycling of zinc, course notes from short course presented before 4th international symposium on recycling of metals and engineered materials. TMS.

Jha, M.K., Kumar, V., Singh, R.J., 2001. Review of hydrometallurgical recovery of zinc from industrial wastes. Resources, Conservation and Recycling 33, 1–22.

Koch, K., Janke, D., 1984. Schlacken in der Metallurgie 1. Verlag Stahleisen mbH, Düsseldorf.

Kozlov, P.A., 2003. The Waelz Process. Ore and Metals Publishing House, Moskau.

Krajewski, W., Krüger, J., 1984. Schlacken und Steinbildung bei der thermischen Blei-und Blei-Zink-erzeugung, Schlacken in der Metallurgie 1. Verlag Stahleisen mbH, Düsseldorf.

Ullmann's Encyclopedia of Industrial Chemistry: Lead, 2007. John Wiley and Sons.

Litz, J.E., 1999. Flue dusts: an ideal feed for resource recovery, residues and effluents – processing and environmental considerations. TMS, 223–239.

Martin, D., et al., November/December 2002. Extending zinc production possibilities through solvent extraction. The Journal of the South African Institute of Mining and Metallurgy, 463–467.

Best Available Techniques Reference Document for the Non-Ferrous Metal Industries, Draft 3, European Commission, 2013.

Meurer, U., 2000. Gewinnung von Zinkoxid aus sekundären Rohstoffen – Neue Entwicklungen in Wälzprozess. In: Liese, F. (Ed.), Intensivierung Metallurgischer Prozesse. GDMB, Clausthal-Zellerfeld, Germany, pp. 183–196.

Muchova, L., Eder, P., Villanueva, A., 2011. End-of-waste criteria for copper and copper alloy scrap. JRC.

Olper, M., Maccagni, M., 2008. From C.Z.O. to zinc cathode without any pretreatment. The EZINEX process, lead and zinc 2008, Durban, South Africa, pp. 85–98.

Pawlek, F., 1983. Metallhüttenkunde. Walter de Gruyter Verlag, Berlin.

Piret, N.L., 1995. Criteria for optimization of recycling processes of primary and secondary copper smelter dusts. In: Third International Symposium on Recycling of Metals and Engineered Materials, Point Clear, Alabama, pp. 189–214.

Püllenberg, R., Höhn, R., 1999. Schlacken in der Bleimetallurgie. GDMB-Verlag, 83.1, 81–94.

Rentz, O., Krippner, M., Hähre, S., 1999. Report on BAT in German Copper Production. University Karlsruhe.

Rumpold, R., Antrekowitsch, J., 2013. Zinc in the copper secondary production and potential recycling strategies. Copper.

Rütten, J., 2009. Ist der Wälzprozess für EAF-Staub noch zeitgemäß? Stand der Technik und Herausforderungen. 2. Seminar – Vernetzung von Zink und Stahl. GDMB, Heft 118, pp. 137–149.

Schlag, S., 23.02.2011. Presentation: Global ZnO supply/demand. In: International Zinc and Zinc Oxide Conference, Cancun, Mexico.

Schlumberger, S., Bühler, J., 2012. Urban Mining: Metal Recovery from Fly and Filter Ash in Waste to Energy Plants. Ash Utilization 2012, Stockholm, Sweden.

Schneider, W.D., Romberg, T., Schwab, B., 2000. Recycling von Zn-Pb-haltigen Reststoffen-die notwendige Weiterentwicklung des IS-Prozesses. In: Liese, F. (Ed.), Intensivierung Metallurgischer Prozesse. GDMB, Clausthal-Zellerfeld, Germany, pp. 171–182.

Steinlechner, S., Antrekowitsch, J., 2013. Amelioration of secondary zinc oxide. In: Presentation at 2013 International Zinc Oxide Industry Conference, Cancun, Mexico.

Sundström, O., 1969. Erzmetall 22, 123–131.

UCA Lanka Pvt Ltd, http://www.ucalanka.com (01.08.11.).

99th World Bureau of Metal Statistics 2001–2011, Ware England, (30.10.12.).

Yamane, L.H., et al., 2011. Recycling of WEEE: characterization of spent printed circuit boards from mobile phones and computers. Waste Management 31, 2553–2558.

Yanlong, W., Longda, W., 2010. Presentation: sunlight through dense fog – the status quo and future outlook for the indirect method of ZnO in China. In: China International Lead & Zinc Conference Proceedings.

http://www.zinc.org/applications/news/zinc_environmental_profiles_published (30.10.13.).

Zinc REACH Consortium, http://www.reach-zinc.eu/pg_n.php?id_menu=31 (14.08.13.).

www.zinc.org/sustainability/resourceserve/zinc_recycling_closing_the_loop, (30.10.13.).

CHAPTER 10

Recycling of Rare Metals

Elinor Rombach, Bernd Friedrich

IME Process Metallurgy and Metal Recycling, RWTH Aachen University, Intzestraße 3, Aachen, Germany

Conventional recycling processes for rare metals are often based on the process routes of mass metals (e.g. Cu, Pb, Zn, Al). Because of low metal contents and production volumes of rare metals, it is not always economical to operate a specific recycling process. In these cases pretreatment or material conditioning steps are used to produce anthropogenic (recycling) concentrates, which are introduced in conventional extraction processes. Such concentrates do not necessarily have very high concentration levels of rare metals, but should be minimized in specific impurities, which are known for disturbing the ongoing process units. In this context it should be noted that geogenic (primary winning) and anthropogenic (secondary recycling) process chains widely overlap; i.e. they are not always clearly separated from each other. This is shown for indium and tellurium, for example (Figure 10.1).

Usually (pretreated) anthropogenic concentrates, recycling raw materials or intermediate products are introduced in pyro- or hydrometallurgical metal winning process routes, to increase their trace element amounts (parts per million ranges of In, Te) up to ranges of a few percent. Byproduct processes, i.e. extraction steps for indium and tellurium, need higher adequate metal contents (>0.1% In, >1% Te). Corresponding materials are, for example, anode slime (copper winning), lead/copper dross (lead winning) or leach residue (zinc winning) as well as slightly contaminated pre-consumer photovoltaic scraps (wafer scrap, sputter targets). For end-of-life (EOL) scrap, i.e. complex composite materials with high impurity concentrations as present in printed circuit boards, purely hydrometallurgical processes are difficult to realize, liable to technical limitations and therefore not recommended (Hagelüken and Meskers, 2010). Especially in the field of rare metals recycling, the selection of suitable process modules for metal concentration is of particular importance.

Most of the rare metals, described in this chapter, are used as key metals in numerous technical applications and were classified as "critical raw materials for the EU" by the European Commission in 2010 (see European Commission-Enterprise and Industry, 2010). For these rare metals main end-use-markets and a selection of developed and especially commercial practiced recycling routes are illustrated. However, since a large number of process alternatives are used,

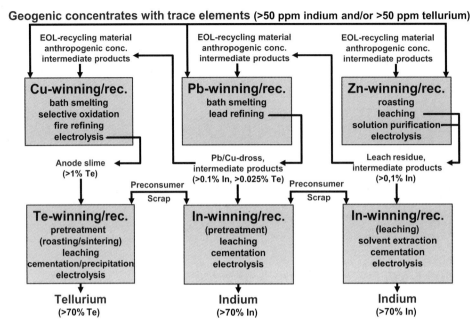

FIGURE 10.1 Summary of current reclamation and recycling routes for indium and tellurium. EOL, end-of-life.

which are also partly highly complex, no claim of completeness can be raised in this document.

10.1 PRECIOUS METALS

The group of eight precious metals (PMs) can be basically distinguished as silver (Ag), gold (Au) and platinum group metals (PGMs). The PGMs comprise six transition metals, three light PGMs (atomic numbers 44–46): ruthenium (Ru), rhodium (Rh) and palladium (Pd) and three heavy PGMs (atomic numbers 76–78): osmium (Os), iridium (Ir) and platinum (Pt). All these metals are of high economic value and have similar chemical and physical properties, such as high melting point, low vapor pressure, high temperature coefficient of electrical resistivity and low coefficient of thermal expansion. Moreover, all PGMs have strong catalytic activity. Main end-use markets for PMs are as follows (European Commission-Enterprise and Industry, 2010; Loferski, 2012; U.S. Geological Survey, 2013):

catalysts: Pt, Pd and Rh are used for automotive catalytic converters and diesel particulate filters to reduce air pollution. PGMs are also used as catalysts in the chemical industry and for petroleum refining. Ag is used as a catalyst in oxidation reactions, e.g. the production of formaldehyde from methanol. For 2012, the PM distribution in this end use was estimated as Pd: >71%; Pt: >31%; and Rh: 69%.

Electric/electronics: In this sector PGMs are used in a variety of applications, such as computer hard discs (Pt, Ru), multilayer ceramic capacitors (Pd) or hybridized integrated circuits. Iridium crucibles were used in the electronics industry to grow high-purity single crystals for use in various applications, including single-crystal sapphire, which was used in the production of backlit light-emitting diode displays in televisions and other electronic devices. The high conductivity of Ag and Au makes them an important component in electrical and electronic equipment. For 2012 the PM distribution in this end use was estimated as Ag: 32%; Au: 5%; Pd: >16%; Pt: <23%; Ru: 62%; and Ir: 55%.

Coins/jewelry/silverware: In 2011 the largest proportion of gold use (66%) as well as approximately 30% of silver and 20% of PGM use went into these nonindustrial, decorative uses.

Photography/mirrors/glass industry: As silver has the highest optical reflectivity of all metals, its use in photography and mirrors is self-evident. However, demand for Ag in photographic equipment has been on the decrease since the introduction of digital cameras since the late 1990s. Platinum equipment was used in the glassmaking industry because of its high melting point and resistance to corrosion. Rhodium is used for flat-panel glass. For 2012 the PM distribution in this end use was estimated as: Ag: >10%; Pt: 7%; Rh: 9%.

Other applications: These applications include dental alloys, solar panels, water treatment, batteries, Radio-frequency identification (RFID) tags and investment tools.

Global mine production of PMs was estimated for 2012 (U.S. Geological Survey, 2013): Ag: 24,000 t/year; Au: 2700 t/year; Pd: 200,000 kg/year; and Pt: 179,000 kg/year. Global consumptions for other PGMs were reported for 2011 (Loferski, 2012): Rh: 28,200 kg/year; Ru: 25,100 kg/year; and Ir: 9360 kg/year. Because of the high economic value of the PMs and their noble (electro-)chemical properties, recycling activities are principally well established, especially for preconsumer, photography, special catalyst and coins/jewelry scraps. But because of the open character of their lifecycles, the recovery of PMs is also limited, especially for postconsumer scrap with dissipative uses, as e.g. Pt/Ru in computer hard disks. Therefore the challenge in PM recycling from consumer applications is the collection and channeling through the recycling chain to different metal recovery processes. Some sector-specific EOL-recycling rates are reported as (UNEP, 2011):

90—100%	For Ag, Au, Pd, Pt in jewelry/coins/silverware.
80—90%	For Pd, Pt, Rh in industrial applications, (including process catalysts/electrochemical, glass, mirror, batteries).
>30—<60%	For Pd, Pt, Rh in vehicles, (including automotive catalysts, spark plugs, Ag-pastes but excluding car-electronics).
	For Ag, Ir, Ru in industrial applications, (including process catalysts/electrochemical, glass, mirror, batteries).
	For Ag, Rh in other applications, (including decorative, medical sensors, crucibles, photographic, photovoltaics).
0—15%	For Ag, Au, Ir, Os, Pd, Pt, Rh, Ru in electronics.

10.1.1 Scrap Metallics, Alloys, Sweeps Minerals and Photographic Materials

PM winning and recycling processes are common (Figure 10.2). They exploit the chemical properties of these metals (e.g. reactivity and oxidation) and use a variety of separation techniques. Typically there is much more silver and gold than PGMs. Many of the processes use very reactive reagents or produce toxic products, and these factors are taken into account by using containment, fail-safe systems and sealed drainage areas to minimize losses. This is further driven by the high value of the metals. Many of the processes are commercially confidential and only outline descriptions are available. One feature of the industry is that, generally, the PMs are recovered on a toll basis, which can be independent of the metal value.

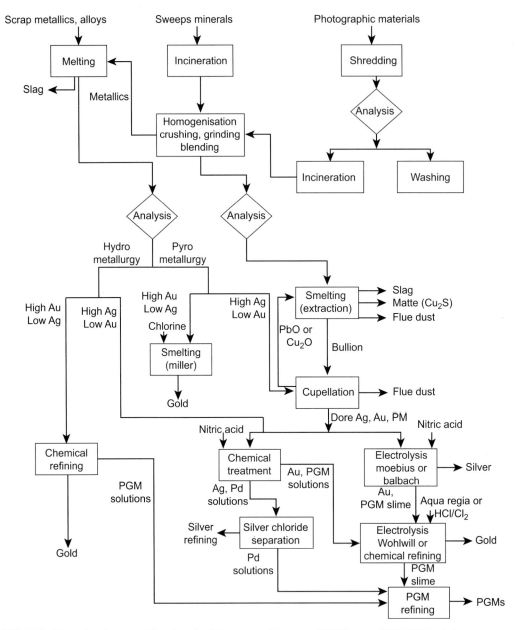

FIGURE 10.2 Example of a general flowsheet for PM recovery (European IPPC Bureau, 2013). PGM, platinum group metal.

Much of the processing is therefore designed to accurately sample and assay the material as well as recover it. Sampling is carried out after the material has been processed physically or from side-streams during normal processing (European IPPC Bureau, 2013).

The individual process steps and technologies used in practice are designed for possible recycling materials, product quality requirements and specific frame conditions in a given location. The main stages in the recovery of PMs can be summarized:

1. Pretreatment and preconcentration of the feedstock, sampling and assay.
2. Concentration, extraction and separation of PMs by pyro- and hydrometallurgical techniques (melting, volatilization, chemical dissolution, precipitation, liquid−liquid extraction, distillation of tetraoxides, ion exchange, electrolytic processes, etc.).
3. Refining (purification, pyrolysis, reduction, etc.) to PM-rich residues or pure metals.

Important recycling sources for Rh are used automotive catalysts and catalysts from the chemical industry. In the case of Ru, recycling of preconsumer scrap plays an important role. This results from the fabrication of Ru sputter targets, which are used in the electronic industry mainly for manufacturing of hard disk drives; usually only 10% of the Ru ends up on the substrate. The biggest portion of Ir-containing recycling materials originates from electrochemical applications (Kralik et al., 2011).

10.1.2 WEEE

Complex EOL scraps like waste of electrical and electronic equipment (WEEE) are commercially integrated in adapted conventional smelter−refinery processes. Mainly copper and lead cycles are used to collect PMs:

- Integrated "primary smelters" like Boliden, Rönnskär (Sweden) or Aurubis, Hamburg (Germany) are focusing on copper concentrates, but the upgrading of the flowsheet and the off-gas treatment enables them to recover PMs as byproducts.
- "Secondary smelters" like Umicore, Hoboken (Belgium) were focused on the recovery of PMs and special metals from scraps, using copper, lead or nickel as collector metals. In this case

the base metals (Cu, Pb, Ni) have, although high in tonnage, a more byproduct character.

At Umicore Precious Metals Refining, Hoboken (Belgium), printed wiring boards (PWBs) or PWB-containing fractions, ICs, processors, connectors and small electronic devices like mobile phones (after removal of the battery) containing typically approximately up to 350 ppm Au, 1500 ppm Ag and 200 ppm Pd together with 20% Cu, 2% Pb, 1% Ni, 10% Fe, 5% Al, 3% Sn and 25% organic compounds are directly treated in integrated copper and PM smelter−refinery operations after mixing with other PM-containing materials (catalysts, byproducts from the nonferrous industries, primary ores) (Hagelüken, 2006, 2009). The organic compounds of the feed material are used as reducing agents and converted to energy; copper acts as a PM collector. The main processing steps of the precious metals operations (PMO) are IsaSmelt furnace, leaching and electrowinning and PMs refinery" (Figure 10.3, see also Figure 10.4).

Feed materials are smelted in a Cu-ISA reactor (ISASMELT™ technology) at about 1200 °C to separate the PMs in a Cu bullion from mostly all other metals concentrated in a Pb-rich copper slag, which is further treated at the base metals operations (BMO). After leaching out the copper in the leaching and electrowinning plant, the PMs are collected in a residue. This PM residue is further refined with a combination of classical methods (cupellation) and unique in-house processes (Ag refinery) developed to recover all possible variations of the separated PMs.

The main processing steps of the BMO are blast furnace, lead refinery and special metals refinery. The Pb-rich copper slag of the Cu-ISA reactor is smelted in a Pb blast furnace together with further Pb-containing raw materials to impure Pb bullion, Ni speiss, Cu matte and depleted Pb slag. PMs collected in the impure Pb bullion and the Ni speiss are separated in form of further PM/Ag residues via

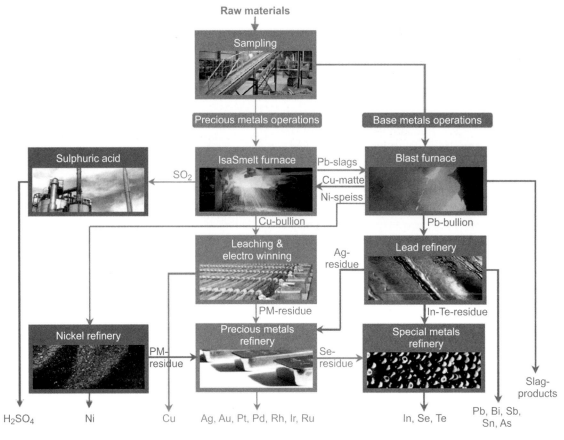

FIGURE 10.3 Mass flow of Umicore's PM-integrated smelter-refinery facility (Hagelüken, 2009).

the Harris process (lead refinery of impure Pb bullion) or via selective leaching (Ni refinery of Ni speiss) to enter the described precious metals refinery.

10.1.3 Catalysts

Specific processes have been developed for recycling of different catalysts (European IPPC Bureau, 2013; Bartz and Wippler, 2005; Hagelüken, 1996; Rumpold and Antrekowitsch, 2012):

- Carbon-based catalysts: These (bearings: C; depletions: 0.5–5% Pd, Pt, Rh, Ru, Pd/Pt) are processed using incineration prior to the dissolution stage.
- Powder-based catalysts: These (bearings: $CaCO_3$, SiO_2, TiO_2, ZrO_2; depletions: 0.1–5% Pd, Pd/Au, Pt, Ir, Ru) and sludges are treated in batches, often in box section furnaces. Direct flame heating is applied to dry and then ignite the catalyst, which is allowed to burn naturally. The air ingress to the furnace is controlled to modify the combustion conditions and an afterburner is used.
- Reforming or hydrogenation catalysts: These (bearings: Al_2O_3, zeolite $M_{x/n}^{n+} \cdot [(AlO_2)_x^- \cdot (SiO_2)_y] \cdot zH_2O$ with M: alkaline (earth) metal) are used in the petrochemical industry and for

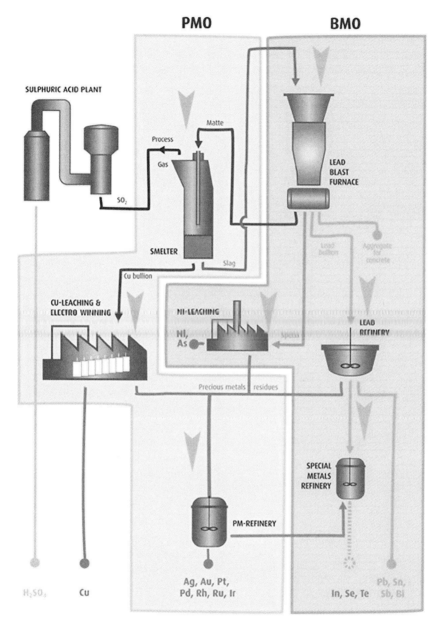

FIGURE 10.4 Flowsheet of Umicore's precious metals recycling loop (PMO: precious metals operation, BMO: base metals operation) (Umicore, 2013a).

hydrocracking; depletions: 0.3–2% Pt, Pt/Re, Pt/Ir, Pd); they can be treated by dissolution of the ceramic base in sodium hydroxide or sulfuric acid. Prior to leaching, the excess carbon and hydrocarbons are burnt off.
- Automotive catalysts: These (bearings: Al_2O_3, cordierite ($2MgO \cdot 2Al_2O_3 \cdot 5SiO_2$); depletions: 0.03–0.3% Pd/Pt(/Rh), Pt, Pt(/Pd)/Rh, Pd/Rh, Pd; content per catalyst: 1.8 g Pt + 0.4 g Rh) can be integrated in Cu, Fe or Ni melts in plasma, electric or converter furnaces, where PGMs can be collected separately. Small operators use open trays to burn off catalysts by self-ignition or roasting; these processes can be dangerous, and fume collection and afterburning can be used to treat the fume and gases. Actually, new hydrometallurgical recycling proposals (hydrochloric acid in combination with hydrogen peroxide as leaching agent) are made to lower energy consumptions and process time as well as to recover rare earth metals (REMs , especially cerium (Ce)) simultaneously with PMs (Rumpold and Antrekowitsch, 2012)
- Organic-based homogenous spent catalysts: These catalysts from, e.g. chemical or pharmaceutical industries, can be treated by distillation and precipitation. The gaseous emissions are treated in an afterburner.

10.2 RARE EARTH METALS

REMs are a moderately abundant group of 17 elements comprising the 15 lanthanides (also referred to as lanthanoids), which comprise elements with atomic numbers 57–71; scandium; and yttrium. REMs can be classified as either light rare earth elements (LREE: 57 (lanthanum) through 64 (gadolinium)) or heavy rare earth elements (HREE: 65 (terbium) through 71 (lutetium)). Yttrium (atomic number 39) is included as an HREE even though it is not part of the lanthanide contraction series. Scandium (atomic number 21), a transition metal, is the lightest REM, but it is not classified as one of the group of LREE nor one of the HREE (Gambogi and Cordier, 2012).

Just as there are many different elements among the REMs, there are many different uses as well; the most important ones are as follows (European Commission-Enterprise and Industry, 2010):

Catalysts: Lanthanum (La) is used in catalytic cracking in oil refineries, and Ce is necessary in catalytic converters for cars.

Magnets: Neodymium (Nd)–iron–boron magnets are the strongest known permanent magnets, which are used in the context of electromobility and for wind power generators. Other REMs used in comparable applications are dysprosium (Dy), samarium (Sm), terbium (Tb) and praseodymium (Pr).

Polishing and glass: Ce oxide is a widely used polishing agent.

Battery alloys: Nickel metal hydride batteries containing La are the first choice for portable tools and are extensively used in hybrid vehicles.

Metallurgy: Ce, La and Nd are used to improve mechanic characteristics of alloyed steel, for desulfurization and to bind trace elements in stainless steel. Smaller shares are also used for magnesium and for aluminum alloys.

Other applications: These applications include the processing of phosphors and pigments and the manufacturing of capacitors and ceramics. A number of merging technologies rely on the properties of REMs, for example: The anodes of solid state fuel cells use either scandium (Sc) or yttrium (Y), which is also necessary for high-temperature superconductors and is used in lasers.

Regardless of the (quite dissipative) end use, REMs are not recycled in large quantities, but could be if recycling became mandated or very high prices of REMs made recycling feasible (Goonan, 2011).

10.2.1 Lanthanides

Lanthanides (La, Ce, Pr, Nd, Pm, Sm, Eu, Gd, Tb, Dy, Ho, Er, Tm, Yb, Lu): Lanthanide statistics are usually reported as rare earth elements (REEs) or rare earth oxides (REOs) equivalents; the REE-to-REO ratio for each of the lanthanides is about 1:0.85. The distribution of REO consumption by type is not homogeneous among market sectors (Goonan, 2011): lanthanides are used in mature markets (catalysts, glassmaking, lighting and metallurgy), which account for 59% of the total worldwide consumption, and in newer, high-growth markets (battery alloys, ceramics and permanent magnets), which account for 41% of the total worldwide consumption. In mature market segments, La and Ce constitute about 80% of REMs used, and in new market segments, Dy, Nd, and Pr account for about 85% of lanthanides used. The estimated 2012 distribution of REOs by end use was as follows (U.S. Geological Survey, 2013): catalysts, 62%; metallurgical applications and (battery) alloys, 13%; glass polishing and ceramics, 9%; permanent magnets, 7%; phosphors, 3%; and other, 6%. The world consumption of lanthanides grew rapidly in the last 25 years; in 2012 world mine production was estimated to be 110,000 Mt (Goonan, 2011; U.S. Geological Survey, 2013).

The recycling of REEs could be considered a very uncommon issue until today and focused mostly on small quantities of preconsumer NdFeB-permanent magnet scrap (U.S. Geological Survey, 2013) as well as postconsumer NiMH battery and phosphors scrap (Umicore, 2013b; News release, 2012). Actual statistics report EOL recycling rates below 1% and an average recycled content (fraction of secondary metal in the total metal input to metal production) between 1% and 10% for La, Ce, Pr, Nd, Gd and Dy rsp. below 1% for Sm, Eu, Tb, Ho, Er, Tm, Yb, Lu; for Pm no data are available (UNEP, 2011).

10.2.1.1 Permanent Magnet Scrap

Preconsumer permanent magnet scrap (swarf and residue) is estimated to represent 20–30% of the manufactured starting alloy (Schüler et al., 2011; U.S. Geological Survey, 2013). Corresponding neodymium–iron–boron ($Nd_2Fe_{14}B$) magnets mostly consist of approximately 65–70% Fe, 1% B, 30% mixture of Nd/Pr, <3% Dy and sometimes Tb.

In-plant recycling activities are reported from NEOMAX group (Hitachi Metals Ltd) including plants in Japan and other countries (Tanaka Oki et al., 2013) (see Figure 10.5):

Cutting sludge (metallic powder coated with oxides) is roasted to completely oxidize the

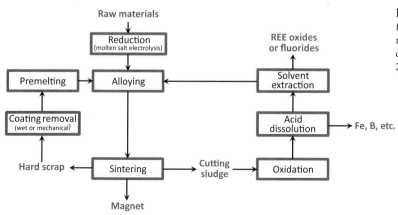

FIGURE 10.5 Simplified flowsheet for in-plant recycling of permanent magnet scraps ($Nd_2Fe_{14}B$ hard scrap, cutting sludge) (Tanaka Oki et al., 2013 and completions).

whole powder. The roasted product is dissolved in acid followed by solvent extraction, precipitation and calcination to obtain Nd and Dy as oxides or fluorides. Both REEs are recovered from these oxides or fluorides by conventional molten-salt electrolysis or thermal reduction process. Solid scrap is generally recovered as alloy via high-frequency vacuum melting. Nickel- or aluminum-chromate plating on the magnet surface will cause problems in magnetic properties of the Nd−Fe−B sintered magnets produced from solid scrap and thus should be removed in advance by wet or mechanical processes. However, these processes are not economical at present and therefore are seldom applied.

Further possible recycling methods, which are under development, are summarized in Table 10.1. Despite these and numerous other recycling proposals, there are no current commercial recycling activities worldwide for post-consumer permanent magnets (Schüler et al., 2011; Tanaka Oki et al., 2013).

10.2.1.2 NiMH Battery Scrap

The recycling of REMs from NiMH battery scrap is in its infancy. Mostly these battery types, consisting of approximately 36−42% Ni, 22−25% Fe, 8−10% REMs and 3−4% Co, are recycled with assumption of REMs lost in slags. In 2011, Umicore (Hoboken, Belgium) and Rhodia (La Rochelle, France) jointly developed a unique pyro-/hydrometallurgical process for REMs recycling from rechargeable battery scrap (News release, 2011, see also VAL'EAS™-process for Li-battery scrap), which represents best available technology standard nowadays (International Resource Panel, 2013): Battery modules are fed directly into an ultra high-temperature smelter without any pretreatment (except for the dismantling of large battery cases). Battery production scrap and slag forming agents are added as well to create three output fractions (see Figure 10.6):

- Metal alloy—Co, Ni, Cu, Fe
- Slag fraction—Al, Li, Mn, REM
- Gas emissions—flue dust (only fraction landfilled)

TABLE 10.1 Selected Recycling Methods for Permanent Magnet Scrap ($Nd_2Fe_{14}B$)

Recycling Methods	Process Characteristics	Comments
Hydrogen decrepitation (HD)	HD, jet milling, aligning and pressing, vacuum sintering, magnet remanufacturing (addition of 1% new Nd) (Zakotnik et al., 2009)	Pilot plant for magnets from disk drivers (Zakotnik et al., 2009); contaminated scraps need further refinery steps (Saguchi et al., 2006)
Whole leaching	Acid leaching and generation of Fe-chloride, Nd-oxide and B-acid (no further details) (News release, 2013)	Pilot plant Loser Chemie GmbH, Germany (News release, 2013)
Selective roasting—leaching	Oxidizing roasting, acid leaching of Nd with H_2SO_4 or HCl, precipitation ($Nd_2(SO_4)_3$) or solvent extraction of neodymium (Tanaka Oki et al., 2013) Chlorinating with NH_4Cl ($T = 350\ °C$), selective dissolving of $NdCl_3$ into water (Itoh et al., 2009)	Laboratory scales
Selective extraction in molten phases	Molten salt: $MgCl_2$ ($T = 1000\ °C$) (Shirayama and Okabe, 2008) Molten metal: Mg (Xu et al., 2000; Okabe et al., 2003); Ag (Takeda et al., 2004) Slag (oxides): melting with LIF-(REM)F_3-fluxes (Takeda et al., 2009)	Laboratory scales

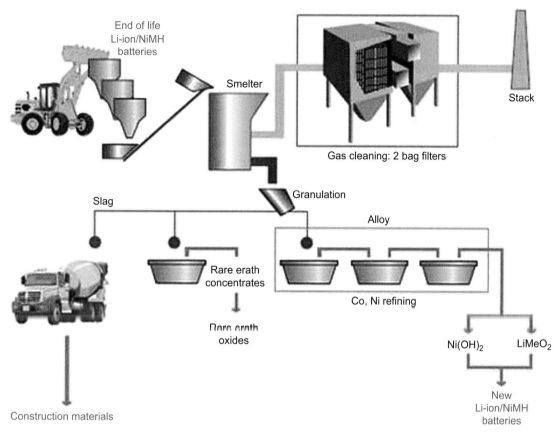

FIGURE 10.6 Flowsheet for Umicore/Rhodia's VAL'EAS™ battery scrap recycling process (Umicore, 2013b) (RE concentrates are hydrometallurgically processed by Rhodia).

The batteries themselves fuel the smelter as their combustible compounds heat the smelter to a high enough temperature that, in combination with a gas cleaning system, ensures no volatile organic compounds (VOCs) or dioxins are emitted. The alloy is further refined in an existing hydrometallurgical process to produce a variety of Co- and Ni-containing materials for use in new batteries and other applications. Recently a hydrometallurgical process was developed to extract a REM concentrate from the slag for further refining rare earth oxides by Rhodia (France) to recover these critical elements. The rest of the slag is valorized by use in construction materials, including Li, whose recovery is currently uneconomical. The recovery of REOs is a business secret; Rhodia patented these process steps together with corresponding waste categories. In different solvent extraction units, rare earths are separated comparable to conventional REMs winning processes for geogenic resources. In summary, the battery recycling process described exceeds the 50% recycling

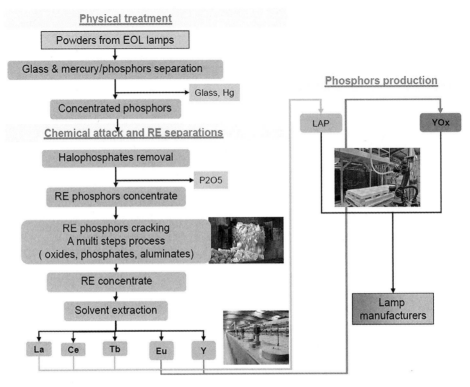

FIGURE 10.7 Flowsheet for Rhodia's planned phosphors recycling process (Rollat, 2012). EOL, end-of-life; LAP, (La, Ce, Tb) PO4; RE, rare earth metal.

efficiency standard imposed by the EU Batteries Directive.

10.2.1.3 Phosphors Scrap

Industrial recycling of REMs from phosphors scrap will start in 2013 by Rhodia (France) (News release, 2012) (see Figure 10.7): EOL energy saving and fluorescent lamps will be collected and pretreated by different recycling companies, which will separate the various components of glass, plastics and metal (with Hg). In two facilities of Rhodia (Saint-Fons and La Rochelle), which is part of the Solvay Group (Belgium) since 2011, separated phosphor powders will be processed to recover La, Ce, Tb, Y, Eu and Gd. No process details are known, but it is reported that the concentrated REMs should be extracted and then recycled, preserving 100% of their original properties. Finally, the REMs will be successively separated and treated to be reused as luminescent materials in the manufacture of new bulbs.

10.2.2 Scandium

The principal use for scandium (Sc) is in aluminum alloys for aerospace components and sporting equipment (baseball and softball bats). Other uses for scandium include analytical standards, electronics, high-intensity metal halide lamps, lasers, metallurgical research and oil-well tracers. Demand for scandium increased slightly in 2012. Global scandium consumption was estimated to be less than 10 Mt. No scandium was mined in the

United States in 2012; foreign mine production data were not available. There are no recycling activities of scandium reported (Gambogi and Cordier, 2012; U.S. Geological Survey, 2013). Because of the absence of data, EOL recycling rates are estimated below 1% (UNEP, 2011).

10.2.3 Yttrium

Yttrium (Y) is consumed mainly in the form of high-purity oxide compounds for phosphors (Gambogi and Cordier, 2012). Smaller amounts are used in ceramics, electronic devices, lasers and metallurgical applications. Principal uses are in phosphors for color televisions and computer monitors, temperature sensors, trichromatic fluorescent lights and X-ray-intensifying screens. Yttria-stabilized zirconia is used in alumina—zirconia abrasives, bearings and seals, high temperature refractories for continuous-casting nozzles, jet-engine coatings, oxygen sensors in automobile engines, simulant gemstones, and wear-resistant and corrosion-microwave radar to control high-frequency signals. Yttrium is an important component in YAl—garnet laser crystals used in dental and medical surgical procedures, digital communications, distance and temperature sensing, industrial cutting and welding, nonlinear optics, photochemistry and photoluminescence. Yttrium also is used in heating-element alloys, high-temperature superconductors and superalloys. The approximate distribution in 2012 by end use was as follows (U.S. Geological Survey, 2013): phosphors, 44%; metallurgical, 13%; and other, 43%.

The world consumption of yttrium, associated with most rare earth deposits, grew rapidly in the last 25 years; in 2012 world mine production was estimated to be 8900 Mt (U.S. Geological Survey, 2013). In the United States only small quantities, primarily from laser crystals and synthetic garnets, are recycled (U.S. Geological Survey, 2013). Actual statistics report no recycling activities (UNEP, 2011).

10.3 ELECTRONIC METALS

The group of electronic metals has not yet been officially defined. In the context of the present document, it will include the following metals: gallium (Ga), In and Te. All are used as key metals in numerous electronic devices and were classified as "critical raw materials for the EU" by the European Commission in 2010 (see European Commission-Enterprise and Industry, 2010).

10.3.1 Gallium

With a share of above 99%, most gallium globally consumed is used as a compound with arsenic (GaAs) for optoelectronic devices (laser diodes, LEDs, photodetectors, solar cells) and integrated circuits (defense applications, high-performance computers, telecommunications) or with nitrogen (GaN) for optoelectronic devices (laser diodes, LEDs). Integrated circuits account for 71% of domestic consumption, optoelectronics for the remaining 29%. Owing to the strong growth of LEDs, laser diodes, smartphones, photodetectors and solar cells, Ga consumption increased rapidly in the last 10 years; in 2012 world primary gallium production was estimated to be 273 Mt (European Commission-Enterprise and Industry, 2010; Jaskula, 2012b; U.S. Geological Survey, 2013).

Recycling activities for gallium in Canada, Germany, Japan, the United Kingdom and the United States are mainly focused on new scrap generated in the manufacture of GaAs-based devices. It was estimated that 50% of gallium consumed worldwide in 2010 came from recycled sources (Jaskula, 2012b); according to UNEP the EOL-recycling rate is very small (<1%) but the recycled content, the fraction of secondary metal in the total metal input to metal production, reaches values between 25% and 50% (UNEP, 2011). Procedural hydrometallurgical routes are usual for new scrap, e.g. via dissolution of crushed GaAs residues in sodium

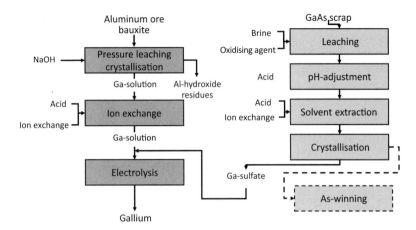

FIGURE 10.8 Simplified flowsheet of potential routes for gallium recovery.

hydroxide solution with hydrogen peroxide (Figure 10.8). According to the abstracts of a Taiwanese patent (TW 2003-92132798, November 20, 2003), the process conditions for the recycling are adjusted so that gallium is enriched as a complex gallium sulfate. It is conceivable that this intermediate product can be introduced into the conventional electrolysis for Ga recovery. The process details are, however, mostly a business secret.

10.3.2 Indium

Production of indium tin oxide (ITO) for flat-panel display devices (74%) and for architectural glass (10%) is the leading global end use of indium, followed by solders (10%) for temperature indicators in fire-control systems; minor alloys (3%) used for surface coatings of optical lenses, bonding agents between nonmetallic materials and for dental- resp. white gold alloys; other applications (3%) include, for example, intermetallic compounds that are used as semiconductors for laser diodes or indium in nuclear reactor control rods. World refinery production of indium has increased more than 5-fold in the last 20 years, when world production amounted to 70–120 t/year compared to today's 670 t/year (European Commission-Enterprise and Industry, 2010; Elsner et al., 2010; Tolcin, 2012).

Recycling possibilities for indium are limited. Indeed the EOL recycling rate is very small (<1%) but the recycled content, the fraction of secondary metal in the total metal input to metal production, reaches values between 25% and 50% (UNEP, 2011). A very large portion of global secondary indium was produced from pure (production) scrap as sputter targets from ITO thin film deposition, which occurs obviously in oxidized form and represents a loss share of more than 70% of deposition material input. Appropriate recycling activities are concentrated on countries like China, Japan and the Republic of Korea, where ITO production and sputtering takes place. They consist of multistage (thermo-)physical/hydrometallurgical processes, with e.g. crushing, leaching, precipitation, cementation, filtration, solvent extraction and/or electrolytic refining units. Indium can also be recovered from copper indium gallium diselenide solar cells (CIGS) to be used in the manufacture of new CIGS solar cells or may be reclaimed directly from old liquid crystal display (LCD) panels. The panels are crushed to millimeter-sized particles and then soaked in an acid solution to dissolve the ITO from which the indium is recovered. Indium recovery from tailings was thought to have been insignificant, as these wastes contain small percentages of the metal and can be difficult to process. However, improvements to the process technology have made indium recovery from tailings feasible when the price of indium is high (Alfantazi and Moskalyk, 2003; Kang et al., 2011; Barakat, 1998; Bihlmaier and Völker, 2011; Tolein, 2012; Merkel and Friedrich, 2010).

At Umicore Precious Metals Refining, Hoboken, Belgium, indium is recovered from electronic scrap (crushed flat-panel displays, solders, etc.) and residues from historical zinc refinery residues among other numerous rare metals (Figure 10.4).

Indium compounds that are charged to the (ISA-)smelter will account to the lead-bearing slag that is subsequently reduced in the lead blast furnace to indium-containing lead bullion. During lead refining it is oxidized selectively into the lead refinery slag via the Harris process, in which a Na_2NO_3–$NaOH$ melt is penetrated by the impurity-containing lead. The product is a salt slag with the oxidized metals. No information is available which special metals refinery process is used to recover indium metal at Umicore, but most likely it is reduced from a leach liquor of the lead refinery slag via cementation, solvent extraction or electrolysis.

10.3.3 Tellurium

Tellurium is increasingly used in cadmium–tellurium-based solar cells (40%). In thermoelectrics (30%), e.g. semiconducting, bismuth telluride is used in cooling devices. In metallurgy (15%), tellurium serve as a free-machining additive in steel, is used to improve machinability while not reducing conductivity in copper, to improve resistance to vibration and fatigue in lead, to help control the depth of chill in cast iron and in malleable iron as a carbide stabilizer. In rubber formulation (5%), tellurium is used as a vulcanizing agent and as an accelerator. Other applications (10%) include the use in catalysts for synthetic fiber production, in blasting caps and as a pigment to produce blue and brown colors in ceramics and glass. World refinery consumption of tellurium was estimated to be about 500–550 t/year in 2011 (European Commission-Enterprise and Industry, 2010; George, 2012).

Tellurium recycling is still embryonic but growing steadily (<10% of supply); recovery of industrial scrap from the photovoltaic (PV) industry provides a growing stream of secondary tellurium expected to represent about 7% of total tellurium in 2010, decreasing though over time as deposition processes for photovoltaics become more efficient and its growth is leveling off (European Commission-Enterprise and Industry, 2010). A plant in the United States recycled tellurium from cadmium–tellurium-based solar cells; however, most of this was new scrap because cadmium–tellurium-based solar cells were relatively new and had not reached the end of their useful life (U.S. Geological Survey, 2013). Different promising recycling methods for (new) PV scrap, with tellurium contents below 1%, are under development (Table 10.2).

For traditional uses, there is little or no old scrap from which to extract secondary tellurium, because these uses of tellurium were nearly all dissipative. A very small amount of tellurium was recovered from scrapped selenium–tellurium photoreceptors employed in older plain paper copiers in Europe. The global EOL recycling rate of tellurium was estimated to be very small (<1%, (UNEP, 2011)). For tellurium recycling from electronic scrap, e.g. a combined pyro/hydro-metallurgical process alternative

TABLE 10.2 Selected Recycling Methods for Photovoltaic Scrap (Cd/Te-Based Solar Cells)

Recycling Method	Process Steps	Document
Hydrometallurgical: sulfuric acid with hydrogen peroxide	Crushing, leaching, ion exchange separation, precipitation (Te), electrolysis (Cd)	US patent: US 2006/0275191 A1 (December 7, 2006)
Hydrometallurgical: acidic/alkaline	Crushing, oxidative acidic leaching with sulfuric acid and hydrogen peroxide, neutralization and filtration, alkaline residue leaching with electrolysis (Te)	US patents: US 6129779 (October 10, 2000) US 6391165 B1 (May 21, 2002)
Hydrometallurgical: nitric acid	Crushing, leaching, electrolysis (Te), "decomposing" of Cd to CdO	US patent: US 5897685 (April 27, 1999)
Pyrometallurgical: solid–gas reactions	Crushing, pyrolysis, hot chlorination and condensation (Cd, Te)	EU patent: EP 1 187 224 A1 (March 13, 2002)

via copper-lead route is working commercially (cf. Umicore's flowsheet in Figure 10.4).

10.4 REFRACTORY METALS (FERRO-ALLOYS METALS, SPECIALTY METALS)

Refractory metals (RMs) are high melting point metals that are characterized by other special physical and chemical properties, such as high density, inertness, corrosion and acid resistance. They are produced both as metal ingot (buttons) using electron beam furnaces and as metal powder that serves as raw material for powder metallurgical treatments like pressing and sintering. The definition of which elements belong to this group differs. As defined at the EU level (European IPPC Bureau, 2013), this group comprises 11 metals, the elements of the fourth to the seventh transition group of the periodic table. Due to their main applications (see U.S. Geological Survey, 2013; European IPPC Bureau, 2013; European Commission-Enterprise and Industry, 2010) they can be subdivided as:

- *Ferroalloy metals (RMs for steel production):* chromium (Cr), manganese (Mn), molybdenum (Mo), niobium (Nb), vanadium (V), and
- *Specialty metals (RMs for special applications):* hafnium (Hf), tantalum (Ta), titanium (Ti), rhenium (Re), tungsten (W), and zirconium (Zr).

Ferro-alloys: Ferro-alloys are mainly used as master alloys in the iron, foundry and steel industry, because it is the most economical way to introduce an alloying element into the steel melt. Besides this, special ferroalloys are also needed for the production of aluminum alloys and as a starting material in specific chemical reactions. As an additive in steel production, ferroalloys improve steel's properties, especially tensile strength, wear and corrosion resistance. In 2012, more than 90% of the produced Cr, Mn, Mo and V (mine productions: 24 Mt Cr, 16 Mt Mn, 250,000 t Mo, 73,000 t V) as well as about 80% of the produced Nb (mine production: 69,000 t) were used in this sector.

Superalloys and special alloys: In 2012 approximately 70% of Re (mine production: 52 t) was used as an important component in high-temperature superalloys for blades in turbine engines, in thermocouples and for electrical contacts that stand up well to electric arcs.

Hard metals and metal carbide powder that can further be treated by powder metallurgical methods to produce hard metal tools are, with a proportion of 50%, the main application fields for W (in 2012 mine production: 73,000 t).

Catalysts: Up to 25% of Re is used in petroleum-reforming Pt/Re catalysts for producing lead-free gasoline.

Aerospace applications: In 2012 an estimated 72% of Ti (in 2012 sponge production: 190,000 t) was used for high-performance aircraft engines and airframes.

Pigments: With 60% of the main use for Ti in dioxide form in white pigments that are nontoxic, it therefore is useful in many applications like cosmetics, food industry and paint (in 2012 sponge production: 190,000 t).

Capacitors: About 60% of total Ta consumption (in 2010 mine production: 765 t/year) is used in the form of metal powder for electrolytic capacitors, which are basic components of modern IT and telecommunication devices (mobiles, notebooks, digital cameras).

Ceramic and refractories: Ceramics and refractories are the main application fields for Zr (in 2012 mine production: 1420 t).

Nuclear energy and chemical process industries: Nuclear energy and chemical process industries are the leading consumers of Hf (in 2012 mine production: not available).

Because of the large number of available secondary raw materials, especially metal oxides from the production of stainless steel (dusts), the recovery of ferroalloys, mainly ferrochrome, has become an important part of the ferroalloy industry (European IPPC Bureau, 2013). But the recovery of RMs is also limited, especially for some described specialty metals, as demonstrated by reported EOL recycling rates (UNEP, 2011):

Ferroalloy metals (RMs for steel production)

>50%	For chromium, manganese and niobium.
>25–50%	For molybdenum.
<1%	For vanadium.

Specialty metals (RMs for special applications)

>50%	For rhenium and titanium.
>10–25%	For tungsten.
<1%	For hafnium, tantalum and zirconium.

10.4.1 Spent Petroleum Catalysts (Molybdenum, Vanadium)

Spent petroleum catalysts are regarded as the most important catalysts for recycling of Mo and V owing to the large volume and value of metals they contain. These also named hydroprocessing catalysts, account for about one-third of total worldwide catalyst consumption and are widely used in the petroleum refining industry for mild hydrogenation and removal of heteroatoms such as sulfur, nitrogen and oxygen, as well as metals like nickel and vanadium. Typically they contain: 2–10% Mo, 0–13% V, 0.5–4% Co, 0.5–10% Ni, 10% S, 10% C on a porous Al_2O_3 support (<30% Al).

The recovery of molybdenum and of the vanadium content can involve the following process steps (European IPPC Bureau, 2013):

1. Thermal pretreatment with initial heating in air at 600 °C (roasting) to remove the residual sulfur, carbon and hydrocarbons and to oxidize the metals to soluble molybdate and vanadate rsp. pretreatment with organic solvents for S, C, C_xH_x removal.

2. A (pressure) leaching step resulting in preferential solubilization of molybdate and vanadate, leaving the nickel cobalt alumina as a solid.

3. Separation of the molybdenum and vanadium.
4. Treatment of the Ni/Co alumina residue to recover the nickel and cobalt content.

The technical and economic effort depends on the type of catalyst (metal content, type of carrier, chemical compounds, impurities, etc.) and quality requirements for the recycled products. Table 10.3 gives an overview of established reclamation facilities in the world for spent hydroprocessing catalysts.

10.4.2 Steelwork Slags (Vanadium)

Besides the fact that vanadium recycled from spent petroleum catalysts is significant (U.S. Geological Survey, 2013), small V contents of recycled steel and ferroalloys are lost to slag during processing eletric arc furnace (EAF) smelting and are not recovered. Because of high mass flow rates and the wide range of applications (V use in steel industry in 2011: 93% of domestic V consumption), steel recycling thus represents the world's largest source of V loss from the vanadium raw materials cycle. Mostly, the generated steelwork slags with 1% V content are used for road and dike construction. Experimental investigations with pyrometallurgical methods (reduction melting followed by aluminothermic slag reduction) were not successful: economic production of a V-rich ferroalloy from corresponding slags seems not possible (Antrekowitsch et al., 2009). Actually the IME (Process Metallurgy and Metal Recycling—Department and Chair

TABLE 10.3 Characteristics of Selected Recycling Processes for spent Hydroprocessing Catalysts (Marafi et al., 2010; Scheel, 2010)

Process (country of business)	Pretreatment	Leaching and Separation Steps	(Intermediate) Products
Gulf GCMC (USA, Canada)	Roasting with Na_2CO_3	Water leaching and precipitation, solvent extraction	MoO_3, V_2O_5, solids for ferroalloy production
CRI-Met (USA)	None	1. Pressure leaching with $NaAlO_2$ and air injection (catalysts) 2. Pressure leaching with $NaOH$/Na_2CO_3 (residues from 1.) and precipitation	MoO_3, V_2O_5, Ni/Co concentrate, $Al(OH)_3$
EURECAT (France)	Roasting (500 °C) with NaOH	Water leaching and ion exchange, solvent extraction, electrolysis	MoO_3, $VOSO_4$, $(NH_4)_2$-/Na_2-MoO_4, Ni-, Co-metal
Taiyo Koko (Japan)	Oxidizing roasting (850 °C) with Na_2CO_3	Wet grinding in a ball mill and precipitation, ion exchange	MoO_3, V_2O_5, solids for ferroalloy production
Full yield (Taiwan)	Catalysts mixing, degreasing	Na_2CO_3 leaching with H_2O_2 and precipitation, ion exchange	MoO_3, V_2O_5, SiO_2 and Al_2O_3 for building industry
AURAMET	Roasting, calcination (1100 °C)	H_2SO_4 leaching and solvent extraction	NH_4VO_3, Co/Ni sulfate solution, Al_2O_3

Moxba-Metrex (The Netherlands), Quanzhuo Jing-Tai Industry (China), Metallurg Vanadium (USA, UK, Germany), H. C. Starck (Germany), Nippon Catalysts Cycle Co. (Japan).

of RWTH Aachen University, Germany) is developing a recycling proposal for the V extraction from steelwork slags. First results can be summarized as follows:

- Hydrometallurgical methods provide ecological/economic advantages compared to pyrometallurgical.
- V content of the slag is dissolved by direct alkaline pressure leaching under oxidizing conditions.
- V can be recovered as ammonium vanadate from the leach liquor by conventional neutralization and precipitation steps.

10.4.3 Tantalum Scrap

Tantalum secondary raw materials are mostly preconsumer scrap that was generated during the manufacture of Ta-containing electronic components as well as from cemented carbide and superalloy scrap. Figure 10.9 demonstrates the considerable amounts of those internal material circuits comparing to external flows (EOL scrap). The unoxidized tantalum scrap (e.g. ingot scrap, sintered parts) can be remelted in an electron beam furnace or treated by dehydrogenation in a vacuum furnace to produce tantalum powder. The second type of scrap represents the oxidized tantalum and fine-grained unoxidized tantalum scrap. The last one has to be oxidized by roasting, before treating with nitric or hydrochloric acid. Both types of scrap result in a residue that contains oxidized tantalum (European IPPC Bureau, 2013).

The biggest handicap for increasing the very low EOL recycling rate for tantalum ($<1\%$) lies in the main application field of Ta in the form of capacitors ($>60\%$), which are in fact not recycled (Gille and Meier, 2012). In this context it has to be noted that Ta in WEEE reaches only trace amounts (<200 ppm) and is lost by slagging and dispersing during conventional WEEE recycling in pyrometallurgical copper process routes.

10.4.4 Titanium Scrap

Growing titanium production has also increased the availability of *titanium secondary*

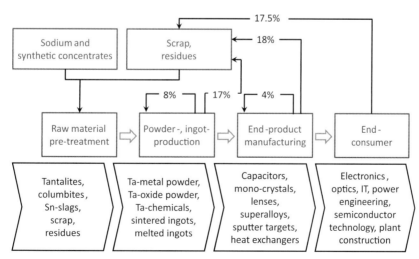

FIGURE 10.9 Internal and external material flow for tantalum recycling (see Gille and Meier, 2012).

raw materials. Especially within the main application field of Ti, the aerospace sector, there is a high level of scrap generation during the production of final-use parts (>80% of input material). In generally, the recirculation of titanium alloys focuses selected and classified scrap, whereas the contaminated and inhomogeneous scrap is mostly downgraded to the ferrotitanium production line; untreated titanium scrap can also be used directly as an additive to steel, nickel, copper, aluminum or other metals.

Clean and sorted scrap is usually introduced into the remelting step of the primary route to produce titanium ingots: while loose scrap can be used for cold hearth melting without further preparation, vacuum arc remelting in specially designed furnaces mix the scrap with titanium sponge and compress it to electrodes. Batches of titanium (mostly preconsumer) scrap and titanium sponge are mixed and pressed to form blocks. The blocks are welded together to produce a consumable electrode. The electrode is then installed in the furnace chamber in a manner where a cooled copper crucible that collects the molten titanium encloses the bottom end of the electrode. An arc is struck between the lower end of the electrode and the bottom of the crucible and the electrode is moved downward as it is consumed (European IPPC Bureau, 2013). Nonetheless, those conventional recycling routes exhibit a very limited refining potential with regard to oxygen contamination.

At IME (Process Metallurgy and Metal Recycling—Department and Chair of RWTH Aachen University, Germany), intensive research was carried out to develop and assess a closed loop recycling process for titaniumaluminide scrap (γ-TiAl), which is currently downgraded as a deoxidation agent in steel production. Those alloys receive special attention of the technical community owing to their applications in automotive turbochargers and the low-pressure turbines of the most recent

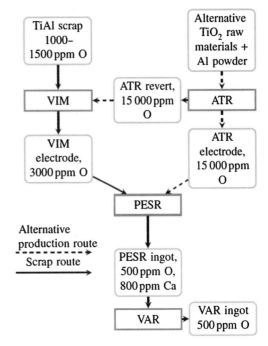

FIGURE 10.10 Integrated concept for alternative production and recycling of γ-TiAl developed at IME, Aachen (Reitz et al., 2011). ATR, aluminothermic reduction; VIM, vacuum induction melting; PESR, pressure electroslag remelting; VAR, vacuum arc remelting.

aero engines. The IME Recycling Process fits their sensitive metallurgical requirements and reduces material processing cost significantly. It comprises vacuum induction melting (VIM) or aluminothermic reduction (ATR) followed by pressure electroslag remelting (PESR) and/or vacuum arc remelting (VAR) (Figure 10.10).

The application of PESR using "active slags" leads to a refinement and in particular to a reduction in oxygen content, which impacts greatly on the mechanical properties. γ-TiAl-scrap remelted by VIM with an oxygen contamination of 3000 ppm could be successfully treated to levels below 500 ppm. In contrast, γ-TiAl obtained through an alternative raw-material route via ATR, and therefore

TABLE 10.4 Ranges of Metal Contents in Different Li Ion Battery Scrap

Li-Ion System (wt%)	Li	Ni	Co	Mn	Cu	Al	Fe	C
Battery cells	~1	7–21[1]			~16	~40	<0.5	~14
Electric vehicle (EV) batteries	<0.5	0–4	0–4	0–4	~11	~25	~21	~9

[1] Sum of Ni, Co and Mn.

contaminated with up to 16,000 ppm oxygen, presents kinetic and technological challenges to the process and could only be treated down to 4000 ppm oxygen. In a techno-economical analysis, the interesting economical potential of the recycling route VIM−PESR−VAR, a triple-melt process that ideally combines flexible scrap melting, chemical deoxidation and final refining, could be highlighted in comparison to alternative processes (Reitz, 2013). This process is currently being transferred to the major Ti alloy $TiAl_6V_4$.

10.5 OTHER METALS

10.5.1 Lithium

Lithium is used in battery production (22–33%), in the glass/ceramic industry (26–30%), in the production of lubricating greases (11–14%), for air treatment (4–5%), as an additive in continuous-casting processes (4%), in molten-salt electrolysis for primary aluminum production (2–4%) and other

TABLE 10.5 Characteristics of Selected Recycling Processes for Li Ion Battery Scrap (Luidold and Antrekowitsch, 2010; Georgi-Maschler, 2011; Friedrich et al., 2012; Elwert et al., 2012; Jaskula, 2012a and Additions)

Company (Place of Business)	Principal Capacity	Characteristics	Comments
TOXCO, Inc. (British Columbia, Canada; Lancaster, Ohio, USA)	Hydro <4000 t/year	Cryogenic process: low-temperature dismantling	+ Recovery of all compounds + High flexibility (all battery types) − Complexity of process
Umicore, S.A. (Hoboken, Belgium; Hofors, Sweden) VAL'EAS™	Combined <500 t/year announced >5000 t/year	Direct smelting (shaft furnace) with subsequent hydrometallurgy (Li recovery from slag in development)	+ Economic process + Also for NiMH batteries − No recovery of Li, Al, electrolyte, graphite, plastic (Only rec. values for Li/Co-oxide)
ACCUREC recycling GmbH (Mühlheim, Germany) ACCUREC-IME	Pyro <300 t/year (mech. pretreatment)	Vacuum pyrolysis (full process with EAF smelting in development)	+ Recovery of Li_2O concentrate + Early separation of valuable components + Co, Ni, Mn as metal alloy + High flexibility − only pilot scale
Xstrata, Ni Corp. (Falconbridge, Ontario, Canada)	Combined (integrated) >5000 t/year	Conditioning (rotary kiln) and introducing into a Co-/Ni winning process (EAF) with subsequent hydrometallurgy	+ Economic process integration + Ni, Co in metallic form − Low recycling efficiency − No recovery of Li, Al, electrolyte, graphite, plastics

Inmetco Inc. (Ellwood City, USA): commercial plant.
Dowa Eco-System Co. Ltd (Japan): >1000 t/year commercial plant.
JX Nippon Mining & Metals Co. (Tsuruga, Japan): commercial plant.
BATREC Ind. AG (Wimmis, Switzerland): <300 t/year only pilot plant for pretreatment.
RECUPYL S.A. (Grenoble, France): <300 t/year only pilot plant for pretreatment.

applications (9–13%). Other applications are, for example, the use of Al–Li alloys in airplane construction or the medical application of lithium to treat depression. Lithium end-use markets have all increased (5.6% per year between 2000 and 2011), and world consumption was estimated to be approximately 26,000 Mt in 2011 (Jaskula, 2012a).

The widespread and constantly increasing use of Li ion batteries in the last two decades leads to an increased battery scrap generation, which is the only one of interest for lithium recycling up to now. Li ion batteries contain high amounts of valuable metals, but since all battery producers sell their own specific types it is difficult to specify exact metal contents in battery scrap mixtures (Table 10.4). Cobalt especially has a strong influence on the economic efficiency of a suitable recycling process for (small and mid-size) battery cells as well as (large-size) electric vehicle battery systems.

Various battery recycling projects under development can basically be divided into pyrometallurgical, hydrometallurgical and hybrid (combined) processes. Besides utilization of specialized battery recycling processes, the addition of spent batteries to existing large-scale processes, which are not dedicated to battery recycling (e.g. extractive cobalt or nickel metallurgy), is common practice and very often an economical advantage (Table 10.5).

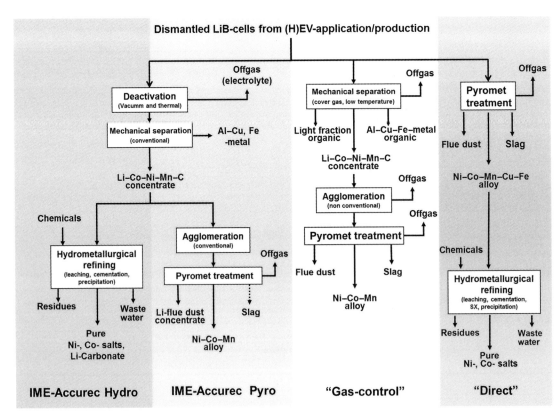

FIGURE 10.11 Potential recycling routes for electric vehicle battery scrap (Friedrich et al., 2011). (H)EV, (hybrid) electric vehicle.

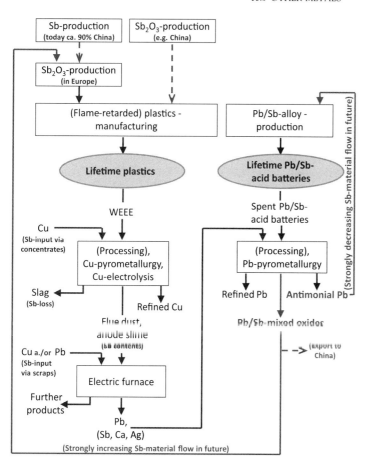

FIGURE 10.12 Antimony circulation in Europe (today and in future). WEEE, waste of electrical and electronic equipment.

Simplified flow charts of the potential recycling routes for (hybrid) electric vehicle battery scrap ((H)EV) are illustrated in Figure 10.11. Until now, only the "direct" route is realized on an industrial scale. In summary, Li ion battery recycling is in its infancy, and only a little EOL recycling in any form is occurring today (EOL recycling rate: <1% (UNEP, 2011)). Increased Li prices could change the current recycling schemes to focus more on the Li content.

10.5.2 Antimony

Most antimony is used in form of trioxide (Sb_2O_3), mainly to enhance the flame-retardant properties of plastics, rubber, textiles and other combustibles (72–75%) as well as a decolorizing and refining agent in the manufacture of glass and ceramics (9%), and furthermore as alloying element for grid metal in lead—acid battery production (19%) (European Commission-Enterprise and Industry, 2010; Angerer et al., 2009). Other applications for lead—antimony alloys are, for example, ammunition, antifriction bearings, cable sheaths, corrosion-resistant pumps and pipes, roof sheet solder and tank lining (Carlin, 2012). In the United States, the three main categories of consumption (metal products, nonmetal products, flame retardants) increased by 15% between 2010 and 2011; the use in flame retardants is expected to remain the principal global use for antimony, while the battery sector loses importance due to new battery technologies (Carlin, 2012; European Commission-Enterprise and Industry, 2010). World mine production was estimated to be approximately 178,000 Mt of antimony in 2011 (Carlin, 2012).

Traditionally antimony is recycled from lead—acid battery scrap, alloy scrap or WEEE and is recovered as antimonial lead to be again consumed by the battery industry with EOL recycling rates between 10 and 25% (European Commission-Enterprise and Industry, 2010; UNEP, 2011): In a first smelting/reduction step the scrap is charged into blast, reverberatory or rotary furnaces where antimony is dissolved in an impure lead bullion or a lead alloy. During subsequent refining to pure lead (mostly in a pyrometallurgical selective oxidation step) it is selected as low-quality Pb/Sb mixed oxide (see Figure 10.12). Changing trends in lead—acid battery production (calcium additive instead of antimony) have generally reduced the amount of secondary antimony:

1. The increasing use in flame retardants is a dissipative application without possible recycling.
2. The availability of recyclable EOL scrap (spent lead—acid batteries with Sb) is decreasing.
3. The traditionally incoming low-quality Pb/Sb mixed oxide is not suitable as an ingredient for flame retardants.

This trend will continue in future and thus the recycling rate will decline.

References

Alfantazi, A.M., Moskalyk, R.R., 2003. Processing of indium: a review. Minerals Engineering 16, 687—694.
Angerer, G., et al., 2009. Rohstoffe für Zukunftstechnologien. IZT/ISI Studie im Auftrag des BMWi. http://www.isi.fraunhofer.de/isi-media/docs/n/de/publikationen/Schlussbericht_lang_20090515_final.pdf.
Antrekowitsch, H., et al., 2009. Experimental investigations on V-containing steelwork slags with a low V_2O_5 content. Berg- und Hüttenmännische Monatshefte 154, 328—333.
Barakat, M.A., 1998. Recovery of lead, tin and indium from alloy wire scrap. Hydrometallurgy 49, 63—73.
Bartz, W., Wippler, E. (Eds.), 2005. Autoabgaskatalysatoren. Expert Verlag, Renningen.
Bihlmaier, A., Völker, M., 2011. Thermisch-mechanische Anreicherung von Indiumzinnoxid aus Displayeinheiten gebrauchter Flachbildschirme. World of Metallurgy — ERZMETALL 64, 79—83.
Carlin, J.F., 2012. USGS, 2011 Minerals Yearbook-Antimony. http://minerals.usgs.gov/minerals/pubs/commodity/antimony/myb1-2011-antim.pdf.
Elsner, H., Melcher, F., Schwarz-Schampera, U., Buchholz, P., 2010. Commodity Top News Nr. 33: Elektronikmetalle-zukünftig steigender Bedarf bei unzureichender Versorgungslage? Bundesanstalt für Geowissenschaften und Rohstoffe.
Elwert, T., Goldmann, D., Schirmer, T., Strauß, K., 2012. Recycling von Li-Ionen Traktionsbatterien-Das Projekt LiBRi. In: Thomé-Kozmiensky, K.J., Goldmann, D. (Eds.), Recycling und Rohstoffe-Band 5. TK Verlag, Neuruppin, pp. 679—690. ISBN: 978-3-935317-81-8.
European Commission-Enterprise and Industry, 2010. Annex V to the Report of the Ad-hoc Working Group on Defining Critical Raw Materials. http://ec.europa.eu/enterprise/policies/raw-materials/files/docs/annex-v_en.pdf.
European IPPC Bureau, 2013. Best Available Techniques (BAT) Reference Document for the Non-ferrous Metal Industries-Draft 3 (February 2013). http://eippcb.jrc.es/reference/BREF/NFMbw_17_04-03-2013.pdf.
Friedrich, B., Vest, M., Wang, H., Weyhe, R., 2011. Processing of Li-based Electric Vehicle Batteries for Maximized Recycling Efficiency. In: Proc. of 16th ICBR September 21—23rd, 2011, Venice, Italy.
Friedrich, B., Träger, T., Weyhe, R., 2012. Recyclingtechnologien am Beispiel Batterien. In: Pinnekamp, J. (Ed.), Die Recycling-Kette — Erfassung, Aufbereitung und Rohstoffrückgewinnung, Band 38. ARA, pp. 10\1—10\12, ISBN: 978-3-938996-93-5.
Fthenakis, et al., 2006. System and method for separating tellurium from cadmium waste. US Patent, Pub. No. US 2006/0275191 A1; December 7, 2006.
Gambogi, J., Cordier, D.J., 2012. USGS, 2010 Minerals Yearbook — Rare Earths. http://minerals.usgs.gov/minerals/pubs/commodity/rare_earths/myb1-2010-raree.pdf.
George, M.W., 2012. USGS, 2011 Minerals Yearbook — Selenium and Tellurium. http://minerals.usgs.gov/minerals/pubs/commodity/selenium/myb1-2011-selen.pdf.

References

Georgi-Maschler, T., 2011. Entwicklung eines Recyclingverfahrens für portable Li-Ion Gerätebatterien (Dissertation am IME der RWTH Aachen). Shaker-Verlag.

Gille, G., Meier, A., 2012. Recycling von Refraktärmetallen. In: Thomé-Kozmiensky, K.J., Goldmann, D. (Eds.), Recycling und Rohstoffe-Band 5. TK Verlag, Neuruppin, pp. 537–560. ISBN: 978-3-935317-81-8.

Goonan, T.B., 2011. Rare Earth Elements – End Use and Recyclability. USGS. http://pubs.usgs.gov/sir/2011/5094/pdf/sir2011-5094.pdf.

Hagelüken, C., 1996. Edelmetalleinsatz und – Recycling in der Katalysatortechnik. World of Metallurgy – ERZMETALL 49, 122–133.

Hagelüken, C., 2006. Recycling of electronic scrap at Umicore's integrated metals smelter and refinery. World of Metallurgy – ERZMETALL 59, 152–161.

Hagelüken, C., 2009. Recycling of precious metals – current status, challenges, developments. In: Proc. of EMC 2009. GDMB, pp. 473–486. ISBN: 978-3-940276-17-9.

Hagelüken, C., Meskers, C.E., 2010. Complex life cycles of precious and special metals. In: Graedel, T.E., et al. (Eds.), Linkages of Sustainability – Strüngmann Forum Report. MIT Press, pp. 163–197. ISBN-10: 0-262-01358-4.

International Resource Panel, Reuter, M., et al., 2013. Metal Recycling: Opportunities, Limits, Infrastructure. UNEP http://www.unep.org/resourcepanel/Publications/MetalRecycling/tabid/106143/Default.aspx.

Itoh, M., Miura, K., Machida, K.-I., 2009. Novel rare earth recovery process on Nd–Fe–B magnet scrap by selective chlorination using NH$_4$Cl. Journal of Alloys and Compounds 477, 484–487.

Jaskula, B.W., 2012a. USGS, 2011 Minerals Yearbook – Lithium. http://minerals.usgs.gov/minerals/pubs/commodity/lithium/myb1-2011-lithi.pdf.

Jaskula, B.W., 2012b. USGS, 2011 Minerals Yearbook – Gallium. http://minerals.usgs.gov/minerals/pubs/commodity/gallium/myb1-2010-galli.pdf.

Kang, H.N., Lee, J.Y., Kim, J.Y., 2011. Recovery of indium from etching waste by solvent extraction and electrolytic refining. Hydrometallurgy 110, 120–127.

Kralik, J., Schapp, J., Voß, S., 2011. Recycling of precious metals at Heraeus: a sustainable contribution to resource efficiency. In: Proc. of EMC 2011. GDMB, pp. 1077–1086. ISBN: 978-3-940276-38-4.

Loferski, P.J., 2012. USGS, 2011 Minerals Yearbook – Lithium. http://minerals.usgs.gov/minerals/pubs/commodity/platinum/myb1-2011-plati.pdf.

Luidold, S., Antrekowitsch, H., 2010. Lithium – Rohstoffquellen, Anwendung und Recycling. World of Metallurgy – ERZMETALL 63, 68–76.

Marafi, M., Stanislaus, A., Furimsky, E., 2010. Handbook of Spent Hydroprocessing Catalysts. Elsevier B.V.

Merkel, C., Friedrich, B., 2010. Rückgewinnung von Sondermetallen für die Elektronikindustrie aus Produktionsabfällen. In: Symp. Rohstoffeffizienz und Rohstoffinnovationen, 04./05.02.2010, Ettlingen, pp. 109–110. ISBN: 978-3-8396-0097-9.

News release, 2011. Umicore and Rhodia Develop Unique Rare Earth Recycling Process for Rechargeable Batteries. http://www.rhodia.com/en/news_center/news_releases/Umicore_rare_earth_160611.tcm (visited 01.03.13.).

News Release, 2012. Solvay recycelt seltene Erden in Frankreich-Zunächst Konzentration auf Energiesparlampen. http://www.gtai.de/GTAI/Navigation/DE/Trade/maerkte,did=680842.html (visited 01.03.13.).

News release, 2013. Recycling von seltenen Erden gestartet. http://institut-seltene-erden.org/recycling-von-seltenen-erden-gestartet-2/ (visited 01.03.13.).

Okabe, T.H., Takeda, O., et al., 2003. Direct extraction and recovery of neodymium metal from magnet scrap. Materials Transaction 44, 798–801.

Reitz, J., 2013. Elektroschlackedesoxidation von Titanaluminiden, Verlag Shaker, ISBN-10: 3844020225 (Dissertation am IME der RWTH Aachen), in press.

Reitz, J., Lochbichler, C., Friedrich, B., 2011. Recycling of gamma titanium aluminide scrap from investment casting operations. Intermetallics 19, 762–768.

Rollat, A., 2012. How to Satisfy the Rare Earths Demand – Rhodia Rare Earth Systems Initiatives. http://www.seii.org/seii/documents_seii/archives/2012-09-28_A_Rollat_Terres_rares.pdf.

Rumpold, R., Antrekowitsch, J., 2012. Recycling of platinum group metals from automotive catalysts by an acidic leaching process. In: Proc. Platinum. The Southern African Institute of Mining and Metallurgy. http://www.saimm.co.za/Conferences/Pt2012/695-714_Rumpold.pdf.

Saguchi, A., et al., 2006. Recycling of rare earth magnet scraps: carbon and oxygen removal from Nd magnet scraps. Journal of Alloys and Compounds 408–412, 1377–1381.

Scheel, E., 2010. Recycling von Nickel/Cobalt/Molybdän und Vanadium/Wolfram/Molybdän-Sekundärmaterialien. In: Proc. 44. Met. Sem.-Sondermetalle und Edelmetalle. GDMB. ISBN: 978-3-940276-23-0.

Schüler, D., Buchert, M., et al., 2011. Study on Rare Earths and Their Recycling, Final Report for the Greens/EFA Group in the European Parliament. Öko-Institut e.V.

Shirayama, S., Okabe, T.H., 2008. In: Proc. of Annual Meeting of MMIJ, March 27–29. The Mining and Materials Processing Inst. of Japan, pp. 69–70.

Takeda, O., Okabe, T.H., Umetsu, Y., 2004. Phase equilibrium of the system Ag–Fe–Nd, and Nd extraction from magnet scraps using molten silver. Journal of Alloys and Compounds 379, 305–313.

Takeda, O.K., Nakano, Sato, Y., 2009. Molten salts. High Temperature Chemistry 52, 63–70.

Tanaka, M., Oki, T., et al., 2013. Recycling of rare earths from scrap. In: Handbook on the Physics and Chemistry of Rare Earths, vol. 43. Elsevier, pp. 159–211.

Tolein, A.C., 2012. USGS, 2010 Minerals Yearbook – Antimony. http://minerals.usgs.gov/minerals/pubs/commodity/indium/myb1-2010-indiu.pdf.

Umicore, 2013a. http://www.preciousmetals.umicore.com/PMR/Process/ (visited 02.05.13.).

Umicore, 2013b. http://www.batteryrecycling.umicore.com/UBR/process/ (visited 02.28.13.).

UNEP, 2011. Recycling Rates of Metals – A Status Report. http://www.unep.org/resourcepanel/Portals/24102/PDFs/Metals_Recycling_Rates_110412–1.pdf.

U.S. Geological Survey, Reston, Virginia, 2013. Mineral Commodity Summaries. http://minerals.usgs.gov/minerals/pubs/mcs/2013/mcs2013.pdf.

Xu, Y., Chumbley, L.S., Laabs, F.C., 2000. Liquid metal extraction of Nd from Nd–Fe–B magnet scrap. Journal of Material Research 15, 2296–2304.

Zakotnik, M., Harris, I.R., Williams, A.J., 2009. Multiple recycling of Nd–Fe–B-type sintered magnets. Journal of Alloys and Compounds 469, 314–321.

CHAPTER 11

Recycling of Lumber

Ambrose Dodoo, Leif Gustavsson, Roger Sathre

Sustainable Built Environment Group, Department of Building and Energy Technology, Linnaeus University, Växjö, Sweden

11.1 INTRODUCTION

Wood from sustainably managed forests can play important roles both as material and as fuel in a transition to a low-carbon society. Wood is widely used as an energy source and as a physical and structural material in diverse applications, including furniture and joinery, pulp and paper, and construction material. There is large potential to improve resource efficiency and thereby reduce greenhouse gas (GHG) emissions through efficient management of post-use wood materials (IPCC, 2007). This chapter explores post-use management of wood products from resource efficiency and climate perspectives. Primary energy and GHG balances are important metrics to understand the resource efficiency of climate change mitigation strategies involving post-use wood products. This chapter describes the mechanisms through which post-use management of recovered wood materials can affect primary energy use and GHG impacts of wood products. To further understand the implications of different post-use management options for wood products, we then explore several quantitative case-studies.

11.2 BACKGROUND

In contrast to materials such as steel and concrete, which are manufactured through technological processes in factories, wood is produced through natural biological processes occurring in growing trees. By dry weight, wood has an elemental composition of about 50% carbon, 44% oxygen, 6% hydrogen, and trace amounts of several minerals (Pettersen, 1984). These elements return to the environment when a wood product is burned or decayed at the end of its service life. Carbon, oxygen and hydrogen generally return in the form of CO_2 and H_2O. The elements thus become bio-available for other trees to use in their growth, leading to continual cycling of materials.

The lifecycle of wood products begins with forest management activities, e.g. seedling cultivation, tree planting and forest thinning. This is followed by harvesting and processing of logs into lumber, and the manufacture, use and end-of-life management of the finished wood products. In addition to the principal flows of round wood and primary wood products, considerable coproducts are generated, e.g. residues from silviculture, harvesting, primary processing when logs are sawn into lumber, and in secondary processing to make products such as doors and windows (Gustavsson and Sathre, 2011). The use of wood as material or fuel has feedback mechanisms that affect total energy use and GHG emissions. Relatively little energy is needed for the manufacture and processing of wood-based materials compared to non-wood alternatives such as concrete and steel (Gustavsson and Sathre, 2006; Perez-Garcia et al., 2005; IPCC, 2007). Typically, wood-based products use mainly biomass residues for processing energy and have lower climate impacts than non-wood alternatives (Gustavsson et al., 2006).

Post-use management options for wood products include reuse, recycling, energy recovery and landfilling with or without the capture of landfill gas (LFG). Reuse of end-of-life products involves the further use of a recovered product in a similar application without reprocessing, while recycling entails reprocessing it to produce a new type of product. For example, large wood frames may be reused in similar structural applications or be remilled (and recycled) into wood flooring. Recovered wood products may be used in different applications, including as raw material for production of particleboard, oriented strand board, medium density fiberboard and animal bedding and mulches. In some areas, deposition in landfills is the most common fate for post-use wood material. For example, in North America demolition waste including wood material is typically disposed in landfills (Salazar and Meil, 2009). This, however, is prohibited in the EU and in some states in the United States (Defra, 2012). While landfilling has typically been the default baseline from which recycling benefits may be measured, in many European countries the default practice may now be to burn untreated wood in conventional energy plants and treated wood in specific incineration plants.

End-of-life management is the single most significant variable for the full lifecycle energy and carbon profiles of wood products (Gustavsson and Sathre, 2009; Sathre and O'Connor, 2010). Post-use wood products contain significant amounts of energy stored in chemical bonds that can be recovered and used to substitute fossil fuels, avoiding fossil emissions (Gustavsson et al., 2006). Currently, woody biomass provides 9% of the global total primary energy, which is more than the share from all other renewable energy sources or nuclear energy (FAO, 2010; IEA, 2009). The share of woody biomass in the global energy mix is projected to double in the coming decades (Mead, 2005). Energy recovery from post-use wood will be an increasingly important component of these renewable energy sources.

Post-use wood products are a potentially important resource in many countries, and Falk and McKeever (2004) observed that up to 90% of solid post-use wood may be recovered. Incomplete data make it difficult to know precisely how much is currently recovered in different countries (Defra, 2012), although a detailed inventory compiled the sources and quantities of recovered wood in 20 countries in the EU (COST Action E31, 2007). About 30 Mt of solid wood was recovered annually in these countries together, corresponding to about 13% of their annual round wood use. Falk and McKeever (2004) reported that 62.5 Mt of solid wood waste was generated in the United States in 2002, most of which was landfilled. They observed that 43% of the generated solid wood waste was suitable for recovery and reuse. Post-use wood may be recovered from construction and demolition sites, municipal and industrial waste, furniture and joinery manufacture, packaging and pallets. Other sources of recovered wood include post-use railroad ties and utility poles, which are often treated with chemical preservative. Incomplete data make it difficult to estimate how much recovered wood is used globally as bioenergy or as raw material. In the EU, 9.1 and 9.7 Mt of recovered wood were used as bioenergy and as raw material in 2007, respectively (Mantau et al., 2010). The share of recovered wood used as bioenergy or as raw material varies significantly among EU countries. For example, 90% of recovered wood in Sweden is used as bioenergy, while 70% of recovered wood in France is used as raw material for further wood processing (Mantau et al., 2010). Typically, post-use wooden materials are transported to material sorting or recycling sites where they are sorted according to size and quality, screened for contaminants, cleaned and designated for different end-use markets. Sorted clean and large wooden materials are typically used in higher value added applications while small wood may be used for low-value purposes (CWC, 1997).

Few lifecycle studies provide comprehensive analysis of the implications of different end-of-life management options for wood products. Salazar and Meil (2009) assessed the energy and carbon balances of typical and wood-intensive buildings and explored scenarios where end-of-life wood materials are either disposed in landfill with recovery of LFG or recovered of energy by combustion, replacing fossil gas and coal for electricity production. The results show that diverting the post-use wood materials from the landfill for combustion significantly improved the energy and carbon balances of the buildings. Petersen and Solberg (2002) analyzed the lifecycle GHG emissions and cost-efficiency of structural beams made with steel or glue laminated (glulam) wood, including the impacts of different end-of-life management scenarios for the demolished wood and steel. They found that the greatest GHG and energy benefits are achieved when the wood is burned for energy to replace fossil fuels. Landfilling the wood resulted in large atmospheric GHG emissions due to the gradual anaerobic decomposition of the wood, releasing methane. They concluded that the differences in impacts between the glulam wood beam and steel beam depend strongly on how the post-use materials are managed.

Dodoo et al. (2009) analyzed the effects of post-use material management on the lifecycle carbon balance of wood- and concrete-frame buildings. The analysis included scenarios where demolished wooden material is used for energy to replace fossil fuels and demolished steel and concrete are recycled to replace virgin raw materials. They found that replacing fossil fuel with the recovered wooden material gives the greatest carbon benefit in the post-use phase of the buildings. Sathre and Gustavsson (2006) analyzed the effects of different post-use management options on the energy and carbon balances of wood lumber. Post-use options included reuse as lumber, reprocessing as particleboard, pulping to form paper products and burning for energy recovery. They compared energy and carbon balances of products made of recovered wood to the balances of products obtained from virgin wood fiber or from non-wood material. They found that several mechanisms affect the energy and carbon balances of recovery wood: direct effects due to different properties and logistics of virgin and recovered materials, substitution effects due to the reduced demand for non-wood materials when wood is reused, and land use effects due to alternative possible land uses when less timber harvest is needed because of wood recovery. They concluded that land use effects have the greatest impact on energy and carbon balances, followed by substitution effects, while direct effects are relatively minor.

Studies on solid waste management scenarios have also included the impacts of post-use wood products. Carpenter et al. (2013) assessed the environmental impacts of different end-of-life management options for construction and demolition waste in a lifecycle perspective. They analyzed scenarios where wood waste is either combusted with and without energy recovery or disposed in landfill with and without LFG recovery. They found that all the impact categories were significantly lower when the wood waste is combusted with energy recovery, compared to the other scenarios. Bolin and Smith (2011a,b) explored the environmental implications of landfilling or energy recovery of preservative treated wood. The impacts analyzed include energy use, GHG emissions, acidification, eutrophication, smog and ecological toxicity. The authors found that energy recovery of the preservative-treated wood results in lower impacts for all categories except for eutrophication and water use. They concluded that appropriate combustion of preservative-treated wood for energy recovery should be permitted. Jambeck et al. (2007) compared the environmental and economic tradeoffs associated with scenarios where treated wood waste is landfilled or combusted for electricity production in a waste-to-energy facility. The ash from the wood combustion was assumed to be landfilled. The economic analysis considered the cost of waste collection, transport, treatment and disposal, and the revenue generated from the sale of electricity for the combustion for energy scenario.

Rivela et al. (2006) analyzed the system-wide environmental impacts and trade-offs associated with the use of recovered wood for particleboard production or for bioenergy. When the recovered wood is recycled into particleboard, energy is assumed to come from natural gas; when the recovered wood is used for bioenergy, particleboard is assumed to be produced from virgin wood. Merrild and Christensen (2009) analyzed the energy and global warming impacts of recycling wood into particleboard or producing particleboard from virgin wood. They found that recycling post-use wood into particleboard results in significant energy and GHG savings compared to the particleboard production from virgin wood, primarily because of the avoided energy for drying virgin wood. The study did not include impacts from upstream activities and processes, e.g. the fate of the forest if virgin wood is not harvested.

11.3 KEY ISSUES IN POST-USE MANAGEMENT OF WOOD

11.3.1 Post-use Wood in Integrated lifecycle Flows

Post-use wood products can be managed as part of an integrated flow of material and energy within and between the forestry, manufacturing, construction, energy and waste management sectors (see Figure 11.1). This integration, which valorizes the post-use materials, can bring energetic, economic and environmental advantages (Sathre and Gustavsson, 2009). Recovery and recycling of wood from demolished buildings is becoming increasingly common. The percentage of end-of-life wood materials that is recoverable is variable, and depends on the practical limitations linked to the building design and whether material recovery is facilitated through deconstruction. A high recovery percentage of demolition wood could be achieved in future as the value of wood as an energy source is more widely recognized, and as more buildings are designed and constructed in ways that facilitate deconstruction to allow greater recycling and reuse of building materials (Kibert, 2003). This may involve the "design for disassembly" of buildings to facilitate the removal of wood products with minimal damage, to maintain their potential for further re-use as a material. Such optimization of end-of-life product recovery and recycling systems may become increasingly important, to gain additional value from the wood as a

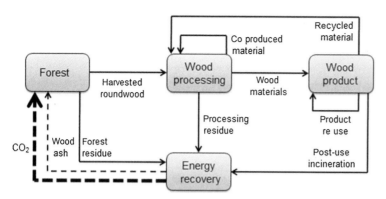

FIGURE 11.1 Schematic diagram of system-wide integrated material flows of wood products.

material before it is ultimately burned to recover its feedstock energy.

11.3.2 Wood Cascading

Additional use of recovered wood material, such as reusing as lumber, reprocessing as particleboard or pulping to form paper products, can improve the environmental performance of the material. Wood products are well suited for material cascading, which has been suggested as a strategy to increase the efficiency of resource use (Haberl and Geissler, 2000). Cascading is the sequential use of a resource for different purposes as the resource quality degrades over time. The cascade concept includes four dimensions of resource economy: resource quality, utilization time, salvageability and consumption rate (Sirkin and ten Houten, 1994). In terms of these four characteristics, optimal utilization of wood resources is achieved by matching the resource quality to the task being performed, so as not to use a high-grade resource when a lower-grade one will suffice; increasing the total utility gained from a resource through prolonging the time during which it is used for various purposes; upgrading a resource through salvaging and reprocessing, where appropriate, for additional higher-grade uses; and balancing the usage rate of a resource with the capacity of forest land to regenerate lost resource quality. Effective cascading use of post-use wood materials may further improve resource efficiency. In some cases, particularly when forest resources are limited, it will be beneficial to employ a more complex cascade chain involving multiple material uses before final burning (Sathre and Gustavsson, 2006). The advantages of such wood cascading depend strongly on the relative quantities of wood products entering service and new wood biomass produced by forests. When forest biomass production is limited, cascading may be beneficial by allowing greater usage from the limited primary biomass production. However, when the biomass production is larger than the amount of wood products made and used, the benefits of material cascading are questionable. At least two conditions can be imagined in which post-use wood cascading, besides energy recovery, could be beneficial: (1) total use of woody biomass increases significantly and the primary harvest is limited, and (2) designation of more forest land as protected reserves to increase biodiversity benefits, together with a limited primary harvest. In the future, if more material and energy services are provided by biomass and fewer by fossil resources, wood cascading is likely to become more important by allowing more intensive use of limited biomass resources.

Reprocessing of recovered wood in a cascade chain may require altered levels of specific energy use for material processing. There are differences in processing energy required because of the different physical properties of virgin and recovered wood, mainly the lower moisture content of recovered wood. Slightly more energy is needed to saw or chip the dry recovered wood, which is harder than green wood. Substantially less energy is required to kiln-dry the recovered wood than the green wood during production of lumber or particleboard. Drying has the largest single demand for energy in the manufacture of lumber and particleboard made from green wood (FAO, 1990). Moisture content, through its effect on heating value, is also important in the comparison of biofuels, e.g. dry recovered wood versus green, freshly harvested biofuel.

11.3.3 Chemical Preservatives

As a biologically produced material, wood is part of natural material cycles and can be decomposed by a variety of organisms such as fungi and insects. To prevent the deterioration of wood products while still in service, some wood is treated with chemical wood preservatives that kill decay organisms. Two main categories of chemical treatments exist: oil-borne preservatives such as creosote and pentachlorophenol, and water-borne preservatives such as copper- and boron-based solutions (Lebow, 2010). Regulations in many countries define the allowable uses of different types of preservatives, which differ between, e.g. residential and industrial applications. The landscape of chemical wood preservatives has changed significantly in the last decades toward safer materials, and continues to change. The use of arsenic in wood preservative solutions, such as the once common chromated copper arsenate (CCA), has been phased out, particularly in residential applications. In the European Union, the Biocidal Products Directive (98/8/EC) covers many common wood preservatives including CCA and creosote, leading to increasing restrictions on their use. Nevertheless, significant quantities of chemically treated wood are currently in service and will require post-use management in the coming years. Opportunities for recycling of preservative treated wood are more limited than for untreated wood (Felton and de Groot, 1996). Particular concerns include worker exposure to emissions from recycling processes, and interference by preservatives with the bonding of adhesives. Energy recovery from treated wood is also restricted, although treated wood can be incinerated under suitable combustion conditions with flue gas cleaning and appropriate ash disposal (Townsend et al., 2008).

11.3.4 Nutrient Cycling

Wood has very small quantities of mineral elements such as Ca, Mg, K and P, although tree leaves and needles typically have higher concentrations of these elements. To avoid loss of these nutrients from forest ecosystems over the long term, ashes from combusted biomass can be applied to growing forests to ensure that nutrient cycles are closed (Stupak et al., 2007). In the absence of ash recycling, the continued export of nutrients contained in the biomass could lead to nutrient deficiency and reduced forest production. In Sweden, for example, the National Board of Forestry has published recommendations regarding the appropriate manner in which ash recycling should be done (Swedish National Board of Forestry, 2002). The dosage of ash application is calculated in such a way as to balance the removal of nutrients in wood, bark and foliage with the return of nutrients in ash. Quality standards are specified for ashes, including minimum content of Ca, Mg, K and P. To avoid the long-term build-up of heavy metals and other contaminants that can be concentrated in the ash, maximum content of trace elements including several heavy metals is also specified. Before wood ashes are applied to the forest, they must be stabilized to slow their dissolution and avoid damage to sensitive flora and fauna. Stabilization can take place both chemically and physically, with the goal that the ashes dissolve slowly over a period of 5–25 years in the field. Ash processing can be done in centralized facilities, or can be done with mobile equipment at the locations where the ash is produced. Ashes can be spread in the forest using ground equipment such as converted tractors, or by helicopter.

11.4 CASE STUDY SCENARIOS

Here, several case study scenarios are explored and analyzed to quantify the primary energy use and GHG implications of different end-of-life management options for recovered wood.

11.4.1 Case Study Method: Primary Energy Use and GHG Balances

This case study is based on a four-story wood-frame building with 16 apartments and a total heated floor area of 1190 m^2, located in Växjö, Sweden. Further details of the characteristics of the building are reported in Persson (1998) and in Dodoo et al. (2012). The mass of major materials contained in the building is shown in Table 11.1.

A method developed by Gustavsson et al. (2006) is used to calculate the primary energy and GHG balances of the scenarios studied. The primary energy used to extract, process, transport and assemble the materials is calculated, and the lower heating values of the logging and processing residues and of the recovered demolition wood used as fuel. The

TABLE 11.1 Mass of Major Materials Contained in the Building

Material	Mass (t)
Lumber	59
Particleboard	18
Plywood	21
Concrete	223
Steel	16
Plasterboard	89

GHG emissions from fossil fuel combustion and cement process reactions are calculated, as well as the potential emissions avoided by replacing fossil fuel with recovered biofuels, and the carbon stock changes in materials and forests. The recovery and use of LFG from landfilled wood products is considered in scenarios involving landfilling. The LFG emission is estimated based on the default IPCC methodology (IPCC, 2006). Recoverable forest residues at harvest are based on data from Lehtonen et al. (2004) and Sathre and Gustavsson (2006). We assume 70% recovery of available harvest residues, and 100% of processing and construction residues. The recovery percentage of post-use wood varies with scenario. The assumed lower heating values for the recovered biomass are 4.25 kWh/kg dry biomass for bark and harvest residues, 4.62 kWh/kg dry biomass for processing residues and 5.17 kWh/kg dry biomass for recovered post-use wood (Gustavsson and Sathre, 2006). The amount of diesel fuel used for biomass recovery, expressed in terms of the heating value of the biomass, is assumed to be 1% for processing residues and post-use wood and 5% for harvest residues (Gustavsson and Sathre, 2006). Specific final energy for building material production is based on Swedish conditions (Björklund and Tillman, 1997; Sathre and Gustavsson, 2006). Electricity is assumed to be produced in coal-fired condensing plants with a conversion efficiency of 40% and distribution loss of 2%. Conversion efficiency is assumed to be unchanged when recovered biofuels replace coal.

The biogenic carbon storage sequestered or released from wood materials is not included in the inventory as the wood is assumed to come from sustainably managed forestry, where carbon flows out of the forest are balanced at the landscape level by carbon uptake by growing trees.

11.4.2 Particleboard Production from Recovered Lumber or Virgin Wood

We analyze the primary energy and GHG implications of producing particleboard from either recovered lumber or virgin wood. Figure 11.2 summarizes the scenarios. The analysis considers particleboard in a 100-year lifecycle perspective and assumes that the particleboard is combusted for energy at the end of its service life. In *Scenario A*, 90% of the lumber in the case-study building, corresponding to 53.1 t, is assumed to be recovered and used as raw material for the particleboard production. In *Scenario B*, particleboard production is from virgin wood, and biomass residues are obtained from forest management and processing activities for the virgin wood. The fuel used to recover and transport the wood material is assumed to be diesel and is calculated with data from Sathre and Gustavsson (2006). When recovered lumber is reprocessed as particleboard (*Scenario A*), the forest is assumed to either remain unharvested and continue to grow by either 0% or 20% over the 100-year lifecycle, or to be harvested at year 0 and used to displace coal. When the forest is harvested to produce particleboard (*Scenario B*), the recovered lumber is combusted, substituting coal. An estimated 9% more electricity and 60% less thermal energy are needed for particleboard production with recovered lumber compared to the particleboard made from virgin wood (Sathre and Gustavsson, 2006).

The primary energy balances of all the scenarios are negative, meaning that more energy is available for external use than is used during

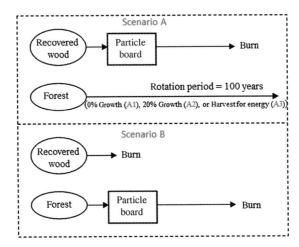

FIGURE 11.2 Comparison of two alternatives for producing particleboard. In *Scenario A* recovered lumber is used as raw material for particleboard, allowing the forest to be used for other purposes. In *Scenario B* the recovered lumber is burned, and the forest is harvested to produce particleboard.

TABLE 11.2 Primary Energy and GHG Implications of Scenarios where Particleboard is Produced from Recovered Lumber or Virgin Wood

Description	Primary Energy (MWh)	Greenhouse Gas (t CO_2e)
A1. RECOVERED LUMBER FOR PARTICLEBOARD, 0% FOREST GROWTH		
Material recovery and production	60	22
Heating value of residues	−215	
Substitution of fossil coal by residues		−85
Product carbon stock changes		0
Forest carbon stock changes		0
Total	−155	−63
A2. RECOVERED LUMBER FOR PARTICLEBOARD, 20% FOREST GROWTH		
Material recovery and production	60	22
Heating value of residues	−215	
Substitution of fossil coal by residues		−85
Product carbon stock changes		0
Forest carbon stock changes		32
Total	−155	−95
A3. RECOVERED LUMBER FOR PARTICLEBOARD, HARVEST FOREST FOR ENERGY		
Material recovery and production	72	25
Heating value of residues	−477	
Substitution of fossil coal by residues		−189
Product carbon stock changes		0
Forest carbon stock changes		0
Total	−405	−164
B. VIRGIN WOOD FOR PARTICLEBOARD, BURN RECOVERED LUMBER		
Material recovery and production	74	26
Heating value of residues	−491	
Substitution of fossil coal by residues		−195
Product carbon stock changes		0
Forest carbon stock changes		0
Total	−417	−169

the product lifecycle (Table 11.2). The GHG balances are negative for all scenarios except when the forest remains standing without growth. The primary energy and GHG balances are lowest when virgin wood is used for particleboard and the recovered lumber is burned. The difference in primary energy and GHG balances is small between the scenario where the recovered lumber is used for particleboard and the forest for biofuel and the scenario where virgin wood is used for particleboard and the recovered lumber for biofuel. The difference in process energy between making particleboard from recovered lumber and virgin wood is small in relation to the total energy flow in the production systems.

11.4.3 A Complex Cascade Chain for Recovered Wood Products

Here we present a more complex material management scenario. Recovered lumber is used for building-frame construction and then recycled as particleboard before combustion with energy recovery (*Scenario C*). This is compared to a scenario where non-wood alternatives including reinforced concrete frame material and gypsum panelboard are used (*Scenario D*). Here we assume that the forest resources are limited, so the building must be constructed either with recovered lumber or alternate non-wood materials. When non-wood materials are used for construction, the recovered lumber is burned in place of coal. A schematic diagram of these scenarios is shown in Figure 11.3. The analysis is based on the same amount of recovered lumber and specific energy data and fuel cycle emission data, as in Section 11.4.2.

The primary energy and GHG balances are lowest when recovered lumber is used for the building frame (Table 11.3). This is due mainly to the fossil fuel used for the production of concrete and steel, compared to the reuse of the recovered lumber, which is assumed to require no additional processing energy. Significant amounts of biomass residues are recovered at the end of the lifecycle of wood product, in contrast to the non-wood alternative materials. The GHG balance is substantially higher for the reinforced concrete materials, owing to the calcination emission of CO_2 during cement production as well as greater fossil fuel use for manufacture of the non-wood materials.

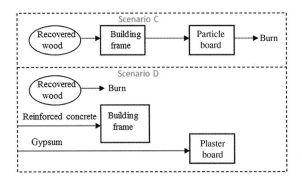

FIGURE 11.3 Comparison of two alternatives for providing building materials, assuming limited forest. In *Scenario C*, recovered lumber is used for building materials. In *Scenario D*, reinforced concrete and plasterboard are used for building materials and the recovered lumber is burned.

TABLE 11.3 Primary Energy and Greenhouse Gas Implications of Scenarios where Recovered Lumber or Non-wood Materials are Used for Building Frame and Panel Products

Description	Primary Energy (MWh)	Greenhouse Gas (t CO_2e)
C. RECOVERED LUMBER FOR PRODUCTS		
Material recovery and production	58	17
Heating value of residues	−233	
Substitution of fossil coal by residues		−92
Product carbon stock changes		0
Total	−175	−75
D. NON-WOOD MATERIALS FOR PRODUCTS, BURN RECOVERED WOOD		
Material recovery and production	171	68
Heating value of residues	−233	
Substitution of fossil coal by residues		−92
Product carbon stock changes		0
Total	62	24

11.5 SUMMARY

In this chapter we have explored the implications of end-of-life management options for wood products and have described several mechanisms through which post-use management of recovered wood materials can affect the lifecycle resource efficiency and climate performance of wood products. This analysis shows how efficient management of post-use wood products can contribute to a sustainable, resource-efficient, low-carbon society. Recovering energy from post-use wood material gives significant primary energy and GHG benefits. These benefits of post-use wood materials may be further optimized when wood is cascaded, in which post-use wood is reused and recycled for use in a sequence of applications and afterward burned to recover the heat content. The benefits of additional material use in complex cascade chains, however, depend largely on the relative abundance of primary forest resources.

Lifecycle and system perspectives of wood products are needed, so that all the lifecycle phases—acquisition of raw material, manufacture, use and end-of-life—are considered and optimized as a whole, including the energy and material chains from natural resources to final services. Primary energy and GHG balances are important metrics when analyzing the resource efficiency and climate mitigation effectiveness of post-use wood management options. Primary energy use largely determines natural resource efficiency and steers the environmental impacts of material production. More so than with other common materials, appropriate post-use management of wood products is important because in addition to its structural use as a physical material, wood can also be used as a sustainable bioenergy source. The increased use of wood products from sustainably managed forests can play an important role in our transition to a low-carbon economy, and post-use management is a critical phase of the wood product lifecycle that should be thoughtfully optimized.

References

Bolin, C.A., Smith, S., 2011a. lifecycle assessment of ACQ-treated lumber with comparison to wood plastic composite decking. Journal of Cleaner Production 19 (6–7), 620–629.

Bolin, C.A., Smith, S., 2011b. lifecycle assessment of borate-treated lumber with comparison to galvanized steel framing. Journal of Cleaner Production 19 (6–7), 630–639.

Björklund, T., Tillman, A.-M., 1997. LCA of Building Frame Structures: Environmental Impact over the lifecycle of Wooden and Concrete Frames. Technical Environmental Planning Report 1997:2. Chalmers University of Technology, Sweden, 156 p.

Carpenter, A., Jambeck, J.R., Gardner, K., Weitz, K., 2013. lifecycle assessment of end-of-life management options for construction and demolition debris. Journal of Industrial Ecology 17 (3), 396–406.

COST Action E31, 2007. Management of Recovered Wood. Accessed at: http://www.ctib-tchn.be/page.php?m=12&s=107&c=36&l=EN (on 29.01.13.).

CWC (Clean Washington Center), 1997. Wood Waste Recovery: Size Reduction Technology Study. Final Report. Accessed at: http://infohouse.p2rlc.org/ref/13/12639.pdf (on 04.05.13.).

Defra, 2012. Wood Waste Landfill Restrictions in England Call for Evidence. Accessed at: http://www.defra.gov.uk/consult/files/consult-wood-waste-document-20120808.pdf (on 29.01.13.).

Directive 98/8/EC of the European Parliament and of the Council of 16 February 1998 Concerning the Placing of Biocidal Products on the Market. Accessed at: http://ec.europa.eu (on 18.02.13.).

Dodoo, A., Gustavsson, L., Sathre, R., 2009. Carbon implications of end-of-life management of building materials. Resources, Conservation and Recycling 53 (5), 276–286.

Dodoo, A., Gustavsson, L., Sathre, R., 2012. Effect of thermal mass on lifecycle primary energy balances of a concrete- and a wood-frame building. Applied Energy 92, 462–472.

FAO (Food and Agricultural Organization of the United Nations), 1990. Energy Conservation in the Mechanical Forest Industries. FAO Forestry Paper 93, Rome.

FAO (Food and Agricultural Organization of the United Nations), 2010. What Wood Fuels Can Do to Mitigate Climate Change. Accessed at: http://www.fao.org/docrep/013/i1756e/i1756e00.htm (on 29.01.13.).

Felton, C.C., de Groot, R.C., 1996. The recycling potential of preservative-treated wood. Forest Products Journal 46 (7–8), 37–46.

Falk, R.H., McKeever, D.B., 2004. Recovering Wood for Reuse and Recycling, a United States Perspective. European COSTE31 Conference. Accessed at: http://www.woodweb.com (on 29.01.13.).

Gustavsson, L., Sathre, R., 2009. A State-of-the-Art Review of Energy and Climate Effects of Wood Product Substitution. Report No. 57. School of Technology and Design, Växjö University, Sweden. ISBN: 978-91-7636-692-9.

Gustavsson, L., Sathre, R., 2011. Energy and CO_2 analysis of wood substitution in construction. Climatic Change 105 (1–2), 129–153.

Gustavsson, L., Pingoud, K., Sathre, R., 2006. Carbon dioxide balance of wood substitution: comparing concrete- and wood-framed buildings. Mitigation and Adaptation Strategies for Global Change 11 (3), 667–691.

Gustavsson, L., Sathre, R., 2006. Variability in energy and carbon dioxide balances of wood and concrete building materials. Building and Environment 41 (7), 940–951.

Haberl, H., Geissler, S., 2000. Cascade utilization of biomass: strategies for a more efficient use of a scarce resource. Ecological Engineering 16, S111–S121.

IEA (International Energy Agency), 2009. Bioenergy – a sustainable and reliable energy source: a review of status and prospects. Paris.

IPCC (Intergovernmental Panel on Climate Change), 2007. Climate Change 2007: Synthesis Report. Contribution of Working Groups I, II and III to the Fourth Assessment Report of the IPCC. IPCC, Geneva, Switzerland.

IPCC (Intergovernmental Panel on Climate Change), 2006. Guidelines for National Greenhouse Gas Inventories. Web-accessed at: http://www.ipcc-nggip.iges.or.jp/public/2006gl/vol5.html (on 18.02.13.).

Jambeck, J., Weitz, K., Solo-Gabriele, H., Townsend, T., Thorneloe, S., 2007. CCA-treated wood disposed in landfills and lifecycle trade-offs with waste-to-energy and MSW landfill disposal. Waste Management 27 (8), 21–28.

Kibert, C.J., 2003. Deconstruction: the start of a sustainable materials strategy for the built environment. UNEP Industry and Environment 26 (2–3), 84–88.

Lebow, S.T., 2010. Wood preservation. Chapter 15. In: Wood Handbook – Wood as an Engineering Material. General Technical Report FPL-GTR-190. Forest Products Laboratory, US Department of Agriculture, Forest Service, Madison, WI.

Lehtonen, A., Mäkipää, R., Heikkinen, J., Sievänen, R., Liski, J., 2004. Biomass expansion factors (BEFs) for Scots pine, Norway spruce and birch according to stand age for boreal forests. Forest Ecology and Management 188 (1–3), 211–224.

Mantau, U., et al., June 2010. EU Wood — Real Potential for Changes in Growth and Use of EU Forests. Methodology Report. Hamburg, Germany, 165 p.

Mead, D.J., 2005. Forests for energy and the role of planted trees. Critical Reviews in Plant Sciences 24 (5), 407—421.

Merrild, H., Christensen, T.H., 2009. Recycling of wood for particle board production: accounting of greenhouse gases and global warming contributions. Waste Management and Research 27 (8), 781—788.

Persson, S., 1998. Wälluddenträhusi fem våningar: Erfarenheter och lärdomar (Wälludden Wooden Building with Five Stories: Experiences and Knowledge Acquired). Report TVBK-3032. Department of Structural Engineering, Lund Institute of Technology, Sweden (In Swedish).

Petersen, A.K., Solberg, B., 2002. Greenhouse gas emissions, lifecycle inventory and cost-efficiency of using laminated wood instead of steel construction. Case: beams at Gardermoen airport. Environmental Science and Policy 5 (2), 169—182.

Pettersen, R.C., 1984. The chemical composition of wood. Chapter 2. In: Rowell, R. (Ed.), The Chemistry of Solid Wood. ACS Publications.

Perez-Garcia, J., Lippke, B., Briggs, D., Wilson, J.B., Boyer, J., Meil, J., 2005. The environmental performance of renewable building materials in the context of residential construction. Wood and Fiber Science 37 (CORRIM Special Issue), 3—17.

Rivela, B., Moreira, M.T., Munoz, I., Rieradevall, J., Feijoo, G., 2006. lifecycle assessment of wood wastes: a case study of ephemeral architecture. Science of the Total Environment 357 (1—3), 1—11.

Salazar, J., Meil, J., 2009. Prospects for carbon-neutral housing: the influence of greater wood use on the carbon footprint of a single-family residence. Journal of Cleaner Production 17 (17), 1563—1571.

Sathre, R., Gustavsson, L., 2006. Energy and carbon balances of wood cascade chains. Resources, Conservation and Recycling 47 (4), 332—355.

Sathre, R., Gustavsson, L., 2009. Using wood products to mitigate climate change: external costs and structural change. Applied Energy 86 (2), 251—257.

Sathre, R., O'Connor, J., 2010. Meta-analysis of greenhouse gas displacement factors of wood product substitution. Environmental Science and Policy 13 (2), 104—114.

Sirkin, T., ten Houten, M., 1994. The cascade chain: a theory and tool for achieving resource sustainability with applications for product design. Resources, Conservation and Recycling 10 (3), 213—276.

Stupak, I., Asikainen, A., Jonsell, M., et al., 2007. Sustainable utilization of forest biomass for energy-possibilities and problems: policy, legislation, certification, and recommendations and guidelines in the Nordic, Baltic, and other European countries. Biomass and Bioenergy 31 (10), 666—684.

Swedish National Board of Forestry, 2002. Recommendations for the Extraction of Forest Fuel and Compensation Fertilizing. Meddelande 3—2002. Accessed at: http://www.svo.se/forlag/meddelande/1545.pdf (on 18.02.13.).

Townsend, T., Dubey, B., Solo-Gabriele, H., 2008. Disposal management of preservative-treated wood products. Chapter 33. In: Schultz, T., et al. (Eds.), Development of Commercial Wood Preservatives. American Chemical Society, Washington, DC.

CHAPTER 12

Paper Recycling

Harald Grossmann[1], Toni Handke[1], Tobias Brenner[2]
[1]Technische Universität Dresden, Germany; [2]Papiertechnische Stiftung (PTS), Germany

12.1 IMPORTANT FACTS ABOUT PAPER RECYCLING

12.1.1 Introduction

The principal raw materials used for papermaking are wood and recovered paper or board. Wood has to be cut in logs or chips, which can then be converted into chemical or mechanical fiber pulps depending on which effect dominates the disintegration process. Chemical pulps, which currently represent 33% of all fibers used for papermaking world-wide, contain only long, strong and flexible cellulose fibers from which all other wood constituents (lignin and hemicellulose) have been reduced. Mechanical pulps (8%) consist of weaker and shorter fibers, which by and large contain all wood components in the same ratio as the wood itself. Both chemical and mechanical pulps are called virgin pulps because they are directly manufactured from wood.

Recovered paper or board is converted into recycled pulp in increasingly complex process chains, the main purpose of which is to remove all unsuitable or potentially detrimental substances from the fibers. All fibers of the resulting recycled (or secondary) pulp therefore have at least once before been part of a paper or board product. They represent 57% of the total fiber consumption (2% are other fiber sources) of the paper industry in the world.

During the last few decades, recovered paper and board has become the principal raw material of the paper industry. The reasons are quite obvious: recovered paper is significantly cheaper and its conversion into pulp requires far less energy than virgin raw materials. Recycling paper and board, however, not only is economical but also contributes to the energy and resource efficiency of the entire industry. In particular, highly developed countries with limited forest resources like Germany or France as well as other countries with emerging economies, such as China and India, are highly dependent on recovered paper.

12.1.2 Utilization and Collection of Recovered Paper

The extent to which used paper and board products are recycled is usually characterized by two key indicators: the recovery rate and the utilization rate.

Recovery rate (%) is the amount of paper recovered for recycling in a certain region, divided by the amount of paper consumed in this very region, on an annual basis, multiplied by 100 (Göttsching and Pakarinen, 2000).

Utilization rate (%) is the amount of recovered paper that is used as raw material in the paper industry in a certain region, divided by the paper production of this region, on an annual basis, multiplied by 100 (Göttsching and Pakarinen, 2000).

These definitions, however, although quite simple and seemingly straightforward, in many cases do not allow a meaningful comparison between regions. A quite obvious—although not the only—reason for this is the fact that they usually do not take into account the amounts of packaging papers that come into a region together with imported goods or which leave the region together with exported goods. The packaging paper used for the latter would of course be included in the production statistics, but it would no longer be available for recovery and utilization in this region—and vice versa. Although the interpretation of the key indicators requires some care, they at least can help to provide a rough idea of paper recycling in a certain region.

Figure 12.1 shows such data for selected countries. China or India, for example, utilize more recovered paper for their domestic paper production than they collect; i.e. they are net importers of recovered paper, while countries like Sweden or Canada collect more paper than they use for their own production, which is based primarily on virgin fibers. Japan, France and Germany recycle almost all available recovered paper and produce most of their papers from this resource.

12.1.3 Paper Consumption Per Capita

Figure 12.2 shows that the annual per capita consumption of paper and paperboard varies significantly between the major regions of the world. The world average in 2011 was close to 57 kg (VdP, 2013). North America is the region with the highest annual consumption per capita of 229 kg, while this figure is as low as 8 kg for Africa.

The consumption levels of Western Europe and North America will remain constant or—even more likely—decline slightly in the long run. But consumption in Asia will increase further, although starting from a rather low level. Because of the huge population in this region, it is assumed that the worldwide average paper consumption will also increase. Asia already today accounts for 40% of global consumption, while EU and North America account for about one-quarter each.

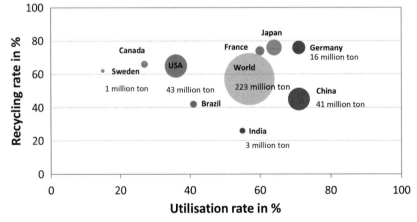

FIGURE 12.1 Recovered paper utilization rate, recycling rate and collection for selected countries in 2011. Bubble size is proportional to recovered paper collection. *Data from VdP, (2013).*

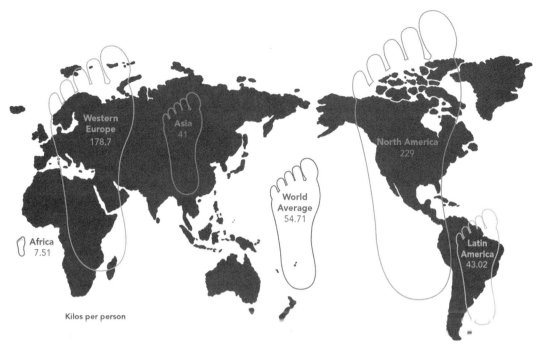

FIGURE 12.2 Annual paper consumption per capita 2009 (Network, 2011).

12.1.4 Collection and Sorting Systems for Major Recovered Paper Grades

Supplying paper mills with used fibrous raw material requires the collection and sorting of recovered paper. The collection of recovered paper can be broken down into industry, trade and administration and private household collections. Collection is organized in various systems, the most important of which are container collection (dropoff system, pickup system), curbside collection (pickup system) and underground collection (dropoff system) (Stawicki and Read, 2010). In addition to the separate collection of recovered paper, a method known as "commingled" collection is practiced in which paper and other waste or metal are collected together (Apotheker, 1990). Commingled collection is not considered appropriate in cases where high-quality recovered paper is required.

The collected paper can either be prepared in sorting factories according to the subsequent use or delivered directly to the paper mill. The recovered paper that has been collected and sorted if necessary is then subdivided into various classes based on the composition and content of unusable or contaminating materials. The classification system for recovered paper in the United States is contained in the "Guidelines for Paper Stock: PS 2008". This system defines the allowable shares of prohibited and contaminating materials for individual paper grades and contains issues of the contract (grade definitions, arbitration, fulfillments by the buyer and seller, purchase agreement). In Europe, recovered paper is commonly classified according to the European List of Standard Grades of Recovered Paper and Board (EN 643) comprising 67 grades subdivided in five classes. The most important grades, quantitatively speaking, are grades 1.02 (mixed papers and boards (sorted)), 1.04 (supermarket corrugated paper and board) and 1.11 (sorted graphic paper for deinking). In addition, another classification for statistical and commercial purposes has four main groups (Stawicki and Read, 2010):

- Mixed grades (mixed paper and boards),
- OCC (old corrugated containers),
- ONP & OMG (old newspaper and magazines) and
- HG & PS (high grade deinking and pulp substitutes)

Initially, recovered paper was sorted manually only. In the meantime, the development of efficient sensors and mechanical separators has made it possible to use sufficiently selective semi automatic or fully automatic systems.

The sorting process for recovered paper consists of the following operations (Blasius et al., 2011):

- delivery of the recovered paper in loose form or as bales, entry inspection, bale dewiring,
- mechanical sorting based on differences in physical properties such as shape, size, specific gravity, stiffness,
- sensor-based sorting and separation by automatic picking, chiefly by blowing,
- manual sorting.

The recovered paper can be delivered to the paper mill as loose material, which is typical of household collections and paper for deinking. All other recovered paper is traded in bales that are typically 1.2 m × 1.0 m × 0.8 m and 550 kg.

The fibrous raw material suitable for producing packaging paper includes packaging paper or board as well as used graphic paper. Only used graphic paper or higher grades can be used as fibrous raw material for the production of graphic paper or tissue paper (deinking process). This distinction is necessary because of the different preparation processes used for stock preparation in the paper mill.

12.1.5 Utilization of Recovered Paper in Different Paper Products

Products from the paper industry can be subdivided into four classes:

- Packaging paper (case materials, carton board, wrapping papers and other packaging papers),
- Graphic paper (newsprint and other graphic paper),
- Tissue paper (household and sanitary) and
- Paper for technical and other purposes.

The utilization rates of recovered paper for the production of these paper grades vary considerably. The highest utilization rates are usually found for corrugated case materials and newsprint (Figure 12.3). On the one hand, these paper grades do not have a sophisticated optical appearance and their product life time is rather short. On the other hand, high strength values cannot be achieved with recovered paper and the high ash content prevents it from being used to a greater extent in hygiene paper. Therefore, if high brightness, strength or softness are required, recovered paper can only be used to a limited degree. The higher the demands placed on the paper, the more energy must be invested in preparing it. Packaging paper requires less energy in stock preparation than a graphic paper that has to undergo extensive cleaning.

12.1.6 Raw Material Efficiency

The average energy consumption for the production of 1 t of paper greatly depends on the furnish composition. Table 12.1 shows some selected products and their specific energy demands.

In Central Europe, packaging paper or newsprint is usually made from 100% recovered paper, thus with a lower energy consumption than paper products containing virgin fibers, e.g. wood-containing coated paper. The

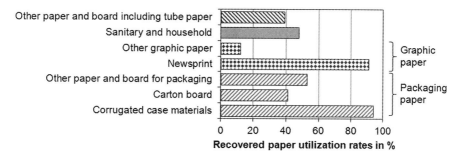

FIGURE 12.3 Recovered paper utilization rates for different paper and board products in Confederation of European Paper Industries (CEPI) countries in 2011 (CEPI, 2012).

difference of about 1000 kWh/t (Blechschmidt, 2010) is due mainly to the different processes used for raw material treatment. The use of recovered paper is always less expensive than the use of virgin fibers.

The water demand of paper mills using recovered paper has been considerably reduced in the last few decades and today ranges between 0 l/kg paper and 10 l/kg paper. In what are termed zero-liquid effluent mills, only the amount of water that is lost through evaporation along the production chain in the dryer section of the paper machine has to be replaced by fresh water.

12.2 STOCK PREPARATION FOR PAPER RECYCLING

12.2.1 Stock Preparation in General

In addition to the fibrous raw material, recovered paper also contains other substances that are introduced by the paper product itself (printing inks, stickies, mineral particles) or during its use and subsequent collection (scraps and leftovers, sand, glass, etc.).

Hence, when producing new paper from recovered paper, the pulp first has to be cleaned intensely, irrespective of the later product.

The objective of recovered paper treatment is to produce a homogeneous substance from which contraries (depending on the product) have been eliminated and which provides fibers that enable a stable production process.

12.2.1.1 Packaging Paper

As far as packaging paper is concerned, it is particularly important to remove coarse contaminants such as films and foils, adhesive and consumer residues, and to produce a homogeneous fiber pulp.

Figure 12.4 shows the main operations of a stock preparation line for the production of

TABLE 12.1 Specific Energy Consumption for Various Types of Paper (Bayerisches Landesamt für Umweltschutz, 2003)

	Heat Demand	Energy Demand	Total
	kWh/t	kWh/t	kWh/t
Sanitary paper	1900	1150	3050
Wood-containing coated	1600	1450	3050
Wood-free coated	1800	1000	2800
Wood-containing uncoated	1250	1370	2620
Wood free uncoated	1600	770	2370
Newspaper	1000	1200	2200
Packaging paper	1500	500	2000

"brown" packaging paper. In older mills, the recovered paper is pulped in medium (10%) to high consistency (up to 20%) batch pulpers with a vat design that includes a dumping system for coarse rejects like huge foils or solid matter as well as a high-consistency cleaner for the removal of heavy impurities such as glass, sand or metal pieces. State-of the-art technology, however, continues to use systems called drum pulpers. After pulping, the fiber suspension is stored in a dump chest in some cases before it is fed to the first screening section equipped with pressurized screens to remove additional contaminants with the help of perforated or slotted screen baskets.

Packaging papers (e.g. testliner) typically consist of two layers with different strength properties. If the pulps for these layers are produced from the same recovered paper mixture, the entire pulp has to be divided into two fractions, one containing significantly longer fibers than the other. The long fiber fraction is then used for the production of the high strength layer.

12.2.1.2 Graphic Paper

The use of recovered paper for graphic paper production is more expensive than for packaging paper because it requires more extensive treatment of the raw material, which consists exclusively of used (or unsold) graphic products. One reason for this is the fact that the printing ink has to be removed from the pulp by a separate energy-intensive process to achieve high brightness. However, not all printing inks can be removed equally well. While the inks used in "old" printing technologies like coldset offset, heatset offset and rotogravure present no major problems, virtually no solution has yet been found for water-based flexography or nonimpact printing inks. This might become particularly problematic in the future as these technologies rapidly gain importance and are already about to replace the older techniques in a number of applications.

Figure 12.5 shows a typical two-loop system for the production of graphic paper. Pulping and sorting are comparable to the stock preparation process used for brown paper, the major difference being the deinking step, which is not needed for pulps used for the production of brown packaging paper. The printing inks are removed by flotation.

Printing ink particles—once detached from fiber surfaces during pulping—are usually removed from the pulp by flotation at very low stock consistencies of about 1%. The removal efficiency is a function of the size of the ink particles. The larger they are the smaller

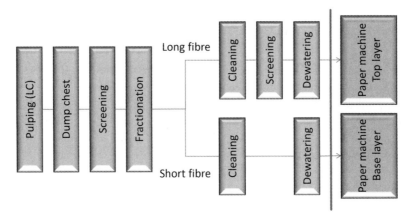

FIGURE 12.4 Concept of stock preparation for the production of packaging paper (testliner). (LC, low consistency.)

is the probability that they become attached to small air bubbles, which, however, is the prerequisite for being floated.

Ink particles that—due to their size—have not been removed by flotation cause dirt specks in the final paper, which is normally not accepted. This can be avoided by breaking them into smaller pieces no longer visible to the naked eye or floatable in a subsequent flotation step. Comminution of the ink particles is done in a disperger after dewatering the suspension to a consistency of 25—35%. Dewatering also serves to separate the first process water loop, which is highly loaded with impurities, from subsequent loops closer to the paper machine. The high stock consistencies and temperatures during dispersing are beneficial to a number of additives, in particular bleaching chemicals, which should therefore be added prior to dispersing.

The most important process steps will be explained in detail in the following section.

12.2.2 Pulping and Slushing

State-of-the-art paper manufacturing still starts with the formation of a wet web from a highly diluted suspension containing fibers and other components such as minerals, starch and chemical additives. The first step in recovered paper treatment, therefore, aims at disintegrating the raw material in water and rendering it into a pumpable suspension. This process is called pulping or slushing. In addition, the process can also remove coarse solid contaminants like wooden parts, plastic films or metal pieces, and mix chemicals into the fiber suspension, especially deinking or bleaching chemicals (Holik, 2000).

With respect to stock consistency, pulping processes can be divided into three categories:

- low-consistency (LC) pulping (below 6% stock consistency),
- medium-consistency (MC) pulping (6—12% stock consistency),
- high-consistency (HC) pulping (beyond 12 and up to about 25% stock consistency).

With respect to the technology used, two different basic processes have become established in the industry:

- conventional pulpers in the form of cylindrical vessels with impellers (rotors) at the bottom or the wall and baffles to avoid solid body rotation, suitable for consistencies between 6% and 19%, and
- drum pulpers, i.e. huge slowly rotating pipes, suitable for consistencies of up to 20%.

Drum pulping is always a continuous process. Conventional pulpers can be performed as batch processes as well. In MC and HC pulping, dilution is required to remove the pulp

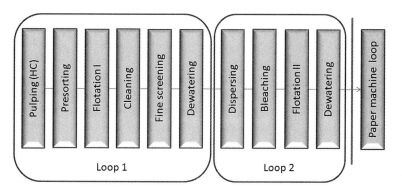

FIGURE 12.5 Concept of stock preparation for the production of graphic paper (newsprint). (HC, high consistency.)

from the pulper. The specific energy demand for pulping is in the range of 15—30 kWh/t paper for conventional pulpers, and 15—20 kWh/t paper for drum pulpers.

Recovered paper always requires—during or after pulping—the separation of coarse impurities that inevitably come with the raw material. Unlike pulpers for virgin fibers, pulpers for recovered paper therefore need a reject handling system.

12.2.2.1 Pulping in Cylindrical Vessels

Vat-shaped pulpers for the disintegration of recovered paper are equipped with a dumping and reject handling system. For economic reasons, this system includes a primary pulper that creates a pumpable suspension of flakes and single fibers and a secondary pulper or disk screen downstream of the primary pulper that crushes the flakes into single fibers. For different ranges of stock consistency, special rotors are available in the primary pulper—flat rotors for LC and spiral rotors for MC and LC. The coarse contaminants are removed from the primary pulper by a ragging device with a winch. Smaller contaminants are removed as light and heavy rejects in the secondary pulper. The light and heavy rejects are treated in downstream devices to recover the adherent fibers and eventually dewater them to the greatest possible extent (Holik, 2000).

12.2.2.2 Drum Pulper

Drum pulpers are primarily used for recovered paper with low wet strength—typically recovered graphic papers. The horizontally aligned system has the shape of a large pipe (up to 4 m in diameter) that rotates at a moderate speed (up to 10 rpm). It consists of two zones (Figure 12.6). The first is a slushing zone where the wet stock (consistency of approximately 20%) is transported upward with the help of baffles aligned perpendicular to the inner wall of the pipe, to repeatedly fall back again. The stock subsequently enters a pipe with a screening zone where it is diluted to 3—5% and the fibers pass through a screen while all contaminants larger than the screen holes are retained. These coarse contaminants are transported to the end of the screen and removed (Holik, 2000).

12.2.3 Screening and Cleaning

Two separation techniques have proved valuable for removing contaminants such as shives, stickies and other solid particles: screening and centrifugal separation. Pressure screens retain particles that are larger or different in shape than the openings of the screens. Centrifugal cleaners accumulate and remove particles that are heavier or lighter than water.

The objective of any separation process is to separate mixed substances into two or more distinct mixtures, at least one of which is enriched with one or more of the mixture's constituents. Recovered paper contains a very wide variety of constituents—many of them in very small quantities. Separation processes generally have three basic mass flows. The "inlet" mass flow, the "accept" mass flow that contains the accepted pulp, and the "reject" with impurities. A separation process is never capable of classifying pulp and impurities with an efficiency of 100%. The smaller the differences between pulp particles and impurities, the poorer will be the separation performance for the two fractions. Therefore, some fibers will always end up in the reject mass flow. Individual separation devices can be interconnected to form circuits known as cascade circuits to ultimately reduce the amount of pulp lost into the reject flow.

The specific energy demand for a screening process is 5—20 kWh/t paper, and for a cleaning process 5—15 kWh/t paper.

12.2.3.1 Screening

The contaminants removed by pressurized screens include shives, stickies, waxes, wet strength paper or plastics. The devices currently

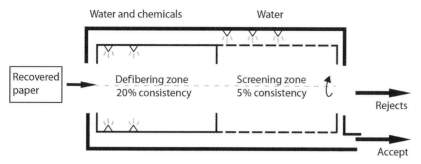

FIGURE 12.6 High-consistency drum pulper.

used are primarily pressure screens, where the screen has the shape of a basket through which the fiber suspension is pumped. A revolving rotor induces pressure pulses to prevent the screen from clogging. Screening devices are distinguished according to the geometry of their openings. The small holes of perforated screens predominantly retain plane particles, whereas the narrow slots of slotted screens are suitable for removing cubical particles (Hanecker, 2006) (Figure 12.7).

12.2.3.2 *Centrifugal Cleaning*

Centrifugal cleaners include a hydro cyclone that creates a spiral flow within the suspension by means of the tangential inlet flow or a rotor. The swirling motion results in a centrifugal field whose forces push heavy particles toward the wall and light particles toward the center. Centrifugal cleaning can thus be used to remove heavy impurities such as pieces of glass, metal or sand as well as light-weight particles such as foam plastics. The removal of different impurities requires specific cleaner designs. Cleaners can be subdivided in heavy particle cleaners, light particle cleaners and combination cleaners for heavy and light contaminants. Another classification is based on the operating stock consistency. The maximum stock consistency for cleaners is approx. 4.5% (Figure 12.8).

12.2.4 Flotation and Washing

Flotation and washing are processes primarily used to reduce the quantity of minerals in a pulp suspension or—even more frequently—to remove ink particles from the suspension (known as deinking).

Washing is a filtration process for removing from the suspension debris particles smaller than the mesh size of a filter cloth (Holik, 2000). Simultaneously, dissolved and colloidal contaminants are removed with the filtrate. The objective in both cases is to remove substances negatively affecting the papermaking process or finished product quality. Materials removed by washing include fillers and coating particles, fines, microstickies and ink.

Flotation is a separation process for cleaning suspensions. It is based on the use of air bubbles that attach themselves to hydrophobic dirt particles and transport them to the surface of the suspension (see Figure 12.9).

Flotation is performed at very low stock consistencies of 0.8–1.2%. It usually requires the addition of soap or fatty acids that collect the hydrophobic ink particles and eventually agglomerate them. The impact caused by an ink-particle agglomerate hitting an air bubble of matching size is a function of the collision velocity and the mass of the agglomerate—it must be sufficiently high to penetrate the laminar boundary layer of the bubble. As a result, the

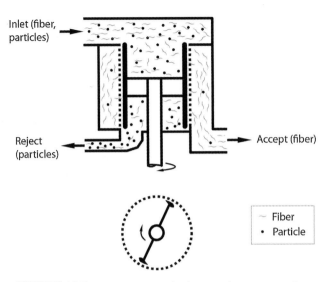

FIGURE 12.7 Pressure screen for low consistency-screening.

agglomerate adheres to the bubble. If the density of the new ink particle bubble agglomerate is still lower than that of water, it will float, i.e. rise to the foam layer at the top of the flotation cell, where it can be discharged by a rotating foam scraper. The accept, i.e. cleaned pulp, is then fed to a number of downstream flotation cells, each of which serves to further increase the brightness of the pulp. The foam of all these cells is collected and fed to a secondary flotation line to recover the fibers and minerals unintentionally floated in the primary cells.

The flotation result is controlled primarily by two parameters: the size of the air bubbles and the energy dispersion. With tiny bubbles, rather small particles are discharged, with larger bubbles larger particles. Particles smaller than 10 μm or larger than 100 μm cannot be efficiently removed by flotation. These particles include printing inks, stickies, fillers, coating pigments and binders. Flotation also removes fillers in some cases. It has a certain deashing effect that is often desirable.

The specific energy demand for flotation is 20–50 kWh/t paper, and for washing 20 kWh/t paper (Holik, 2000).

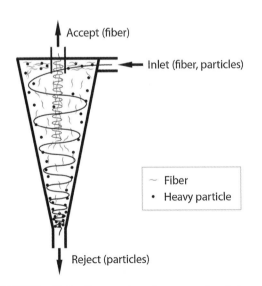

FIGURE 12.8 Cleaner for the removal of heavy contaminants.

12.2.5 Dewatering

Dewatering increases the stock consistency of the pulp. Compared to the washing process, where a solid–solid separation occurs

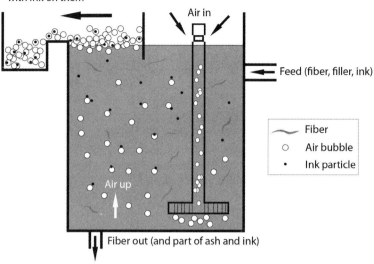

FIGURE 12.9 Flotation principle.

(e.g. separation of mineral particles from fibers), dewatering separates liquids from solids. Different technologies are suitable for this: disk filters as well as devices with belts or screw presses, all of which are of major importance.

Dewatering is aimed at:

- Ensuring the optimal stock consistency for process steps such as bleaching or dispersing,
- Closing water loops to confine the diffusion of contaminants or additives,
- Saving space in storage vessels and
- Compacting rejects or sludge for the economical recovery of these materials.

Belt filters and disk filters remove the water up to a discharge consistency of about 10–12% and have an average specific energy demand of 5 kWh/t paper. Belt filter presses and twin wire presses achieve stock consistencies of more than 30% with a specific energy demand of less than 10 kWh/t paper, while screw presses allow a thickening up to 65% consistency with a specific energy demand of 10–15 kWh/t paper. Screw presses require an inlet consistency of at least 10%. These devices are therefore usually combined with belt filters or disk filters (Figure 12.10).

12.2.6 Dispersing and Bleaching

Dispersing and bleaching are used for products that must meet high optical quality demands, e.g. magazine paper, newspaper and writing paper. Dispersing and bleaching are not commonly used in the case of brown grades.

The objective of dispergers is to comminute any printing ink particles present in the suspension and separate any printing ink particles still attached to the fibers to eliminate them later by flotation. Comminution and separation serve two aims: producing floatable particles or reducing the size of particles to such an extent that they are below the visibility threshold of the human eye, i.e. <50 μm. This gives the paper product a homogeneous appearance, since the printing ink particles will no longer produce large dirt specks, although the brightness diminishes—the paper turns gray, with typical losses amounting to two to three brightness points. The functional principle of a disperger is actually quite simple: a rotor rotates against a stator. The rotor/stator plates are equipped with many large cutting surfaces, much like refiners. The energy consumption is high: if dispersing is viewed as a single process step, it requires most of the electric energy used in stock preparation—between 50 and 90 kWh/t paper at a temperature of 60–100 °C.

These losses can be compensated in the bleaching step. Bleaching agents are typically added to the pulp just upstream of or in the disperger itself, since the disperger operates at approximately 30% stock consistency and mixes them thoroughly. A typical combination is using hydrogen peroxide in the first step after preflotation and sodium hydroxide in the second step at the end of the process. What is important in this context is to couple an oxidative with a reductive bleach in order to achieve maximum effect.

Problems may occur in hydrogen peroxide bleaching due to high amounts of iron and catalase. Iron ions can be rendered harmless by using sodium silicate, and catalase formation is prevented by high temperatures in the bleaching tower. Once both contaminants are under control, the hydrogen peroxide concentration can be kept at low levels. It is advisable to conduct sodium hydrosulfite bleaching in the absence of oxygen and at high temperatures (70 °C) in a bleaching tube. If the reaction were carried out in a batch tower, too much oxygen would greatly inhibit bleaching and the effect would be negligible.

12.2.7 Limits and New Trends in Stock Preparation

12.2.7.1 Selectivity

The processes currently used for the treatment of recovered paper are predominantly separation processes designed to eliminate undesired substances such as printing inks, stickies, sand or plastics. In most cases, these substances are used for energy recovery or put in landfills. In selective separation, individual substances can continue to be recovered and utilized. For instance, calcium carbonate or kaolin can be recovered in order to be reused as coating pigment or internal sizing aid. Separated fines and plastics can be used in biopolymers, wood plastic composites or paper plastic composites. The selectivity of state-of-the-art separation processes, however, is not yet good enough to achieve this without considerable problems. If their selectivity was increased, however, the products of tomorrow could be manufactured from today's rejects.

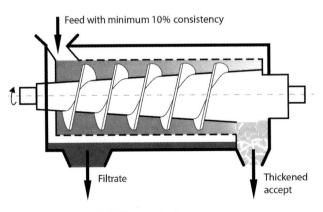

FIGURE 12.10 Screw press.

12.2.7.2 Energy Demand

State-of-the-art stock preparation processes must greatly dilute the suspension for deinking in order to effectively remove the printing ink particles. The large water volumes required for this, however, result in high energy consumption for pumping. To prevent this, the stock consistency must be as high as possible in stock preparation, which cannot be achieved with current flotation technology. Several ongoing research projects are devoted to this theme, and are developing and testing alternative slushing and deinking technologies, among other.

The latter include adsorption deinking (Handke et al., 2012) and ultrasound deinking (Großmann et al., 2010), for example. In adsorption deinking, adsorbers take up the printing ink particles at high stock consistencies of 15% and, at the same time, have a dispersing effect (ball mill effect); 90% less water needs to be pumped in the facility owing to the high stock consistency. Ultrasound deinking attempts to separate the printing ink particles from the fibers by cavitation. This is possible in particular when stock consistencies are high.

12.2.7.3 New Printing Inks

Current flotation technology can remove only hydrophobic printing inks like conventional offset and gravure printing inks. Digital printing inks or water-soluble inks cannot be deinked, or only with great difficulty. Because they offer advantages to printers in some cases, their market share continues to grow. Flexo ink in particular is being used to an ever greater extent in the graphic sector, and causes problems in deinking. The printing ink colloidal dissolves in the water and discolors the entire pulp. This is no serious problem for brown paper, but disturbing in the case of graphic paper. The removability of problematic printing inks by conventional technology is a problem that must yet be solved by printing ink manufacturers. Because they lack any incentive to do so, new removal technologies must be developed that can deal with problematic printing inks.

12.2.7.4 Water Consumption

The specific volume of effluents produced during papermaking has declined in Germany from approximately 45 m^3/t paper in the 1970s to 10–11 m^3/t in 2010. The level has stagnated at approximately 10 m^3/t paper since 2000. Some paper mills have completely closed the water circuits in their production facilities, which means that no effluents accumulate and only the water (approximately 1 m^3/t paper) evaporated during production must be replaced by fresh water. A completely closed water circuit, however, gives rise to a variety of problems due to gradually increasing concentrations of dissolved matter and contaminants. These problems can be solved only with great technical effort. The potential of current papermaking processes for reducing the water demand and effluent volume is therefore limited (Jung et al., 2011).

12.2.7.5 The Impact of Climate Change

The internationally agreed upon goals to reduce greenhouse gas emissions in view of the expected climate change can only be realized in the short term in many cases by the increased use of regenerative energy sources, in particular wood. This is reflected by the drastically intensified global competition for this resource noticeable today, and the impressive rise in wood prices.

The lack of forest resources in some emerging countries and the widely accepted necessity to use far more bio-based than fossil products and fuels are bound to increase the competition for wood all over the world. Against this background, it is very likely that the demand for recovered paper and board as well as alternative fibers will increase further at least in the midterm.

However, the utilization rate of recovered paper cannot be further increased in many countries. For this reason, the global paper industry will become increasingly dependent on wood or alternatives fibers, even in the midterm. The use of recovered paper or wood for energy generation will be promoted by subventions in some cases. Current studies show, for example, that more wood will be used for energy generation than as a material source in future (Mantau, 2012). This development is detrimental to the environment and the economy, since the material recovery from recovered paper and wood saves fossil fuels, and the longer process chains involved mean more jobs. The important task here is to take political countermeasures and, above all, ensure the recycling of recovered paper and wood.

References

Apotheker, S., October 1990. Curbside collection: complete separation versus commingled collection. Resource Recycling.

Bayerisches Landesamt für Umweltschutz, 2003. Klimaschutz durch effiziente Energieverwednung in der Papierindustrie – Nutzung von Niedertemperaturabwärme. Augsburg: s.n., ISBN 3-936385-36-x.

Blasius, K., Escabasse, J.-Y., Farreira, B., 2011. Verbesserte Ausbeute. Recycling Magazine 2.

Blechschmidt, J., 2010. Taschenbuch der Papiertechnik. Carl Hanser, München. ISBN 978-3-446-41967-4.

CEPI, 2012. Key Statistics 2011, European Pulp and Paper Industry. CEPI, Confederation of European Paper Industries, Brussels.

Göttsching, L., Pakarinen, H., 2000. Recycled Fibre and Deinking. Helsinki: s.n. In: Papermaking Science and Technology. Book 7.

Großmann, H., Fröhlich, H., Wanske, M., 2010. The potential of ultrasound assisted deinking. Norfolk. In: Proceedings of the 2010 TAPPI PEERS Conference.

Handke, T., Schrinner, T., Großmann, H., 2012. Adsorption Deinking – A New Approach for Higher Energy Efficiencies in Paper Recycling, vol. 9, 1, pp. 32–38.

Hanecker, E., 2006. Altpapierrohstoff: Rohstoff Altpapier und seine Aufbereitung. In: Strauß, J. (Ed.), Papierherstellung im Überblick, Manuskript EK 697. PTS Papiertechnische Akademie, München.

Holik, H., 2000. Unit operations and equipment in recycled fiber processing. In: Pakarinen, L., Göttsching, H. (Eds.), Papermaking in Science and Technology, Recycled Fiber and Deinking, Part 7. Fapet Oy, Finnland.

Jung, H., et al., 2011. Wasser- und Abwassersituation in der deutschen Papierindustrie. Wochenblatt für Papierfabrikation 9, 478–481.

Mantau, U., 2012. Holzrohstoffbilanz Deutschland, Entwicklungen und Szenarien des Holzaufkommens und der Holzverwendung von 1987 bis 2015. Hamburg: s.n., p. 65.

Network, Environmental Paper, 2011. The State of the Paper Industry: 2011 [PDF] Asheville, NC 28801: s.n.

Stawicki, B., Read, B., 2010. COST Action E48 – the Future of Paper Recycling in Europe: Opportunities and Limitations. The Paper Industry Technical Association (PITA), Dorset, Great Britain.

VdP, 2013. Papier 2013 ein Leistungsbericht. Verband deutscher Papierfabriken e.V., Bonn.

CHAPTER 13

Plastic Recycling

Li Shen, Ernst Worrell

Copernicus Institute of Sustainable Development, Utrecht University, Utrecht, The Netherlands

13.1 INTRODUCTION

Plastics are synthetic organic polymers, mainly made from petrochemical feedstocks. Since the invention of the first plastic or polymer in the early 1900s (i.e. Bakelite), and the development of polyethylene in the 1930s, the number of plastics has increased dramatically, and so has their use. As society has steadily increased its use of plastics, plastic waste management has become a growing concern around the world. Today, about 280 Mt of plastics are produced annually. The key producers of plastics are China (23%), Europe (21%), North America (20%) and the rest of Asia (excluding China; 21%).

Plastics offer many advantages for specific applications. Plastics are easy to shape, do not corrode or decompose only slowly and the characteristics can be adapted to the specific needs by using composites or adding specific layers or additives. These characteristics are also the source of some of the problems associated with plastics. From the association of chlorinated plastics with dioxin formation in waste incinerators to the plastic soup, the durability and wide use of plastics contributes to significant waste management problems. Recycling of plastics is one of the key strategies to reduce the environmental problems associated with plastic waste. Moreover, plastics are mainly made from petrochemical feedstocks, which have increased in price over the past decades, are concentrated in a relatively small number of locations and will be in short supply within the next decades. Hence, plastic recycling will reduce reliance on fossil fuels.

Recycling of plastic wastes, especially post-consumer, has had a slow start. Compared to other commonly used materials such as paper, glass and metals, recovery and recycling rates are generally low. Even in countries with advanced waste management systems and long experience in recycling, plastic recycling rates are typically much lower than rates of other materials. This is also partly caused by the huge variety of plastics, additives and composites used. This variety is one of the key advantages of plastics, and one of the reasons for its versatility. Yet, this variation is a problem for recycling.

In this chapter, we will first discuss trends in plastic consumption, and types of plastics used. This is followed by a description of the current situation with respect to plastics in the waste stage and recycling. We will primarily focus on post-consumer waste streams, as little data are available on pre-consumer plastic waste. Then, the typical processes in plastic recycling are discussed, followed by a discussion of the environmental benefits, using one of the key recycled plastics (PET) as an example. We end with some concluding remarks on the future challenges and opportunities of plastics recycling.

13.2 USE OF PLASTICS

There is a huge variety of plastics, and the applications similarly are extremely wide. Today, plastics are found in virtually all the things we do and use throughout all aspects of life. Still, a few uses dominate current plastics consumption. In the EU-15 member states, packaging was the dominant user of plastic (38%), followed by building and construction, household wares, automotive, electronics and a variety of applications. Figure 13.1 depicts the distribution of applications of plastics.

There are many types of plastics. Plastics are subdivided in thermoplasts and thermosets. Thermoplasts do not undergo chemical changes when heated and can be molded again. The main thermoplasts are polyethylene (PE), polypropylene (PP), polystyrene (PS) and polyvinyl chloride (PVC). PE can be subdivided in high-density polyethylene (HDPE), low-density polyethylene (LDPE) and linear low-density polyethylene (LLDPE), based on the way the polymer chains are distributed in the plastic. Thermosets can melt and take shape once; i.e. after solidification they stay solid. In the thermosetting process, a chemical reaction occurs that is irreversible. Polyurethane (PUR) is one of the most used thermosets. Recycling of thermosets is harder, and can only be done in a chemical process (see below).

Figure 13.2 depicts the key plastic types used in the EU-27. The global distribution of plastic types is comparable to the distribution found in the European Union. National distribution of uses and plastic types may vary, depending on specific circumstances. Some plastic types are used predominantly in specific applications. For example, the use of PVC in packaging has been reduced in many countries, but PVC, together with PUR and PS, is one of the key plastics used in the building and construction sector. In the packaging markets, the polyolefins (i.e. PE, PP), PET and PS dominate. Table 13.1

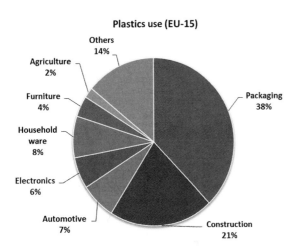

FIGURE 13.1 Key applications of plastics. *Distribution is based on data from the EU-15 member states.*

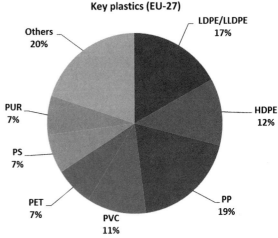

FIGURE 13.2 Key plastic types as used in the European Union (EU-27) in 2011. *Based on data from Plastics Europe.*

TABLE 13.1 Typical Applications of Common Plastic Types

Number	Abbreviation and Name	Typical Applications
1 PETE	PET: polyethylene terephthalate	Bottles and flasks for soft drinks, mineral water, detergents and pharmaceutical products; blister packs; packaging for ready meals
2 HDPE	HDPE: high-density polyethylene	Thick-walled applications such as bottles and flasks, barrels, jerry cans, crates and jails; films for refuse bags; packaging for carpets and instruments
3 V	PVC: polyvinyl chloride	Blister and press-through packs for medication; films for perishables
3 V	PC: polycarbonate	Refillable milk bottles; specific refillable packaging for liquids
4 LDPE	LDPE: low-density polyethylene	Foil and film, such as shrink wraps, tubular film, sacks and covering wraps for bread, vegetables, fruit and carrier bags
4 LDPE	LLDPE: linear low-density polyethylene	Ultra-thin films: elastic wrap foil or stretch films
5 PP	PP: polypropylene	Buckets, crates, boxes, caps for bottles or flasks, transparent packaging for flowers, plants, confection products; yogurt and dairy product cups; industrial adhesive tapes

(*Continued*)

TABLE 13.1 Typical Applications of Common Plastic Types (cont'd)

Number	Abbreviation and Name	Typical Applications
6 PS	PS: polystyrene	Food service disposables; boxes and dishes for meat products and vegetables; boxes for ice; boxes for video tapes
6 PS	EPS: expanded polystyrene	Buffer packaging for household devices, electronics and instruments; flasks and pipettes for the medical industry; egg packaging and fast food packaging
7 OTHER	Other	Other packaging

The numbers refer to the code used to sort plastics for recycling.

summarizes some of the key uses of the main plastic types used.

Plastics are an attractive material for many applications because plastics are easy to shape and material characteristics can be adapted or even tailored to the specific application. This is often done by adding additives or forming composites through adding layers of special materials (with e.g. special barrier properties for oxygen, carbon dioxide, ultraviolet light) in the plastic. While this affects the application of the plastic positively (e.g. reduced material use), this can become a barrier in recycling, effectively limiting or even blocking recycling of the plastic. Typically, additives and fillers are about 20% of the plastic weight, and can be even more for some applications. The distribution of key plastics as depicted in Figure 13.2 may, hence, contain other materials. Figure 13.2 is based on the main plastic in the product application.

Currently, bio-based plastics are still a minute fraction of the total volume of plastics used. However, in the last decade, the emerging bio-based plastics experienced a rapid growth. The global capacity of the emerging bio-based plastics has increased from 0.1 Mt in 2003, to 1.16 Mt by 2011. The global production of bio-based plastics is likely to grow strongly in the next decade and to reach 2.3 Mt in 2013 and 5.0 Mt in 2016 (Shen et al., 2010a; European Bioplastics, 2013). The key bio-plastics at this moment are bio-based PET (39%), PLA (polylactic acid) and blends (16%), bio-based PE (17%) and starch plastics and blends (11%) (European Bioplastics, 2013).

13.3 PLASTIC RECYCLING

The 2011 plastic use in the European Union is estimated at 47 Mt (based on the consumption of plastic convertors and processors), and post-consumer wastes are estimated at 25 Mt. These figures demonstrate that besides the production of industrial wastes, there is also a considerable stock buildup in society. For example, many plastics currently used in construction (21% of total plastic consumption in the EU) end up in buildings with long lifetimes. These plastics will become available as waste in the future as these buildings are renovated or demolished.

This stock buildup also explains why there are considerable differences between the plastic uses and plastic wastes. Figure 13.3 depicts the distribution of use categories in consumption and plastic wastes. Packaging is by far the dominant factor contributing to plastic waste (76%), while it is about 38% of plastic use. Especially, plastic waste from construction, automotive, electronics and household wares trail behind consumption figures, because of retention of the products in stocks in society.

In 2012, based on figures of PlasticsEurope, 25.2 Mt of post-consumer plastic waste was collected in the European Union. Of this, more than 60% (15.6 Mt) was recovered and 40% (9.6 Mt) was disposed of with municipal solid waste (MSW). Of the 15.6 Mt of recovered plastic waste, about 6.6 Mt was actually recycled, while the remainder was likely used as refuse-derived fuel (RDF) or incinerated in MSW incinerators with energy recovery (about 9 Mt). While Europe can be considered a leader in plastics recycling, only about 26% of plastic waste is recycled. This is much lower than that of other materials (e.g. the recycling rate of paper and board is nearly 72% in Europe).

Large differences in recovery and recycling rates can be observed across countries. For example, in 2012, Switzerland, Germany, Austria, Luxembourg, Belgium, Sweden, Denmark, the Netherlands and Norway had very high recovery rates. These countries had a ban on landfilling, which may have contributed to high recovery rates. The recovery rates for other member states varied between lows of 12–15% and about 60%. The recycling rates also varied considerably, but are all far below the recovery rate. The highest rate of recycling was seen in Norway (about 37%) and the lowest in Malta at 12%. Some countries do not have energy recovery systems and therefore all recovered wastes were recycled (e.g. Lithuania, Cyprus and Malta), while Germany, Sweden, Belgium, the Netherlands and Norway have recycling rates of 30–37% compared to very high recovery rates (92–98%), meaning that a large fraction of the recovered plastic is actually incinerated.

Typically, plastics are recycled mechanically (see below). However, some plastic cannot be mechanically recycled because of the characteristics of the material (e.g. thermosets) or because of the low purity (caused by mixing with other plastics, being composites, being laminated with multiple layers or presence of additives and fillers). Feedstock recycling is then an option. Here, the low-purity mixed plastics are converted into syngas or liquid fuel via a pyrolysis process and to be used as a reducing agent (as carbon monoxide) in blast furnaces in the iron and steel industry to replace coke or mineral oil. For plastic-containing electrical and electronic waste, feedstock recycling also recovers precious metals. In 2008, only 0.07 Mt

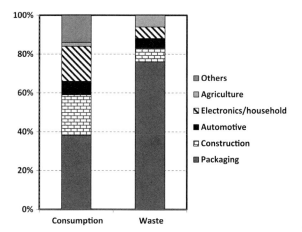

FIGURE 13.3 Distribution of plastic applications in consumption and waste in the European Union.

of plastic was processed by feedstock recycling in the European Union. In Europe, currently about 0.22 Mt of capacity has been installed for this type of feedstock recycling to provide an alternative reducing agent for the iron and steel industry.

Next to mechanical recycling and feedstock recycling, some plastics such as PET can be chemically recycled. In this technology, used plastics are depolymerized into monomers, which can be repolymerized to produce virgin polymer. Currently, chemical recycling is hardly used on a commercial scale.

13.4 MECHANICAL RECYCLING

Today, mechanical recycling is the main technology used to recycle plastics. Mechanical recycling of plastics in the European Union slowly increased from just below 5 Mt in 2006 to about 6.6 Mt in 2012 (comparable to a recycling rate of 26%). Mechanical recycling typically includes four steps. The collected material is first sorted (step 1). The sorted material is shredded (step 2), and then washed and dried (step 3). The material can then be melted and reprocessed to make pellets (that can be used by a manufacturer or convertor) or products directly (step 4). Below we discuss the key steps in more detail.

Sometimes, relatively pure streams of waste plastic are recovered (e.g. through dedicated collection systems for plastic bottles through a refund system). These streams can deliver high-quality recycled material with desired properties, requiring only minimal sorting to remove any impurities (e.g. bottle caps, labels).

Sorting. The first step in a recycling process is the collection of the waste and transporting it to a sorting plant to sort the plastic mix. The collected mix of plastics usually consists of various types of plastics, especially for post-consumer waste. Pre-consumer (production) waste can be (relatively) pure. In the case of post-consumer waste, there are usually still non-plastic impurities in the material, such as labels and little pieces of metal. These are first removed. Separation into various types of plastics is required to improve material quality. The separation of mixed plastics is challenging, and various techniques are applied in varying combinations. The process and the order in which they are used are defined by the composition of the mixed waste stream. Eddy current separator, sink–float separation, drum separators/screens, induction sorting, X-ray technology and near infrared (NIR) sensors are the most commonly used separation techniques. Most companies use a combination of different techniques to obtain sufficiently pure streams. The design of a sorting installation may be tailored to the incoming stream of plastic waste, to optimize sorting efficiency. The achievable purity level is a trade-off between (energy) costs and market requirements. This, by definition, will result always in impurities. The maximum achievable purity by separating mixed plastics waste is 94–95%. High-quality recycled material should have at least a purity of 98% to be used as input into manufacturing processes. This means that in subsequent steps further refining is necessary if high-quality recycled material is to be produced. The key sorting technologies are as follows:

- *Induction sorting*. Material is sent over a conveyor belt with a series of inductive sensors underneath. These sensors locate different types of metal, which are then separated by fast air jets.
- *Eddy current separator*. An "eddy current" is an electric current that occurs when changing the magnetic field within a conductor, and is used to separate non-ferrous metals.
- *Drum separator/screen*. These separate materials on the basis of the particle size. Waste is fed into a large rotating drum, which is perforated with holes of different sizes. Materials smaller than the diameter of the

holes drop through, and larger particles remain in the drum.
- *Sink—float separation.* The plastic waste is separated based on the specific weight of the material relative to the fluid. In water, some plastics (e.g. PET, PVC and PS) will sink, and others will float (e.g. PE, PP and EPS). After the sink—float separation, the fractions still need to go through another process to separate the different polymers.
- *X-ray technology.* X-rays can be used to distinguish between different types of material based on density.
- *Near infrared sensor.* When plastics are illuminated, they mostly reflect light in the near infrared (NIR) wavelength spectrum. The NIR sensor can distinguish between different materials based on the way they reflect light. The fractions are then blown with an air jet to separate them from the mixed stream. This is currently the preferred method by the industry to accurately identify the many different polymers.

Shredding. The next step is to reduce the size of the scrap, to enable processing larger pieces of plastic waste and to improve the density of the material for more efficient storage and transport. A shredder consists of rotating blades driven by an electric motor, some sort of grid for size grading and a collection bin. Materials are added to the shredder by a hopper. The product of shredding is a pile of plastic flakes.

Washing. After the plastics are shredded into small flakes they are washed. Although most post-consumer waste is washed, this is not true for all input material used for recycling. Some regrind or even agglomerate is processed instantly. Cold or hot water, up to 60 °C, may be applied. Cold water use may result in increased use of chemicals (e.g. sodium hydroxide) and mechanical energy. The waste water from the washing is often treated internally for internal reuse. The washed plastic flakes are dried until they contain less than 0.1 wt% moisture and are ready for reprocessing.

Reprocessing. There are different techniques for reprocessing, with the most common ones being as follows:

- *Agglomeration.* This process is mainly applied for recycling plastic films. Film is cut into small pieces, heated by friction (to allow for agglomeration) and subsequently cooled down by injecting water. This is usually carried out in a single machine. The product is referred to as crumbs or agglomerates and is not ideal for further processing. The agglomerates can be mixed with plastic flakes for extrusion. The agglomeration process is very energy intensive (approximately 300—700 kWh/t of plastic). Agglomeration can be avoided for most injection and extrusion grade plastics.
- *Extrusion.* The most commonly applied technique for reprocessing recycled plastic is extrusion. It is commonly used to manufacture pellets from virgin plastics, and also used to produce pellets from recycled material. The material is blended and then injected in the extruder from a hopper. It comes into contact with a rotating screw that forces the plastic flakes forward into a heated barrel at the desired melt temperature of the molten plastic (ranging from 200 to 275 °C). The pressure allows the plastic (beads) to mix and melt gradually as they are pushed through the barrel. The melt is degassed to remove oils, waxes and lubricants. Finally the molten plastic is pushed through a sieve to remove impurities, cooled and pelletized.

After the agglomeration or extrusion phase, the agglomerates or pellets are ready for the final processing step, the choice of which is determined by the final product:

- *Injection molding.* The first stage of this manufacturing process is identical to that of extrusion; i.e. the pellets are molten again, but then the polymer is pressed with high pressure into a split mold. The mixture is pressed into the mold until it is full and, after cooling to allow the plastic to solidify, the mold is opened and the product can be removed.

- *Blow molding.* The spiral screw of the extruder forces the plasticized polymer through a die. A short hollow tube and compressed air is used to expand the tube until it fills the mold and obtains its required shape. This manufacturing technique is used for manufacturing bottles and other containers.
- *Film blowing.* Film blowing is a process used to manufacture items such as plastic bags. It is a technically more complex process and requires high-quality raw materials. The process involves blowing compressed air into a thin tube of polymer to expand it to the point where it becomes a thin film tube.
- *Fiber extrusion.* The melt extruded polymer (polyester) is sent to the spinneret where the spinning of filament takes place. The filaments then pass through a denier setter before they enter the finishing steps where the spun filaments are drawn, dried, cut into staple fiber and finally baled for sale.

Today, in practice, high-quality recycled material can be made from containers made from HDPE, PP and PET (especially bottles). Specialized streams can also be recycled to provide high-quality recycled material. For example, in several countries, including The Netherlands, PVC window frames are collected separately and recycled into material for new window frames. However, other plastic products, especially films and foils, represent a large challenge for the recycling process. Films are hard to sort, and the large use of plasticizers and other additives makes it impossible to guarantee a high purity of the recycled material. Hence, this material may currently be used to replace other materials than plastics (e.g. in construction) or may be incinerated as RDF. The environmental and economic impact of recycling may hence vary considerably from case to case, needing careful analysis.

13.5 IMPACT OF RECYCLING

The environmental impact of recycling depends on many factors, ranging from the energy used for collecting the plastic waste to the type of material and application being replaced by the recycled plastic (which is partly determined by the quality of the recycled material). The markets for recycled plastic are still limited, but growing. Applications of the recycled material vary and will affect the overall environmental benefits and economics of recycling.

In this section, we discuss the results of a life-cycle analysis (LCA) of PET bottle recycling (Shen et al., 2010b). Globally, almost three times as much PET is used for textile production as for packaging. In Europe, the amount of collected post-consumer PET bottle waste increased from 0.2 Mt in 1998 to 1.6 Mt in 2011. About 52% of all used PET bottles in Europe were collected for recycling in 2012. It is expected that PET bottle waste collection in Europe will continue to increase.

Figure 13.4 shows the flowsheet of the production of recycled PET flakes. After the baled bottles are opened, loose bottles are sorted by color and material type. Transparent (uncolored) bottles have a higher economic value than blue and green ones. The unwanted color fractions and unwanted materials (e.g. paper and metal) are either sold as by-products or disposed of in local MSW management facilities. MSW can be incinerated with or without energy recovery, or landfilled, depending on the available local infrastructure. Next, the bottles are sorted. The plant in Europe uses automated sorting (through color recognition technology), while most Asian producers use manual sorting. Some producers include a step using hot water washing to remove labels before

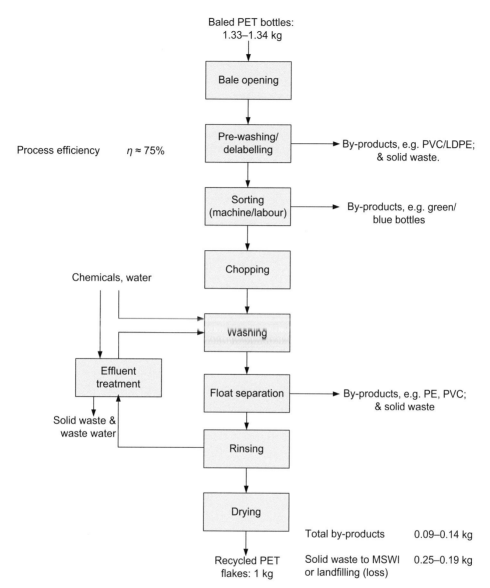

FIGURE 13.4 Schematic presentation of PET bottle recycling (including mass balance). *Source: Shen et al. (2010b).*

the sorting process. The plastics labels are either sold as by-products (mainly consisting of LDPE and/or PVC) or sent to local MSW management. The bottles are then chopped into flakes, followed by a float separation step to separate PET from other plastics (e.g. HDPE caps) based on density differences. PE obtained from this step is sold as a by-product. The PET flakes are then washed in a cleaning solution, rinsed and dried. In some production lines, a second chopping step (also called "fine crushing") is required to ensure that the PET flakes meet the quality requirements. Finally, the dried PET flakes are ready to be transported to a pellet plant or a fiber plant.

Mechanical recycling is the physical conversion of flakes into fiber or other products (e.g. bottles or sheets) by melt extrusion. Currently, there are two ways to produce recycled fiber from mechanical recycling. In flake-to-fiber recycling (1), flakes are off-loaded and they are dried in a column dryer before they are melt extruded. The extruded polymer is filtered before it passes through the spinneret where filament spinning takes place. After the filaments pass a denier setter, they enter the finishing process where the spun filaments are drawn, dried, cut into staple fiber and baled. The entire process has 1% process solid waste. In many other mechanical recycling plants (2), flakes are first extruded into pellets and then converted into fiber and other products. PET flakes are dried prior to the melt extrusion step. The extruded polymer is further purified through a filtration step. After cooling, the polymer is pelletized and dried. The PET pellets are then delivered to the fiber spinning plant where they are melt-spun into filament fiber. A small amount of ethylene glycol (EG) may be added to meet the final quality requirements.

In chemical recycling, PET polymer is broken down into monomers or oligomers via various depolymerization technologies. Chemical recycling has a higher cost than mechanical recycling. It usually requires a large scale in order to become economically feasible. The important advantage of chemical recycling is that the quality of virgin polyester can be achieved. Current commercially available chemical recycling technologies include glycolysis, methanolysis and alkaline hydrolysis. The glycolysis of PET yields the oligomer bishydroxyl ethylene terephthalate. The process is usually conducted in a wide range of temperatures, 180–250 °C, with excess EG and in the presence of catalysts. After the glycolysis process, the oligomer passes through a fine filtration step before it is repolymerized into PET. The recycled polymer is then spun into fiber. The entire process creates about 5% process solid waste. In methanolysis, PET is depolymerized with methanol to DMT (dimethyl terephthalate) and EG in the presence of catalysts under 2–4 MPa pressure and 180–280 °C. The reaction mix is cooled and DMT is recovered from the mix via precipitation, centrifugation and crystallization. The recycled polymer is then converted into fiber via spinning and finishing processes. The methanolysis route is used commercially. The depolymerization of the DMT is technically identical with the polymerization process leading to virgin PET. The recycled amorphous PET polymer is sent to the fiber production plant or upgraded to bottle grade resin for bottle production.

The LCA of the different processes has shown that recycled polyester fibers produced from mechanical recycling have lower environmental impact than virgin polyester. The results (Shen et al., 2010b) show that recycled polyester fibers offer important environmental benefits over virgin polyester. Depending on the allocation of the benefits of open loop recycling, energy savings of 40–85% and reductions in greenhouse gas (GHG) emissions of 25–75% can be achieved.

However, PET fiber used in textiles is a product that cannot be further recycled via mechanical recycling. Chemical recycling is technically possible, but the economic feasibility of large-scale operation is still to be proven. Recycled fiber produced from chemical recycling offers lower impacts compared to virgin polyester. While mechanical recycling has a better environmental profile than chemical recycling, chemically recycled fibers can be applied in a wider range of applications than mechanically recycled fibers. Another important way of recycling PET bottles is bottle-to-bottle recycling. In this case, a close-loop recycling system is formed. In theory, PET can be recycled multiple times before it is finally converted into fiber.

The environmental impact of such "cascading" recycling systems has been studied by Shen et al. (2011). The results show that based on the current global demand of PET bottles and fiber, the recycling system, which includes both bottle-to-fiber recycling and bottle-to-bottle recycling, can offer 20% of impact reduction in both primary energy demand and GHG emissions. Multiple recycling trips can further reduce the environmental impact by maximally six percentage points, due to the lower share of bottle demand (35%) compared to that of the fiber demand (65%).

13.6 CONCLUSIONS AND OUTLOOK

Today, plastic recycling is still limited compared to most other bulk materials. Recycling rates for plastics are increasing in many countries around the world, while an international market for recycled plastics is developing. To further increase the recycling rate, the industry faces a number of challenges. The key challenge is the quality of the recovered and recycled material. The use of fillers, additives and composites on one hand and contamination with other materials in the recovery process on the other hand are key issues to come to a cleaner recycled plastic product. This can be achieved by good integration of collection, recovery and separation technology. The solution will consist of an effective combination of these three steps, and may vary for different product streams and waste management systems. This is observed in relatively high recycling rates of PET and HDPE, while the recycling rates of films are still very low (typically not exceeding 15%). Dedicated collection and recovery systems will help increase raw material quality. Furthermore, increasing the recycling rate will also need to include the design stage of products. While common in certain areas, in the key uses of plastics (e.g. packaging), design for recycling is still limited. The use of different plastics for different parts of packaging (e.g. caps, labels, containers) and the use of a variety of additives and fillers are currently barriers for high recycling rates. Improving design practices may reduce this variety and enable better recycling.

New technology will be needed to produce high-purity plastic from recovered material, to improve separation efficiency and effectiveness (e.g. separating the growing number of bio-based plastics), and to handle mixed plastic wastes to still produce a high-quality recycled product. Chemical recycling may be part of this, and will need further development to decrease costs and energy use.

Finally, currently a large part of the recovered plastic is "down cycled", or used as fuel to generate energy in industrial processes or incinerators. Better monitoring is needed to track actual recycling rates and the fate of the recovered material, and to allow optimization and "quality cascading" of recycled plastic to generate the highest economic and environmental gains. Today's information in most countries does not enable such analysis and optimization. Such a system, and the growing international market, will need an international, widely accepted and used quality certification system, integrated into a transparent monitoring system.

References

European Bioplastics, 2013. Bioplastics: Facts and Figures. European Bioplastics, Berlin, Germany.

Shen, L., Worrell, E., Patel, M.K., 2010a. Present and future development in plastics from biomass. Biofuels, Bioproducts & Biorefining 4, 25–40.

Shen, L., Worrell, E., Patel, M.K., 2010b. Open-loop recycling: a LCA case study of PET bottle-to-fibre recycling. Resources, Conservation and Recycling 55, 34–52.

Shen, L., Nieuwlaar, E., Worrell, E., Patel, M.K., 2011. Life cycle energy and GHG emissions of PET recycling: change-oriented effects. International Journal of Life Cycle Analysis 16, 522–536.

Further Reading

Gutowski, T., Dahmus, J., Albino, D., Branham, M., 2007. Bayesian Material Separation Model with Applications to Recycling. MIT, Cambridge, MA.

Joosten, L.A.J., Hekkert, M.P., Worrell, E., 2000. Assessment of the plastic flows in The Netherlands using streams. Resources, Conservation and Recycling 30, 135—161.

Kuczenski, B., Geyer, R., 2010. Material flow analysis of polyethylene terephthalate in the US, 1996—2007. Resources, Conservation and Recycling 54, 1161—1169.

Patel, M.K., Jochem, E., Radgen, P., Worrell, E., 1998. Plastic streams in Germany — an analysis of production, consumption and waste generation. Resources, Conservation and Recycling 24, 191—215.

Patel, M.K., von Thienen, N., Jochem, E., Worrell, E., 2000. Recycling of plastics in Germany. Resources, Conservation and Recycling 29, 65—90.

PlasticsEurope, 2012. Plastics — the Facts 2012. An Analysis of European Plastics Production, Demand and Waste Data for 2011. PlasticsEurope, Brussels, Belgium.

Welle, F., 2011. Twenty years of PET bottle to bottle recycling—an overview. Resources, Conservation and Recycling 55, 865—875.

Welle, F., 2013. Is PET bottle-to-bottle recycling safe? Evaluation of post-consumer recycling processes according to the EFSA guidelines. Resources, Conservation and Recycling 73, 41—45.

CHAPTER 14

Glass Recycling

Thomas D. Dyer

Division of Civil Engineering, University of Dundee, Dundee, UK

14.1 INTRODUCTION

Glass is one of the oldest synthetic materials, with evidence of its production in ancient Egypt dating back to at least 3000 BC (Martin and MacFarlane, 2002). Its optical clarity, hardness and strength mean that it has made its way into every aspect of human activity, with significant roles in glazing, packaging, vehicles, housewares, electrical equipment and fibers in insulation products and composite materials. Moreover, the material has played an instrumental role in many advances in science and technology.

Whilst the recycling of glass has been conducted almost since its discovery, and *reuse* of glass containers has been carried out throughout the nineteenth and twentieth century(Emmins, 1991), large-scale recycling in recent times started only in the 1970s. This has partly been driven by legislation aimed at diverting waste away from landfill. Glass is a readily recycled material, in that it can be remelted and reformed into articles with the same characteristics as the original material—"closed-loop" recycling. However, to maximize recycling levels, it has been found that alternative "open-loop" routes must also play a role. This chapter examines the issues relating to the practicability and viability of closed-loop glass recycling, and examines open-loop options.

14.2 TYPES OF GLASS

The different applications for glass require different chemical compositions. The largest glass product output is container glass, which is based on a fairly consistent composition of about 15% sodium oxide (Na_2O), 9% calcium oxide (CaO) and slightly more than 70% silicon dioxide (SiO_2). The only significant variation in composition results from the different colors that are produced. These are most commonly clear (known as "flint"), brown ("amber") and green, although other colored products are manufactured in smaller quantities. Green glass is achieved through the addition of small quantities of chromium, whilst amber glass contains both chromium and iron. In the case of flint glass, there is a requirement for very low concentrations of other metals in the raw materials.

Like flint container glass, flat glass for glazing is required to be free from metallic impurities to achieve optical clarity. However, the composition is usually not dissimilar to container glass. There is typically a proportion of magnesium oxide (MgO), which is largely absent from container glass.

Glass fiber can take two forms. *Continuous-filament glass fiber* can be used as reinforcement in composite materials as well as, in woven form, in protective clothing and electrical insulation (Edgar et al., 2008). *Glass wool* is finer and tends to be used in applications such as building insulation. Glass wool is usually based on soda-lime—silicate formulations. However, continuous-filament glass fiber can have a range of compositions depending on the purpose the fiber is to be put to. High stiffness fibers ("E-glass") are normally composed of alumino-borosilicate glass. Raising the alumina content leads to higher strengths ("S-glass"), whilst reducing it leads to enhanced corrosion resistance (Shelby, 1997).

The type of glass used in vehicles depends on where it is used, although it is normally soda-lime—silicate. Windshields are composed of laminated glass, whereas side and rear windows are normally tempered glass.

The composition of glass components used in electronic equipment also varies. Conventional light bulbs are normally made from soda-lime—silicate glass, although fluorescent lighting tubes and low-energy bulbs are usually made from borosilicate glass. Cathode ray tubes (CRTs), although becoming less common, are of a similar composition, but contain quantities of metals such as barium, strontium, zirconium and lead (Edgar et al., 2008). CRTs are progressively being replaced by alternative screen technologies such as liquid crystal display (LCD) screens, which usually contain borosilicate glass substrates.

Glass used in housewares is again dependent on the application. However, one important group of products is glass cookware that is resistant to thermal shock. These articles are made from borosilicate glass, which is typically 80% silica, 13% boron trioxide oxide (B_2O_3), 4% sodium oxide and 2% alumina (Al_2O_3) (Edgar et al., 2008). Borosilicate glasses are also used in scientific and medical glassware.

14.3 GLASS MANUFACTURE

Glass is manufactured by melting a suitable combination of raw materials in a furnace and using various processes to form it before it is cooled. A wide variety of raw materials are used, but, in soda-lime—silicate glass, the most commonly encountered materials are silica sand (SiO_2), soda ash (sodium carbonate, Na_2CO_3) and calcium carbonate ($CaCO_3$). In borosilicate glasses, boron trioxide is obtained from borax ($Na_2B_4O_7 \cdot 10H_2O$). Al_2O_3 is obtained from feldspar minerals or alumina derived from refinement of bauxite.

A wide variety of techniques are available for manufacturing glass articles. This includes the automated blowing of glass containers in split molds, the manufacture of flat glass through the float glass method (molten glass is floated over a bath of molten tin) and the manufacture of fibers through rotary wool forming (where centrifugal force is used to force molten glass through very fine holes in a rotating drum) and mechanical drawing, where molten glass is allowed to flow under the action of gravity through holes in a platinum plate (known as a "bushing") and the resulting filaments wound together into a strand that is, in turn, wound onto a drum (Bralla, 2007).

14.4 GLASS RECOVERY FOR REUSE AND RECYCLING

Glass is a material that, in many of its forms, is reusable. This is particularly true of bottles and other glass vessels, which are still able to satisfy their role as a container long after their original use is over. This has led to manufacturers and governments recognizing that savings in terms of energy and resources could be achieved through the recovery and reuse or recycling of glass containers.

As mentioned previously, one of the earliest examples of initiatives devised to recover glass was container deposit schemes. Soft drink advertising in Ireland and England provides clear evidence of manufacturers' efforts to recover containers through such means from as early as 1800 (Emmins, 1991).

In many parts of the world, container-deposit legislation requires consumers to pay a small deposit on containers at the point of sale, refunded when the container is returned at a redemption point. Redemption points can take a number of forms, including establishments selling deposited products and automated "reverse vending machines" where a user disposes of an end-of-life container and is presented with a receipt that can be exchanged for cash.

Container deposit legislation has been put into effect in The Netherlands, Norway, Denmark, Sweden and Germany, as well as parts of the United States, Canada and Australia. The materials involved need not be glass, but glass has, in most cases, been the starting point of many of these initiatives.

Whilst the first deposit schemes had refill of glass articles in mind, and refilling still continues, there has been a shift toward recycling the materials back into manufacture. Whether this is a favorable development is debatable—as will be discussed later, the benefits of recycling glass are considerable, but it is unlikely that they exceed those of reuse.

Financial incentives to return glass containers are not always necessary. Bottle banks began to appear in the 1970s, and numbers have grown: the first UK bottle bank was installed in 1977, and growth in numbers has been almost exponential—there are now more than 50,000.

In some instances, particularly in the catering and drinks industry, sorting glass containers in terms of color presents practical problems. In the UK, this problem has been addressed by permitting collection of "comingled" glass containers. Curbside glass collection is now carried out by a number of local authorities in the UK. These schemes include both color-sorted and mixed collection (Waste and Resources Action Programme).

Much of the recent growth in the recovery of container glass has been driven by legislation. In Europe the driver has been the Packaging and Packaging Waste Directive (1994), which is discussed in more detail later. However, other EU directives have driven the growth in recycling of glass from other sectors. In particular, the Waste Electrical and Electronic Equipment Directive (2003) and the End-of-Life Vehicles Directive (1997) both set targets for recycling of materials from these sources.

Currently recycling of glass from electrical and electronic equipment is relatively limited. This is partly because glass from these products is often contaminated with other substances. Moreover, many electrical and electronic products contain scarce elements whose high market value detracts from the recovery of glass. For instance, the interior of fluorescent tubes is coated in a layer of phosphor, which itself will also have become contaminated with toxic mercury. This renders the quality of glass unsuitable for recycling, and emphasis is thus placed on recovering valuable mercury and rare earth phosphors from the lighting elements. Nonetheless, technologies are being developed to efficiently remove the contaminated phosphor layer, which may realize better opportunities to recover the glass (Rey-Raap and Gallardo, 2013).

Tempered glass from vehicles is readily recovered for recycling. Laminated windscreens require slightly more specialized techniques to separate the glass from the polymer layer. This usually involves breaking up the glass using a roller or similar mechanical action, followed by removal of the glass.

The recovery of flat glass for recycling must be conducted in a manner that prevents it from being mixed with other construction materials. This is because the physical characteristics of glass make it indistinguishable from the bulk of the other inorganic, nonmetallic materials in demolition waste. Therefore, much of the glass arising from construction for recycling tends to come from the replacement of glazing during maintenance and refurbishment. Additionally, contamination of glass with putty and other materials has acted as an obstacle to recycling, but progress is now being made (Tandy and Way, 2004). The construction industry is now moving toward a philosophy of designing structures for deconstruction—ensuring that at the end of a structure's life it can be systematically dismantled, with optimal separation of materials.

14.5 REUSE OF GLASS

Before discussing recycling, it is worth briefly discussing the alternative option of reuse. Reuse of returnable glass containers is a practice that has become well established in many parts of the world. This approach is often a highly sustainable one, since the material undergoes no reprocessing prior to return into the system. A recent study in the UK on delivered-to-doorstep containers for milk found that returnable glass containers compared well in environmental impact terms with competing products (Fry et al., 2010). The shift toward globalization has meant that glass containers for food and drink are diverse in size and shape and often originate from remote locations, making container return uneconomical and impractical in terms of logistics. Thus, in many countries, recovery of intact glass bottles has seen a decline, magnifying the need for recycling.

14.6 CLOSED-LOOP RECYCLING OF GLASS

The introduction of waste glass ("cullet") back into the glassmaking furnace will cause it to melt and mix with the other raw materials. This recycling process—sometimes referred to as "remelt"—can be repeated indefinitely without any loss of performance.

The quality of the glass product required defines whether glass recycling can be employed for a given manufacturing process. For instance, the inclusion of cullet in the manufacture of flat glass is not always possible, because of the strict requirements in terms of raw material purity needed to achieve the desired optical properties. However, return of cullet to container manufacture is eminently possible, and can tolerate soda-lime—silicate glass from other sources.

For cullet to be recycled back into container manufacture, it is essential that color sorting of glass is conducted. This is because each color of glass can tolerate limited levels of contamination with other colors. As a result of this, quality requirements limit the level of cullet in green containers to 90%. Amber glass and flint glass are more sensitive to contamination, and so technical limits on cullet content are 70 and 60%, respectively. Color sorting can be achieved by collection schemes that require the public to sort their returned bottles. Collection of mixed-color glass from commercial sources has led to the introduction of automated "electronic eye" sorting equipment that is able to identify rogue particles and remove them through air jets, or similar.

Automated color sorting of glass will make up just one stage of several in the cullet processing scheme. These are required because glass from containers is likely to be present alongside other materials. These include materials used for labeling and closure, such as metals, plastics and paper. There may also be other materials present in the recovery streams, including stone, ceramics and other types of glass, such as borosilicate materials, which cannot be recycled in soda-lime—silicate glass manufacture. Processes include:

- Hand sorting—to remove obvious contamination;

- Crushing;
- Sieving—to also remove some foreign materials such as metals and plastics;
- Magnetic sorting—to remove ferrous metals
- Eddy current sorting
- Vacuum sorting

14.7 ENVIRONMENTAL BENEFITS OF CLOSED-LOOP RECYCLING OF GLASS

The closed-loop recycling of soda-lime–silicate glass has a number of environmental benefits. First, it is returning a mineral resource back into the loop—a practice that is an essential aspect of sustainable development. This is arguably of less importance than for other materials, since none of the raw materials used to make glass are notably scarce: silica sand and limestone are mineral resources of which there are large mineral reserves. Soda ash either is produced using the Solvay process, which uses readily available raw materials, or is produced from the mineral *trona* ($Na_3(CO_3)(HCO_3) \cdot 2H_2O$), which is present as large deposits around the globe (United States Geological Survey, 2012). However, despite this apparent abundance, it should be stressed that not all of these minerals are readily accessible or suitable for glass manufacture. For instance, only a small proportion of available sand deposits are suitable for glass manufacture—only six UK sites produce silica glass suitable for the manufacture of flint container glass (British Geological Survey, 2009).

There are other more significant environmental benefits to recycling. One of these is the reduction of waste. Glass is an inert waste, since it is insoluble and chemically unreactive under normal conditions. From the perspective of the contemporary philosophies with regard to landfill disposal, this is of lesser concern when compared to noninert wastes—glass will not decompose to produce greenhouse gases and is unlikely to leach harmful substances that could threaten groundwater quality. Nonetheless, the disposal of glass in this way represents a wasteful practice that utilizes land which could otherwise be put to more productive use. Landfill space in many areas of the world is at a premium, and reducing disposal is considered an essential part of moving toward sustainable land use.

Disposal of waste containing glass through combustion is also an undesirable fate for the material. Glass is clearly noncombustible, and so where combustion is carried out in a waste-to-energy plant with the aim of generating electricity and local heating, no benefit is gained from its presence. However, glass has a further detrimental effect, in that its high specific heat capacity acts as a sink for heat that would otherwise be put to beneficial use (Edwards and Schelling, 1999).

The most significant benefit to the environment deriving from the return of cullet to the glassmaking process is the reduction in energy demand and the consequent reductions in pollution.

The processes for extracting and processing raw materials for glass manufacture all require energy. More significantly, melting these raw materials in the glass furnace requires large quantities: temperatures between 1400 and 1600 °C are required. Whilst the temperature requirement is not affected by the inclusion of cullet, energy savings are realized as a result of the behavior of cullet at elevated temperatures in contrast to other raw materials. To increase the temperature of cullet, energy is required exclusively to raise the temperature. In the case of soda ash, calcium carbonate and dolomite, additional energy is required to thermally decompose them. The resulting energy saving is shown in Figure 14.1. Similarly, a study by flat glass producers determined that for each 10% increase in cullet used in glass manufacture, there was a 2–3% saving in energy, with

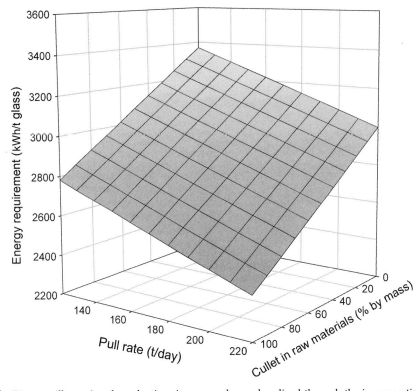

FIGURE 14.1 Diagram illustrating the reductions in energy demand realized through the incorporation of cullet in the raw material feedstock (UNID, 1993).

a consequent 230 kg reduction in CO_2 emissions (Glass, 2010).

14.8 THE GROWTH OF GLASS RECYCLING

Archaeological evidence of the recycling of glass has been found in sites deriving from the earlier days of the Byzantine Empire and the later Roman Empire (Degryse et al., 2006). However, in contemporary times, glass recycling started on a major scale in the 1970s, driven partly by concerns about energy security.

Levels of recycling have steadily increased since then in most of the more developed world. In many European countries, levels of recycling were high—77% by 1994 in the Netherlands—although performance was varied. The UK is a useful example, since it has moved from a relatively low glass recycling rate to a currently more respectable level. This has in part been driven by legislation, namely the Producer Responsibility Obligations (Packaging Waste) Regulations (1997), which implemented the EU Packaging and Packaging Waste Directive (1994) in the UK. The regulations placed responsibilities on the producers and handlers of packaging to achieve periodically updated recycling targets.

The Directive initially set targets to recover 50—65% of packaging waste overall by 2001, with 25–45% being used for recycling. This target was subsequently revised in 2004 to recover a minimum of 60% by 2008, with recycling levels at 55 and 80% by the same date. Specific targets for recycling container glass were also set at 60%.

The Producer Responsibility Obligations (Packaging Waste) Regulations have undergone periodic amendments and revisions, each increasing UK recovery targets for container glass from 38% in 1999 to 81% in 2010. This highest target is not set to change until at least 2017, the result of indications that this level is optimal. This fits well with theoretical calculations that have shown, when the energy requirement of collection of glass is compared to the energy savings of closed-loop recycling, the optimum benefit is obtained at this level (Edwards and Schelling, 1999).

The overall effect of these developments is shown in Figure 14.2, which plots the rise in levels of UK glass recycling relative to the total amount of glass in the waste stream and container glass production. The influence of the Directive on levels of recycling after 1997 is clearly evident.

Not all of this recycling is closed-loop recycling. The major reason for this is that, in the UK, there exists a barrier to total absorption of cullet arisings into remelt. This is the result of significant differences between container glass products made in the UK and those on sale to consumers.

The majority of container glass output in the UK is flint, whereas the majority of container glass entering the waste stream is green. This is due to the importation of products such as wine, and the exportation of flint glass-contained products. This disparity led to concerns that a green cullet surplus would arise as recycling rates increased. This did not occur, and the reason is evident from Figure 14.3,

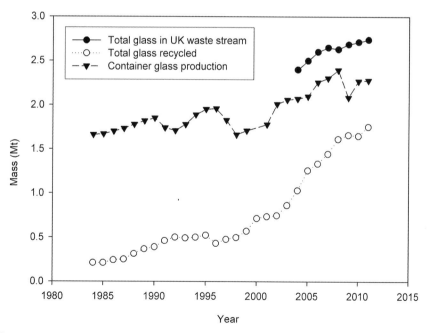

FIGURE 14.2 Total glass recycled in the UK alongside the total amount of glass in the waste stream (British Glass Manufacturers' Confederation, written communication).

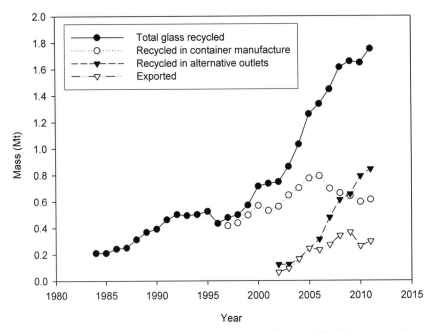

FIGURE 14.3 Recycling routes for glass in the UK (British Glass Manufacturers' Confederation, written communication).

which shows the emergence of two outlets for the surplus material: export and recycling in alternative applications—open-loop recycling. Open-loop recycling is also often necessary for the recycling of borosilicate glass products and soda-lime–silicate glass with more exotic chemical additions.

The development of alternative outlets for glass has involved considerable research and development work to match the physical and chemical characteristics of the material with the needs of a wide range of different sectors. These outlets are discussed in detail in the next section.

14.9 OPEN-LOOP GLASS RECYCLING

A number or alternative outlets for glass exist. Some of these also involve remelting glass, but involve forming into very different products. Other applications involve melting glass in the presence of other materials, with different effects depending on the manufacturing conditions. Additionally, there are a number of applications that exploit the properties of granular broken glass. Finally, the possibility of exploiting the chemistry of glass in the manufacture of commercially useful compounds has also been explored by researchers. All of these outlets for glass are examined below.

Whilst in some cases, the application is an established or growing practice, in other instances the outlet has been explored only from a research and development perspective or at a pilot scale. Nonetheless, such uses have been included on the grounds that feasibility has been demonstrated.

14.9.1 Alternative Glass Products

One alternative glass product in which cullet can be recycled by remelting is in the manufacture of *ballotini*—small glass spheres (typically

1–60 μm in diameter) that are highly reflective and used in applications that include road markings and signs, reflective safety clothing and projection screens. Ballotini are normally required to be colorless, and so only flint cullet can be used in their manufacture.

Glass wool insulation consisting of short glass fibers can also be manufactured from cullet. Tolerances for recycled glass are relatively high, and there is no color requirement. One manufacturer is currently using up to 80% cullet. Continuous filament glass fiber cannot currently be made using cullet, as a result of strict manufacturing requirements (Rodriguez Vieitez et al., 2011).

Glass wool insulation comes under the broader category of mineral wools. Research has been carried out recently into means of converting inorganic byproducts into mineral wools using waste glass as a melting aid and to reduce the viscosity of the melt. Such materials include incinerator bottom ash, sludge from dredging (Scarinci et al., 2000) and wastes from hydrometallurgy (Pelino, 2000).

Foamed glass is an established commercial product that can tolerate 98% cullet in its manufacture. The manufacture of foamed glass normally involves the introduction of foaming agent particles into glass, which is then heated to a temperature of between 700 and 900 °C, where the glass becomes a viscous fluid and the foaming agent evolves gas that forms bubbles (Hurley, 2002).

Foamed glass can take the form of loose aggregate, pelletized aggregates and blocks and shapes. Lightweight aggregates are discussed in a later section. The foaming agents used include silicon carbide (SiC), manganese oxide (MnO_2), hematite (Fe_2O_3) calcite or other forms of calcium carbonate ($CaCO_3$) and calcium sulfate ($CaSO_4$) (Hurley, 2002; Marceau, 1967). Foamed glass blocks and shapes are mostly used for insulation purposes, since they have low thermal conductivities coupled with relatively high compressive strengths.

14.9.2 Ceramics

The use of glass cullet as an ingredient in ceramic products is also a possible outlet. Two different approaches to using the material can be employed: use as a sintering additive in fired ceramics, and in the manufacture of "glass ceramics".

Cullet plays two roles when used as a sintering additive. Firstly, it becomes fluid at lower temperatures than the other constituents, acting to bind the solid particles together, reducing porosity. Additionally, solid particles dissolve into the liquid glass, leading to sintering at relatively low temperatures (Figure 14.4).

Research has been conducted into the use of cullet in conventional clay-based fired ceramics, with much success. Trials of the manufacture of sanitary ware containing 4.5% glass have been shown to reduce the energy requirement (compared to the conventional raw material composition) by reducing the firing temperatures (Hancock, 2011). Another study has examined the use of glass as a replacement for feldspars in porcelain stoneware tiles (Tucci et al., 2004). The optimal level of replacement was 10% of the feldspar (1.8% of the total). Similar results were obtained for tiles made using up to 10% cullet (Brusatin et al., 2005).

Laminated windscreen glass has been employed in brickmaking, with the polymer layer used as a combustible component that decomposes during firing to leave pores in the brick (Mörtel and Fuchs, 1997). This gives the product a higher strength at reduced density.

Glass ceramics are ceramic materials comprising both crystalline and amorphous phases. They are manufactured by holding glassy materials within temperature ranges at which conversion of some of the amorphous material to crystalline phases occurs. Much research has been conducted into the use of cullet in glass ceramics containing other waste products and byproducts, as outlined in Table 14.1.

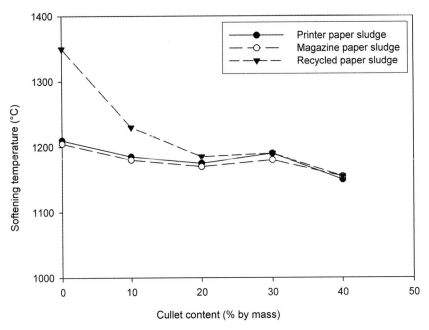

FIGURE 14.4 Reduction in softening temperature with cullet content in ceramic materials manufactured using mixtures of cullet and sludge from paper production (Asquini et al., 2008).

TABLE 14.1 Conditions for the Production of Glass Ceramics Incorporating Glass Cullet from Various Studies

Other Inorganic Material	Cullet Content, % by Mass	Nucleation Temperature, °C	Crystallization Temperature, °C	Reference
NUCLEATION-CRYSTALLIZATION				
Incinerator fly ash/ pollution control residue	35	560, 35 min	1000, 120 min	Romero et al. (1999)
Incinerator bottom ash	50, 90	1000, up to 8 h		Barbieri et al. (2000)
SINTER CRYSTALLIZATION				
Dredging sludge	20	940, 5 h		Brusatin et al. (2005))
Sunflower husk ash	30, 40, 50	1400, 1300, 1250 (for 30, 40 and 50% cullet respectively), no time specified		Quaranta et al. (2011)
PETRURGIC METHOD				
Coal-derived bottom ash	0–60%	Cooling from 1500 °C to ambient at 1–10 °C/ min		Francis et al. (2002)

All of the studies that have examined the possibility of using cullet in this way cite the manufacture of ceramic tiles as the most likely application for such a product.

14.9.3 Glass as Abrasive Media

Glass is a hard material, which makes it suitable for use as abrasive media. This normally takes the form of using crushed cullet in a conventional shot-blasting configuration. However, more advanced use of recycled glass in this area has also been initiated, including the use of particles of aluminum oxide embedded in "sponge media" made from recycled glass to clean metal surfaces (Anonymous, 2005). Successful results have also been achieved with media made from molded and sintered glass powder used for vibratory mass finishing in place of polyester resin bonded ceramic particles (Benjarungroj et al., 2012).

14.9.4 Glass as Unbound Aggregate

The strength, stiffness and hardness of glass make it highly suited for use as an aggregate, once crushed into a granular form. Some of the properties of glass with respect to aggregate performance are shown in Table 14.2.

One of the higher value routes to using loose recycled glass aggregate is as granular sub-base material in highway construction. In the UK, requirements for these materials are defined in the Highways Agency document, "Specification for Highway Works" (Highways Agency, 1998). The document defines various types of granular material, although Types 1 and 2 are the most commonly used. Type 1 is the coarser of the two types suitable for use in most sub-base applications, whereas Type 2 is finer and used only in the construction of lightly trafficked pavements.

Glass satisfies the requirements for both types of material, namely high abrasion resistance (measured using the Los Angeles test), resistance to freezing and thawing (a result of the absence of pores in the material) and non-susceptibility to "frost heave" (Henry and Morin, 1997).

In many cases a minimum California bearing ratio (CBR) value may also be required, which is a measure of strength. The minimum requirement will normally be at least 15%, and values for glass (Table 14.2) come relatively close to this. The most effective solution is to combine recycled glass with crushed rock. Typically a glass content of up to 15% by mass of rock yields CBR values identical to that of the rock alone (Clean Washington Center, 1998; Younus Ali et al., 2011).

The direct shear strength of recycled glass is also provided in Table 14.2, indicating behavior typical of a well-graded angular sand or gravel, making the material suitable for unbound sub-base applications, as well as for embankment fills, structural and nonstructural fill and backfill (Disfani et al., 2012; Wartman et al., 2004).

14.9.5 Glass in Bituminous Highway Pavements

The use of recycled glass aggregate in bituminous mixtures for highway construction is becoming an increasingly accepted practice. Glass can be used in the base, binder course and surface course in highway construction, although the levels used in the surface course tend to be lower (10–15%) (Huang et al., 2007) than in the other layers (up to 40% for one UK example) (Cemex). When used in the surface course, since large glass particles can produce glare, limit skid resistance and possess the potential to damage tires and cause injury to people, particle sizes are normally limited to less than 4.75 mm (Cemex; USDTFHA, 1997).

One of the main disadvantages of glass over conventional aggregates in this application is that glass particles are essentially free of

TABLE 14.2 Aggregate Characteristics of Recycled Glass of Three Different Maximum Particle Sizes (Disfani et al., 2011, 2012)

Characteristic	Particle Size		
	Coarse, 19 mm Down	Medium, 9.5 mm Down	Fine, 7.75 mm Down
Specific gravity	2.48	2.50	2.50
Flakiness index, %	94.7	85.4	N/A
Modified Proctor test values			
$\gamma_{d,max}$, kN/m^3	N/A	19.5	17.5
w_{opt}, %	N/A	8.8	10.0
LA abrasion value, %	24.8	25.4	27.7
California bearing ratio (CBR), %	—	31–32	18–21
Direct shear test			
σ_n (30–120 kPa)	N/A	52–53°	45–47°
σ_n (60–240 kPa)	N/A	50–51°	42–43°
σ_n (120–480 kPa)	N/A	—	40–41°

porosity, making it impossible for bitumen to form a strong bond. This results in a susceptibility to "stripping"—the deterioration of the bond between aggregate and bitumen as a consequence of contact with water, leading to a loss of strength. However, the addition of lime at a level of about 2% rectifies this problem (Su and Chen, 2002).

Benefits of using glass in bituminous highway pavements include improved skid resistance and light reflectivity in surface courses (Su and Chen, 2002). The permeability of pavements containing glass tends to be higher, which is potentially a benefit to safety since it allows water to drain more freely from road surfaces in rainy weather. It has also been found that the stiffness modulus of asphalt increases with glass content, and that asphalt mixes containing glass become less susceptible to fatigue (Arabani, 2011; Arabani et al., 2012).

14.9.6 Glass in Concrete

The use of glass aggregate in concrete is also possible, although concerns regarding damaging alkali–silica reaction (ASR) have limited this. ASR is a reaction that occurs between aggregates containing reactive silica and hydroxide ions associated with alkali–metal ions from the cement. The reaction leads to the disruption of the glass structural network, leading to the formation of an open gel (Glasser, 1992). This undergoes hydration, leading to swelling, the development of stresses and, eventually, cracking.

The best means of avoiding ASR is to include additions in the cement fraction of the concrete that are capable of controlling ASR. These include fly ash, ground granulated blast furnace slag (GGBS) and metakaolin (Hansen et al., 1999; Dhir et al., 2009; Ling et al., 2011). Lithium admixtures are also effective (Topcu et al., 2008).

Another feature of the ASR behavior of glass is that it displays a "pessimum" particle size, below which expansion is not observed. For soda-lime–silicate glass, this is typically about 1 mm, and so by using glass as fine aggregate (sand) below this size, ASR is avoided (Figure 14.5).

The effect of glass aggregate on the fresh properties of concrete depends on particle size. Glass used as coarse aggregate generally has the effect of improving workability (de Castro and de Brito, 2013; Topcu and Canbaz, 2004), whilst finer particle sizes reduce workability (Tan and Du, 2013; Park et al., 2004).

Enhanced characteristics include reduced thermal conductivity, increased heat capacity (Alani et al., 2012) and improved retention of strength after exposure to fire (Ling et al., 2012a).

Similar research has also been conducted using CRT glass and glass from LCD screen substrates (Ling and Poon, 2012a,b; Ling et al., 2012b; Chen et al., 2011). The lead content of CRT glass is normally considered an environmental concern. However, research has attempted to exploit this by using it as fine aggregate in concrete and mortar used for radiation shielding in structures where X-rays are used (Ling and Poon, 2012a; Ling et al., 2012b). Pretreatment in dilute nitric acid to remove lead at the glass surface permitted it to be used as 100% of the fine aggregate (Ling and Poon, 2012b).

14.9.7 Light-Weight Aggregate

In a previous section, the foaming of molten glass using gas-evolving particles has been discussed. This process can also be employed to produce light-weight aggregate, and often employs glass cullet. Such aggregates are now well-established products and have applications

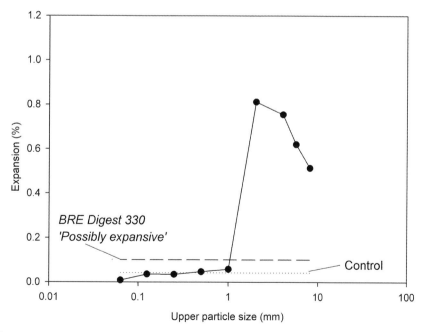

FIGURE 14.5 The 52-week expansion of concrete tested in accordance with BS 812-123 containing glass fine aggregate present as narrow particle size fractions. The dashed line represents the threshold of concern in UK guidance.

in concrete construction as well as in horticultural applications.

Light-weight aggregate can take two different forms—granulated aggregate and pelletized aggregate. Granulated aggregate is one possible product of the manufacture of continuous foamed glass sheets, which are broken up into suitable particle sizes (Hurley, 2002).

Pelletization involves forming granules comprising powdered cullet and a foaming agent. These granules are then passed through an inclined rotary furnace during which melting and gas evolution occur.

Whilst pelletized aggregates can be made with just glass and foaming agent, it is increasingly common for other materials to be included. Research into such products has included fly ash from coal-fired power generation (Kockal and Ozturan, 2011; Kourti and Cheeseman, 2010), sludges from aggregate and stone processing (Ducman and Mirtič, 2009) and dredging sludges (Wei et al., 2011; Chiou and Chen, 2013).

Low density coupled with high strength and relatively low water absorption are desirable characteristics, which can be achieved through finer glass particle sizes and higher glass content (Kockal and Ozturan, 2011; Wei et al., 2011; Chiou and Chen, 2013).

As for normal-weight glass aggregate, there exists a concern regarding ASR, when lightweight aggregate containing glass is used in concrete. Mixed results have been obtained with regard to this in the literature (Limbachiya et al., 2012; Mladenović et al., 2004; Ducman et al., 2002), and it probably necessary to establish ASR reactivity for a given product where this reaction may be an issue.

14.9.8 Glass as a Cement Component in Concrete

Soda-lime—silicate glass is a pozzolanic material: in a finely divided form, it will undergo a reaction with calcium hydroxide to form calcium silicate hydrate (CSH) gel (Dyer and Dhir, 2001). CSH is the main contributor to strength development in hydrated Portland cement, and calcium hydroxide is another of its hydration products, meaning that glass combined with Portland cement is a viable cement. The pozzolanic reaction is identical to ASR, the difference being that the particles are sufficiently small to prevent the formation of large volumes of gel before it dissolves to form CSH gel (Douglas and El-Shamy, 1967; Idir et al., 2013).

The possibility of using glass powder as a means of controlling ASR (in the same way as GGBS and fly ash) has been suggested (Idir et al., 2010). The ASR-controlling capability of glass powder has been confirmed, although it is less than that of fly ash (Shi et al., 2005). Moreover, testing indicates that the duration over which the controlling effect is observed is finite (Dyer and Dhir, 2010).

Investigation of borosilicate glass (from lighting tubes) as a filler in concrete has identified the material as being particularly pozzolanic, exceeding the performance of silica fume, which is generally a highly reactive material (Shakhmenko et al., 2012).

14.9.9 Filtration Media

The ability of volumes of granular glass to permit rapid percolation of fluid has meant that the material has found application in filtration applications. Studies examining the ability of recycled glass as a filter medium have found that it is potentially extremely effective at removing suspended solids from water.

A study examining the effect of particle size on the use of glass filter beds on the tertiary treatment of wastewater identified that a particle size of 0.5—1.45 mm was most appropriate (Horan and Lowe, 2007). This represented a compromise that avoided the rapid blocking (or "blinding") by trapped solids of finer glass particles and the low particle-capture efficiency of coarser particle sizes. The amount of media required was 10% less compared to sand. This approach has been developed into a two-stage filtration system with the first filter containing medium-grade (0.5–1 mm) glass particles and the second containing a finer grade (0.2–1 mm) (Lavender, 2009).

Other researchers have found that the smooth surface of glass appears to limit the extent to which solid particles become attached, although this can be rectified through the use of coagulants (Soyer et al., 2010), or through the use of dual-media beds containing glass and coarser anthracite (Soyer et al., 2013). Similarly, glass in combination with larger clay balls in a pebble matrix filter has been shown to be suitable for pretreatment of drinking water (Rajapakse and Fenner, 2011).

14.9.10 Zeolites and Other Silicate Products

It has already been seen that solubilization of silicate ions from glass occurs at high pH (Figure 14.6). This is further enhanced at elevated temperatures (Mavilia and Corigliano, 2001). For this reason, a number of researchers have examined the possibility of producing chemical products through hydrothermal reaction of glasses in high pH conditions. A range of chemical compounds can be produced, including double-layer silicates and zeolites, which have applications as absorbents, ion exchange media and desiccants. The conditions used for the synthesis of three such substances from cullet are shown in Table 14.3.

Similar conditions have also been used to produce articles with relatively high tensile strength (Veloza et al., 1999). The process involved using hydrothermal hot pressing to

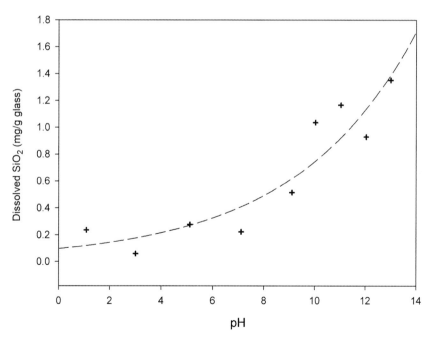

FIGURE 14.6 Influence of pH on the dissolution of SiO_2 from soda-lime–silicate glass into water (Bacon, 1968).

TABLE 14.3 Conditions Used to Produce Various Products from Glass Cullet Using Hydrothermal Processes

Product	Temperature	Other Reactants	References
Rhodesite, zeolite Na—P1	150 °C	NaOH	Grutzeck and Marks (1999)
Tobermorite	100 °C	NaOH, CaO	Coleman (2011)
Zeolite A	Extraction at 105 or 140 °C, subsequent reaction temperature with $Na_2O_2Al_2O_3$ not specified, but probably 20—175 °C	Extraction of silica using NaOH followed by reaction with $Na_2O \cdot Al_2O_3$	Mavilia and Corigliano (2001)

shape mixtures of glass and water at temperatures of about 200 °C.

14.10 CONCLUSIONS

The closed-loop recycling of glass into glassmaking is clearly a sustainable practice, but barriers exist to the complete return of all recovered glass to glassmaking. For this reason a number of alternative outlets for cullet have become established, and further outlets are being developed. Whether the alternatives under development will become established depends on the associated economic and practical benefits. The issue of economic viability of alternatives is one which has not been explored in this chapter. In some instances, researchers have included economic analyses. However, a critical influence on viability is the market price for cullet, and this varies considerably around the world and can fluctuate considerably over relatively short periods of time. For this reason, the economics of open-loop recycling of glass are probably best explored on an individual case basis.

References

Alani, A., MacMullen, J., Telik, O., Zhang, Z.Y., 2012. Investigation into the thermal performance of recycled glass screed for construction purposes. Construction and Building Materials 29, 527—532.

Anonymous, 2005. Shipyard takes softer, greener approach to surface prep. Journal of Protective Coatings and Linings 22, 12—15.

Arabani, M., 2011. Effect of glass cullet on the improvement of the dynamic behaviour of asphalt concrete. Construction and Building Materials 25, 1181—1185.

Arabani, M., Mirabdolazimi, S.M., Ferdowsi, B., 2012. Modeling the fatigue behaviours of glasphalt mixtures. Scientia Iranica 19, 341—345.

Asquini, L., Furlani, E., Bruckner, S., Maschio, S., 2008. Production and characterization of sintered ceramics from paper mill sludge and glass cullet. Chemosphere 71, 83—89.

Bacon, F.R., 1968. The chemical durability of silicate glass — part one. Glass Industry 49, 438—446.

Barbiori, L., Corradi, A., Lancellotti, I, 2000. Bulk and sintered glass-ceramics by recycling municipal incinerator bottom ash. Journal of European Ceramic Society 20, 1637—1643.

Benjarungroj, P., Harrison, P., Vaughan, S., Ren, X.J., Morgan, M., 2012. Investigation of thermally treated recycled glass as a vibratory mass finishing media. Key Engineering Materials 496, 104—109.

Bralla, J.G., 2007. Handbook of Manufacturing Processes. Industrial Press.

British Geological Survey, 2009. Mineral Planning Factsheet: Silica Sand.

Brusatin, G., Bernardo, E., Andreola, F., Barbieri, L., Lancellotti, I., Hreglich, S., 2005. Reutilization of waste inert glass from the disposal of polluted dredging spoils by the obtainment of ceramic products for tiles applications. Journal of Material Science 40, 5259—5264.

Cemex. Glasphalt product datasheet. http://www.cemexliterature.co.uk/pdf/Asphalt_Glasphalt_Data_Sheet.pdf. (accessed 2013.).

Chen, S.-H., Chang, C.-S., Wang, H.-Y., Huang, W.-L., 2011. Mixture design of high performance recycled liquid crystal glasses concrete (HPGC). Construction and Building Materials 25, 3886—3892.

Chiou, I.J., Chen, C.H., 2013. Effects of waste-glass fineness on sintering of reservoir-sediment aggregates. Construction and Building Materials 38, 987–993.

Clean Washington Center, 1998. A tool kit for the use of post consumer glass as a construction aggregate. Report No.GL-97-5.

Coleman, N.J., 2011. 11 Å tobermorite ion exchanger from recycled container glass. International Journal of Environment Waste Management 8, 366–382.

de Castro, S., de Brito, J., 2013. Evaluation of the durability of concrete made with crushed glass aggregates. Journal of Cleaner Production 41, 7–14.

Degryse, P., Schneider, J., Haack, U., Lauwers, V., Poblome, J., Waelkens, M., Muchez, P., 2006. Evidence for glass 'recycling' using Pb and Sr isotopic ratios and Sr-mixing lines: the case of early Byzantine Sagalassos. Jounal of Archaeological Science 33, 494–501.

Dhir, R.K., Dyer, T.D., Tang, M.C., 2009. Alkali-silica reaction in concrete containing glass. Materials and Structures 42, 1451–1462.

Disfani, M.M., Arulrajah, A., Bo, M.W., Hankour, R., 2011. Recycled crushed glass in road work applications. Waste Management 31, 2341–2351.

Disfani, M.M., Arulrajah, A., Bo, M.W., Sivakugan, N., 2012. Environmental risks of using recycled crushed glass in road work applications. Journal of Cleaner Production 20, 170–179.

Douglas, R.W., El-Shamy, T.M., 1967. Reactions of glasses with aqueous solutions. Journal of American Ceramic Society 50, 1–8.

Ducman, V., Mirtič, B., 2009. The applicability of different waste materials for the production of lightweight aggregates. Waste Management 29, 2361–2368.

Ducman, V., Mladenović, A., Šuput, J.S., 2002. Lightweight aggregate based on waste glass and its alkali-silica reactivity. Cement and Concrete Research 32, 223–226.

Dyer, T.D., Dhir, R.K., 2001. Chemical reactions of glass cullet used as a cement component. Journal of Materials in Civil Engineering 13, 412–417.

Dyer, T.D., Dhir, R.K., 2010. Evaluation of powdered glass cullet as a means of controlling harmful alkali-silica reaction. Magazine of Concrete Research 62, 749–759.

Edgar, R., Holcroft, C., Pudner, M., Hardcastle, G., 2008. UK Glass Manufacture 2008-A Mass Balance Study. British Glass.

Edwards, D.W., Schelling, J., 1999. Municipal waste life cycle assessment. Part 2: transport analysis and glass study. Transaction of the Institute of Chemical Engineers 77 (Part B), 259–274.

Emmins, C., 1991. Soft Drinks. Shire Publications.

Francis, A.A., Rawlings, R.D., Boccaccini, A.R., 2002. Glass-ceramics from mixtures of coal ash and soda lime glass by the petrurgic method. Journal of Material Science Letters 21, 975–980.

Fry, J.M., Hartlin, B., Wallén, E., Aumônier, S., 2010. Life Cycle Assessment of Example Packaging Systems for Milk. Waste and Resources Action Programme.

Glass for Europe, 2010. Recyclable waste flat glass in the context of the development of end-of-waste criteria.

Glasser, F.P., 1992. Chemistry of the alkali–aggregate reaction. In: Swamy, R.N. (Ed.), The Alkali Silica Reaction in Concrete. Blackie, pp. 30–53.

Grutzeck, M.W., Marks, J.A., 1999. Synthesis of double-layer silicates from recycled glass cullet: a new type of chemical adsorbent. Environmental Science and Technology 33, 312–317.

Hancock, P., July/August 2011. Fluxing issues – container glass in sanitaryware applications. Asian Ceramics, 50–53.

Hansen, M.R., Wang, M., Collins, T., Nielsen, M.L., 1999. Recycled glass concrete for pavement patching – alkali-silica reactivity. In: Bank, L.C. (Ed.), 5th ASCE Materials in Engineering. ASCE, pp. 847–851.

Henry, K.S., Morin, S.H., 1997. Frost susceptibility of crushed glass used as construction aggregate. Journal of Cold Regions Engineering 11, 326–333.

Highways Agency, 1998. Manual of contract documents for highway works. Specification of Highway Works 1.

Horan, N.J., Lowe, M., 2007. Full-scale trials of recycled glass as tertiary filter medium for wastewater treatment. Water Research 41, 253–259.

Huang, Y., Bird, R.N., Heidrich, O., 2007. A review of the use of recycled solid waste materials in asphalt pavements. Resources, Conservation and Recycling 52, 58–73.

Hurley, J., 2002. A UK Market Survey for Foam Glass. Waste and Resources Action Programme.

Idir, R., Cyr, M., Tagnit-Hamou, A., 2010. Use of fine glass as ASR inhibitor in glass aggregate mortars. Construction and Building Materials 24, 1309–1312.

Idir, R., Cyr, M., Tagnit-Hamou, A., 2013. Role of the nature of reaction products in the differing behaviours of fine glass powders and coarse glass aggregates used in concrete. Materials and Structures 46, 233–243.

Kockal, N.U., Ozturan, T., 2011. Characteristics of lightweight fly ash aggregates produced with different binders and heat treatments. Cement and Concrete Composites 33, 61–67.

Kourti, I., Cheeseman, C.R., 2010. Properties and microstructure of lightweight aggregate produced from lignite coal fly ash and recycled glass. Resources, Conservation and Recycling 54, 769–775.

Lavender, P., 2009. Filter media: treating chemical wastewaters. Filter Separator 45, 16–18.

Limbachiya, M., Meddah, M.S., Fotiadou, S., 2012. Performance of granulated foam glass concrete. Construction and Building Materials 28, 759–768.

Ling, T.-C., Poon, C.-S., 2012a. Feasible use of recycled CRT funnel glass as heavyweight fine aggregate in barite concrete. Journal of Cleaner Production 33, 42–49.

Ling, T.-C., Poon, C.-S., 2012b. A comparative study on the feasible use of recycled beverage and CRT funnel glass as fine aggregate in cement mortar. Journal of Cleaner Production 29–30, 46–52.

Ling, T.-C., Poon, C.-S., Kou, S.-C., 2011. Feasibility of using recycled glass in architectural cement mortars. Cement and Concrete Composites 33, 848–854.

Ling, T.-C., Poon, C.-S., Kou, S.-C., 2012a. Influence of recycled glass content and curing conditions on the properties of self-compacting concrete after exposure to elevated temperatures. Cement and Concrete Composites 34, 265–272.

Ling, T.-C., Poon, C.-S., Lam, W.-S., Chan, T.-P., Fung, K.K.-L., 2012b. Utilization of recycled cathode ray tubes glass in cement mortar for X-ray radiation-shielding applications. Journal of Hazardous Materials 199–200, 321–327.

Marceau, W.E., 1967. Method of making a molten foamed glass. US Patent 3325264. United States Patent Office.

Martin, G., MacFarlane, A., 2002. The Glass Bathyscaphe: How Glass Changed the World. Profile Books.

Mavilia, L., Corigliano, F., 2001. High added value products from off-quality waste glass. In: Dhir, R.K., Limbachiya, M.C., Dyer, T.D. (Eds.), Recycling and Reuse of Glass Cullet. Thomas Telford, pp. 63–73.

Mladenović, A., Šuput, J.S., Ducman, V., Škapin, A.S., 2004. Alkali-silica reactivity of some frequently used lightweight aggregates. Cement and Concrete Research 34, 1809–1816.

Mörtel, H., Fuchs, F., 1997. Recycling of windshield glasses in fired bricks industry. Key Engineering Materials 136 (Part 3), 2268–2271.

Park, S.B., Lee, B.C., Kim, J.H., 2004. Studies on mechanical properties of concrete containing waste glass aggregate. Cement and Concrete Research 34, 2181–2189.

Pelino, M., 2000. Recycling of zinc-hydrometallurgy wastes in glass and glass ceramic materials. Waste Management 20, 561–568.

Quaranta, N., Unsen, M., López, H., Giansiracusa, C., Roether, J.A., Boccaccini, A.R., 2011. Ash from sunflower husk as raw material for ceramic products. Ceramics International 37, 377–385.

Rajapakse, J.P., Fenner, R.A., 2011. Evaluation of alternative media for pebble matrix filtration using clay balls and recycled crushed glass. Journal of Environmental Engineering 137, 517–524.

Rey-Raap, N., Gallardo, A., 2013. Removal of mercury bonded in residual glass from spent fluorescent lamps. Journal of Environmental Management 115, 175–178.

Rodriguez Vieitez, E., Eder, P., Villanueva, A., Saveyn, H., 2011. End-of-Waste Criteria for Glass Cullet: Technical Proposals. European Commission Joint Research Centre.

Romero, M., Rawlings, R.D., Rinćon, J.Ma, 1999. Development of a new glass-ceramic by means of controlled vitrification and crystallisation of inorganic wastes from urban incineration. Journal of European Ceramic Society 19, 2049–2058.

Scarinci, G., Brusatin, G., Barbieri, L., Corradi, A., Lancellotti, I., Colombo, P., Hreglich, S., Dall'Igna, R., 2000. Vitrification of industrial and natural wastes with production of glass fibres. Journal of European Ceramic Society 20, 2485–2490.

Shakhmenko, G., Korjakins, A., Kara, P., Bumanis, G., 2012. Low-carbon concrete using local industrial by-products. In: Jones, M.R., Newlands, M.D., Halliday, J.E., Csetenyi, L.J., Zheng, L., McCarthy, M.J., Dyer, T.D. (Eds.), Concrete in the Low Carbon Era. University of Dundee, pp. 978–988.

Shelby, J., 1997. Introduction to Glass Science and Technology. Royal Society of Chemistry.

Shi, C., Wu, Y., Riefler, C., Wang, H., 2005. Characteristics and pozzolanic reactivity of glass powders. Cement and Concrete Research 35, 987–993.

Soyer, E., Akgiray, Ö., Eldem, N.Ö., Saatçi, A.M., 2010. Crushed recycled glass as a filter medium and comparison with silica sand. Clean – Soil, Air, Water 38, 927–935.

Soyer, E., Akgiray, Ö., Eldem, N.Ö., Saatçi, A.M., 2013. On the use of crushed recycled glass instead of silica sand in dual-media filters. Clean – Soil, Air, Water 41, 325–332.

Su, N., Chen, J.S., 2002. Engineering properties of asphalt concrete made with recycled glass. Resources Conservation and Recycling 35, 259–274.

Tan, K.H., Du, H., 2013. Use of waste glass as sand in mortar: part I – fresh, mechanical and durability properties. Cement and Concrete Composites 35, 109–117.

Tandy, B.C., Way, A.G.J., 2004. R&D to Improve Site Practices for Collection and Clean Separation of Composite (Glass) Materials in the Construction and Demolition Industry. Waste and Resources Action Programme.

Topcu, I.B., Canbaz, M., 2004. Properties of concrete containing waste glass. Cement and Concrete Research 34, 267–274.

Topcu, I.B., Boga, A.R., Bilir, T., 2008. Alkali–silica reactions of mortars produced by using waste glass as fine aggregate and admixtures such as fly ash and Li_2CO_3. Waste Management 28, 878–884.

Tucci, A., Esposito, L., Palmonari, C., Rambaldi, E., 2004. Use of soda-lime scrap-glass as a fluxing agent in porcelain stoneware tile mix. Journal of European Ceramic Society 24, 83–92.

United Nations Industrial Development Organization, Ministry of Internal Trade and Industry, 1993. Output Seminar of Energy Conservation in Glass Energy.

United States Geological Survey, 2012. Mineral Commodities Summaries 2012.

US Department of Transportation Federal Highway Administration, 1997. User guidelines for waste and byproduct materials in pavement construction. Report FHWA-RD-97–148.

Veloza, Z.M., Yanagizawa, K., Yamasaki, N., 1999. Recycling waste glass by means of the hydrothermal hot pressing method. Journal of Material Science Lett. 18, 1811–1813.

Wartman, J., Grubb, D.G., Nasim, A.S.M., 2004. Select engineering characteristics of crushed glass. Journal of Materials Civil Engineering 16, 526–539.

Waste and Resources Action Programme. Case study: color-separated collection of glass for re-melt: bath and North East Somerset Council. http://www.wrap.org.uk/sites/files/wrap/4379%20WRAP%20Case%20Study%20-%20Glass%20Bath%20_%2008.09.09.pdf (accessed 05.13.).

Wei, Y.-L., Lin, C.-Y., Ko, K.-.W., Wang, H.P., 2011. Preparation of low-water-sorption aggregates from harbour sediment added with waste glass. Marine Pollution Bulletin 63, 135–140.

Waste and Resources Action Programme. Case study: colour-mixed collection of glass for re-melt: Hart District Council. http://www.wrap.org.uk/sites/files/wrap/4545%20WRAP%20Case%20Study%20-%20Glass%283%29%20_%2009.09.09.pdf (accessed 05.13.).

Younus Ali, M.M., Newman, G., Arulrajah, A., Disfani, M.M., 2011. Application of recycled glass – crushed rock blends in road pavements. Australian Geomechanics 46, 113–121.

CHAPTER 15

Textile Recycling

Jana M. Hawley

Textile and Apparel Management, University of Missouri, Columbia, MO, USA

15.1 INTRODUCTION

Textile recycling impacts many entities and contributes significantly to the social responsibility of today's society. By recycling, companies can realize larger profits because they avoid charges associated with dumping in landfills; at the same time, textile recycling contributes to goodwill associated with environmentalism, employment for marginally employable laborers, contributions to charities and disaster relief, and the movement of used clothing to areas of the world where inexpensive clothing is needed.

Because textiles are nearly 100% recyclable, nothing in the textile and apparel industry should be wasted. According to the Environmental Protection Agency, United States consumers annually consume nearly 84 pounds of textile products per capita and dispose of 4.3 pounds/capita daily (Environmental Protection Agency, 2011). To compound the problem, the global supply of fibers increased from 52.6 Mt in 2000 to 70.5 Mt in 2010 (Oerlikon Textile GmbH & Co, 2010), with developing countries leading this trend. A recent report shows that China has surpassed the United States, becoming the number one consumer of fiber in the world. This report points out that China will continue to have the fastest-growing fiber consumption market for the next 10 years (Bharat Textiles, 2013). Fiber consumption growth is a double-edged sword in that while increased fiber consumption does stimulate the economy (projected to add 10–20 new factories to meet the world demand), it also contributes significantly to the problem of disposal.

As consumers continue to buy at a rate that meets wants rather than needs, the problem of what to do with waste is compounded. Furthermore, clothing in today's marketplace is different from that of several decades ago, not only in that it is more rapidly produced and distributed, but also in fiber content. Textile waste is composed of both natural and synthetic materials such as cotton, wool, polyester, nylon and spandex. After synthetic fibers came onto the market in the mid-twentieth century, textile recycling became more complex for two distinct reasons: (1) increased fiber strength made it more difficult to shred or "open" the fibers, and (2) fiber blends made it more difficult to purify the sorting process. Given that the recycling industry must cope with everything that enters the recycling stream, it would be more efficient if clothing were manufactured in a single-material system in order to facilitate the recycling process.

Recently, a variety of strategies and technologies have been implemented for recycling textiles, including the Council for Textile Recycling's (CTR) call for a zero-waste goal by 2037. This robust goal can be accomplished if the industry and consumers embrace a holistic approach, form strategic partnerships, and heighten conscientious consumption. Zero-waste focuses on a closed-loop industrial/societal system whereby waste is considered residual raw material or a resource for value-added products. This may involve redesigning both products and processes in order to allow waste to be remanufactured into new products. Zero-waste concepts consider the entire life cycle of products in the context of a comprehensive systems understanding (Hawley, 2009).

The process of textile recycling occurs when postindustrial or post-consumer waste enters the recycling pipeline through manufacturing waste collection, donations to charitable organizations, or passing on to another person for reuse. Unfortunately, this process does not preclude textile waste from ending up in landfills.

15.2 THE RECYCLING EFFORT

It is well established that recycling is economically beneficial, yet much of the discarded clothing and textile waste fails to reach the recycling pipeline. The United States textile recycling industry annually recycles about 3.8 billion pounds of post-consumer textile waste (PCTW). This accounts for approximately 15% of all PCTW, leaving 85% in the landfills (Council for Textile Recycling, 2013). Furthermore, the total textile per capita consumption has increased over time. When statistics for emerging markets are added, the worldwide plethora of PCTW becomes daunting.

Textile recycling can be classified as postindustrial (also referred to as pre-consumer) or post-consumer. Postindustrial waste consists of by-product materials from the textile, fiber and cotton industries that are remanufactured for the automotive, aeronautical, home building, furniture, mattress, coarse yarn, home furnishings, paper, apparel and other industries. Post-consumer waste, on the other hand, is defined as any type of garment or household article made from manufactured textiles that the owner no longer needs and decides to discard. These clothing items are donated because they are worn out, are damaged, no longer fit or have gone out of fashion. Used clothing and household textiles are often donated to charities such as Goodwill or Salvation Army. Charity agencies then sort the clothes and select items for their retail shops, with the surplus sold to textile graders (also known as rag dealers) for pennies on the pound. The price per pound of used clothing is dependent on current market value but often ranges from 3–6 cents/pound. At regularly scheduled routes, trucks are dispatched to pick up the surplus. The clothes are then taken to recycling warehouses, where the sorting process begins. The primary goal for the rag dealers is to earn profits, but often the business owners are also committed to making a positive environmental impact.

The sorting process starts as clothes are loaded onto a conveyor belt. The first step is to perform a crude sort that removes heavy and large items such as coats, curtains and blankets. The second sort separates other readily identifiable items, such as jeans, shirts and household. As the process proceeds, the sorts become more and more refined. For example, once all trousers are picked, they are further sorted based on criteria such as women's or men's, woolens or chinos, condition (e.g. tears, missing buttons, stains) or brands and style (Levi's, Dickies or Wranglers). As post-consumer clothing is sorted, it is also graded to meet the requirements of specific markets. It is not uncommon for a fully integrated rag dealer to sort over 400 grades at any given time. It is often the quality of grading that distinguishes a competitive advantage of one rag dealer over another. Sorting includes used clothing exported to developing countries, vintage collectibles, wipers, items for upcycling

FIGURE 15.1 Pyramid model for textile recycling categories, based on volume-to-value ratio (Hawley, 2006).

and shoddy (to be remanufactured into other products, including yarns for apparel). For the most part, volume of used clothing is inversely proportional to value (Hawley, 2006). Figure 15.1 illustrates the recycling pyramid.

15.3 EXPORT OF SECONDHAND CLOTHING

The largest volume of recycled textiles (approximately 48%) is sorted for secondhand exports, primarily for export to developing countries or disaster relief. In 2011, the United States exported more than 761 million kg of worn clothing to places around the world, with much of it being sent to developing countries (US Department of Commerce, Office of Textiles and Apparel, December 2012). Once sorted, the goods are compressed into large bales (usually 275–500 kg), wrapped and warehoused until an order is received. Several things are considered when sorting for used clothing markets: climate, relationships between exporters and importers and trade laws for used apparel. Exporters of used clothing and textile products are faced with several trade barriers and restrictions, including restrictions based on sanitation, phytosanitary, national security, human/animal health and deceptive practices (Office of Textiles and Apparel, 2012).

15.4 CONVERSION TO NEW PRODUCTS

Recycled clothing and textiles that are not wearable or appropriate for wipers are either mechanically or chemically converted to fiber and processed into value-added products. Mechanical operations include cutting, shredding, carding and processing the fabric. Chemical processes involve enzymatic, thermal, glycolysis or methanolysis methods (Hawley, 2009). Once the post-consumer textiles are converted, they can be further processed into new products for consumption. These value-added products include stuffing, automotive components, carpet underlays, building materials and blankets.

In the textile and apparel recycling industry few value-added conversion products exist; however, more and more efforts are focused on new ways to treat the old. Long-interview data were obtained from four companies during the summer of 2010. It was found that a variety of products are being made from recycled fiber, most of which are mechanically processed from postindustrial and post-consumer waste. These products are used for a variety of purposes. For example, as part of the cleanup after the Deepwater Horizon oil spill of 2010, GeoHay partnered with Polyester Fibers to support cleanup. The strategic partnership resulted in the production of oil booms that proved to effectively filter tar from the water as it washed onto shore. GeoHay™ outperformed competitive brands as well as natural hay for removal of particulate and turbidity in the water (Hillegass, 2010). Table 15.1 illustrates a variety of products that are currently manufactured from mechanically recycled postindustrial and

TABLE 15.1 Products Made from Mechanically Recycled Post-consumer and Postindustrial Waste

Product	Description	Uses
PRODUCTS MANUFACTURED BY LEIGH FIBERS OF WELLMAN, SOUTH CAROLINA		
QuietLeigh™	Acoustical product made from shoddy and custom blended	• Tarpaulins • Ropes • Padding & stuffing • Sound deafening
SafeLeigh™	Flame-retardant product made from shoddy and custom blended	• Upholstery backing • Headliners • Interior engine components • Interior acoustical components
SafeLeigh Premium™	Made from recycled meta-aramids. Provides environmentally friendly properties	• Mattresses • Substitute for asbestos
Greenloop™	Processes and products that provide integrated solutions from PCTW	
PRODUCTS MANUFACTURED BY MARTEX FIBERS OF SPARTANBURG, SOUTH CAROLINA		
Jimtex for ECO2Cotton®	Yarns made from 75–80% recycled postindustrial cotton with acrylic, polyester, or recycled polyester added for strength	• Blankets • Fleece knits • Crafts • Home textiles • Sweaters • T-shirts • Hosiery • Insulation • Pads & mattress fill
PRODUCTS MANUFACTURED BY GEOHAY OF INMAN, SOUTH CAROLINA		
GeoHay™	Erosion and sediment control products manufactured from post-consumer textile waste	• Stabilization • Drainage • Erosion control • Roadway separation • Landfill leachate • Surface retention • Wildlife habitat restoration

post-consumer waste. Figure 15.2 further illustrates the breadth of products being developed.

15.5 CONVERSION OF MATTRESSES

Mattress waste poses a unique challenge. Not only have landfill fees for mattresses skyrocketed in recent years; a single mattress can take up to 23 cubic feet in the landfill, and oftentimes the mattresses cause damage to landfill equipment. In the United States alone, between 20 and 40 million mattresses are disposed of annually. Mattresses have high transportation costs associated with the recycling process, contain chemical flame retardants, and are labor-intensive to dismantle.

At the Hutchinson Correctional Facility (HCF) inmates dismantle old mattresses that come from suppliers all over the Midwest. Mattress recycling involves dismantling the mattress to fundamental parts: springs, wood, cotton, foam and fabric covers. Inmate labor is one of the few cost-effective ways to manage mattress recycling. Figure 15.3 illustrates mattresses recycled at HCF.

15.6 CONVERSION OF CARPET

A discussion of textile recycling would be remiss if carpet recycling were not included. Approximately 4 billion pounds of carpet are disposed of each year in the United States, but recycling carpet is a complex process that often requires extensive processing to convert it to new products (Wang et al., 2003). Recycling of both residential and commercial carpet has been undertaken on a national scale in the United States, with a variety of end products including high-energy fuel, shingles, flooring, auto parts and new carpet. Carpet that is collected is tested for fiber type (e.g. polyester, nylon, olefin), the surface is separated from the backing, and components are processed into a variety of end uses. Because carpet is heavy and bulky, transportation costs are a major factor in the cost efficacy of the recycling. To facilitate the process, carpet is now strategically designed to allow for easier capture of component parts during both mechanical and chemical recycling processes. In addition, a strategic infrastructure is being implemented to help reduce transportation costs.

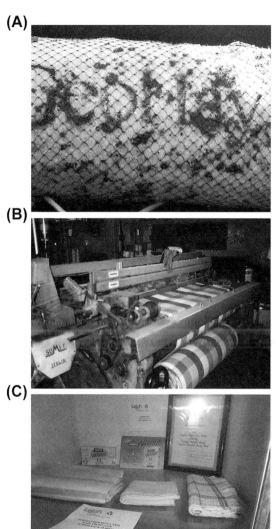

FIGURE 15.2 Products made from post-consumer waste. (A) GeoHay™, used for Deepwater oil spill cleanup. *(Photo courtesy of GeoHay.)* (B) Acrylic blankets made from post-consumer sweaters, Prato, Italy. *(Photo courtesy of author.)* (C) Products made from postindustrial waste, Leigh Fibers. *Photos courtesy of author.*

15.7 WIPERS

Clothing that has seen the end of its useful life as clothing may be turned into wipers for industrial use. T-shirts are a primary source for

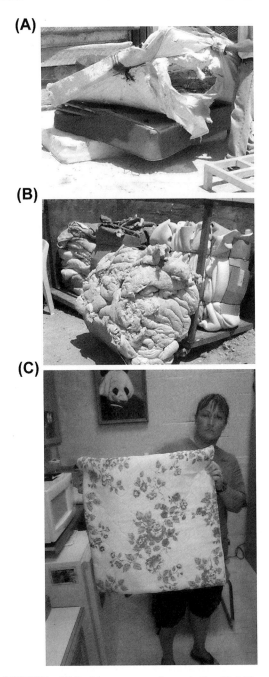

FIGURE 15.3 Mattress recycling at the Hutchinson Correctional Facility. (A) Inmates at HCF dismantle mattresses. (B) Mattress components are baled to be sold to recyclers. (C) Pet crate pads are made from recycled mattress components at the HCF. *Photos courtesy of author.*

this category because the cotton fiber content makes an absorbent rag and polishing cloth. Wipers are then sold to a variety of industries including automotive, housekeeping, furniture making and other industrial users.

15.8 LANDFILL AND INCINERATION

Although rag sorters work hard to avoid landfill, post-consumer waste eventually reaches the end of its useful life. Recent research has explored the use of textiles for fuel. Although emission tests of incinerated used fibers are above satisfactory and the British thermal unit (BTU) value is respectable, the process of fueling boiler systems in North American power plants is not feasible. However, in Europe, it is not uncommon to find textiles used as an alternative fuel source.

15.9 DIAMONDS

The *diamond* category in the recycling pyramid refers to the collectible clothing found on the conveyor belts. Although the volume of collectible clothes is approximately 1% of the total volume, the value of these items makes it worth the mining effort. For example, some collectible Levi's jeans have been sold at auction for more than $10,000 per pair.

Categories of diamonds include vintage collectibles, luxury fibers, Harley-Davidson and other highly collectible branded goods, couture clothing, or special event T-shirts. Many of these mined goods have global markets and the number of consumers who seek collectibles is on the rise. For example, Japanese consumers seek vintage Americana items such as Ralph Lauren, Levi's, Coach leather goods, or Harley-Davidson, whereas American consumers seek Italian leather and French couture items.

Many owners of vintage shops are members of the National Association of Resale and Thrift Shops. This trade association promotes public education about the used clothing industry and has more than 1000 members that serve thrift shops, resale vendors and consignment shops.

15.10 SUMMARY

As more attention is paid to environmental issues, including value-added options for recycled textiles, companies must continue to innovate options not only for environmental reasons, but also for economic. McDonough and Braungart (McDonough and Braungart, 2002) point out that the textile industry ranks among the worst industries on environmental impact. In order to improve this standing and make the best of textile recycling processes, several considerations must be met. As we move toward a zero-waste goal, we must give attention not only to downcycling, which means that products lose integrity through each cycle, but also to design thinking that allows for a biomimetic approach. Both manufacturers and consumers must be challenged to send all postconsumer and postindustrial waste into the recycling pipeline. To improve recycling, holistic attention to the supply chain must be implemented. The environmental footprint can be reduced if both science-based and societal-based approaches across the textile supply chain are considered.

References

Bharat Textiles. http://www.bharattextile.com/ (accessed 29.01.13).

Council for Textile Recycling, 2013. Wear Donate Recycle. http://www.weardonaterecycle.org/ (accessed 29.01.13).

Environmental Protection Agency, 2011. Municipal Solid Waste Generation, Recycling, and Disposal in the United States: Facts and Figures for 2010. United States Environmental Protection Agency, Washington, DC.

Hawley, J.M., 2006. Digging for diamonds: a conceptual framework for understanding reclaimed textile products. Clothing and Textiles Research Journal 24 (3), 262–275.

Hawley, J.M., 2009. Understanding and improving textile recycling: a systems perspective. In: Blackburn, R. (Ed.), Sustainable Textiles: Life Cycle and Environmental Impact. Woodhead Publishing, Cambridge, pp. 179–199.

Hillegass, R., 2010. Polyester Fibers and GeoHay Support Oil Containment Effort in Gulf of Mexico. http://www.benzinga.com/press-releases/10/05/b299257/polyester-fibers-and-geohay-support-oil-containment-effort-in-gulf-of-m (accessed 01.02.13).

McDonough, W.,, Braungart, M., 2002. Cradle-to-Cradle. North Point Press, New York.

Oerlikon Textile GmbH & Co, 2010. The Fiber Year 2009–2010: A World Survey on Textile and Nonwovens Industry. Oerlikon Textile, Remscheid.

Office of Textiles and Apparel, 2012. Worn (Used) Clothing and Textile Products. http://web.ita.doc.gov/tacgi/eamain.nsf/b6575252c552e8e28525645000577bdd/401dbca37c2d025985256efb0068f875?OpenDocument#Exporting%20Used%20Clothing%3A%20Foreign (accessed 30.01.13).

US Department of Commerce, Office of Textiles and Apparel, December 2012. US Exports of Textiles and Apparel: Worn Clothing Schedule B: 63090. http://otexa.ita.doc.gov/scripts/tqexp_ads.exe/htspage (accessed 30.01.13).

Wang, Y., Zhang, Y., Polk, M., Kumar, S., Muzzy, J., 2003. Recycling of carpet and textile fibers. In: Andrady, A.L. (Ed.), Plastics and the Environment. John Wiley & Sons, New York.

CHAPTER 16

Cementitious Binders Incorporating Residues

Yiannis Pontikes[1], Ruben Snellings[2]

[1] KU Leuven, High Temperature Processes and Industrial Ecology Research Group, Department of Metallurgy and Materials Engineering, Kasteelpark Arenberg 44, 3001 Heverlee, Belgium; [2] Ecole Polytechnique Fédérale de Lausanne (EPFL), Laboratory of Construction Materials, Institute of Materials, Station 12, 1015 Ecublens, Switzerland

16.1 INTRODUCTION

Cements or cementitious binders are materials that can set and harden to a rock-like solid material from an initial paste and are typically used to bind other materials together. In concrete, cement acts as the binder phase between the sand and gravel aggregates. The binding in cements occurs through a chemical reaction that leads to the establishment of physicochemical bonding between the solids. Most cement produced today is based on so-called Ordinary Portland Cement (OPC). The success of OPC-based construction materials (mortars, concrete) is based on their comparatively low cost, ease of use, robustness against misuse and flexibility in terms of transport and shaping of structural elements.

OPC is composed of finely ground clinker with a small amount of gypsum. Clinker is produced at high temperatures, typically around 1450 °C in a rotary kiln, from blends of limestone with small quantities of other materials such as clays. In terms of phases, clinker is predominantly composed of alite (Ca_3SiO_5; 50–70%), and smaller amounts of belite (β-Ca_2SiO_4; 5–25%), tricalcium aluminate ($Ca_3Al_2O_6$; 5–10%), and tetracalcium aluminoferrite ($Ca_4(Al,Fe)_4O_{10}$; 5–15%), a typical OPC chemical composition is (in wt%) CaO: 65 ± 3; SiO_2: 21 ± 2; Al_2O_3: 5 ± 1.5; Fe_2O_3: 3 ± 1; MgO < 5. Blending additional constituents, such as coal fly ash, limestone, natural pozzolans (e.g. volcanic ashes and tuffs), and ground granulated blast-furnace slag with the clinker creates blended cements with properties depending on the materials added.

Considering that cement is primarily composed of only five common major elements (Ca, Si, Al, Fe, O), it comes as no surprise that cement production is geographically widespread. China is a leading country, producing approximately 58% of total cement whereas

India, second-ranked producer at 7%, is expected to become a major player in the years to come (USGS, 2013a; IEA, 2009). Over the past decades, cement production has been increasing steadily and reached a total of 3.7 Gt in 2012 (USGS, 2013a). Considering a cement content in concrete of 10–15 wt%, this results in an estimate of 25–37 Gt of concrete production, or about 1.5 m^3 of concrete produced per person annually. The massive scale of production positions concrete as the second most used material by humankind after water.

Nevertheless, both the need of high temperature processing and the use of limestone contribute to the environmental footprint of OPC production. Although values vary depending on kiln technology, one tonne of cement clinker production currently requires about 3800 MJ and emits approximately 0.87–0.92 t of CO_2 (Habert et al., 2010; WBCSD, 2009). Nonetheless, considering the massive volumes produced, the total footprint of the cement sector is rather substantial. Data suggest that 5–8% of the total anthropogenic CO_2 emissions derive from the cement sector, depending on how the calculations were performed and the reference year (Olivier et al., 2012; Barcelo et al., 2013). A set of actions is undertaken by the cement sector itself, improving the energy efficiency, using alternative raw materials and fuels, developing new cement types, increasing the use of supplementary cementitious materials or decreasing the cement-binder quantities used in concrete by more sophisticated blending (OECD/IAI and the World Business Council for Sustainable Development, 2009). In addition to the above, a range of novel binders have been proposed, offering potential alternative paths to OPC. In all these paths, the use of residues and by-products is integrated and can contribute to more closed material loops. It is the scope of this chapter to describe all the above, emphasizing on the use of residues and by-products.

16.2 CLINKER PRODUCTION: PROCESS FLOW, ALTERNATIVE FUELS AND ALTERNATIVE RAW MATERIALS

The use of alternative fuels and raw materials (AFR) in cement production is taking place already from the early 1970s since it suppresses the production cost of cement while promoting a more efficient waste management. Indeed, seen in a wider framework of energy and material flows, the cement sector contributes by using a variety of household and industrial residues, preventing them from being land-filled. The latter is a dimension often neglected when the cement sector is evaluated only on the basis of CO_2 footprint. Still, as the notions of "co-processing" and "industrial ecology" become more acknowledged and endorsed, cement plants are more widely regarded and operate as "co-processing industries of alternative fuels and raw materials for sustainable cement production" (The GTZ-Holcim Public Private Partnership, 2006). For the guiding principles concerning co-processing of AFR, a comprehensive list of principles has been published (The GTZ-Holcim Public Private Partnership, 2006). For instance, it is expected that co-processing takes place by respecting the waste hierarchy, without affecting the environment, humans or cement quality, by companies that are qualified and in-line with country-specific conditions.

16.2.1 Alternative Fuels

The use of alternative fuels is common practice in many countries, with the thermal substitution levels of alternative fuels for conventional fossil fuels able to exceed 80% (WBCSD, 2005). In general, the trend of using alternative fuels is increasing and nowadays more than 64% of the cement plants in Europe use residues as alternative fuels (Rootzén, 2012). The list of candidate alternative fuels is quite wide, with meat and bone animal meal and sewage sludge being the most widely used. Table 16.1 summarizes the alternative fuel options; note that substantial differences are documented on the profile of the alternative fuels used per country (Aranda Usón et al., 2012). It is obvious that other alternative fuels can be used per case, however, it is also obvious that limitations do exist. The choice of alternative fuels has to meet different levels of considerations, ranging from business oriented aspects to more technical ones. Certain preconditions exist, for instance, the cost of the alternative fuels is expected to be lower than that of traditional fuels, with a calorific value higher than 14 MJ/kg, low water content and physical properties that permit efficient handling and processing. In terms of chemistry, there are strict quality controls and attention is placed to Cl (<0.2 wt%), S (<2.5 wt%), PCB (<50 ppm), and heavy metals (<2500 ppm, out of which Hg <10 ppm and Cd + Tl <90 ppm); threshold values are indicative, after (Mokrzycki et al., 2003). The volatile metals, like alkalis, as well as P, Zn and F are also closely monitored and controlled. In general, the chemistry of the fuel should be well controlled because it may have an impact on kiln operation, clinker and cement quality as well as emissions (Aranda Usón et al., 2012; Chinyama, 2011).

16.2.2 Alternative Raw Materials

When it comes to alternative raw materials, again many possibilities exist, although the chemistry of OPC clinker poses a certain ceiling on what can be introduced. Since OPC clinker is primarily composed of CaO, Ca-rich residues can be introduced up to higher levels. Despite this constraint, the high temperature clinkering process offers an important degree of freedom since it permits the use of materials with little consideration of their actual phase composition. Indeed, at the processing temperatures the system approaches thermodynamic equilibrium and the raw materials disintegrate readily to form the stable clinker phase assemblage (Noirfontaine et al., 2012).

Therefore, a grouping of alternative raw materials can be made based on their major constituent, Table 16.2. Additional residues, such as construction and demolition debris, refinery spent catalysts as well as spent aluminum potliners should also be considered; they are not listed in Table 16.2 as they may also have a heating value. Besides major oxides, minor oxides and heavy metals in particular (e.g. Cr) are controlled, although this is not standard practice globally (Huang et al., 2012). Depending on the type of material, different entry points in the process are selected. As with the alternative fuels, the list is not exhaustive and in principle other materials can be used, provided they meet business, logistic, handling and chemical quality criteria. Nonetheless, unlike the alternative fuels, the percent substitution does not typically exceed 5 wt%.

TABLE 16.1 Alternative Fuels Typically Used in the Cement Industry

Category	Fuels
Gaseous fuels	Refinery waste gas, landfill gas, pyrolysis gas, natural gas.
Liquid fuels	Tar, chemical wastes, coal slurries/distillation residues, waste solvents, used oils, used oil + oiled water, wax suspensions, petrochemical waste, asphalt slurry, paint waste, oil sludge, paper fiber sludge.
Solid fuels	Petroleum coke (petcoke), fine/anodes/chemical cokes, oil-bearing soils, shale/oil shales, paper/cardboard waste, rubber residues, pulp sludge, sewage sludge, used tyres, battery cases, plastics residues, wood waste, impregnated saw dust, domestic refuse, meat and bone animal meal, animal fat, rice husks, refuse derived fuel, agricultural waste, automobile shredder residue.

Modified after Chinyama (2011).

TABLE 16.2 Alternative Raw Materials that are/can be Used in OPC Clinker Production

Major Substitution for	Examples
Al_2O_3	Coating residues, aluminum recycling sludge.
$CaCO_3$	Industrial lime from neutralization processes, lime sludge from sewage treatment.
SiO_2	Foundry sand, contaminated soil from soil remediation.
Fe_2O_3	Roasted pyrite from metal surface treatment, mechanical sludge from metal industry, red sludge from industrial waste water treatment, bauxite residue "red mud", non-ferrous slags, mill scales from iron and steel production.
Ca-Si-Fe-(Mg)-oxides	Steel slags.
Ca-Si-Al-oxides	Blast-furnace slag.
Si-Al-Ca-Fe-oxides	Fly ashes from coal combustion, bottom and fly ashes from incineration plants after treatment.

Adapted from The GTZ-Holcim Public Private Partnership (2006) with data from Verein Deutscher Zementwerke (Association of German Cement Manufacturers), U.S. EPA (2008) and private communications.

The above-mentioned small substitution rate has been challenged by Eco-Cement, a family of cements developed in Japan in the 1990s that has gone to full industrial implementation since 2001, complying with national standards (Hanehara, 2005; Katagiri, 2012). These cements use approximately 50 wt% incinerator bottom and fly ash from the combustion of municipal urban solid waste (MUSW) and sewage sludge as raw materials. The final cements are comparable to Portland cement but small differences in the phase assemblage and properties do exist. The key element in this process is to address the chlorides and heavy metals typically found in incinerator ashes and to that extent, the high temperature processing during clinkering is crucial. In an alternative path for ash utilization, washing systems are installed in existing cement facilities for an efficient ash pre-treatment (Katagiri, 2012).

16.3 FROM CLINKER TO CEMENT: RESIDUES IN BLENDED CEMENTS

The substitution of Portland cement clinker by so-called supplementary cementitious materials (SCMs) is generally being considered as the most effective way of reducing the environmental impact and CO_2 emissions associated to the production of cement and concrete (Habert et al., 2010; Snellings et al., 2012). The reduction of the clinker content scales proportionally with the embodied CO_2 of the blended cement and enables to valorize massive quantities of industrial by-products and residues. By-products that are conventionally used to replace clinker in either blended cements or in concrete are blast-furnace slags from pig iron production, fly ashes from coal-fired electricity production, and silica fume from silicon refineries. The level of substitution generally depends on the CaO content of the SCMs. High-CaO blast-furnace slag, Figure 16.1, can be blended into cement up to levels of 95 wt%, Figure 16.2. Materials containing less CaO need $Ca(OH)_2$ liberated by the hydration of the clinker component in order to react and can therefore be blended into cements only up to 30—40 wt% without significant loss in binding properties.

In general, the substitution of clinker by SCMs results in a lowered early strength that gradually recovers and often surpasses the strength of the reference cement at the long term. Not only the strength development, but also the durability of the cement is changed by clinker substitution. In general, blended cements develop a finer microstructure which leads to a much reduced permeability and enhanced protection against the ingress of deleterious agents, especially chloride-bearing solutions (e.g. de-icing salts or seawater). On the other hand, an overall reduction in the CaO

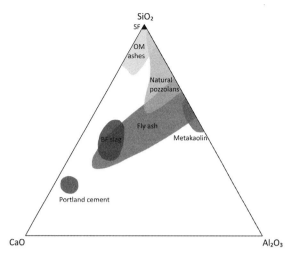

FIGURE 16.1 Chemical composition of Ordinary Portland cement (CEM I) compared to the compositional ranges of conventional SCMs such as blast-furnace slags (BF slag), fly ashes, natural pozzolans, kaolinite-rich thermally activated clays (metakaolin), silica fume (SF) and organic material (OM) ashes. *Adapted from Snellings et al. (2012).*

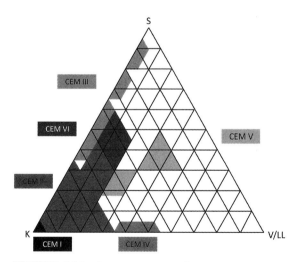

FIGURE 16.2 Cement type specifications according to EN 197-1 (European Committee for Standardization, 2000) extended with low-clinker ternary cement types CEM VI (Delort, 2013). Cement types are defined based on the substitution levels of clinker (K) by blast-furnace slag (S), fly ash (V), and limestone (LL).

level of the cement leads to a lower buffer capacity to carbonation and the resulting rebar corrosion. However, the lowered binder permeability is instrumental in mitigating the carbonation process to some extent.

Global substitution levels, usually expressed as clinker factor—the proportion of clinker in the cement—were estimated to average around 0.77 in 2010 (Snellings et al., 2012). Considering a global cement production of 3700 Mt in 2012 (USGS, 2013a) of which an estimated 180 Mt of gypsum, the use of SCMs in cement and concrete amounts up to 515 Mt. As blast-furnace slag reserves (2012 production of 290 Mt (USGS, 2013b)) are more or less entirely exploited, only an increased use of fly ash (2012 production estimate of 670 Mt (World Coal Association, 2013)) can be instrumental in reducing clinker factors further down to 0.56, well surpassing the goal set by the International Energy Agency of 0.71 for 2050 and leading to a further 25% reduction in embodied CO_2 from current levels (32.5% compared to 1990). Major issues in an increased usage of fly ash are the material heterogeneity and the quality or overall low reactivity, leading to practical limitations in substitution levels or in extreme cases to incompatibility with the cement. Improved quality control and adjustments of the coal raw materials and the combustion process in an industrial symbiosis approach as well as beneficiation of the fly ash residues will have to be considered to increase fly ash recycling into blended cements. On a different level, a more efficient organization of supply lines to limit transport costs is also needed to achieve lower clinker factors at reasonable economic and environmental costs.

Future reductions in clinker factors cannot be based on the use of conventional SCMs alone as it can be expected that future developments in steelmaking and electricity generation will decrease blast-furnace slag and fly ash supplies (Damtoft et al., 2008). Moreover, local availabilities of conventional by-product SCMs are variable and limited in many developing regions.

Other materials will need to step in and a wide range of industrial and societal residues have been the subject of investigation. Many materials under consideration face issues such as low reactivity (e.g. copper slags), the presence of potentially expansive phases (e.g. free lime and periclase in steel slags), or the presence of problematic elements (e.g. in some metallurgical slags and municipal solid waste ashes (van Oss and Padovani, 2002)), and need further investigation to establish their potential use in cement-based construction materials. Special attention should be directed to the potential leaching of toxic elements to the environment. Cementitious materials have often been advocated as good alternatives for hazardous waste disposal because of the immobilization of many heavy metals in the high pH environment in the cement. However, the fate of construction materials after service life and the potential remobilization of heavy metals upon carbonation is often ignored and the potential proliferation of toxic waste should be a matter of concern. As already presented in the case of Eco-Cement and ash utilization in Japan, pre-treatment of these resources by upgrading their quality will facilitate their integration into final cementitious binders. In combination with the above, actions taken in the residue-producing-installations, e.g. adjusting slag chemistry after metal separation while still at high temperature or controlling slag solidification (USGS, 2013b; World Coal Association, 2013), can also contribute.

Another issue is that potentially useful and environmentally safe residues such as waste-glass (Shi et al., 2005) are often only available in volumes that are relatively small compared to the global clinker production. Therefore, it is expected that the extended use of residues will depend on local availabilities and that in many situations primary resources such as natural pozzolans, thermally activated clays and limestone will find increasing application in blended cements (Scrivener and Nonat, 2011). Alternatively, spurred by local conditions and needs, new approaches may emerge as is the case in the Enhanced Landfill Mining concept (Jones et al., 2012). In this approach, residues already in landfills are excavated, separated for the recovery of materials and subsequently thermally processed in a plasma furnace coupled with a gasification unit, generating syngas and a fully vitrified slag. The latter, considering that its starting composition is in the vicinity of pozzolanic materials, can be used as such or can be compositionally modified at high temperatures toward known glassy hydraulic binders (e.g. blast-furnace slag) or other compositions (Pontikes et al., 2013a). In view of the plasma technology, metals can be fumed or become part of a metal alloy, leaving the slag purified. In essence, this approach and similar ones (Yamaguchi et al., 2013) put forward a cradle to cradle scheme for the production of hydraulic and other novel binders (Pontikes et al., 2013b). Interestingly, all processes developed on MUSW, whether "fresh", already incinerated as ash, or smelted (vitrified) as slag, could lead to binders produced on-site, at volumes reflecting to some extent the local needs. In practice, this can be an example of closed material loops.

In terms of cement industry standards a clear trend can be discerned toward the development of ternary or multiple blended cements that combine various residues or, more commonly one residue and limestone. Limestone has been shown to play a double role: on the one hand it can act as an inexpensive filler that accelerates the hydration of the clinker and that can reduce the water demand of the cement (Gutteridge and Dalziel, 1990), on the other hand, limestone can react in synergy with Al-bearing SCMs to form more space-filling hydration products (De Weerdt et al., 2011; Antoni et al., 2012). Binary and ternary cements containing high amounts of limestone have been adopted recently into cement standards around the world, Figure 16.2, and it can be expected that they will constitute an important share of future cement production. Further extensions of cement specifications and definitions of new cement types are currently being investigated and will provide ample opportunity for producers to diversify and adapt their products to the local market conditions.

16.4 ALTERNATIVE CEMENTS FOR THE FUTURE: REDUCING THE CO_2 FOOTPRINT WHILE INCORPORATING RESIDUES

Next to OPC based binders, a range of alternative cements has been developed or is being investigated (Gartner and Macphee, 2011; Schneider et al., 2011; Juenger et al., 2011; Shi et al., 2011). Many of the already existing alternatives have found niche applications, e.g. calcium aluminate cements for refractory bricks, but out of technical (performance or durability) or economic reasons (raw materials of low cost and economy of scale for OPC) have not replaced OPC on a wider scale. The discussion herein cannot be exhaustive but two alternative cement families will be presented as they appear to be promising in terms of (extended) industrial implementation in the near future, partially driven by their reduced CO_2 footprint and the substantial amounts of residues incorporated: (1) calcium sulfoaluminate (CSA) containing cements; and (2) alkali-activated cements. There is also revived interest for calcium aluminate cements as well as supersulfated cements, the latter being particularly interesting from a residue recycling perspective, and more information can be found elsewhere (Juenger et al., 2011).

16.4.1 Calcium Sulfoaluminate Cements

Calcium sulfoaluminate based cements contain ye'elimite ($Ca_4Al_6O_{12}(SO_4)$) as a major component next to other major phases such as belite and ferrite. The hydration of ye'elimite in combination with added calcium sulfate is responsible for the early strength development while the hydration of belite and ferrite contributes largely to later age strength gain. Compared to Portland cement, CSA cements can be produced at lower temperatures of about 1250 °C and involve much lower CO_2 emissions from raw materials. CSA cements were initially largely produced in China as specialty cements for high-early strength or self-stressing applications and contained high levels of ye'elimite (60—70%). Bauxite is needed as raw material to achieve high ye'elimite contents and as a result these cements are relatively expensive; in the future secondary alumina-rich resources could be explored. More recent developments have been aiming at producing CSA cements that contain less ye'elimite (about 30%) while maintaining a strength development comparable to Portland cement. Different approaches have been claimed to compensate for the relatively slow strength development of the remaining clinker components. One is the doping of the belite component with minor elements such as boron to stabilize the more reactive α-polymorph (Gartner and Macphee, 2011), another is the formation and activation of the ternesite ($Ca_5Si_2O_8(SO_4)$) clinker phase (Dienemann et al., 2013). CO_2 reductions of 20—30% were reported compared to Portland cement production. Currently, the use of residues such as fly ashes and slags both as raw materials and as SCMs is being tested to further reduce both cost and CO_2 footprint. In addition, bauxite residue ("red mud") and gypsum from other processes can be used (Pontikes and Angelopoulos, 2013; Arjunan et al., 1999), considering the higher levels of Fe_2O_3 and $CaSO_4$ in the raw meal. While preliminary data are encouraging, more data on concrete durability are needed, especially regarding carbonation and corrosion behavior (Barcelo et al., 2013).

16.4.2 Alkali-Activated Cements

Unlike the cements presented so far, alkali-activated cements usually do not necessitate a high temperature processing step for the synthesis of the solid reactants or precursors. In most cases, alkali-activated cements are developed with by-products and residues from other industrial processes, and in fact, these materials constitute the actual reactive components. This characteristic makes alkali-activated cements undoubtedly attractive for further investigation, yet, at the same time, they are more vulnerable to the supply and characteristics of the residues. Moreover, a number of scientific and technological questions still remain that are typically investigated per case.

The term alkali-activated cements is applied to an extensive range of formulations. The classification of alkali-activated cements can be based on the nature of the solid precursors, i.e. glassy, amorphous (like metakaolin), or crystalline; on the chemistry of the solid precursors, i.e. alkaline binding systems ($Me_2O-Al_2O_3-SiO_2-H_2O$) and "alkali–alkaline-earth" binding systems ($Me_2O-MO-Al_2O_3-SiO_2-H_2O$, where Me = Na, K, ... and M = Ca, Mg, ...) (Glukhovsky, 1967); on the origin of solid precursors, i.e. slag-based, fly ash-based, OPC-based, and so on (Shi et al., 2006); or on the nature of the resulting binding phase, i.e. C–(A)–S–H and/or N–A–S–(H).

Conceptually, these cements are formed after mixing two components, a solid precursor and an alkaline activating solution. The resulting paste can set and harden usually at room temperature. Most often used solid precursors are fly ash from coal combustion, metakaolin, a product of thermal treatment of kaolinite clays, as well as ground granulated blast-furnace slag. Other residues explored are phosphorous slag, ferrous slags (from steel and stainless steel production), non-ferrous slags (from lead, zinc, copper, nickel production), as well as a range of other materials (e.g. Shi et al., 2006). These components can be used alone or in binary/ternary mixtures. With respect to the alkaline activators, these are usually concentrated solutions of (Na,K)-hydroxides and (Na,K)-silicates, with the possibility of -carbonates and -sulfates being used (Shi et al., 2006).

The phenomena taking place after mixing are difficult to follow analytically as they run concurrently, yet, a model is suggested where dissolution, speciation equilibrium, gelation, reorganization and polymerization are described as the main governing stages (Duxson et al., 2007 and references therein). The final binding phase depends on the starting composition, in particular the level of Ca in the $Na_2O-CaO-Al_2O_3-SiO_2$ system. For high-calcium compositions, an amorphous to partially crystalline C–(A)–S–H material typically forms, relatively cross-linked, with a moderate degree of Al substitution and a low C/S ratio (Wang and Scrivener, 1995). On the other hand, for low Ca compositions, an amorphous N–A–S–(H) material develops, resembling to zeolites (Bell et al., 2008), and highly cross-linked. This material contains low levels of chemically bound water (Rahier et al., 1996). For intermediate levels of Ca in the precursor blend, recent work in the area demonstrated the decisive role of Ca (Garcia-Lodeiro et al., 2011) and revealed that an interaction between C–(A)–S–H and N–A–S–(H) phases takes place over time (García-Lodeiro et al., 2013; Ismail et al., 2014) leading to a hybrid N–(C)–A–S–H phase. Apart from amorphous phases crystalline phases can also precipitate, depending on the activator and the processing. For instance, hydroxide activators can lead to zeolite formation which may be intensified if hydrothermal curing is applied.

Besides compositions within the $Na_2O-CaO-Al_2O_3-SiO_2$ diagram, additional chemistries are also explored. Fe–Al-rich amorphous silicates are interesting precursors (Pontikes et al., 2013a), in addition to other works where authors tried to synthesize magnesium-containing analogues of aluminosilicate binders (MacKenzie et al., 2013) and to substitute boron for aluminum in the inorganic polymer structure (Williams & van Riessen, 2011). The role of Fe is also a topic of interest (Lemougna et al., 2013), with materials originating from non-ferrous slag typically demonstrating competitive physical and mechanical properties (e.g. Maragkos et al., 2009).

16.5 CONCLUSIONS

Cement plants today can operate as recipients of materials for energy and product enabling the "co-processing" of large volumes of residues. Depending on economic, logistic and legislation aspects, "ad-hoc" solutions emerge. Factors such as variability, homogeneity and volumes of these secondary streams in conjunction with maximum acceptance levels for certain elements, pose limitations on what can be used eventually. Next to the materials used for clinker production, impressive volumes of about 500 Mt/y of residues are already being incorporated downstream into a wide range of blended cements. Ground granulated blast-furnace slags as well as fly ashes from coal combustion are used to a significant extent. Thus, despite constraints, cement plants can be seen as interwoven, interacting elements in the social/industrial symbiotic matrix.

Still, apart from the established and proven processes, new ideas emerge, aspiring to reduce the CO_2 footprint of cement today and often "problem driven" by other industrial sectors or societal needs. In most cases, these ideas do not reach the level of industrial implementation. Yet, a certain culture is developing that is embracing and considering these alternatives, often for local or niche applications. The beneficiation of these residues, securing a proper and environmentally sound chemistry, a desirable microstructure and an overall quality that fluctuates only in a minor, reasonable way, will facilitate their use. This would necessitate also a mentality shift, where residue producing industries become actively involved in the valorization of their residues, without allocating the liability to end-users. Industrial ecology provides the conceptual framework for these actions to emerge but there is a need for support by legislation, regulations and society. For the binders in particular, a challenge would be to move from prescriptive to performance-based standards, with no compromise on their long-term durability and no illusions on the fact that these binders will be used in real-life conditions where control is not always exceptional.

In view of the above, a subtle transformation is taking place where the industrial processes of today become more sustainable and where new processes and materials emerge as alternatives. It remains an open question if the kinetics of this transformation are fast enough to meet the technical and non-technical challenges of the materials themselves as well as the need for a smaller environmental footprint of humanity in total.

Acknowledgments

YP is thankful to the Research Foundation — Flanders and RS acknowledges support of the European Community under FP7-Marie Curie IEF grant 298337. Prof. H. Rabier and Prof. J. Wastiels from VUB as well as Dr. P.T. Jones from KU Leuven and Dr. M. Kazemi-Kamyab from EPFL are also acknowledged for their constructive comments.

References

Antoni, M., Rossen, J., Martirena, F., Scrivener, K., 2012. Cement substitution by a combination of metakaolin and limestone. Cement and Concrete Research 42, 1579—1589.

Aranda Usón, A., López-Sabirón, A.M., Ferreira, G., Llera Sastresa, E., 2012. Uses of alternative fuels and raw materials in the cement industry as sustainable waste management options. Renewable and Sustainable Energy Reviews 23, 242—260.

Arjunan, P., Silsbee, M.R., Roy, D.M., 1999. Sulfoaluminate-belite cement from low-calcium fly ash and sulfur-rich and other industrial by-products. Cement and Concrete Research 29, 1305—1311.

Barcelo, L., Kline, J., Walenta, G., Gartner, E., 2013. Cement and carbon emissions. Materials and Structures, (Published Online).

Bell, J.L., Sarin, P., Driemeyer, P.E., Haggerty, R.P., Chupas, P.J., Kriven, W.M., 2008. X-ray pair distribution function analysis of a metakaolin-based,

KAlSi$_2$O$_6$·5.5H$_2$O inorganic polymer (geopolymer). Journal of Materials Chemistry 18, 5974–5981.

Chinyama, M.P.M., 2011. Alternative fuels in cement manufacturing. In: Manzanera, M. (Ed.), Alternative Fuel. InTech.

Damtoft, J.S., Lukasik, J., Herfort, D., Sorrentino, D., Gartner, E.M., 2008. Sustainable development and climate change initiatives. Cement and Concrete Research 38, 115–127.

De Weerdt, K., Haha, M.B., Le Saout, G., Kjellsen, K.O., Justnes, H., Lothenbach, B., 2011. Hydration mechanisms of ternary Portland cements containing limestone powder and fly ash. Cement and Concrete Research 41, 279–291.

Delort, M., 2013. Low-clinker ternary cements: performance and standardization. In: 7th International VDZ-Congress.

Dienemann, W., Ben Haha, M., Bullerjahn, F., Schmitt, D., 2013. Belite calciumsulfoaluminate ternesite (BCT): a new low carbon clinker technology. In: Proceedings of the 7th International VDZ-Congress.

Duxson, P., Fernández-Jiménez, A., Provis, J.L., Lukey, G.C., Palomo, A., van Deventer, J.S.J., 2007. Geopolymer technology: the current state of the art. Journal of Materials Science 42, 2917–2933.

European Committee for Standardization, 2000. EN 197–1, Cement – Part 1: Composition, Specifications and Conformity Criteria for Common Cements.

Garcia-Lodeiro, I., Palomo, A., Fernández-Jiménez, A., Macphee, D.E., 2011. Compatibility studies between N-A-S-H and C-A-S-H gels. Study in the ternary diagram Na$_2$O-CaO-Al$_2$O$_3$-SiO$_2$-H$_2$O. Cement and Concrete Research 41, 923–931.

García-Lodeiro, I., Fernández-Jiménez, A., Palomo, A., 2013. Variation in hybrid cements over time. Alkaline activation of fly ash-Portland cement blends. Cement and Concrete Research 52, 112–122.

Gartner, E.M., Macphee, D.E., 2011. A physico-chemical basis for novel cementitious binders. Cement and Concrete Research 41, 736–749.

Glukhovsky, V.D., 1967. Soil Silicate Articles and Structures. Budivelnyk Publisher, Kiev.

Gutteridge, W.A., Dalziel, J.A., 1990. Filler cement: the effect of the secondary component on the hydration of Portland cement. Part I. A fine non-hydraulic filler. Cement and Concrete Research 20, 778–782.

Habert, G., Billard, C., Rossi, P., Chen, C., Roussel, N., 2010. Cement production technology improvement compared to factor 4 objectives. Cement and Concrete Research 40, 820–826.

Hanehara, S., 2005. Eco-cement and eco-concrete. Environmentally compatible cement and concrete technology. In: COE Workshop on "Material Science in the 21st Century for the Construction Industry – Durability, Repair and Recycling of Concrete Structures", Hokkaido University Conference Hall, Sapporo, Japan.

Huang, Q., Yang, Y., Wang, Q., 2012. Potential for serious environmental threats from uncontrolled co-processing of wastes in cement kilns. Environmental Science & Technology 46, 13031–13032.

IEA, 2009. Cement Roadmap. http://www.iea.org/publications/freepublications/publication/Cement_Roadmap_Foldout_WEB.pdf.

Ismail, I., Bernal, S.A., Provis, J.L., San Nicolas, R., Hamdan, S., Van Deventer, J.S.J., 2014. Modification of phase evolution in alkali-activated blast furnace slag by the incorporation of fly ash. Cement and Concrete Composites 45, 125–135.

Jones, P.T., Geysen, D., Tielemans, Y., Van Passel, S., Pontikes, Y., Blanpain, B., Quaghebeur, M., Hoekstra, N., 2012. Enhanced landfill mining in view of multiple resource recovery: a critical review. Journal of Cleaner Production 55, 45–55.

Juenger, M.C.G., Winnefeld, F., Provis, J.L., Ideker, J.H., 2011a. Advances in alternative cementitious binders. Cement and Concrete Research 41, 1232–1243.

Katagiri, M. Experiences of incineration ash recycling by Japan's cement industry. In: 2012 R3C International Symposium. Waste Resource Management and Incineration Ash Utilization, Nangyang Executive Centre, NTU, Singapore, 23 and 24 October 2012.

Lemougna, P.N., MacKenzie, K.J.D., Jameson, G.N.L., Rahier, H., Chinje Melo, U.F., 2013. The role of iron in the formation of inorganic polymers (geopolymers) from volcanic ash: a 57Fe Mössbauer spectroscopy study. Journal of Materials Science 48, 5280–5286.

MacKenzie, K.J.D., Bradley, S., Hanna, J.V., Smith, M.E., 2013. Magnesium analogues of aluminosilicate inorganic polymers (geopolymers) from magnesium minerals. Journal of Materials Science 48, 1787–1793.

Maragkos, I., Giannopoulou, I.P., Panias, D., 2009. Synthesis of ferronickel slag-based geopolymers. Minerals Engineering 22, 196–203.

Mokrzycki, E., Uliasz-Bocheńczyk, A., Sarna, M., 2003. Use of alternative fuels in the Polish cement industry. Applied Energy 74, 101–111.

Noirfontaine, M.N., Tusseau-Nenez, S., Girod-Labianca, C., Pontikis, V., 2012. CALPHAD formalism for Portland clinker: thermodynamic models and databases. Journal of Materials Science 47, 1471–1479.

OECD/IAI and the World Business Council for Sustainable Development, 2009. Cement Technology Roadmap 2009-Carbon Emission Reductions up to 2050.

Olivier, J.G.J., Janssens-Maenhout, G., Peters, J.A.H.W., 2012. Trends in Global CO$_2$ Emissions; 2012 Report. PBL Netherlands Environmental Assessment Agency; Ispra: Joint Research Centre, The Hague.

van Oss, H.G., Padovani, A.C., 2002. Cement manufacture and the environment: part I: chemistry and technology. Journal of Industrial Ecology 6, 89–105.

Pontikes, Y., Angelopoulos, G.N., 2013. Bauxite residue in cement and cementitious applications: current status and a possible way forward. Resources, Conservation and Recycling 73, 53–63.

Pontikes, Y., Machiels, L., Onisei, S., Pandelaers, L., Geysen, D., Jones, P.T., Blanpain, B., 2013a. Slags with a high Al and Fe content as precursors for inorganic polymers. Applied Clay Science 73, 93–102.

Pontikes, Y., Machiels, L., Jones, P.T., Blanpain, B., 6–9 October 2013b. Cradle to cradle cementitious binders and building materials from municipal wastes. In: Workshop on Sustainable Waste Management for Ecocities and Rural Areas. Venice International University.

Rahier, H., Van Mele, B., Biesemans, M., Wastiels, J., Wu, X., 1996. Low-temperature synthesized aluminosilicate glasses: part I. Low-temperature reaction stoichiometry and structure of a model compound. Journal of Materials Science 31, 71–79.

Rootzén, J., 2012. Reducing Carbon Dioxide Emissions from the EU Power and Industry Sectors. An Assessment of Key Technologies and Measures. Department of Energy and Environment, Chalmers University of Technology, Gothenburg, Sweden.

Schneider, M., Romer, M., Tschudin, M., Bolio, H., 2011. Sustainable cement production—present and future. Cement and Concrete Research 41, 642–650.

Scrivener, K.L., Nonat, A., 2011. Hydration of cementitious materials, present and future. Cement and Concrete Research 41, 651–665.

Shi, C., Wu, Y., Riefler, C., Wang, H., 2005. Characteristics and pozzolanic reactivity of glass powders. Cement and Concrete Research 35, 987–993.

Shi, C., Krivenko, P.K., Roy, D., 2006. Alkali-Activated Cements and Concretes. Taylor & Francis, NY, USA.

Shi, C., Jiménez, A.F., Palomo, A., 2011. New cements for the 21st century: the pursuit of an alternative to Portland cement. Cement and Concrete Research 41, 750–763.

Snellings, R., Mertens, G., Elsen, J., 2012. Supplementary cementitious materials. Reviews in Mineralogy and Geochemistry 74, 211–278.

The GTZ-Holcim Public Private Partnership, 2006. Guidelines on Co-processing Waste Materials in Cement Production.

U.S. EPA, 2008. Cement Sector Trends in Beneficial Use of Alternative Fuels and Raw Materials.

USGS, 2013a. Mineral Commodities Summary: Cement.

USGS, 2013b. Mineral Commodities Summary: Iron and Steel Slag.

Wang, S.-D., Scrivener, K.L., 1995. Hydration products of alkali activated slag cement. Cement and Concrete Research 25, 561–571.

WBCSD, 2005. Guidelines for the Selection and Use of Fuels and Raw Materials in the Cement Manufacturing Process.

WBCSD, 2009. Cement Industry Energy and CO_2 Performance. "Getting the Numbers Right".

Williams, R.P., van Riessen, A., 2011. Development of alkali activated borosilicate inorganic polymers (AABSIP). Journal of the European Ceramic Society 31, 1513–1516.

World Coal Association, 2013. Coal Statistics.

Yamaguchi, N., Nagaishi, M., Kisu, K., Nakamura, Y., Ikeda, K., 2013. Preparation of monolithic geopolymer materials from urban waste incineration slags. Nippon Seramikkusu Kyokai Gakujutsu Ronbunshi/Journal of the Ceramic Society of Japan 121, 847–854.

CHAPTER 17

Industrial By-products

Jaana Sorvari[1], Margareta Wahlström[2]

[1]Finnish Environment Institute, Helsinki, Finland; [2]VTT Technical Research Centre of Finland, VTT, Finland

17.1 WHAT IS A BY-PRODUCT?

Besides the desired products, industrial processes generate by-products: that is, unintentional wastes and residues that may be recyclable either as such or after processing. What is understood by the term "industrial by-product" varies internationally, however. Formerly, the European legislation did not recognize this concept. Hence, the outcomes of an industrial process were either products or waste, in which the term "product" generally refers only to the main product that the process intends to produce. Process conditions are mainly adjusted to attain the highest quality of the product; usually, no attention is paid to the properties of the residues even though they can have a demand as substitutes for virgin materials or manufactured new products. Whether a residue is classified as waste or a by-product has implications for its recycling and reuse because any professional treatment, disposal or reuse of waste is regulated by waste legislation and requires a permit. As for a product, only product specifications need to be followed. The producer is responsible for any problems arising during the life time of the product whereas anyone in the possession of waste is also liable for its due and safe management. The liability extends to the final application: for example, maintenance of a road where such material has been utilized. At present, the by-product concept has been adopted within the European Union (EU). The European Waste Framework Directive (2008/98/EC) also includes the generic End of Waste (EoW) criteria, which define when a residue ceases to be waste and turns into a product, potentially regulated by the Registration, Evaluation, Authorisation and Restriction of Chemicals (REACH) directive and existing specific legislation, such as the regulation for construction products or fertilizers. The generic EoW criteria include requirements for recovery operations, usefulness for specified purposes, existing demand or market, technical compliance and acceptability from the viewpoint of adverse impacts to the environment and human health. These criteria and the methodology developed by Joint Research Centre (JRC) (Delgado et al., 2009) serve as the basis for defining waste-specific criteria. By the end of 2012, specific EoW criteria existed for only a few materials, e.g. iron, steel and aluminum scrap and glass. Thus, from the European regulatory perspective, most residues generally referred to as by-products are in fact waste.

In this chapter, no stand is taken regarding whether the material at hand is to be defined a by-product from a regulatory viewpoint. Therefore, an industrial by-product is understood to be a solid inorganic residue that is formed in the metal and mining industry, energy production, forest industry, chemical industry or building industry, and has reuse or recycling potential. Thus, waste or residues that are generally considered to be a part of municipal waste (e.g. tires), organic residues (e.g. wastewater treatment sludge) and liquid by-products from chemical industry, as well as some specific residues that are currently classified as hazardous wastes within the EU or in the United States (US) (for example, jarosite and automobile shredder residue) are excluded. Some of the by-products, e.g. those generated in construction and demolition, are discussed in more detail in other chapters of this book.

17.2 MAJOR BY-PRODUCTS AND THEIR GENERIC PROPERTIES

Inorganic industrial by-products are formed in several industries (Table 17.1). The process whereby a specific by-product is formed determines its generic properties and composition. However, process conditions, equipment and raw materials eventually dictate its specific technical and environmental properties and, hence, suitability for recycling. Generally speaking, waste concrete from construction and demolition, reclaimed asphalt from road maintenance and slag from crude iron and steel production are recycled to a large extent all over the world. The recycling rate of different by-products varies in different countries, however, and finding exact information on the rates is difficult. Inadequacy in documenting the volumes and final destination of by-products as well as variation in classifying waste and its management practices (e.g. whether filling mines is considered recycling) are probable causes of the latter problem.

17.3 WHERE AND HOW TO USE BY-PRODUCTS

Many by-products have beneficial properties that make them suitable for applications in which they can replace virgin materials. Table 17.2 summarizes potential applications for the major by-products presented in Table 17.1. Prevailing applications are found in the construction/building industry and, to a lesser extent, in agriculture and horticulture. In some cases, recovery of valuables or use as a raw material of new products is an option. Practically, regardless of application, the use of a by-product generally requires detailed knowledge of its composition, physical properties and variations in quality, as well as of the specifications of the virgin material it would replace. In most cases, some pretreatment, processing, or both are required to render a by-product usable.

The characteristics of the material and quality requirements of the application determine the processing needs of a by-product. Treatment and processing options include sorting to improve the quality of the material, e.g. removal of impurities or unwanted fractions through selective demolition; conditioning and weathering to stabilize the material—for example, to enhance the hydration of the free lime and magnesia of steel slag used in earth construction; crushing or grinding to remove porous particles and to convert granular monolithic materials; sieving to separate different particle size fractions; filtration; sintering to reduce porosity and enhance the strength or electrical and thermal conductivity of fine materials; pelletizing, granulation and briquetting to homogenize heterogeneous materials, to avoid turning fine materials to dust

TABLE 17.1 By-products Generated in Different Industries and Their Generic Properties, Relevant From the Viewpoint of Recycling

Industrial Sector	By-product	Formation	Main Characteristics and Remarks
Construction and demolition	Concrete	1. Dismantling of roads, runways and structures, e.g. bridges, buildings (⇒ reclaimed concrete) 2. Generated continuously in the manufacture of hollow-core slabs 3. Washing of tank trucks that transport concrete, and concrete mixers in concrete plants. Forms in the settling of washing liquid	1. Can include subbase soil material and bitumen (asphalt-concrete pavement in roads) or sealants (buildings). Upgrading can be done on-site or off-site. The quality depends on the type and use of the structure, and the accuracy of the selective demolition system. Potential contaminants include PCBs (from sealants used in concrete buildings), asbestos, dioxins, heavy metals (cadmium, chromium, mercury, lead), molybdenum, CFC, oil, PAH, phenols. 2. Typically almost pure concrete containing only minor residues of mineral and plant oil used in molding. 3. Granularity and water content varies. Can contain some hardened concrete. Can be recycled back to the manufacturing process.
	Bricks	Dismantling of structures, building	Consists of clay and mortar. Can include some impurities originating from the raw material or from the construct (e.g. chimneys). Generated and used often together with waste concrete from demolition, difficult to clean due to formation of dust that is harmful to human health (barrier to reuse as such without processing).
	Mineral wool	1. Dismantling of buildings 2. Manufacture of mineral wool tiles when the raw material is melted and defibrated and further processed to shaped products	Comprises amorphous synthetic, vitreous fibers produced from mineral raw materials. Covers glass wool, rock wool or slag wool, depending on the starting material. Includes 18% of sodium, potassium, calcium, magnesium and barium oxides.
	Reclaimed asphalt	Road maintenance and construction: 1. Cutting of pavement 2. Crushing of asphalt tiles	Properties resemble those of virgin stone materials. Includes mainly stone material and few percentages of bitumen. Good technical properties owing to low fines content. Since the fines are bound into bitumen, the sensitivity to water is lower. In general, recycling does not pose significant risks to the environment or human health. Deterioration of quality is a problem when asphalt is recycled several times.
	Glass	Dismantling of buildings	Main components include silicon oxide, calcium carbonate and sodium carbonate. Amorphous and inert material, and lighter than equivalent mineral soil materials. A part can be reused as such.
	Plasterboard and wallboards	Dismantling of buildings	Consists of gypsum (see gypsum below). Contains impurities, such as nails, screws, wallpaper and other wall coverings.

(Continued)

TABLE 17.1 By-products Generated in Different Industries and Their Generic Properties, Relevant From the Viewpoint of Recycling (cont'd)

Industrial Sector	By-product	Formation	Main Characteristics and Remarks
	Excess soil	Construction in general	Physically similar to virgin soil material. May contain impurities emitted from previous activities at the construction site, such as heavy metals or organic contaminants, which determine the environmental acceptability. Poor technical properties can hinder recycling in earth construction.
Metal industry	Blast furnace slag (BFS)[1]	Manufacture of crude iron. Forms in the blast furnace 1. When the slag is rapidly cooled with water at a rate sufficient to ensure that no crystallization occurs = granulated BFS 2. When the slag is allowed to cool in air = air-cooled BFS	Contains mainly natural rock minerals, with the oxides of silicon, calcium, magnesium and aluminum being the main constituents. Porous material which has low density, low thermal conductivity, high friction and low wear resistance. Water-cooled BFS is vitreous material and its good hydraulicity enables its use as a binder agent. Sulfate (particularly in air-cooled BFS) and vanadium are potential most soluble contaminants.
	Steel slag	1. Primary steelmaking,[2] i.e. manufacture of crude steel by combusting crude iron in a converter in the presence of quicklime, dolomite and oxidizing agents of the ore. 2. Secondary steelmaking, (see EAFS below)	Consists mainly of calcium, silicon, magnesium, and aluminum oxides. Particle size distribution equivalent to gravel, but coarser with higher bulk density, mechanical properties depend on the porosity. Unreacted lime results in heterogeneous and unstable material. Contains usually various additional elements as potential contaminants that can hinder recycling, such as cadmium, chromium, molybdenum, nickel, selenium.
	Ferrochromium (FeCr) slag	Manufacture of ferrochrome (used in the production of stainless steel) from chromite. Forms in the cooling of molten slag.	Consists of silicon, magnesium and aluminum oxides. From the viewpoint of technical properties, can be used as a substitute of sand, gravel and crushed rock. Particles are sharper compared with sand. Chromium content is high (up to ca. 10%) which prevents recycling in some applications, e.g. in applications where dusting and spreading via air is possible.
	EAFS	Manufacture of crude steel from steel scrap by the electric arc furnace (EAF) process. Forms when the liquid slag is first solidified and further cooled with water.	Porous material with a risk of volume expansion. Wear-resistant material with similar or better technical properties than natural materials. Can have a high content of chromium.
	Zinc slag	Galvanization of steel to avoid corrosion. Forms when the slag is skimmed from the molten zinc bath.	Contains more than 50% zinc, which can be recovered.

	Nonferrous slag	Recovery and processing of nonferrous metal from natural ores. Forms in smelting and subsequent cooling of molten slag.	Rocklike or granular materials, such as copper (Cu), nickel (Ni), phosphorus, lead (Pb), and zinc (Zn) slag. Characteristics vary depending on the ore, process and slag cooling method and rate. Some slag are classified as *hazardous wastes* (e.g. Cu slag). Processed air-cooled and granulated Cu, Ni, and phosphorus slag are mechanically sound, resistant to abrasion and stable. Particularly granulated slag generated by rapid quenching tends to be vitreous which, reduces friction and skid resistance and can be a problem in some applications, e.g. pavements.
	Flue dust	Manufacture of ferrous and nonferrous metals. Flue dust forms in the separation of particles from flue gas.	Composition depends on the raw material and process. In those cases where the dust contains several metals as impurities, it is often classified as hazardous waste.
	Foundry sands	Casting of molds. Forms as a residue in both ferrous and nonferrous metal casting process.	Composition varies depending on the sand type (e.g. green sand, furan sand, alkaline phenolic sand). The basic constituent is natural silica sand which can contain traces of metals originating from the metal molded. Contains generally also traces of organic binders. Most of the foundry sands are recycled back to casting process, but some need to be removed after several rounds. Some spent foundry sands are characteristically hazardous, e.g. those from brass and bronze foundries.
Mining and enrichment of ores	Red mud (ilmenite)	Production of bauxite. Forms in the refinement of bauxite to produce aluminum.	Consists of titanium-iron oxide. Finely crushed, alkaline material (pH >12). Contains several heavy metals and metals, e.g. arsenic, chromium, lead, nickel, selenium, zinc, as impurities with varying concentrations.
	Waste rock	Mining and quarrying; part of the ore body, which is separated at the excavation front in order to get access to the ore.	Contains low grades of value minerals, which cannot be mined and processed profitably. Differing characteristics depending on the ore. Waste rock from open-pit mines and quarries are mostly inert natural materials if they do not contain sulfides, which might generate acidic drainage.
	China clay (kaolin) waste	Mining. Forms in the extraction of kaolin mineral when the sand and mica residues are separated from the mineral slurry.	Properties equivalent to primary aggregates used in earth construction. Consists of overburden, waste rock, coarse sand and micaceous residue, coarse sand being the main component. Mica can cause changes in the density. Almost 90% of the coarse sand is quartz. Production of 1 t of china clay results in 9 t of waste.
	Final tailings	Enrichment of metal ores. Forms when the ore is treated chemically to separate the desired metal or mineral.	Differing characteristics depending on the raw material (ore) and process. Includes mineral material, metal residues and chemicals used in the enrichment.

(Continued)

TABLE 17.1 By-products Generated in Different Industries and Their Generic Properties, Relevant From the Viewpoint of Recycling (cont'd)

Industrial Sector	By-product	Formation	Main Characteristics and Remarks
	Lime (calcium oxide) e.g. lime kiln dust (LKD)	Refining of limestone (= calcium carbonate). Forms during filtering of the exhaust gases from heating limestone in a kiln.	Very fine, white powdery material of uniform size, principal constituents are calcium and magnesium carbonates. Chemical and physical properties vary depending on the type of lime being manufactured, feedstock and fuel type, process, type of kiln and dust collection method. Based on reactivity determined by the amount of free lime and magnesia, distinction is made between two types: calcitic (chemical lime, quicklime, etc.) and dolomitic.
Energy production	Fly ash[3] from the combustion of coal, peat, wood or mixture	Combustion of coal, peat, wood or mixture of these. Forms when a portion of light, noncombustible mineral matter is emitted from the furnace along with flue gases and volatilized minerals and separated as fly ash in mechanical or electric particulate control devices, e.g. a dust precipitator.	Fly ashes have generally better thermal insulation compared to natural materials. Being fine-textured can absorb humidity and silt. Frost susceptibility depends on particle size distribution. The composition of fly ashes varies significantly depending on boiler type and used fuels, but main constituents of coal combustion fly ash (CCFA) include silicon, aluminum and iron compounds with smaller amounts of oxidized alkali metals and some heavy metals as impurities. CFFA is fine powdery material with a grain size of 0.002–0.1 mm corresponding to loam. CCFA ash is a consolidating material. Consolidation can be accelerated by activators (e.g. concrete, lime, FGD slag). The properties of co-combustion (peat, wood etc.) fly ashes are in principle similar to coal fly ashes, but the quality variations are usually more severe. The main components in the fly ash from peat combustion (PCFA) are the oxides of iron, magnesium, potassium and organic matter. Concentration of heavy metals is lower compared to ash from wood combustion. Corrosiveness can be a problem in some applications. The high concentration and/or potential leaching of (heavy) metals, such as antimony, arsenic, boron, barium, cadmium, lead, selenium, molybdenum and zinc, can also prevent recycling. Some ashes, e.g. PCFA, can contain radioactive elements.
	Bottom ash[3] and slag from the combustion of coal, peat, wood or mixture	Combustion of coal, peat, wood or mixture of these. Forms when a portion of noncombustible mineral matter is retained in the bottom of the furnace.	The ratio of fly ash and bottom ash depends on the combustion technique. Fixed bed combustion generates more bottom ash than fluidized bed combustion. Granularity of bottom ash is equivalent to coarse or gravelly sand, some particles are glassy. It is non pozzolanic unlike fly ash, therefore does not consolidate as easily. Bottom ash is usually cooled with water and therefore, its water content is high. Water can be removed by storing before use.

17.3 WHERE AND HOW TO USE BY-PRODUCTS

	FGD (flue gas desulphurization) slag	Combustion of coal. Forms in the treatment of flue gases using lime for removing sulfur, based on 1. Wet-scrubbing process (or wet process) which results in a dry or wet end product, generates also a filter cake (waste water treatment) 2. Dry additive injection process: waste collected with a particulate collection device 3. Wet-dry process (or spray-absorption/semidry process): waste collected with a particulate collection device	1. Consists mainly of contaminated gypsum (see phosphogypsum below), content of impurities reported to be up to 6%. 2. Consists of fly ash and reaction products of additive input fed into the furnace (e.g. limestone, lime). The amount of ash in the waste varies between 30% and 80%. 3. A Dry mixture of fly ash and desulphurization waste, consisting mainly of calcium sulfite and calcium sulfate and some unreacted lime and calcium carbonate. The fly-ash content depends on the separation degree of fly ash before desulphurization.
Forest industry	Paper mill sludge (i.e. fiber sludge)	Manufacture of paper from wood pulp. Forms in the separation of the primary sludge generated in the mechanical wastewater treatment.	Quality and composition varies according to the manufacturing process and wastewater treatment method. Sometimes primary sludge is mixed with sludges from other parts of the process. In general, fiber sludge consists of organic material and caoline (around 50/50%). Caoline is very stable material while the organic matter mainly consists of cellulose and wood fibers. Impurities include heavy metals that come from wood, chemicals used in the process, tubes and fillers.
	De-inking sludge	Manufacture of recycled pulp from waste paper. Forms in the removal of impurities by filtering and further removal of ink by flotation.	Includes mostly short fibers, i.e. organic matter and clay (caoline, soapstone, bentonite) with impurities such as, ink, heavy metals and plastics, the final waste has been usually incinerated on site after thickening and drying. Although the organic matter content is high, de-inking sludge has proved suitable for some earth construction applications due to low biodegradability.
	Green liquid precipitate	Manufacture of wood pulp for papermaking. Forms in the soda recovery plant belonging to the chemicals recovery line when the precipitate is removed from green liquid.	Consist of calcium carbonate and different insoluble impurities of green liquid, such as soot, metal oxides, sodium compounds, silicates and impurities of wood. Generally considered as inert waste due to low organic matter content. Granularity equivalent to loam with high water content. Compression strength is too low to be used in earth construction without additives, such as fiber sludge and fly ash.
Chemical industry	Iron calcinate/hematite filler	Manufacture of sulfuric acid. Forms when pyrite is combusted in calcining plant.	Consists of iron (III) oxide, which is chemically stable, durable and resistant to corrosion.

(Continued)

TABLE 17.1 By-products Generated in Different Industries and Their Generic Properties, Relevant From the Viewpoint of Recycling (cont'd)

Industrial Sector	By-product	Formation	Main Characteristics and Remarks
	Kiln dust	Manufacture of cement. Forms in cement kiln, recovered from the dust collection system for the control of air emissions.	Covers two different dusts: cement kiln dust (CKD) and lime kiln dust (LKD). CKD is a fine powdery material equivalent to Portland cement. The principal constituents are compounds of lime, iron, silica and alumina. The free lime tends to be in coarser particles closest to the kiln. Its content in LKD can be significantly higher than in CKD. Calcium and magnesium carbonates are the principal mineral constituents of LKD, the composition varies depending on the product (chemical, hydrated or dolomitic lime or quicklime) manufactured. Part of the dust can be separated and returned to the kiln.
	Gypsum	Forms in various chemical processes, e.g. in the manufacture of titanium pigment, in the reaction between phosphorus mineral and sulfuric acid in the manufacture of phosphorus fertilizer (phosphogypsum), in flue-gas desulphurization (FGD, see above)	Composed of calcium-sulfate-dihydrate ($CaSO_4 \cdot 2H_2O$). Large quantities are formed in the phosphoric acid process of the phosphorus fertilizer production. Gypsum is moderately water-soluble and it is used particularly in soil amendment and in building industry. Depending on the origin of the raw material (phosphate rock), phosphogypsum can include impurities, particularly fluoride and radioactive elements that can prevent its recycling.

The list aims to present the most important materials with recycling potential and it is not exhaustive

[1] Two additional types of BFS, i.e. expanded or foamed BFS and pelletized BFS, are also known; the former forms when the slag is cooled and solidified by controlled quantity of water, air or stream and the latter when the slag is cooled and solidified with water and air quenched in a spinning drum.
[2] Primary steelmaking slag covers slag from Linz-Donawitz process (LD slag), converter slag, slag from continuous casting (CC slag), and slag from basic oxygen furnace (BOF slag).
[3] Ashes often comes from mixed combustion of different raw materials, e.g. wood, peat, coal and, ashes with similar generic physical properties are formed also in the combustion of municipal solid waste; the chemicals involved then vary depending on the waste material.

Source: Data compiled from various sources Recycled Materials Resource Center (RMRC), Mroueh et al. (2000), Mäkelä and Höynälä (2000), Jha et al. (2001), Ramachandra Rao (2006), Orkas (2001), U.S. Environmental Protection Agency (2007), RMT Inc. (2003).

TABLE 17.2 Summary of the Main recycling Options and Examples of By-products and Techniques for Processing Them Suitable for the Specific Application

Application	By-product	Examples of Processing
Construction (civil engineering)		
• As aggregate in different structures: roads[1], landfills, buildings, noise barriers, embankments, fills	Crushed concrete, all ashes, BFS[3], steel slag[4], Ni slag, Cu slag, foundry sand[5] waste rock, concrete sludge, china clay waste, final tailings from limestone mines, excess soil, green liquor precipitate (landfills[6])	Ash: stabilization using cement, FGD, or equivalent Ash: Modification of technical properties using fiber sludges Air-cooled BFS, steel slag, mineral wool: crushing or grinding; Air-cooled BFS: sorting to required fractions Steel slag: recovery of metallic particles, conditioning and weathering Reclaimed concrete: removal of metals and impurities
• Pavement	All ashes, reclaimed asphalt, red mud, steel slag, mineral wool	
• Groundwater protection layers, i.e. landfill liners	Fiber sludges from forest industry, paper mill sludge, de-inking sludge), green sand (foundry sand)	Fiber sludge: refinement[9] (e.g. with bentonite) Concrete sludge: homogenization, drying and crushing China clay sand: grading and washing only (equivalent to primary aggregates)
• Roofing	Ni and Cu slag	
• Masonry aggregate	Phosphorus slag	
• Soil stabilizer or solidifying agent in soft or wet soils	Ashes, lime (LKD), cement and lime kiln dusts, mixture of fiber sludge and ash	
Agriculture and horticulture: as fertilizer or soil amendment	Granulated BFS, FGD gypsum, red mud, steel slag, cement kiln dust, ashes from the combustion of wood[7] and peat, foundry sand[5], some fiber sludges, dusts from steel making, foundries and smelters, residues from galvanizing	Fine materials, e.g. ash: granulation (to avoid dusting) followed by sieving Ashes: addition of boron (if needed)
Reuse[2]	Glass, concrete, bricks and tiles from construction and demolition	Manual dismantling, sorting and cleaning (i.e. separation from mortar, grout and adhesives), crushing, sieving and magnetic separation Mechanical reclamation
	Phenolic sand (to replace new sand in green sand molding system in gray iron foundry)	
Building industry		
• As additive and binder in construction materials (e.g. source of fines): In (Portland) cement and concrete, asphalt concrete	Fly ash,[8] BFS, red mud, FGD gypsum, flue dust form BFS and steel making. Ni slag, Cu slag, foundry sand, fiber sludge	Flue dust: sintering Foundry sands: removal of very fine particles Fiber sludge: drying
• In concrete bricks and tiles	Red mud	
• As constituent or fine aggregate/binder or filler in asphalt/bitumen and bitumen bound materials	Foundry sand, red mud, ashes, cement kiln dust, steelmaking slag, Ni slag, Cu slag, Zn slag, phosphorus slag	
• As constituent in roofing felt	Ni slag	
• As insulator in structures	Ferro-chromium slag	Granulation or crushing (air-cooled FeCr slag)

(Continued)

TABLE 17.2 Summary of the Main recycling Options and Examples of By-products and Techniques for Processing Them Suitable for the Specific Application (cont'd)

Application	By-product	Examples of Processing
Manufacture of new products • Portland cement • Plasterboards • Ceramic tiles • Sandblasting sand • Mineral wool • Lightweight aggregate • Glass aggregate for floor tiles, abrasives, roofing shingles etc.	BFS, foundry sand, phosphorus slag FGD gypsum Zn slag Ni slag Glass, foundry sand Fiber sludge mixed with fly ash Fiber sludge	Foundry sand: briquetting, reduction of silica content Pelletization followed by heating and cooling Melting in high temperature, followed by rapid cooling with water
Metallurgical industry • As auxiliary material to form silicates (slag) in secondary refinement process of copper	Foundry sands	
Mining industry: backfilling of mines	Waste rock, tailings	No processing needed
Waste management and reclamation	Neutralizing (alkaline) materials, stabilizing or solidifying materials, e.g. ashes, FGD gypsum	

[1] Freezing can be a problem.
[2] Has to be free of dangerous substances and undamaged structures, etc.
[3] Used particularly in road construction.
[4] Due to strength and good abrasion and impact resistance particularly well-suited for road constructs used by heavy vehicles.
[5] Particularly sand from iron, steel and aluminum foundries (U.S. Environmental Protection Agency, 2007).
[6] In the water-retaining layer (surface structures), often mixed with ash or fiber sludge in order to increase structural strength and to ensure water-retaining capacity.
[7] Use in forestry has been questioned because of the potential long-term risks associated with cadmium content (Pitman, 2006).
[8] Ash from coal combustion is used more frequently than other ash.
[9] The need of processing depends on the properties of the material and the requirements set by the application, use without processing is also possible.
Source: Recycled Materials Resource Centre (RMRC), Mroueh et al. (2000); Mäkelä and Höynälä (2000); Jha et al. (2001); Ramachandra Rao (2006); Orkas (2001); U.S. Environmental Protection Agency (2007); RMT Inc. (2003); U.S. Environmental Protection Agency (2001).

or to ease storage and use; modification of chemical properties, e.g. to increase the solubility of micronutrients or improve fertilizing properties by adding nutrients; magnetic separation or thermal treatment using a rotary kiln or plasma to recover metals; and vitrification to immobilize soluble constituents. Depending on the material, several processes may be needed to make it suitable for a specific application.

Some of the most common applications of by-products are briefly described below.

17.3.1 Civil Engineering

In civil engineering, by-products are widely used as aggregates in different structures and fills in road and landfill construction and maintenance (Figure 17.1). The basic components of earthworks are pavement structures, load-dissipating structures (base course and subbase) and foundation structures. The type of structure and quality of subsoil determine which components are in fact necessary in a particular construction. Different layers call for different material properties (Table 17.3), the major determinants being stability, hydraulic conductivity, bearing capacity and freeze-thaw resistance (in northern regions).

FIGURE 17.1 Road building using reclaimed concrete.

In road building, by-products can be used as bulk fills, aggregates or binders (Sherwood, 2001). As aggregates, they are used either as unbound, i.e. granular, materials (primarily to substitute gravel) or bound. Bitumen, concrete, lime and their combinations are common binders. Some by-products, such as blast furnace slag (BFS) and fly ash, can also be used for binding. In addition, by-products can be used for thermal insulation and as a drainage layer in filtering layers; in noise barriers; as fillers in landscaping and pavement; as relief materials; and as sealing materials. The properties of metal slag residues, such as BFS, in particular are favorable since their thermal resistance is generally higher compared with virgin materials and their bulk density is lower. Therefore, the volume of material needed to build the desired structure is lower, allowing thinner structure layers compared with natural materials. This is a relevant feature considering freezing. Mineral wool residue does not freeze as easily as soil material, which enables its use during winter in cold regions.

By-products with suitable geotechnical properties for use in landfill structures come from various industries. They have mainly been used in top covers, particularly in the impermeable mineral layer. For example, fly ash mixed with paper sludge or sewage sludge has been used in top cover layers. Use of by-products in bottom layers is less frequent, most probably owing to the technical standards, which require adequate hydraulic conductivity, and environmental requirements. Moreover, the bottom layer rarely can be removed if the by-product turns out unsuitable.

The geotechnical properties of a single by-product are generally improved by mixing it with other materials, which can be either by-products or virgin materials. In the case of the mineral layer, variations in quality and the influence of weather conditions during mixing and compaction at the site are key issues. Furthermore, estimation of the long-term stability and

TABLE 17.3 Technical Requirements for By-products Used in the Different Layers in Earth Construction, per the Technical Guidelines Issued in Finland (Mäkelä and Höynälä, 2000)

Application	Purpose and Technical Requirements	Examples of Suitable By-products
Roads, parking lots, courtyards		
• Pavement	• Should serve as an even and wear resistant foundation for traffic and prevent penetration of water to lower layers. The requirements therefore include: strength, robustness, resistance to cracking, deformation, and wear.	• Metallurgical slag, reclaimed asphalt, ashes from coal and peat combustion, waste rock
• Base course	• Should persist tensions and deformations arising from traffic, distribute the load to wider area of lower layers, form an even base for pavement and serve as drying material for the water penetrating though pavement. Should have: sufficient carrying capacity and water permeability, freeze-thaw resistance, frost resistance, suitable granularity.	• Crushed metallurgical slag, granulated BFS, crushed concrete, waste rock, stabilized fly ash
• Sub-base	• Serves as a dissipating layer and increases carrying capacity and frost resistance, leads away the water penetrating from upper layers and prevents the capillary rise of water. Should possess: sufficient carrying capacity particularly in melting phase, sufficient frost resistance, freeze-thaw resistance, suitable granularity, thermal insulation capacity or good freezing resistance, adequate hydraulic conductivity.	• Like base course, also stabilized bottom ash and concrete sludge
• Protective course	• Should prevent capillary rise of water, separate pavement from ground base and prevent its mixing with upper layers, dissipate water from the structure, prevent or slow down frost penetration, and secure sufficient carrying capacity during melting.	• Crushed metallurgical slag, e.g. FeCr slag, crushed concrete, waste rock, consolidating or stabilized ash, ash from coal and peat combustion, crushed rock wool, crushed bricks and concrete sludge
• Roadbed (embankment)	• Serves as homogenic material that ensures even frost heaves and dislocations during melting. Requirements therefore include load and frost resistance, and sufficient strength during melting.	• Metallurgical slag, waste rock, nonconsolidating ash (to limited extent)
• Embankment relief	• Should reduce loads falling on soft road bed. Used in berms of road and street structures and courtyards and parking areas. Requirements include load resistance, sufficient strength during melting and lightness.	• Granulated BFS

TABLE 17.3 Technical Requirements for By-products Used in the Different Layers in Earth Construction, per the Technical Guidelines Issued in Finland (Mäkelä and Höynälä, 2000) (cont'd)

Application	Purpose and Technical Requirements	Examples of Suitable By-products
Soil structures in house building		
• Foundation pad	• Should ensure sufficiently strong and even bed for foundation and transfer loads to foundations to base course. Requirements include sufficient carrying capacity and frost resistance.	• Crushed concrete, metallurgical slag, e.g. FeCr slag, granulated BFS, consolidating or stabilized fly ash
• Flooring pad	• Forms a sufficiently load-bearing base for upper structures. Should possess adequate frost resistance, sufficiently low compressibility and thermal insulation capacity. Should preferably have low thermal conductivity. It should be possible to compact the material.	• Metallurgical slag, e.g. FeCr slag, granulated BFS, crushed concrete, waste rock, crushed bricks, consolidating and stabilized ash, bottom ash and slag, concrete sludge as such or crushed (to limited extent)
• Side fill of foundations	• Balances the variability in strength due to freeze-thaw phases and prevents detriments caused by freezing. Should have sufficient frost resistance, suitable granularity, thermal insulation capacity or good resistance toward freezing, and adequate water permeability. It should also be possible to compact the material.	• As above (flooring pad)
• Drainage layer	• Keeps other structures dry by leading waters to land drains and cutting capillary rise of water. Granularity is an important determinant of the functioning of this structure. Should prevent capillary rise of water, and have suitable granularity and good water permeability. It should also be possible to compact the material.	• Metallurgical slag, e.g. FeCr slag, granulated BFS
Noise barriers and butts structures		
• Surface layer	• Acts as a growing base and protects lower layers against erosion. Requirements include resistance to erosion particularly during rain and melting of frost, and suitability for a growing base.	• Ash and slag from coal combustion, peat combustion ash mixed with soil material, crushed concrete, crushed bricks, concrete sludge
• Fill	• Acts as a subgrade for upper structures and shall therefore retain its strength during melting and be independent of weather conditions. Requirements include sufficient stability, freeze-thaw resistance and water permeability, condensability, sufficiently low compressibility, and sufficient shearing strength.	• Metallurgical slag, crushed concrete, concrete sludge (limited use), waste rock and final tailings, crushed bricks, consolidating or stabilized fly ash, crushed mineral wool, excess soil

(Continued)

TABLE 17.3 Technical Requirements for By-products Used in the Different Layers in Earth Construction, per the Technical Guidelines Issued in Finland (Mäkelä and Höynälä, 2000) (cont'd)

Application	Purpose and Technical Requirements	Examples of Suitable By-products
• Subgrade	• Should balance any slumps in the structure and act as a foundation to other structures. Should have sufficient rigidity and shearing strength in order to attain stability and adequate carrying capacity.	• Metallurgical slag, waste rock, consolidating or stabilized fly ash, crushed concrete, crushed bricks, concrete sludge (limited use)
Landfill, surface structures		
• Growing base	• Acts as a growing base for vegetation planted after covering of landfill and protects lower layers against erosion, among others. Requirements therefore include resistance to erosion and suitability for a growing base.	• Ash from peat combustion, excess soil containing humus, crushed bricks, crushed mineral wool, concrete sludge (as soil amendment), fly ash with layers of till soil (compacted)
• Topsoil	• Should protect lower structures from mechanical load caused by plant roots, among others, also slows down the percolation of water to underlying drainage fill. Should therefore possess sufficient water permeability and shearing strength (in slope).	• Fly ash with layers of till soil (compacted), excess soil containing humus, concrete sludge, ash from peat combustion
• Drainage fill	• Shall lead the waters from upper layers to collecting drains located outside landfill's surface structure. This requires sufficient water permeability and frost resistance.	• Metallurgical slag, waste rock, bottom ash and slag
• Seal course	• Prevents percolating water from reaching underlying structures and waste fill. Should also avoid gases to reach upper layers and air. Functioning and durability of seal course are vital for the performance of the whole surface structure. Should therefore have suitable granularity to reach low water permeability and frost resistance. Shearing strength (in slope) should also be sufficient and it should be possible to compact the material.	• Fiber sludge and de-inking sludge as such or mixed with ash or bentonite, fly ash + bentonite + stabilized clay in layers, compacted fly ash and FGD gypsum + bentonite, ash from peat combustion
• Gas collection layer	• Collects landfill gases and leads them to collection structures or pipes and outside the cover structure. Requirements therefore include sufficient air permeability and suitable granularity to attain the required water permeability and frost resistance. It should also be possible to compact the material.	• Metallurgical slag, coarse crushed concrete, waste rock, bottom ash and slag

degradation of any organic material in by-products often involves high uncertainty.

17.3.2 Agriculture, Horticulture and Landscaping

Use of a by-product in agriculture or horticulture presumes that it has properties favorable to plant growth. The positive effect can be related to the supply of nutrients or conditioning soil by altering its chemical, physical or biological composition. In the former case, the nutrients need to be soluble or transformable to a form that is available to plants.

By-products that have liming properties and can thus correct soil acidity and thereby increase crop yield have been used for ground improvement. Blast furnace slag, steelmaking slag and ash from wood combustion are the most common by-products used for this purpose. Kiln dust is also suitable for agricultural purposes. In the US, in particular, zinc-bearing hazardous dust from electric arc furnaces has been used in the production of zinc micronutrient fertilizers (U.S. Environmental Protection Agency, 2001). Steelmaking slag and sludge also contain some plant nutrients such as calcium and magnesium, and can thus increase their content in herbage. The EU directive on fertilizers sets the minimum nutrient levels for basic slag from steelmaking (EC 2003/2003).

The motivation for using wood ash is to avoid depleting essential soil nutrients and reduce the harmful effects of acidification of forest soils and surface waters. Wood ash also releases potassium, sodium, boron, and sulfur. Granulated ash fertilizers are generally less soluble than the equivalent powdered fertilizers and products stabilized by self-hardening. The increase in soil pH resulting from the application of fly ash can also decrease the availability of heavy metals to plants (Scotti et al., 1999) and increase the uptake of some nutrients such as silicon, phosphorus, and potassium in paddy soils (Lee et al., 2006).

The use of red mud as a soil amendment has been studied extensively. In Australia, red mud was found to have the potential to reduce eutrophication of rivers and waterways by retaining nutrients, particularly phosphorus, in infertile sandy soils (Summer et al., 1996) and to increase pasture production. The potential release of harmful substances to sensitive waterways and groundwater needs to be studied prior to such use, however. Seawater-neutralized red mud has also shown good capacity to immobilize soluble acid and metals, particularly aluminum, zinc, and copper, from acid sulfate soil solutions. This feature makes it a good alternative for lime if leaching is an issue.

17.3.3 Recovery of Valuables

In particular, slag from metallurgical processes contains useful components that can be recovered. Manganese, silica, magnesia, alumina and niobium of steelmaking slag, and nickel and cobalt of copper slag, are examples of recoverable metals in by-products (Ramachandra Rao, 2006). Steelmaking slag could be used as a secondary source of iron. Dust from the steelmaking process also contains valuable metals, e.g. silver, that can be separated by leaching.

Silica present in ash can be separated and used, for example, in coatings and the manufacture of glass. It is also possible to recover metals from ashes, and their high iron fraction can be utilized for iron ore feed and cement.

Construction and demolition waste contain significant amounts of metals that are generally sold as scrap to be used as raw materials. The metals are first shredded, and then iron is removed by magnetic separation. Finally, nonferrous metals such as copper, aluminum and stainless steel are removed by density separation.

17.3.4 Recycling On-site

Recycling on-site is the preferred option from the viewpoint of the environmental impact arising from storing, transporting and processing or refining residues. On-site recycling is suitable for by-products that are equivalent to manufactured products such as concrete residues generated in the manufacture of hollow core slabs and sludges of mixing plants, and reclaimed asphalt generated in the renewal of pavement.

On-site recycling in the manufacturing process has also been presented as an alternative means of recycling steelmaking slag (Ramachandra Rao, 2006). In this option, the phosphorus content of the slag needs to be reduced before it can be returned to the process.

17.3.5 Use as Raw Material or Additive of New Products

Some by-products have proven to be practical raw materials for new products. For example, BFS has been widely used in the manufacture of concrete owing to properties similar to Portland cement, whereas fiber sludge with high inorganic content has proven a suitable raw material in the manufacture of cement. Fiber sludge has also been used as the base raw material in some industrial sorbent and animal bedding products (RMT Inc., 2003). Blast furnace slag can be combined with silica or alumina and converted to fibers for rock wool. Spent foundry sand can be used as a source of silica in this process.

Fly ash can be mixed with the main raw material to produce ceramics, floor and wall tiles, sound insulation panels, fillers in polymers and rubbers, and zeolites and inorganic fibers (Kumar et al., 2007), among others. Besides ash, foundry sands in particular are used as additives to provide fines in cement and concrete. Some applications, such as the manufacture of cement, may require mechanical activation that enables the use of higher proportions of ash and the attainment of improved quality for the new product. At the same time, the potential effect on the durability of concrete structures has raised some discussion about the suitability of ash for this purpose. Both the EU and ASTM International have issued standards for the quality of fly ash that can be used in concrete (DIN EN 450, ASTM International C618 − 12a).

17.3.6 Use in Waste Management and Wastewater or Flue Gas Treatment

Some by-products have the potential for use in waste or wastewater treatment. Ash and cement kiln dust, among others, have been used to stabilize and solidify waste. The use of ash in composting plants has resulted in acceleration of the composting process, better quality of the end product and improved occupational safety owing to the decreased number of harmful microbes (Ojala, 2008). Ash from combustion generated in the forest industry has also been used to stabilize soil contaminated by lead. Kiln dust, too, has been used in the remediation of contaminated environments and is also suitable for wastewater treatment (RMT Inc., 2003). The use of red mud to treat wastewater (Summer et al., 1996), e.g. acid mine drainage water, has also evoked interest. Green liquid precipitate formed in the manufacture of wood pulp has proven suitable for neutralizing wastewater, although its phosphorus content somewhat restricts its use.

Asokan et al. (2010) studied the potential of a mixture of coal combustion residues to immobilize the toxic elements of jarosite formed in the refinement of zinc ore. The solidified end product could be used in construction as bricks and blocks. Ferrous slag has also been used together with jarosite to make building materials. It is worth noting that jarosite is generally considered hazardous waste, and thus such end products may not be environmentally acceptable.

Cement kiln dust has been demonstrated to be an efficient material for the removal of sulfur dioxide from cement kiln flue gas (RMT Inc.,

2003). The reported removal efficiencies vary from 90% to 95%.

17.4 TECHNICAL AND ENVIRONMENTAL REQUIREMENTS

Uniform quality is perhaps the most important prerequisite that any by-product has to fulfill to be used on a large scale. By-products are also generally required to be technically at least as good as the virgin materials for which they substitute. In particular, by-products used in earth construction should have a service life comparable to virgin materials. The producer is responsible for ensuring that the by-product complies with the defined quality criteria and that quality control is carried out according to the requirements. Then, the user must use the material in defined applications according to the producer's instructions.

Any waste to be recycled generally needs to go through an extensive basic characterization that integrates information about its generic physicochemical properties and long-term behavior as well as the concentration of harmful substances and their environmental fate. A by-product that is used continuously usually does not need to go repeatedly through characterization unless its quality changes significantly: for example, as a result of changes in the production process or raw materials. However, it is very important to regularly control the by-product's quality to ensure that its properties meet specified requirements. A quality control system to ensure conformity with environmental and technical specifications should therefore be established in connection with the characterization. Particular attention should be paid to sampling, to produce representative samples for both characterization and quality control. Here, the sampling method, number of samples and time of sampling are important factors. Any sampling plan should therefore be based on the characteristics of the specific by-product as well as on the objectives of the sampling.

17.4.1 Quality Control

The quality control system and the testing needs included are material- and application-specific (see an example of potential parameters to be measured in Table 17.4). The technical characteristics relevant to the application and the most critical environmental characteristics are selected for periodic checking. Homogeneous materials with well-known characteristics require less frequent monitoring than heterogeneous and less well-investigated materials. The quality control program should include a description of the properties to be monitored, monitoring and sampling methods, sampling and control frequencies, limit values for quality and acceptable errors, as well as reporting methods.

The quality of a by-product can be proved through certification. The certificate can contain a detailed description of the properties that need to be controlled. In addition, use of a by-product as a product requires information about its long-term behavior and suitable applications. For example, in the case of earth construction, this generally requires building and monitoring pilot constructions and preparing design manuals and user guidelines. Therefore, several countries have manuals for the use of by-products in landfill construction, among others. These manuals usually contain a description of geotechnical and environmental requirements to be met and cover work stages from manufacturing to monitoring. In Europe, by-products used in construction are also regulated by the Construction Products Regulation; in the future, they will need to have a CE mark if they are sold as construction products in Europe. A CE mark is a producer's declaration that the product fulfills EU directives or national requirements. Harmonized test methods must be

TABLE 17.4 Examples of the Relevant Technical and Environmental Features and Their Measurement in the Case of a By-product Planned to be Recycled in Earthworks (Mroueh et al., 2000; Arm, 2003)

Material Property	Parameter to be Measured	Measurement/Basis of Determination
TECHNICAL FEATURES		
Chemical aggressiveness, corrosiveness	Several	pH, salt concentration, electrical conductivity
Wear resistance	Ball mill-value	Ability of a material sample to withstand degradation in a ball mill machine (EN 1097-9)
	Los Angeles-figure	Degradation of a material sample that is placed in a rotating drum with steel spheres, several standards (e.g. EN 1097-2, ASTM C131A–D)
Carrying capacity	Modulus Compression	See below
	CBR-value	Force needed to achieve the defined penetration (2.5/5.0 mm) when a steel plunger is pressed into the surface of the material, several standards (e.g. EN 13286-47, ASTM D1883)
Firmness	Friction angle	Maximum angle before which one of the items will begin sliding, describes the grain-to-grain frictional resistance
	Cohesion	Force that holds the material together
Transformation	Modulus	Several measurements can be used
Compression	Modulus	Deformation of material when compressed
Packing	Maximum dry volume weight	Density of dry material
	Optimum water content	Determined from the moisture content/dry density curve
Frost susceptibility	Frost heave	Granularity (particle size and distribution) Water permeability
	Segregation potential	Capillarity: extent of the material to elevate or depress the surface of a liquid that is in contact with it
Thermal properties	Heat conductivity	Heat energy transferred by the material per unit time and per unit surface area, divided by the temperature difference
Stability	Change in granularity	Granularity
Erosion sensitivity	Speed of water current	Granularity
Granularity	Granularity curve Granule size	Separation of different particle size fractions by sieving.
Water permeability	Permeability coefficient	Volume of water that will flow in unit time through a unit volume of a porous material across which a unit pressure difference is maintained
Capillarity	Height of capillary action	Granularity

TABLE 17.4 Examples of the Relevant Technical and Environmental Features and Their Measurement in the Case of a By-product Planned to be Recycled in Earthworks (Mroueh et al., 2000; Arm, 2003) (cont'd)

Material Property	Parameter to be Measured	Measurement/Basis of Determination
Specific gravity	Volume weight of the solid material	Ratio of the weight of a given volume of material to the weight of an equal volume of water
ENVIRONMENTAL CHARACTERISTICS		
Dry matter and water content	Concentration of dry matter	Residue after direct drying at 105 °C or water content determined by titration or distillation (EN 14346)
Composition	Total concentration of substances included in the material	Analysis method is substance-dependent, several standardized alternative methods were issued, e.g. microwave or aqua regia digestion (CEN 13656 or 13657) for metals, gravimetric and gas chromatographic method for hydrocarbons
Organic matter content	Total concentration of organics	Total organic carbon (TOC) using CEN 13137 or ignition loss at 550 °C
Biodegradability	Amount of matter that can be consumed (degraded) by microorganisms	Consumption of oxygen or generation of carbon dioxide (aerobic conditions) or generation of methane (anaerobic conditions) when the material is inoculated a specified time with microorganisms, several standardized tests (e.g. ISO 17556)
Leaching of substances in time from a granular material	Concentration of substances in eluate	Cumulative concentration of substances in the eluate after an upflow during a specified time through a column filled with the material (characterization test, CEN/TS 14405)
Leaching of substances from a monolithic or stabilized material	Concentration of substances in eluate	Concentration of substances in eluate after a specified time when the material is immersed in water (characterization test, NEN7345)
Solubility of substances in different pH conditions, acid/base neutralization capacity	Concentration of substances in eluate	Concentration of substances in the eluates of parallel extractions at different pH values (characterization test, CEN/TS 14429 CEN/TS 14997, CEN/TS 15364)
Leaching of substances from a granular material	Concentration of substances in eluate	Concentration of substances in leachate of a one or two-stage batch leaching test (quality control test, EN12457)
Leaching of substances from the surface of non-permeable materials or material with low permeability	Concentration of substances in solute	Diffusion of substances using a surface solubility test, e.g. Dynamic Monolithic Leaching Test (DSLT) CEN/TS 15863 or NVN 7347 for materials prone to silting
Ecotoxicity	Several alternatives, e.g. LD_{50}/LC_{50}, NOEC	Toxicity response of soil organisms (e.g. Enchytraids, plants) to the material, several standardized tests available

(Continued)

TABLE 17.4 Examples of the Relevant Technical and Environmental Features and Their Measurement in the Case of a By-product Planned to be Recycled in Earthworks (Mroueh et al., 2000; Arm, 2003) (cont'd)

Material Property	Parameter to be Measured	Measurement/Basis of Determination
Ecotoxicity of leachate	Several alternatives, e.g. LD_{50}/LC_{50}, NOEC	Toxicity response of aquatic organisms (e.g. *Daphnia magna*, bacteria, plants) to the leachate from the material, several standardized tests available

CBR = California bearing ratio; LD_{50}/LC_{50} = lethal dose/concentration, concentration where 50% of test animals are killed; NOEC = no-observed effect concentration, highest concentration which does not cause adverse effects to test organism.

used to prove compliance in case testing is required.

17.4.2 Environmental Acceptability

In the recycling of any waste, environmental suitability is often a key issue. Potential hazards to the environment and human health depend on the properties and use of the material, e.g. the existence of harmful substances, the type of application, environmental conditions in the place where it is managed, and the existence and characteristics of receptors that may be exposed or might receive emissions from it. Environmental and health risks are therefore also different in different stages of recycling. During storing; processing, building or construction, maintenance and demolition of a by-product structure; or application of soil amendments and fertilizers, the focus is often on occupational hazards. Inhalation of dust and skin contact are generally the major routes of human exposure at these stages. As for materials used in buildings, e.g. by-product–based gypsum board, the potential release of volatiles and radiation into indoor air can be an important issue. Here, quality standards for indoor air can be used as a reference in assessing the material's acceptability. However, the release of harmful substances in the form of dust, volatiles or dissolved chemicals also poses environmental risks. Leaching of harmful substances is a major issue in the case of by-products used in earth construction; therefore, case-specific information on the long-term leaching behavior of harmful substances is required. Leaching of contaminants should also be taken into account in the case of soil amendments and fertilizers.

Several internationally standardized laboratory tests are available to test the leaching behavior of by-products (see Table 17.4). Benchmarks have been issued to compare against the test results and to assess environmental suitability. In testing by-products to be used in earth construction, the leaching mechanism, which depends on the material properties and environmental conditions, determines the test methods. Typically, there is a difference between granular and monolithic materials. In the case of the former type of material, it is assumed that water percolates through the material layer and solubility of the compounds determines their leaching. The so-called percolation test is then used as a study method. Unlike granular materials, monolithic (or stabilized/solidified) materials have low hydraulic conductivity and maintain their shape when in contact with water. Water is assumed to be in contact only with the surface and not to percolate through it. Leaching behavior is thus surface dependent, often dictated by diffusion, and therefore is studied using the so-called diffusion test.

Several standardized toxicity tests are available for further investigation into the toxicity of the leachate beyond chemical analysis. In some cases, a site-specific risk assessment is required that takes into account the specific application and conditions at the location where the by-product is used. Moreover, life cycle analysis (LCA) is a technique that has been used to compare the overall environmental impact of by-products against virgin materials used in the same application or products manufactured from virgin raw materials. Because the starting point of LCA is a product, it usually relies on generic rather than site-specific data. It is therefore poorly suited for assessing the potential adverse environmental impact at a specific site.

The existence and concentration of harmful and toxic substances in a by-product depend on the raw materials and process conditions during its generation. In addition, any weathering, treatment or processing can alter its physicochemical properties, or the speciation of elements, and consequently change their availability. In the case of fertilizers and soil amendments, the type of application and volume or rate and frequency of application finally determine the probability of their adverse impact.

The environmental and health risks caused by by-products can be reduced by processing, i.e. by eliminating or transforming their toxic substances into nontoxic chemicals or by diminishing their availability—for example, by immobilization using a stabilizer. Emissions to the environment can also be limited by coating or paving. However, in the long term, cover materials require regular maintenance because of wearing and breakage. Another approach to minimizing environmental and health risks is to limit the use of by-products in insensitive areas and applications. In the case of earthworks, this can mean that by-products are not allowed in important groundwater areas and in soils of high hydraulic conductivity, such as gravel and sand. Above all, a by-product structure should not be in contact with groundwater, taking into account the seasonal and weather-related alterations of the water table. Application of by-products in the environment can also require the establishment of safety zones. For example, some studies suggest buffer strip zones between wood ash–treated areas and watercourses and lakes, to avoid potential adverse effects that could arise from the increase in pH and nutrient content and potential mobilization of toxic compounds (Aronsson and Ekelund, 2004).

17.5 CONCLUDING REMARKS

Industrial processes generate substantial amounts of mineral industrial residues that can be technically suitable and environmentally acceptable for use in different applications, particularly within the construction industry. How much of the recycling potential of these by-products is actually used in practice depends on many factors and thus can vary considerably in different countries.

First of all, whether any residue is considered a waste or a product is a crucial issue. Management and treatment or recycling of wastes requires a permit. The considerable time required to attain a permit is in fact considered one of the main obstacles to recycling (Sorvari, 2008). A lack of clear environmental criteria, in particular, can result in time-consuming permit processes because the authorities need to conduct a case-by-case acceptability evaluation.

Road construction, for example, assumes fast deliveries, and the material to be used in the structures is not selected well beforehand. For example in colder regions, by-products from energy production, such as ash, are mainly generated during the cold season, whereas the construction works that could use them are run in the summertime. This assumes the existence of a storage system or some logistic center where ash can be safely stored in such a form that its technical and environmental properties remain acceptable. During storage, any release to the environment, e.g. dusting or leaching of harmful substances, needs to be controlled, which creates extra costs and increases the price of by-products. This is a question of supply, as well, because in some countries the price of virgin materials is so low that it hinders its replacement with by-products.

Logistics is also an issue because the road transport of by-products over long distances is generally not profitable, and owing to the consequent emission of greenhouse gases it can be seen as environmentally disadvantageous. In addition, the heterogeneous quality of some by-products reduces their desirability compared with virgin materials. Despite the many practical barriers to recycling, various environmental strategies and regulations call for material efficiency, saving of natural resources, and recycling of waste. Attainment of these objectives requires removing the barriers and perhaps introducing some policy instruments—for example, economic incentives. Moreover, we need to develop systems that help the demand meet the supply. In fact, material exchange databanks have been established in some countries for this purpose. Process technology, too, could provide some new and more economical solutions for generating by-products of a high and homogeneous technical quality, improving by-products' quality, or changing their properties to render them acceptable and desirable for different applications. New feasible applications could also increase the recycling of by-products. In some cases, a lack of information about the long-term environmental fate of harmful substances present in by-products prevents them from being recycled. Practice has shown that environmental safety needs to be carefully studied to avoid future problems. Therefore, more data should be produced by long-term field tests, because laboratory tests can rarely fully simulate the long-term environmental fate of using by-products.

Whether by-products are or will be recycled on a large scale in the future ultimately depends on the overall benefits attained. Such benefits can be environmental (e.g. saving natural resources), economic (e.g. lower costs of structures) and social (increasing the potential for employment). These benefits are the building blocks of sustainable development. The problem is that not all of the sustainability components are interesting to the different stakeholders involved in by-product recycling and reuse. The main benefit the waste producer or user of by-products seeks is economic benefit. In some cases, technical properties may also argue for using a by-product instead of virgin materials, whereas the environmental authority looks at overall environmental benefits to the society or region. Social aspects such as employment are the interest of regional decision makers. How to combine the interest of these different stakeholders is thus the key question in attaining the sustainable management of by-products.

References

Arm, M., 2003. Mechanical Properties of Residues as Unbound Road Materials — Experimental Tests on MSWI Bottom Ash, Crushed Concrete and Blast Furnace Slag (Doctoral thesis). KTH Land and Water Resources Engineering, Stockholm. Report No 64. Swedish Geotechnical Institute.

Aronsson, K.A., Ekelund, N.G.A., 2004. Biological effects of wood ash application to forest and aquatic ecosystems. Journal of Environmental Quality 33, 1595—1605.

Asokan, P., Saxena, M., Asolekar, S.R., 2010. Recycling hazardous jarosite waste using coal combustion residues. Materials Characterization 61 (12), 1342—1355.

Delgado, L., Catarino, A.S., Eder, P., et al., 2009. End-of-Waste Criteria. Final Report. EUR—Scientific and Technical Research Series. European Commission, Joint Research Centre, Institute for Prospective Technological Studies, Seville. Available at: http://ipts.jrc.ec.europa.eu.

Jha, M.K., Kumar, V., Singh, R.J., 2001. Review of hydrometallurgical recovery of zinc from industrial wastes. Resources, Conservation and Recycling 33, 1—22.

Kumar, R., Kumar, S., Mehrotra, S.P., 2007. Towards sustainable solutions for fly ash through mechanical activation. Resources, Conservation and Recycling 52, 157—179.

Lee, H., Ha, H.S., Lee, C.H., et al., 2006. Fly ash effect on improving soil properties and rice productivity in Korean paddy soils. Bioresource Technology 97, 1490—1497.

Mäkelä, H., Höynälä, H., 2000. Sivutuotteet ja uusiomateriaalit maarakenteissa — Materiaalit ja käyttökohteet (By-products and Recycled Products in Earth Construction — Materials and Applications). Teknologiakatsaus 91/2000. Tekes — the Finnish Funding Agency for Technology and Innovation, Helsinki.

Mroueh, U.-M., Wahlström, M., Mäkelä, E., et al., 2000. By-products in Earth Construction — Assessment of Acceptability. Technology Review 96/2000. Tekes — the Finnish Funding Agency for Technology and Innovation, Helsinki.

Ojala, J., 2008. Evaluation of compost maturity and quality with ash amendment. Pro gradu. In: Finnish with an English Abstract. University of Helsinki, Department of Ecological and Environmental Sciences;, Helsinki. Available at: http://www.tsr.fi/database.

Orkas, J., 2001. Technical and Environmental Requirements for Surplus Foundry Sand Utilization. Helsinki University of Technology Publications in Foundry Technology. TKK-VAL-3/2001. Helsinki University of Technology, Espoo.

Pitman, R.M., 2006. Wood ash use in forestry — a review of the environmental impacts. Forestry 79 (5), 563—588.

Ramachandra Rao, S.R., 2006. Resource Recovery and Recycling from Metallurgical Wastes. In: Waste Management Series, vol. 7. Elsevier B.V.

Recycled Materials Resource Centre (RMRC). Materials. College of Engineering, Madison. http://rmrc.wisc.edu/.

RMT, Inc., 2003. Beneficial Use of Industrial By-products, Identification and Review of Material Specifications, Performance Standards, and Technical Guidance. National Council for Air and Stream Improvement, Inc. (NCASI), Madison.

Scotti, I.A., Silva, S., Baffi, C., 1999. Effects of fly ash pH on the uptake of heavy metals by chicory. Water, Air, and Soil Pollution 109, 397—406.

Sherwood, P.T., 2001. Alternative Materials in Road Construction: A Guide to the Use of Recycled and Secondary Aggregates, second ed. Thomas Telford Ltd, London.

Sorvari, J., 2008. Developing environmental legislation to promote recycling of industrial by-products — an endless story? Waste Management 28, 489—501.

Summer, R.N., Smirk, D.D., Karafilisc, D., 1996. Phosphorus retention and leachates from sandy soil amended with bauxite residue (red mud). Australian Journal of Soil Research 34, 555—567.

U.S. Environmental Protection Agency, 2001. The Micronutrient Fertilizer Industry: From Industrial Byproduct to Beneficial Use. EPA-305-B-01—006. Available at: http://www.epa.gov/nscep/index.html.

U.S. Environmental Protection Agency, 2007. Foundry Sands Recycling. EPA530-F-07-018. Available at: http://www.epa.gov/osw/conserve/imr/foundry/.

CHAPTER 18

Recovery of Metals from Different Secondary Resources (Waste)

Stefan Luidold and Helmut Antrekowitsch
Chair of Nonferrous Metallurgy, Montanuniversitaet Leoben, Leoben, Austria

18.1 INTRODUCTION

Raw materials are very essential for the economy and subsequently for the modern society. They are required for industrial activities as well as for infrastructure and products used in everyday life. The importance of oil, gas, and coal has often been mentioned since the oil crises in the 1970s that lead to sharp price increases. However, the importance of nonenergy materials, such as minerals and metals, was neglected in the western industrialized world due to a long period of stable and low raw material costs. This leads to a considerable concentration of mine production for individual raw materials in a small number of countries. For example, more than 95% of rare earths as well as more than 80% of tungsten and antimony are produced today in China. Therefore, the European Union (EU) has become increasingly more dependent on imports of many raw materials. The resulting high vulnerability was already demonstrated in the past by the rush for tantalum in 2000 due to the boom of mobile phones and more recently by the extreme price increase for rare earths during the period 2010–2011. To manage the complex challenge of ensuring the availability of raw materials, the EU has started the EU Raw Materials Initiative as an integrated strategy in 2008, which is based on multiple pillars. Its first action was the identification of critical nonenergy raw materials on the EU level by an ad-hoc working group after the development of a methodology for the rating of criticality. This activity resulted in a report analyzing a selection of 41 minerals and metals (Luidold, 2013).

Concerning the recovery of critical raw materials from wastes, such as consumer products after their economic lifetimes, the present situation is quite different. Some of them (tungsten, silver, gold and platinum group metals) have been recycled for a long time due to their high prices, which makes an economic reclamation from designated waste streams relatively easy. However, these elements also do not have very high recycling rates due to their strong dilution in many applications. Furthermore, other elements (In, Ga, Ge, etc.) are not yet being recovered from end-of-life products or their recycling activities on an industrial scale are currently under development (rare earths, etc.) (Luidold, 2013).

Simply said, if end-of-life recyclates have sufficient economic value, they will be recycled when the appropriate technological infrastructure exists for recovering their contained elements. However, not all end-of-life goods have a high value. Especially for technological metals, the existing metallurgical infrastructure (Figure 18.1) for their recovery is limited, resulting in low recycling rates of these elements (Reuter et al., 2013).

However, recycling rates have been defined in many altered ways, from different perspectives and for various life stages; sometimes the term is left undefined. Therefore, Graedel et al. defined the metal life cycle and flow annotation, as shown in Figure 18.2 for flows related to a simplified life cycle of metals and the recycling of production scrap and end-of-life products.

Here, the end-of-life recycling rate (Figure 18.3) refers only to functional recycling—that is, recycling in which the physical and chemical properties that made the material desirable in the first place are retained for subsequent use (Graedel et al., 2011).

The recycled content (RC = $(j+m)/(a+j+m)$, Figure 18.4) describes the fraction of secondary (scrap) metal in the total metal input to metal production (Graedel et al., 2011).

Finally, the old scrap ratio (OSR = $g/(g+h)$, Figure 18.5) indicates the fraction of old (postconsumer) scrap in the recycling flow (Graedel et al., 2011).

However, low recycling rates are not only caused by economic reasons and limited metallurgical infrastructure but also by several additional factors. At the base are limitations set by nature, such as physics, chemistry, metallurgy, and thermodynamics. These dictate the outcome of a given input into a recycling process, the constraints of nature tying together the use of energy, metal recovery, and the quality and quantity of the inevitable losses. An important aspect at this stage is the relationship between quality and recovery of a

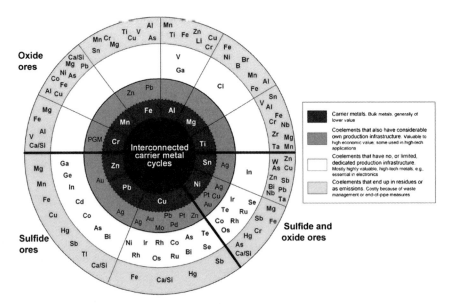

FIGURE 18.1 The metal wheel showing metal linkages in natural resource processing, illustrating the capacity of available metallurgical processes to deal with impurities in their (primary or secondary) feed (Verhoef et al., 2004).

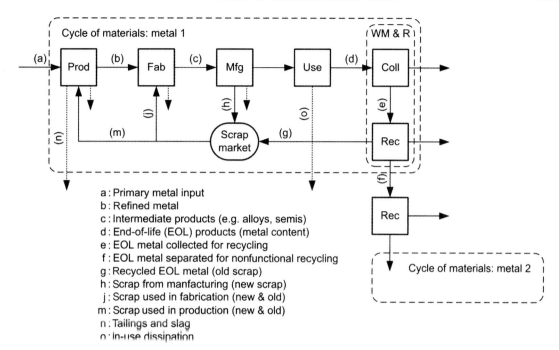

FIGURE 18.2 Metal life cycle and flow annotation (Graedel et al., 2011). Boxes indicate the main processes (life stages): Prod, production; Fab, fabrication; Mfg, manufacturing; WM & R, waste management and recycling; Coll, collection; Rec, recycling. Yield losses at all life stages are indicated by dashed lines (with WM referring to landfills). When material is discarded to WM, it may be recycled (e), lost into the cycle of another metal (f, as with copper wire mixed into steel scrap), or landfilled. The boundary indicates the global industrial system, not a geographical entity.

given metal or metals. Another major issue is the way (consumer) goods are produced—whether product design includes recyclability or not. Another important influence is the structure of the waste collection system, because the collection of postconsumer waste is very much a logistics challenge. There are typically several stakeholders in the logistics chain with different aims and motivations. Often, the minimization of collection costs is more important than the maximization of material efficiency. Another obstacle to recycling is the lack of consumer awareness concerning collection and recycling possibilities. An additional factor affecting the volume and availability of material for recycling is the leakage of waste into the "gray market" of informal, or illegal, low-technology recyclers (Reuter et al., 2013).

On the basis of two examples, the complexity of recycling from waste will be shown.

18.2 PRODUCTION OF FERROALLOYS FROM WASTE

It is already common practice to process secondary raw materials beside primary ores, known as residual materials and byproducts of different operations, for the extraction of pure metals and also ferroalloys. Hence, local shortages of raw materials and delivery problems of the mines can be surmounted by purchasing high-quality materials and by the acceptance of residues. When using secondary raw materials, energy and investment costs for producing

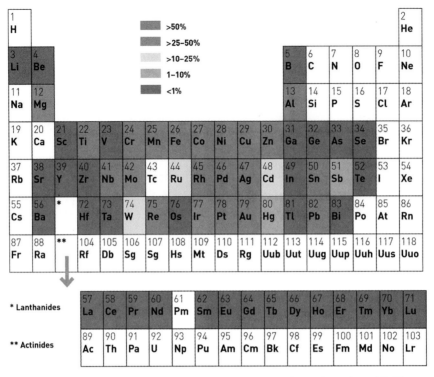

FIGURE 18.3 The end-of-life recycling rate (in %) for various metals (Graedel et al., 2011). Unfilled boxes indicate that no data or estimates are available, or that the element was not addressed in the study. (For interpretation of the references to color in this figure legend, the reader is referred to the online version of this book.)

concentrates are lower than mining the ores and first enrichment steps. Disadvantages of this consideration are extensive processes for the recovery of valuable metals because of the presence of complex compounds and high-melting oxides. The aim is a complete conversion of the residues into commercially saleable products such as pure metals, ferroalloys, or inert material for landfilling and re-use as building materials.

Numerous research activities have been carried out due to the fact that the annual consumption of refractory metals increased considerably in the last 40 years and stricter landfill regulations as well as bans on exports were released. In the meantime, a variety of publications with different recycling processes for the waste material have become well known. A large part of them only deal with substeps, such as roasting or leaching, or techniques that are specialized in individual residues. Subsequently, this chapter gives an overview of possible residues for recovering vanadium (V), nickel (Ni), or molybdenum (Mo) and describes different ways of waste prevention and a complete recycling route for one type of secondary material, spent catalysts. Additionally, characterizations and basic studies of used catalysts of the oil refining are mentioned (Hoy et al., 2011).

A comparison of valuable metal concentration in established ores and two important types of residues (spent catalysts and fly ash) is shown in Table 18.1. Wide ranges of prevalent percentages of vanadium, nickel, and molybdenum are listed together with two special cases in parentheses. Considering the high values of vanadium in spent

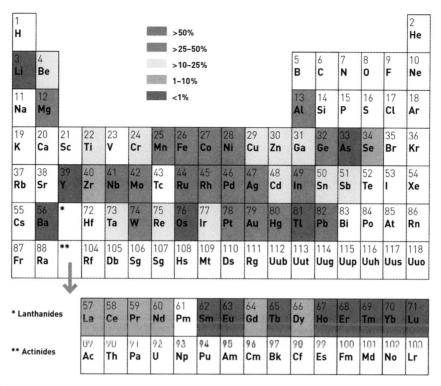

FIGURE 18.4 Recycled content for various metals (Graedel et al., 2011).

catalysts of petroleum refining and in fly ash, it is self-evident to develop a process for recycling. Nickel and molybdenum represent the same issue, especially in spent catalysts (Hoy et al., 2011).

The composition of the ore and the structure of the residues must not be ignored. If valuable metals are present as complex compounds, a selective enrichment and a complete separation will be consequently more difficult (Hoy et al., 2011).

In recent years, studies on the recovery of economically important metals have increased due to rising commodity prices and increasing costs for the deposition of recycling material. Furthermore, an export ban prevents the exportation of hazardous waste from North America or Europe to non-Organisation for Economic Co-operation and Development locations, such as Africa, Asia, and India, which have less strict landfill regulations (Hoy et al., 2011; Marafi and Stanislaus, 2008).

18.2.1 Secondary Raw Materials

The assortment of secondary raw materials for the production of ferroalloys is very extensive, but only a couple of residues can be used. Reasons for this are the low amounts of material or minor contents of contained vanadium, nickel, and molybdenum. In addition, it is complex to develop a recycling process for discontinued production of waste materials with an alternating chemical composition. A literature study offered a couple of interesting types of residues, which are listed in Table 18.2.

These materials are residues for the mentioned industry sectors, but they are valuable and

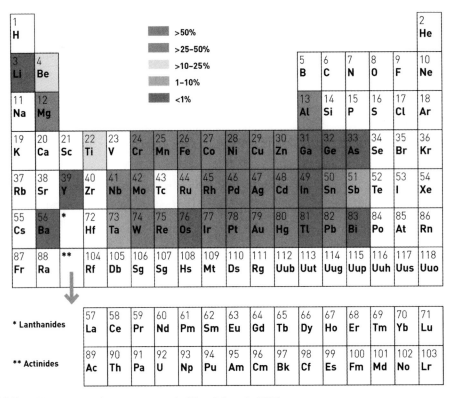

FIGURE 18.5 Old scrap ratio for various metals (Graedel et al., 2011).

represent the feedstock for the production of ferroalloys. The listed residues are partially already secondary raw materials for the production of ferrovanadium, ferromolybdenum, ferronickel, or mixed forms thereof. A brief discussion on these secondary materials follows; however, catalysts are described in more detail because of intensified research work (Hoy et al., 2011).

TABLE 18.1 Comparison of Vanadium, Nickel, and Molybdenum Content in Ores and Residues (Hoy et al., 2011)

Element	Ore	Spent Catalysts	Fly Ashes
Vanadium (wt%)	0.50—11.00	0.05—11.30	0.40—7.00 (20.00)
Nickel (wt%)	0.20—3.50	0.02—9.40	0.80—1.00 (22.00)
Molybdenum (wt%)	0.08—0.25	0.50—18.50	<0.01

18.2.1.1 Acid Production

Catalysts for the production of sulfuric acid must not be supplied to any landfill, as there are problematic eluate values. They are usually used in the catalytic oxidation of SO_2 to SO_3 by oxygen, and with the technology available at present it is not possible to set these types of catalysts aside or to substitute them. For this reason, it requires regeneration with a possible recovery of valuable metals, which usual comprise 2—5% vanadium and <1% nickel (Hoy et al., 2011; Garcia et al., 2001; Ognyanova et al., 2009).

TABLE 18.2 Types of Residues and Industrial Sectors (Hoy et al., 2011)

Type	Industrial Sector	Reference
Catalysts	Petroleum industry	Furimsky (1996), Marafi and Stanislaus (2008)
Catalysts	Production of acid	Garcia et al. (2001), Ognyanova et al. (2009)
Catalysts	Fertilizer industry	Singh (2009), Sadanandam et al. (2008)
Catalysts	Fat hardening/hydrogenation	Hochenhofer (2003)
Combustion residues	Firing plant	Tsai and Tsai (1998), Navarro et al. (2007)
Further residues	Petroleum industry	Holloway and Etsell (2002), Gomez-Bueno et al. (1981)
Slags and dusts	Steel mill	Schreiter (1962)
Byproducts	Aluminum production	Gupta and Krishnamurthy (1992), Cinar et al. (1992)
Byproducts	Fertilizer industry	Schreiter (1962), Rostoker (1958)
Byproducts	Uranium mining	Schreiter (1962), Rostoker (1958)

18.2.1.2 Fertilizer Industry

Other catalysts are applied in the production of fertilizers, in particular for the manufacture of pure hydrogen and nitrogen to generate ammonia. Depending on the process step, these catalysts contain 10–32% nickel oxide, which represents a level of about 8–25% elemental nickel. Meanwhile, because of optimizations, the life cycle could be extended up to 6 years, whereby the amount of used catalysts decreased. On the other hand, a rise in production based on the increased demand is taking place. Therefore, the residual material volume in this division is estimated in China, India, and some developing countries at 3000 t per year on its own (Hoy et al., 2011; Singh, 2009).

18.2.1.3 Fat Hardening

To increase the melting point of oils and fats in the food industry, hydrogenation is carried out with catalysts consisting of about 15–42% nickel. These speed up the so-called fat hardening of vegetable oils, such as corn or coconut oil, where hydrogen is added to double bonds in the chemical structure. The produced hardened oils as well as the saturated fats are used in the manufacture of margarine because of their chemical stability and oxidation resistance (Hoy et al., 2011; Chorfa et al., 2010; Yang et al., 2011; Idris et al., 2010).

18.2.1.4 Cogeneration Plants

Residues of cogeneration plants result from the combustion of oil, coal, or gas and accrue in a solid form directly in the burning boilers (boiler remainder), gas phase (fly ash from the cyclone), or after the off-gas treatment as dry sludge. In the year 2000, the total consumption of heavy oil was estimated to be 14 million tons in Italy. About 27,600 tons of ash was produced, which contained 0.1–2.9% vanadium and 0.32–1.37% nickel, depending on process parameters. This corresponds to a theoretical amount of 30–800 t of vanadium and 90–380 t of nickel whereby a recovery process is possible (Hoy et al., 2011; Seggiani et al., 2003).

The same was described in a publication from Taiwan in 1998. In the burning of $15 \cdot 10^9$ l of fuel oil, 43,000 t of boiler residues occurred. Depending on the level of off-gas treatment, fly ash from electrostatic precipitators (13,000 t) and cyclone residues (30,000 t) accrued. This relates to over 50 t of vanadium and 130 tons of nickel in the fly ash. The residues of the cyclone include about 570 t vanadium and 240 t nickel (Hoy et al., 2011; Tsai and Tsai, 1998).

In addition to boiler and fly ashes, dry sludges from the off-gas treatment are collected. According to a publication of the Austrian Federal Environment Agency and information from power plant operators, vanadium and nickel are included in concentrations, of 0.5% and 1.0% respectively, especially in pressed dry sludges from oil-fired power plants (Hoy et al., 2011).

18.2.1.5 Slags and Dusts

Other possible sources of refractory metals are residues from special steel plants. In addition to dusts, slags with enriched amounts of vanadium, nickel, and molybdenum, usually in oxide form, are produced. The level of useful metals depends on the alloy specifications. It is not possible to determine the exact contents of valuable metals due to different alloys and scraps that are used in the electric arc furnace (Hoy et al., 2011; Navarro et al., 2007; Moskalyk and Alfantazi, 2003).

18.2.1.6 Production of Aluminum

In addition to the mentioned residues, vanadium occurs in a byproduct generated at the extraction of aluminium oxide from bauxite in the well-known Bayer process. The ore contains up to 760 ppm V_2O_5, according to a Turkish publication. Due to the caustic digestion, about a quarter of the vanadium oxide passes into the liquid phase, whereby enrichment to 14–20% V_2O_5 takes place because of circular flow of the leach. From these concentrations on, a separation is necessary whereby vanadium will be obtained as sodium vanadate in the sludge. In the course of different treatments, about 10% of the initial mass of the sludge is converted to pure vanadium oxide (V_2O_5), which is processed in the production of ferroalloys (Hoy et al., 2011; Cinar et al., 1992).

18.2.1.7 Petroleum Industry

One of the first steps of petroleum refining includes fractional distillation for separating different hydrocarbons with varied boiling points. During this process, the residue becomes enriched by vanadium because metals cannot follow the route of light fractions. Subsequently, long-chain alkanes are cracked in order to cover the high fuel demand at which a light oil fraction and residues, which occur mainly in the extraction of oil from oil sands, are obtained. Most of the valuable metals accumulate as nonvolatile compounds in the remaining oil fraction or the so-called petroleum coke. This residual material has to be burned, so a further concentration of vanadium and partially nickel takes place. Depending on the feedstock, 2–10% vanadium, in some cases even 20%, and up to 4.5% nickel are present in the resulting ash (Hoy et al., 2011; Gary and Handwerk, 2001).

After fractional distillation at atmospheric pressure and a rectification of the previously formed residues in vacuum, a cracking of long-chain hydrocarbons is performed. This process is necessary to cover the increased industrial demand for high-grade fuels, due to the fact that crude oil is not as abundant as necessary. The most common process is fluid-bed catalytic cracking in a fluidized bed of distillated fractions together with catalysts. Because purity plays a decisive role in the sale and use of products from petroleum, it is necessary to remove sulfur, nitrogen, oxygen, and metals such as vanadium and nickel from the naphtha in subsequent process steps using the hydrogen process (hydrotreating). Catalysts used for this purpose are roughly divided into groups of hydrodesulfurization, hydrodenitrogenation, and hydrodemetalization. Combined modern processes apply a two-stage procedure: in the first step, impurities are removed (hydrogenation step), and then splitting (hydrocracking step) takes place. If large quantities of isoparaffins and olefins exist, it is possible to convert them with the aid of strong acids (sulfuric or phosphoric acid) in the presence of catalysts to high-quality gasoline with a well-defined octane index. In this case, alkylation and polymerization will be accomplished (Hoy et al., 2011; Furimsky, 1996; Marafi and Stanislaus, 2008; Irion and Neuwirth, 2005).

18.2.2 Treatment and Recycling Activities of Spent Oil Refining Catalysts

The treatment of used catalysts is divided into two concepts. The first option deals with the possibility of reducing the amount of waste material by regeneration, a targeted application, and research for higher activity. The second scope covers a whole recycling process for spent catalysts in which hydro- and pyrometallurgical operations result in pure valuable metals such as vanadium, nickel, or molybdenum or in alloys made of iron and the mentioned metals (Hoy et al., 2011).

18.2.2.1 Reduction of Waste Material Amount

Due to an increasing focus on the residue volume, some challenges are adopting new processes, optimizing existing processes, and developing and using modern types of catalysts. The following methods to reduce the waste material amounts are some common procedures, depending on the refinery (Hoy et al., 2011; Furimsky, 1996; Marafi and Stanislaus, 2008; Zeng and Cheng, 2009):

- Regeneration and rejuvenation
- Cascades and extended process application
- Material recycling for new catalysts
- Catalyst improvement (activity, life cycle)

18.2.2.2 Recycling of Spent Catalysts

In the case of discharging catalysts from the process chain of oil refining due to inactivity and lack of regeneration, the spent material must not be delivered to a landfill immediately, according to environmental conditions. This is caused by adherent toxic refined products (e.g. hydrocarbons), noxious substances (metals, sulfides, sulfates, etc.), and bad leachate characteristics. Consecutively, hydrometallurgical recycling is marginally mentioned because of the widespread topic. The main focus is on pyrometallurgical processes and research work to recycle spent catalysts (Hoy et al., 2011).

18.2.2.2.1 HYDROMETALLURGICAL RECYCLING

Reams of hydrometallurgical ways for recycling have been published. Most of them are designed as a combination of pyro- and hydrometallurgical processes (Figure 18.6). This is caused by adherent residues, such as the remaining hydrocarbons from the refining process, which hinder the handling and lead to some problems in leaching operations such as oil films, conservation, and auxiliary acid consumption. In addition, valuable metals do not exist in their pure form. A conversion into soluble compounds is necessary and handled by roasting at temperatures up to 800 °C under atmospheric conditions. Due to the solubility of the formed oxides in acid, a separation of vanadium, nickel, or molybdenum and alumina is feasible. Another possibility is alkaline roasting with caustic soda to produce water-soluble compounds. The next steps before extraction are purification of the leachate by selective precipitation and cementation followed by enrichment via solvent extraction or ion exchange

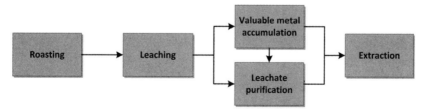

FIGURE 18.6 General flowsheet of combined pyro- and hydrometallurgical processing (Hoy et al., 2011).

(Hoy et al., 2011; Kuck, 2010; Scheel and Stelter, 2010; Dufresne, 2007).

18.2.2.2.2 PYROMETALLURGICAL RECYCLING

Pyrometallurgical processes are rare and relatively difficult to handle due to the high melting points of the components associated with huge energy and refractory demand. For example, the melting point of the major constituent alumina is higher than 2050 °C. Vanadium in its oxidic form V_2O_3 is solid until 1970 °C. In this case, high amounts of lime and silica are required to form a liquid slag at temperatures of approximately 1700 °C (Hoy et al., 2011).

First of all, calcination under oxidizing atmosphere removes adherent carbon, sulfur, and hydrocarbons. By selecting correct process parameters, molybdenum can be volatilized because of the high vapor pressure of MoO_3. After filtering the exhaust gas, a separation process for molybdenum oxide is necessary. The next step starts with dry catalysts that are melted in an electric arc furnace at temperatures around 1500–1700 °C. Either aluminum, carbon, or silicon (FeSi) acts as a reducing agent to produce a liquid metal phase together with molten steel scrap (Hoy et al., 2011).

Another process route works with sulfur from pyrite to produce matte. Therefore, a separation of metal and matte is possible followed by a separate regeneration of the nickel matte. This procedure is realized as part of a whole recycling process at Eurecat Inc. and described in the following chapter (Hoy et al., 2011; Berrebi et al., 1994).

Selective chlorination is possible but difficult to adopt in industrial dimensions because of installation costs and environmental regulations. The gas for the chlorination is pure chlorine gas or a mixture of chlorine, air, nitrogen, and carbon monoxide. Molybdenum and vanadium are converted in the reactor to gaseous chlorides and/or oxychlorides at temperatures below 600 °C. They are recovered from the vapor phase by selective condensation. The solid residue is composed of alumina and nonvolatile or less volatile chlorides of cobalt and nickel. These chlorides are extracted by water leaching (Hoy et al., 2011; Marafi and Stanislaus, 2008; Gaballah and Djona, 1995).

18.2.2.2.3 COMBINED PROCESSES

One of these combined processes for the recycling of spent catalysts is employed at Eurecat Inc. in France and other places of business around the world. The technology of Eurecat Inc. is only one example to explain the interaction of pyro- and hydrometallurgical operations. A couple of proceedings (Marafi and Stanislaus, 2008; Scheel and Stelter, 2010; Dufresne, 2007) are well known and publicly available to examine catalyst recycling. The first step is to eliminate free hydrocarbons (up to 5 wt%) and precipitated carbon (5–20 wt%) in a regeneration furnace at temperatures up to 480 °C. Metals like V, Ni, Mo, W, Co, and Fe exist in their sulfide form in a spent catalyst. During the regeneration process, they are transferred into the corresponding oxides at moderate temperatures of 220–270 °C. After regeneration, alkaline roasting with caustic soda is necessary to produce leachable compounds. Hot water removes molybdenum, vanadium, tungsten, parts of aluminum, arsenic, and phosphorus in the following leaching operation. The insoluble matrix made of alumina forms the residue with nickel and cobalt. After filtration, the leachate is purified by precipitation to get a solution with molybdenum, vanadium, and tungsten as the main components because impurities such as aluminum and phosphorus would influence subsequent steps. Arsenic associates with molybdenum, tungsten, and vanadium and decreases the quality of the product. The separation of molybdenum and vanadium or tungsten and vanadium is done by ion exchange resins. The solid residue of alkaline leaching typically contains up to 4 wt% nickel and/or cobalt with alumina and/or silica. It is accepted as an ore substitute. There, an electric arc furnace fuses the solid feedstock and separates cobalt and nickel in a matte from the aluminum and residual impurities, which are eliminated as slag. Following the pyrometallurgy, a conventional hydrometallurgical treatment of the sulphide phase is executed. Solvent extraction separates nickel and cobalt and electrolysis precipitates the end products (Hoy et al., 2011; Berrebi et al., 1994).

18.3 RECYCLING CONCEPTS FOR RARE EARTH CONTAINING MAGNETS

Some of the current recycling concepts for permanent magnets containing rare earths are explained in the following chapters. Most of the actual operating recycling facilities are designed to treat process scraps such as swarfs and sludges; postconsumer magnets are not currently recycled. Hitachi (Hitachi, 2013) planned to start a postconsumer magnet recycling in 2013 (Kaindl et al., 2013).

18.3.1 Recycling by Hydrogen Processing

The University of Birmingham (Zakotnik et al., 2006, 2008, 2009) has done several investigations for the recycling of scrap magnets by the use of hydrogen. Full-density NdFeB-magnets can be processed by hydrogen decrepitation (HD), which produces a powder as raw material for new magnets. The powder is milled, pressed, and resintered to new magnets. This can be repeated four times, after which the properties get worse due to the increasing oxide content in the magnet. By adding fresh neodymium powder, the quality of the product can be improved. The schematic process is shown in Figure 18.7.

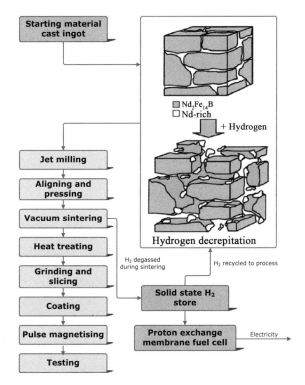

FIGURE 18.7 Hydrogen decrepitation process (Walton, 2011).

The volume of the Nd-rich binder phase increases during the reaction with hydrogen, so the whole magnet collapses into fine powder. A problem can be the layer, which improves the corrosion resistance (Ni, Zn etc.) because the hydrogen cannot react with the magnet material (Kaindl et al., 2013).

After milling the product in the HD process, the powder is aligned and pressed into a new shape. During the vacuum sintering, the contained hydrogen degasses and can be stored for further use in the recycling process or to get electricity out of a proton exchange membrane fuel cell. After several further process steps, the magnet is ready for testing. The produced powder can also be used to manufacture resin bonded magnets (Kaindl et al., 2013; Walton, 2011).

18.3.2 Recycling in Molten Magnesium or Magnesium Chloride

Because of the good solubility of rare earths in liquid magnesium and magnesium chloride, these materials are used for the recycling of rare earth permanent magnets. This section describes two ways to treat NdFeB magnets by magnesium and magnesium chloride (Kaindl et al., 2013).

18.3.2.1 Liquid Magnesium Chloride

Okabe and Shirayama (2011) developed a process that can recover Nd and Dy from sintered NdFeB-magnets by extraction with molten $MgCl_2$. The scrap is charged in a cage and dipped into the molten magnesium chloride. The rare earths are dissolved while the rest of the material remains in solid state. In Figure 18.8, the scheme of the process can be seen. The reaction mass can be filtrated to obtain the liquid salt with the whole rare earths (Kaindl et al., 2013).

The rare-earth chlorides can be extracted by vacuum distillation of the magnesium, respectively $MgCl_2$, and separation. After a reduction of the chlorides, Nd and Dy are obtained in metallic form (Kaindl et al., 2013).

18.3.2.2 Molten Magnesium

A process developed by Takeda et al. (2006) can recover Nd from a mixture of NdFeB-scrap and other scraps. The feed is charged by a basket into the reaction chamber, which can be seen in Figure 18.9. At a temperature of about 1000 °C, magnesium evaporates and builds an alloy with Nd in the scrap. This liquid alloy flows through the slits in the crucible and is collected in a tantalum crucible (Kaindl et al., 2013).

Extraction rates of up to 95% can be achieved. The purity of the metal amounts to 98% for Nd and 99% for Mg after a vacuum evaporation of Mg (Kaindl et al., 2013).

18.3.3 Recycling by a Chemical Vapor Transportation Process

The selective stability of various compounds at different temperatures is used for the recycling by chemical vapor transportation. Murase

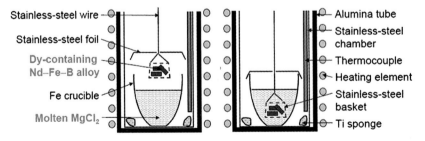

FIGURE 18.8 Recycling by reaction with liquid $MgCl_2$ (Shirayama and Okabe, 2009).

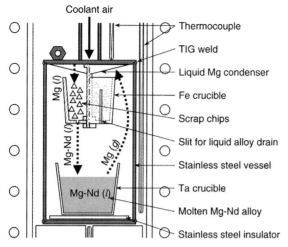

FIGURE 18.9 Recycling by reaction with magnesium (Takeda et al., 2006).

et al. (1995) described a process where the rare earths are transported in form of chemical vapor complexes. These complexes are formed by reaction with chlorine and aluminum chloride and the rare-earth chlorides are collected at higher temperatures (800–1050 °C) than iron chlorides (below 500 °C). At the range of the rare-earth chlorides, $NdCl_3$ is collected at higher temperatures than $DyCl_3$. The purity of the rare-earth chlorides is up to 99% (Kaindl et al., 2013).

18.3.4 Recycling by Processing with Ammonium Chloride

Itoh et al. (2009) developed a process to recycle rare earths from magnets by chlorination of scrap magnets, such as grinding residues. The neodymium of the powdery NdFeB reacts with ammonium chloride (NH_4Cl), while the iron is not affected. The chlorination takes place at about 300 °C for 3 h; approximately 90% of the rare earths could be recovered in form of rare earth chlorides ($RECl_3$). During this time, the surface of the scrap increases from 2.3 to 23.5 m^2/g. The chlorides from the powder can be recovered and separated from the iron by washing with water. The rare-earth chlorides are soluble, while the iron remains solid (Kaindl et al., 2013).

References

Berrebi, G., Dufresne, P., Jacquier, Y., 1994. Recycling of spent hydroprocessing catalysts: EURECAT technology. Resources, Conservation and Recycling 10, 1–9.

Chorfa, N., Hamoudi, S., Belkacemi, K., 2010. Conjugated linoleic acid formation via heterogeneous hydrogenation/isomerization of safflower oil over mesostructured catalysts. Applied Catalysis A: General 387, 75–86.

Cinar, F., et al., 1992. Production of Ferrovanadium from Vanadium Sludge. Düsseldorf/Neuss, Germany, pp. 209–217.

Dufresne, P., 2007. Hydroprocessing catalysts regeneration and recycling. Applied Catalysis A: General 322, 67–75.

Furimsky, E., 1996. Spent refinery catalysts: environment, safety and utilization. Catalysis Today 30, 223–286.

Gaballah, I., Djona, M., 1995. Recovery of Co, Ni, Mo and V from unroasted spent hydrorefining catalysts by selective chlorination. Metallurgical and Materials Transactions B: Process Metallurgy and Materials Processing Science 26, 41–50.

Garcia, D.J., Blanco, L.J.L., Vivancon, M.D.M., 2001. Leaching of vanadium from sulphuric acid manufacture spent catalysts. Revista de Metalurgia (Madrid) 37, 18–23.

Gary, J., Handwerk, G., 2001. Petroleum Refining, Technology and Economics, fourth ed. Marcel Dekker Inc., New York.

Gomez-Bueno, C.O., Spink, D.R., Rempel, G.L., 1981. Extraction of vanadium from athabasca tar sands fly ash. Metallurgical Transactions B 12, 341–352.

Graedel, T.E., Allwood, J., Birat, J.P., Reck, B.K., Sibley, S.F., Sonnemann, G., Buchert, M., Hagelüken, C., 2011. Recycling rates of metals — a status report, A status report of the working group on the global metal flows to the International Resource Panel, UNEP.

Gupta, C.K., Krishnamurthy, N., 1992. Extractive Metallurgy of Vanadium. Elsevier Science Publisher B.V.

Hitachi, Hitachi Develops Recycling Technologies for Rare Earth Metals. Online: http://www.hitachi.com/New/cnews/101206.pdf (Access: 13.03.13.).

Hochenhofer, M., 2003. Schlackenmetallurgie und Entschwefelung bei der Herstellung von Ferronickel aus Sekundärrohstoffen (Diploma thesis). Nonferrous Metallurgy, University of Leoben, Austria.

Holloway, P., Etsell, T.H., 2002. Recovery of vanadium from oil sands fly ash. In: International Symposium on Vanadium; Vanadiumgeology, Processing and Applications, pp. 227–242.

Hoy, C., Luidold, S., Antrekowitsch, H., 2011. Recycling of secondary materials for the production of ferroalloys. In: Proceedings of EMC 2011, vol. 4. GDMB, Düsseldorf, Germany, pp. 1375–1390.

Idris, J., Musa, M., Yin, C.Y., Hamid, K.H.K., 2010. Recovery of nickel from spent catalyst from palm oil hydrogenation process using acidic solutions. Journal of Industrial and Engineering Chemistry 16, 251–255.

Irion, W.W., Neuwirth, O.S., 2005. Oil Refining, Ullmann's Encyclopedia of Industrial Chemistry, VII. Wiley-VCH Verlag GmbH & Co. KGaA, pp. 1–52.

Itoh, M., Miura, K., Machida, K.I., 2009. Novel rare earth recovery process on Nd-Fe-B magnet scrap by selective chlorination using NH4Cl. Journal of Alloys and Compounds 477, 484–487.

Kaindl, M., Poscher, A., Luidold, S., Antrekowitsch, H., 2013. Investigation on different recycling concepts for rare earth containing magnets. In: Proceedings of EMC 3, Weimar, Deutschland, pp. 1151–1165.

Kuck, P.H., 2010. Nickel, Minerals Yearbook, 2008. U.S. Geological Survey (USGS).

Luidold, S., 2013. The situation regarding critical raw materials for the high technological sector in Austria. In: EUMICON – Contributions to a Resilient Raw Materials Policy, pp. 605–627.

Marafi, M., Stanislaus, A., 2008. Spent catalyst waste management: a review. Part I – developments in hydroprocessing catalyst waste reduction and use. Resources, Conservation and Recycling 52, 859–873.

Moskalyk, R., Alfantazi, A., 2003. Processing of vanadium: a review. Minerals Engineering 16, 798–805.

Murase, K., Machida, I.I., Adachi, G.Y., 1995. Recovery of rare metals from scrap of rare earth intermetallic material by chemical vapour transport. Journal of Alloys and Compounds 217, 218–225.

Navarro, R., Guzman, J., Saucedo, I., Revilla, J., Guibal, E., 2007. Vanadium recovery from oil fly ash by leaching, precipitation and solvent extraction processes. Waste Management 27, 425–438.

Ognyanova, A., Ozturk, A.T., De Michelis, I., Ferella, F., Taglieri, G., Akcil, A., Veglio, F., 2009. Metal extraction from spent sulfuric acid catalyst through alkaline and acidic leaching. Hydrometallurgy 100, 20–28.

Okabe, T.H., Shirayama, S., 2011. Method and apparatus for recovery of rare earth element. Patent: US 20110023660 A1.

Reuter, M.A., Hudson, C., van Schaik, A., Haiskanen, K., Meskers, C., Hagelüken, C., 2013. Metal recycling: opportunities, limits, infrastructure, A Report of the Working Group on the Global Metal Flows to the International Resource Panel, UNEP.

Rostoker, W., 1958. Extractive metallurgy of vanadium. In: The Metallurgy of Vanadium. John Wiley & Sons, Inc., pp. 1–32.

Sadanandam, R., Fonseca, M.F., Srikant, K., Sharma, A.K., Tangri, S.K., Suri, A.K., 2008. Production of high purity cobalt oxalate from spent ammonia cracker catalyst. Hydrometallurgy 91, 28–34.

Scheel, M., Stelter, M., 2010. Recycling von refraktärmetallhaltigen Sekundärrohstoffen. Sondermetalle und Edelmetalle, Heft 121, Schriftenreihe der GDMB, pp. 103–116.

Schreiter, W., 1962. Vanadium, Seltene Metalle, Band III. VEB Deutscher Verlag für Grundstoffindustrie, pp. 280–311.

Seggiani, M., Vitolo, S., Narducci, P., 2003. Investigation on the porosity development by CO2 activation in heavy oil fly ashes. Fuel 82, 1441–1450.

Shirayama, S., Okabe, T.H., 2009. Selective extraction of Nd and Dy from rare earth magnet scrap into molten salt. In: Processing Materials for Properties III. The Minerals, Metals & Materials Society (TMS), Bankok, Thailand, pp. 469–474.

Singh, B., 2009. Treatment of spent catalyst from the nitrogenous fertilizer industry – a review of the available methods of regeneration, recovery and disposal. Journal of Hazardous Materials 167, 24–37.

Takeda, O., Okabe, T.H., Umetsu, Y., 2006. Recovery of neodymium from a mixture of magnet scrap and other scrap. Journal of Alloys and Compounds 408–412, 387–390.

Tsai, S.L., Tsai, M.S., 1998. A study of the extraction of vanadium and nickel in oil-fired fly ash. Resources, Conservation and Recycling 22, 163–176.

Verhoef, E.V., Dijkema, G.P.J., Reuter, M.A., 2004. Process knowledge, system dynamics, and metal ecology. Journal of Industrial Ecology 8, 23–43.

Walton, A., 2011. The Use of Hydrogen to Extract and Recycle Rare Earth Magnets. Online: http://www.eu-recycling.com/pdf/University_of_Birmingham_Allan_Walton.pdf (Access: 27.02.13.).

Yang, Q.Z., Qi, G.J., Low, H.C., Song, B., 2011. Sustainable recovery of nickel from spent hydrogenation catalyst: economics, emissions and wastes assessment. Journal of Cleaner Production 19, 365–375.

Zakotnik, M., Devlin, E., Harris, I.R., Williams, A.J., 2006. Hydrogen decrepitation and recycling of NdFeB-type sintered magnets. Journal of Iron and Steel Research International 13 (1), 289–295.

Zakotnik, M., Harris, I.R., Williams, A.J., 2008. Possible methods of recycling NdFeB-type sintered magnets using the HD/degassing process. Journal of Alloys and Compounds 450, 525–531.

Zakotnik, M., Harris, I.R., Williams, A.J., 2009. Multiple recycling of NdFeB-type sintered magnets. Journal of Alloys and Compounds 469, 314–321.

Zeng, L., Cheng, C.Y., 2009. A literature review of the recovery of molybdenum and vanadium from spent hydrodesulphurization catalysts. Part I: metallurgical processes. Hydrometallurgy 98, 1–9.

CHAPTER 19

Recycling of Carbon Fibers

Soraia Pimenta[1], Silvestre T. Pinho[2]

[1]Department of Mechanical Engineering, Imperial College London, South Kensington Campus, London, UK; [2]Department of Aeronautics, Imperial College London, South Kensington Campus, London, UK

19.1 INTRODUCTION

The worldwide demand for carbon fibers (CFs) reached approximately 46,000 t in 2011, and this number is expected to grow at a rate over 13% per year (Roberts, 2011). Carbon fiber reinforced polymer (CFRP) is now used in a widening range of applications, and in growing content in most of them; the aircraft industry is an impressive example, with the new Boeing 787 and Airbus A350 having up to 50% of their weight in CFRPs.

The increasing use of CFRPs generates an increasing amount of CF waste, comprising end-of-life (EOL) prepregs, manufacturing cutoffs, testing materials, production tools and EOL components (Pimenta and Pinho, 2011). Taking the aeronautics sector as an example, the first civil aircraft with a significant amount of CFRP will soon be decommissioned; within 30 years, the same will happen to the new composites generation aircraft, with each vehicle representing more than 20 t of CFRP waste. The automotive and wind industries will be other great sources of CFRP waste, within a similar time frame.

Recycling composites is inherently difficult because of (1) their complex composition (with fibers, matrix and fillers), (2) the cross-linked nature of thermoset (e.g. epoxy) resins (which cannot be remolded) and (3) the combination with other materials (metal fixings, honeycombs, hybrid composites, etc.). Presently, most of the CFRP waste is landfilled (Pickering, 2006), which is unsatisfactory in environmental, economic and even legal terms (EU 1999/31/EC, 1999; EU, 2000/53/EC, 2000). Consequently, turning CFRP waste into a valuable resource and closing the loop in the CFRP life cycle (Figure 19.1) is vital for the continued use of the material in some applications (e.g. the automotive industry).

Consequently, CFRP recycling has received great attention over the last 15 years, not only from researchers (Pickering, 2006; Pimenta and Pinho, 2011) but also from several collaborative industrial entities (e.g. Aircraft Fleet Recycling Association [AFRA] and European Composite Recycling Services Company [ECRC]). The major current challenge is now the successful commercial implementation of recycling operations and the establishment of applications for the recycled composites.

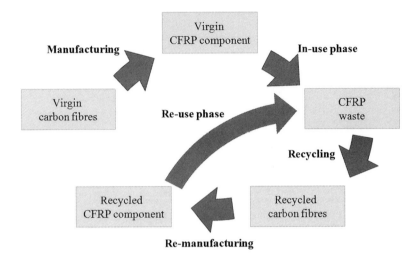

FIGURE 19.1 Closed carbon fiber reinforced polymer (CFRP) life cycle.

This chapter establishes the state of the art in CF recycling in Section 19.2, and analyses remanufacturing methods in Section 19.3. Section 19.4 discusses potential applications for the recyclates, and the environmental impact of recycling is addressed in Section 19.5. Section 19.6 discusses some challenges faced by the CFRP recycling industry, before the main conclusions are summarized in Section 19.7.

19.2 CARBON FIBER RECYCLING PROCESSES

Several methods for recycling CFRP have been investigated, as shown in Figure 19.2; most recycling routes include a fiber reclamation step (to break down the waste and recover fibers/fibrous products), followed by a remanufacturing step (to reimpregnate the reclaimed fibers in a new composite). Recycling processes can be classified as mechanical or thermochemical, depending on the main method used to break down the waste; Table 19.1 compares advantages and drawbacks of the different methods.

19.2.1 Mechanical Recycling

Mechanical recycling involves breaking down the composite by, for instance, shredding, crushing, or milling; the resulting scrap pieces can then be segregated by sieving into powdered products (rich in resin) and fibrous products (rich in fibers) (Kouparitsas et al., 2002; Palmer et al., 2010).

Typical applications for mechanically recycled composites include their reincorporation in new composites (as filler or reinforcement) and use in construction industry (e.g. as fillers for artificial woods or asphalt, or as mineral sources for cement; Conroy et al., 2006). Because these are low-value applications, mechanical recycling is mostly used for glass fiber reinforced polymers (GFRPs).

19.2.2 Thermochemical Fiber Reclamation

Thermochemical fiber reclamation consists on recovering the fibers (with high thermal and chemical stability) from the CFRP by using an aggressive thermal or chemical process to break down the matrix (typically a thermoset).

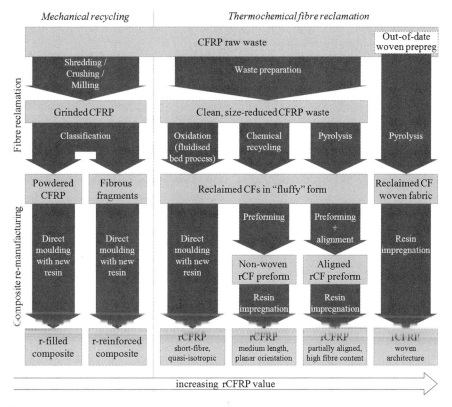

FIGURE 19.2 Overview of carbon fiber reinforced polymer (CFRP) recycling and remanufacturing processes. CF, carbon fiber; rCF, recycled carbon fiber; rCFRP, recycled carbon fiber reinforced polymer.

This is usually preceded by preliminary operations, such as cleaning and mechanical size reduction of the waste (Turner et al., 2011).

Generally, recycled carbon fibers (rCFs) have a clean surface (Figure 19.3(A)) and mechanical properties comparable to the virgin precursors (Figure 19.4). Nevertheless, depending on the reclamation process used, some surface defects (Figure 19.3(B)) and strength degradation have also been reported (Heil et al., 2009, 2010; Pimenta and Pinho, 2012), especially for large-scale processes.

19.2.2.1 Pyrolysis

Pyrolysis, the thermal decomposition of organic molecules in an inert atmosphere (e.g. nitrogen), is one of the most widespread recycling processes for CFRP. During pyrolysis, the CFRP is heated up in the (nearly) absence of oxygen; the polymeric matrix is volatilized into lower-weight molecules, whereas the CFs remain inert and are eventually recovered (Meyer et al., 2009). Existing studies (Alsop, 2009) suggest that reusing matrix products is not economically viable; nevertheless, the resin calorific energy can be recovered, making the pyrolysis process self-sustained.

Pyrolysis is currently the only recycling method with full-size commercial-scale implementations:

- ELG Carbon Fibre—Recycled Carbon Fibre, Ltd (ELG-RCF, formerly Recycled Carbon Fibre Ltd from the Milled Carbon Group), in the UK;

TABLE 19.1 Advantages and Drawbacks of Different Recycling Processes

	Advantages	Drawbacks
Mechanical	• Simple process; • recovery of both fibers and resin; • no use or production of hazardous.	• Modest mechanical performance; • unstructured, coarse and variable fiber architecture; • limited remanufacturing possibilities
Pyrolysis	• High retention of fiber properties for optimized processes; • energy recovery from the resin; • good adhesion between rCFs and epoxy; • existing commercial implementations.	• possible deposition of residual matrix/char on fiber surface; • quality of fibers is sensitive to processing parameters; • need for off-gases treatment unit.
FBP	• High tolerance to contamination; • no residual products on fiber surface; • well established and documented process.	• Large fiber-strength degradation; • fiber length degradation; • unstructured fiber architecture; • no material recovery from the matrix.
Chemical	• Very high retention of fiber properties; • potential for recovering valuable matrix products.	• Fiber adhesion to epoxy resins is commonly reduced; • low tolerance to contamination; • environmental impact if using hazardous solvents; • reduced scalability of most processes.

FBP, fluidized bed process; rCFs, recycled carbon fibers.

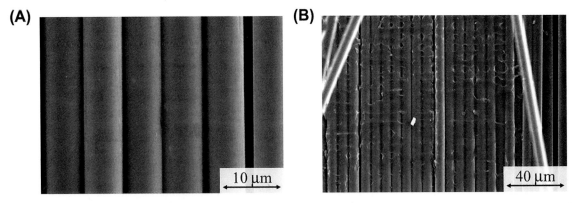

FIGURE 19.3 Recycled (pyrolyzed) carbon fibers, seen under a scanning electron microscope *(Adapted from Pimenta and Pinho, 2012)*. (A) Clean fibers. (B) Fibers with residual matrix.

- MIT-RCF (Reengineered Carbon Fiber, a Materials Innovation and Technologies LLC company), in the United States;
- CFK Valley Stade Recycling GmbH & Co. KG, in Germany;
- Karborek S.p.A., in Italy.

19.2.2.2 Oxidation in Fluidized Bed

Oxidation is another thermal process for CFRP recycling, consisting in combusting the polymeric matrix in a hot and oxygen-rich flow (e.g. air at 450–550°C). The fluidized bed process (FBP), developed at the University of

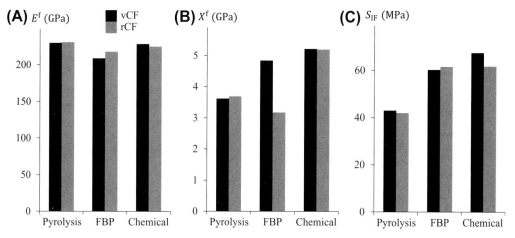

FIGURE 19.4 Mechanical properties of recycled carbon fibers (rCFs) and their virgin carbon fiber (vCF) precursors. *(Data sources: Heil et al. (2009) for pyrolysis; Wong et al. (2007a) and Jiang et al. (2008) for fluidized bed process (FBP); Jiang et al. (2009) for chemical.)* (A) Fiber stiffness. (B) Fiber tensile strength. (C) Interfacial shear strength (with epoxy).

Nottingham, is the most well known implementation of such a process (Yip et al., 2002; Pickering, 2006).

During recycling, CFRP scrap (reduced to fragments approximately 25 mm large) is fed into a bed of silica on a metallic mesh. As the hot air stream passes through the bed and decomposes the resin, both the oxidized molecules and the fiber filaments are carried up within the air stream, while heavier metallic components sink in the bed; this natural segregation makes the FBP particularly suitable for contaminated EOL components. The fibers are separated from the air stream in a cyclone, and the resin is fully oxidized in an afterburner; energy recovery to feed the process is feasible.

19.2.2.3 Chemical Recycling

Chemical methods for CFRP recycling are based on a reactive medium (e.g. catalytic solutions, alcohols and supercritical fluids) under low temperature (typically below 350 °C). The polymeric resin is decomposed into relatively high value oligomers (which can be reused as chemical feedstock), while the CFs remain inert and are subsequently collected.

Some of the currently most mature implementations of chemical CFRP recycling are:

- Adherent Technologies, Inc., in the United States. Allred et al. (2001) developed a proprietary catalytic tertiary recycling process, consisting in a low-temperature liquid catalysis (sometimes complemented by dry pyrolysis);
- Hitachi Ltd., in Japan. Nakagawa et al. (2009) used a benzyl-alcohol and a catalyst in a nitrogen atmosphere, under 200 °C, to recover both glass and CFs from composites.
- Chemical recycling using supercritical fluids (SCFs). These are fluids at temperatures and pressures (typically just) above the critical point, hence presenting liquid-like density and dissolving power, and gas-like viscosity and diffusivity. Several types of SCF (usually coupled with alkali catalysts) have been used for CF recycling, such as acetone (Pinero-Hernanz et al., 2008) and propanol (Jiang et al., 2009).

Despite being a more recent approach, recycling with optimized SCFs is already recognized for producing rCFs with virtually no mechanical degradation.

19.3 COMPOSITES REMANUFACTURING

The second phase of the recycling processes (Figure 19.2) consists in reimpregnating the reclaimed fibers with a new matrix. The typical form of rCFs, with discontinuous, unsized, filamentized and unstructured fibers (Figure 19.5), is unlike the standard form of commercialized virgin fibers. Consequently, several remanufacturing processes have been developed; Table 19.2 summarizes their main advantages and drawbacks, and Figure 19.6 presents typical mechanical properties of the recycled CFRPs (rCFRPs) produced.

FIGURE 19.5 Recycled (through pyrolysis) carbon fibers (CFs) in discontinuous, unsized, filamentized and unstructured (fluffy) form.

TABLE 19.2 Advantages and Drawbacks of Different Remanufacturing Processes

	Advantages	Drawbacks
Direct molding	• Established processes for virgin composites; • low complexity and cost.	• Very low fiber content ($V^f < 20\%$); • subcritical fiber length; • process problems because of rCF form.
Impregnation of nonwoven preforms	• Existing process for virgin materials; • widely used process and well documented; • similar performance to conventional structural materials; • demonstrators manufactured for automotive and aircraft industries.	• Fiber damage and breakage for filamentized architectures; • competing marked is dominated by relatively cheap materials.
Impregnation of aligned preforms	• Improved uniaxial properties; • possibility to tailor laminates; • improved fiber length and content.	• Improving packability requires nearly perfect alignment; • ongoing process development.
Impregnation of woven fabrics	• Structured architecture with continuous reinforcement and high fiber content; • simple and established manufacturing processes for virgin materials; • optimized performance comparable to virgin CFRP.	• Applicable only to out-of-date woven prepreg rolls;

CFRP, carbon fiber reinforced polymer; rCF, recycled carbon fibers.

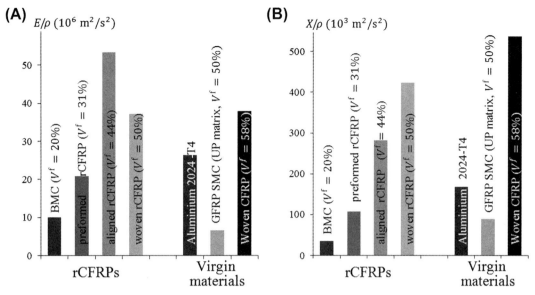

FIGURE 19.6 Comparison between the mechanical properties of typical recycled carbon fiber reinforced polymer (rCFRPs) and conventional virgin structural materials. *(Data sources: Turner et al. (2009) for bulk molding compound (BMC) and aligned rCFRP; Pimenta and Pinho (2013) for preformed rCFRP; Pimenta and Pinho (2012) for recycled and virgin woven CFRP; Aerospace Specification Metals (ASM, Inc.) for aluminum 2024-T4; Menzolit (Sheet Molding Compound [SMC] 1800) for glass fiber reinforced polymer (GFRP).)* (A) Specific tensile stiffness. (B) Specific tensile strength.

19.3.1 Direct Molding

Recycled CFs have been directly incorporated in composites by direct molding processes, such as injection molding (IM) and bulk molding compound (BMC) compression. These processes are readily available for virgin composites (mainly glass reinforced); their application to rCFs requires tailoring processing conditions and compounding formulation, but has been successfully demonstrated by independent researchers:

- Wong et al. (2007b) used IM with FBP fibers and a polypropylene (PP) matrix loaded with coupling agents;
- George et al. (2007), from Boeing/North Carolina State University, used IM with pyrolyzed fibers and a polycarbonate matrix;
- Turner et al. (2009) molded several epoxy BMCs with fibers recovered through the FBP and SCFs.

Because the high molding pressures involved in IM and BMC reduce fiber length to subcritical levels, these recyclates exhibit low stiffness and strength (similar to molded glass fiber composites; Figure 19.6), as well as low electrical conductivity; nevertheless, the mechanical properties are considerably superior to those of the unloaded polymers.

19.3.2 Impregnation of Nonwoven Preforms

The most widely used manufacturing process for rCFRPs includes a preforming step, in which the recycled fibers are converted into a nonwoven dry 2D or 3D product, followed by their subsequent impregnation through resin infusion or resin transfer molding (RTM). Preforming methods include:

- Papermaking technique. This has been used by Wong et al. (2009a), in collaboration with Technical Fibre Products Ltd. Szpieg et al. (2009) used a similar method and manufactured a fully recycled composite, with recycled PP scrap;
- Carding. This method has been applied by Nakagawa et al. (2009), from Hitachi Ltd, as well as Cornacchia et al. (2009, from Karborek S.p.A.);
- Three-dimensional engineered preforming process, developed by Janney et al. (2009) from Material Innovations and Technology, LLC (MIT).

These methods yield composites with discontinuous fibers and random orientation. A typical feature of these materials is the presence of a multiscale reinforcement (Pimenta et al., 2010), with filamentized fibers and bundles held together by minor quantities of residual resin of different sizes. Although the presence of bundles has been traditionally seen as a recycling defect, Pimenta and Pinho (2013) have showed they actually increase the fracture toughness of rCFRPs without significantly affecting stiffness or strength.

The rCF preforms have typically low permeability—especially when using filamentized fibers—thus increasing the required molding pressure and breaking the fibers during manufacture (Pimenta et al., 2010). Consequently, the performance of rCFRPs saturates at intermediate fiber contents ($V^f \approx 30\%$) (Wong et al., 2009a); nevertheless, the specific properties of rCFRPs are typically close to those of aluminum (Figure 19.6).

19.3.3 Impregnation of Aligned Preforms

Fiber alignment is key to improve the mechanical performance of composites manufactured with discontinuous rCFs; this not only improves uniaxial mechanical properties, as it benefits fiber packing thus allowing for higher fiber content in the composites. Alignment methods used to rCFRPs include:

- Nonwoven processes: all processes mentioned in the previous section can be modified so as to induce some degree of fiber alignment in the preforms (e.g. Turner et al., 2009 for papermaking technique);
- Centrifugal alignment rig, developed in the University of Nottingham (Wong et al., 2009b). It uses a rotating drum equipped with a convergent nozzle, which aligns a highly-dispersed suspension of rCFs;
- Fibrecycle (UK research project) has processed dry CF waste into yarns, slivers and tapes. These have been further processed into woven and noncrimp fabrics and reimpregnated with thermoplastic matrices, achieving nearly the same stiffness as the corresponding virgin materials.

The stiffness of state-of-the-art aligned rCFRPs (Figure 19.6) already overcomes that of aerospace-graded aluminum and virgin woven CFRP, both in absolute and specific terms. It is expected that any further improvement in the alignment level will significantly benefit both stiffness and strength because of its side effect on achievable fiber content. Ultimately, well-aligned rCFRPs will have mechanical properties competing only with virgin CFRPs.

19.3.4 Impregnation of Woven Fabrics

As some recycling processes can preserve the reinforcement architecture of the waste, it is possible to recover the structured weave from large woven items (e.g. out-of-date prepreg rolls). These have been subsequently reimpregnated (through RTM or resin infusion) into woven rCFRPs by:

- Meredith (2009, 2012). The recycled composite was used to manufacture noncritical parts of an environmentally sustainable formula 3 car;

- Pimenta and Pinho (2012). The mechanical properties of different recycled woven composites were compared to those of the virgin precursor (see Section 19.6.2 for more details) showing that nearly 100% of the stiffness and 80% of the strength were recovered.

19.4 APPLICATIONS FOR RECYCLED CARBON FIBERS AND COMPOSITES

19.4.1 Structural Applications

One of the most promising applications for rCFRPs consists of noncritical structural components (Pimenta and Pinho, 2011); this would fully exploit the mechanical performance of the fibers, thus increasing the final value of recycled products.

The aeronautics industry is particularly interested in incorporating rCFRPs in the interiors of aircraft, as long as materials are traceable and their properties consistent. Although rCFRPs need to be allowed to mature in nonaeronautical applications first, the involvement of aircraft manufacturers in CFRP recycling activities (e.g. in AFRA) is a good indicator of a very promising market.

There is also scope to manufacture automotive components with rCFRPs; this is motivated, on the one hand, by the improved performance of the recyclates when compared to more traditional materials. On the other hand, the introduction of substantial CFRP components in mass-production cars raises a significant waste management problem; incorporating recycled composites back into production lines would therefore boost green credentials, as required by tightening legislation on recyclability and sustainability (EU 1999/31/EC, 1999; EU, 2000/53/EC, 2000).

A few structural components have been manufactured with rCFRPs as technology demonstrators:

- Crashworthy and secondary components for the automotive industry: wing mirror cover and car door panel (Warrior et al., 2009), wheelhouse (Janney et al., 2009), driver seat (Nakagawa et al., 2009) and rear structure (Meredith, 2009);
- Components for aircraft interiors: seat armrest (George, 2009);
- Other markets have also been identified, such as construction industry, sports and household goods, and wind turbines (Pimenta and Pinho, 2011).

19.4.2 Nonstructural Applications

A few alternative nonstructural applications have also been demonstrated for rCFs:

- Electromagnetic interference shielding veils, as manufactured by Wong et al (2010) through a papermaking technique. Although the electrical conductivity of individual rCFs was similar to that of the virgin precursors ($\sim 1.8 \times 10^3\ \Omega\,cm$), the shielding effectiveness of recycled nonwoven veils was slightly lower than the corresponding virgin veil (-13% in average), because of incomplete fiber dispersion. Nevertheless, the veils with 80 g/m^2 exhibited sufficient electromagnetic shielding (40 dB attenuation) for commercial applications.
- Heating elements. Pang et al. (2012) manufactured rCF nonwoven veils and studied the influence of fiber length, binder type and content on their electrical conductivity and folding endurance, envisaging applications in the food industry. The heating behavior was studied, and overall performance was optimized for veils with 12 mm long fibers and 15% of acrylic thermoset binder.
- High performance ceramic disk brakes. This possibility has been studied within the ReBrake project, led by Surface Transforms PLC and funded by the Technology Strategy

Board, UK (Job, 2010). The concept proved to be successful and is now being commercially exploited by Carbon Ceramics Ltd.

19.5 LIFE-CYCLE ANALYSIS OF CARBON FIBER REINFORCED POLYMERS

Although CFRP recycling methods have been extensively investigated over the past 20 years, only a few life-cycle analyses (LCA) have been performed to assess the effect of the EOL stage on the environmental impact of CFRPs.

When compared to conventional structural materials, such as aluminum and steel, CFRPs offer a considerable environmental benefit during the in-use phase in transport applications, because of weight—and, therefore, fuel—savings. However, the production phase is very energy-intensive for CFRPs, and both metals mentioned are very easily recyclable. Consequently, using composites offers environmental benefits only when the savings during the in-use phase overcome the cost of production and EOL phases.

Considering landfilling or incineration as the current solution for CFRP waste, LCA show that CFRP is environmentally superior to steel for the majority of transport applications, as well as to aluminum in aircraft (Song et al., 2009; Scelsi et al., 2010; Duflou et al., 2012). However, aluminum outperforms CFRP in road transport applications if the EOL composite component is landfilled or incinerated (Song et al., 2009). These studies underline the importance of establishing suitable recycling routes for CFRPs.

Witik et al. (2013) provide a very thorough analysis of the environmental impact of different EOL routes for CFRP: (1) recycling through pyrolysis, (2) incineration with energy recovery and (3) landfilling disposal. For the former option, different re-use possibilities are considered: either replacing GFRPs or virgin CFRPs, and considering either static or transport applications. Incineration with energy recovery was found to have a smaller environmental impact than landfilling. Recycling is preferable to incineration when using the recyclate to replace virgin CFRP and GFRP in automotive applications, but not when replacing GFRP in static applications.

This study clearly shows that recycling operations must be application-driven, and that producing high-value recycled products is critical for their success. Nonsafety critical transport applications (e.g. in automotive or aircraft industries) are clear candidates for re-using the recyclates; in this case, the mechanical performance of recycled composites is of extreme importance as it governs the environmental impact during the in-use phase.

19.6 FURTHER CHALLENGES

19.6.1 Establishment of Successful Carbon Fiber Reinforced Polymer Recycling Chains

With recycling and remanufacturing processes now reaching maturity, the major current challenge to CFRP recycling operations is the establishment of a sound CFRP recycling chain supporting the effective commercialization of recycled products. Some of the main issues to overcome are:

- Global strategy: organized networks for CFRP recycling—bringing together suppliers/users (composite-related industries), recyclers and researchers—must be created, so as to understand the current state of the art and plan for future developments on the topic according to industrial needs.
- Incentives for recycling: governments should consider supporting the option of recycling; this could involve not only penalties for nonrecyclers (e.g. landfilling taxes) but also direct privileges (e.g. carbon credits) for companies recycling their CFRP waste. As recycling companies go commercial, it is necessary to implement specific legislation covering their operations.

- Logistics and cooperation in the supplying chain: waste suppliers must cooperate with recyclers, which includes supplying the waste in a continued and suitable form and providing the recyclers with material certificates whenever possible (e.g. for out-of-date prepreg rolls). Conversely, recyclers must ensure that materials and components supplied will not undergo reverse engineering.
- Market identification and product pricing: this requires that (1) characteristics and properties of different rCFRPs are known, (2) their processing times and costs are assessed and (3) a value for the recycled label is established.
- LCA: the environmental, economic and technical advantages of rCFRPs over other materials and disposal methods can be estimated only through cradle-to-grave analyses of the whole CFRP life cycle.
- Market establishment: ultimately, the major current challenge for the success of CFRP recycling is the establishment of a market for the recyclates. Creating a market requires all previous issues to be overcome, so rCFs are accepted as an environment-friendly and cost-effective material.

19.6.2 Understanding the Mechanical Response of Recycled Carbon Fiber Reinforced Polymers

Given the potential of rCFRPs to be incorporated into noncritical structural applications, it becomes crucial to understand the relations between microstructure, mechanical properties and damage mechanisms of these materials. First, this provides informed guidance for reclaimers and manufacturers to improve their processes toward recyclates with optimal performance. Second, it promotes the effective use of rCFRPs in structural components by making engineers more aware and confident on the response of these materials.

Pimenta and Pinho (2013) investigated the mechanical response of three different rCFRPs manufactured from nonwoven preforms, giving particular attention to the influence of the reinforcement architecture on the macroscopic mechanical properties. It was shown that preserving fiber bundles during reclamation and remanufacturing toughened the recyclates by more than one order of magnitude. Fracture toughness in excess of 40 kJ/m^2 was measured for the rCFRP with the coarsest architecture, which is significantly above the typical value exhibited by aluminum. Strength and stiffness, on the contrary, were similar for all materials tested (Table 19.3).

In another study, Pimenta and Pinho (2012) compared the responses of two woven rCFRPs (with fibers recovered in the ELG-RCF industrial plan using two different pyrolysis cycles) and their virgin precursor. Figure 19.7 shows the relation between (1) strength retention at the filament level, and (2) retention of mechanical properties of composites (normalized for the same fiber content). The stiffness of the composite (and, therefore, of the fibers) was unaffected

TABLE 19.3 Effect of Different Reinforcement Architectures on the Mechanical Properties of Recycled Carbon Fiber Reinforced Polymers Manufactured from Non-woven Preforms (Pimenta and Pinho, 2013)

	Stiffness	Strength	Toughness	Manufacturability
Filamentized/dispersed	●	●●	○	● (Fiber breakage)
Intermediate	●●	●●	●	●●
Bundled/coarse	●	●	●●	● (Voids)

Key: ○-low; ●-intermediate; ●●-high.

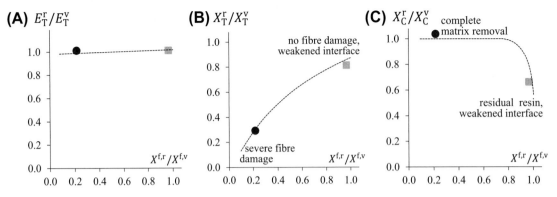

FIGURE 19.7 Retention of mechanical properties at the composite level versus retention of fiber strength: experimental measurements (data points: ● aggressive pyrolysis cycle; ■ gentle pyrolysis cycle) and suggested trends (dashed lines). Each data point corresponds to average properties measured in at least 50 single filaments and 10 composite specimens. *(Data source: Pimenta and Pinho (2012).)*, (A) stiffness, (B) tensile strength and (C) compressive strength.

by the reclamation process, even for the very aggressive cycle, which reclaimed only 25% of fiber strength. Retention of tensile strength was very similar at the filament and composite level, as expected for a woven composite. On the contrary, the compressive strength of the composite was fully recovered after the most aggressive cycle indicating good interfacial adhesion—but decreased with increasing fiber strength because of the presence of residual matrix, which weakened the interface.

Both these studies (Pimenta and Pinho, 2012, 2013) show that the response of rCFRPs is very complex and open to further research. Moreover, they make the case for market-driven recycling in which the specific mechanical requirements of the envisaged application for the recyclates are taken into account when selecting the optimal reclamation and remanufacturing process.

19.6.3 Scalability and Optimization of Recycling Operations

Although many small-scale fiber reclamation processes have been able to recover virgin-like recycled fibers, recent studies by Heil et al (2010) and Pimenta and Pinho (2012) show that commercial-scale recycling is much more challenging and often results in fiber defects—either fiber damage for aggressive processes or residual matrix for gentle cycles. Some of the additional challenges faced by commercially viable reclamation processes are:

- Throughput and processing time. Continuous processes can have running times under 30 min (RCF's/ELG Carbon Fibre's website), whereas typical laboratory batch processes report hours for recycling a few grams of composite (Meyer et al., 2009). To guarantee the same level of matrix removal, the former implies using more aggressive cycles, therefore risking fiber degradation.
- Implementation as a continuous process. In a batch process, atmosphere and temperature can be precisely tuned in time. Continuous processes are necessarily open, and require the cycle to be controlled relatively to position. Guaranteeing ideal and uniform conditions is, therefore, much more difficult.
- Unknown and mixed feedstock. Although most research is done with traceable scrap material, commercial recycling must deal with mixed feedstock and unknown

TABLE 19.4 Defects Expected at Recycled Fibers and Composites From Different Reclamation Cycles

	Defect	Reclamation Cycle		
		Too aggressive	Ideal	Too gentle
Fiber level	Fiber loss	●●	○	○
	Residual matrix	○	○/●	●●
	Stiffness reduction	○	○	○
	Strength reduction	●●	○/●	○
Composite level	Weak interface	○/●	○/●	●
	Voids	○	○	●
	Fiber breakage	●	○/●	○

(Adapted From Pimenta and Pinho, 2012)

specifications. Aiming for optimal reclamation conditions requires identifying and sorting the feedstock in great detail, which may not always be technically feasible or economically viable.

If one of the two goals of recycling—matrix removal and retention of fiber properties—is to be sacrificed (Table 19.4), it is useful to consider the application foreseen for the recycled fibers (as suggested in Figure 19.6 and Table 19.3). Nevertheless, it must be highlighted that aggressive cycles yield a considerable loss of fiber mass and, therefore, of profit.

19.7 CONCLUSIONS

A comprehensive overview on the state-of-the-art and market outlook for CFRP recycling operations was presented; recycling and remanufacturing processes were reviewed, and the commercialization challenges and potential markets for the recyclates were identified.

A critical comparison between recycling processes proved each of them to have specific advantages and drawbacks, suggesting complementarity rather than competition. Optimized batch-scale recycling processes yield rCFs with high retention of mechanical properties; however, it is expected that commercial implementation will operate under nonoptimal conditions.

Although remanufacturing CFRPs is challenging, mainly because of the unusual form of the recycled fibers, the mechanical performance of some rCFRPs overcomes that of some conventional structural materials. A few structural demonstrators for the automotive and aircraft industries have been manufactured, suggesting that rCFs can be used in nonsafety-critical structural applications; LCA studies support that such applications would present an environmental benefit.

References

Adherent Technologies, Inc. Official website at http://www.adherent-tech.com/ (accessed 03.13).
AFRA; Aircraft Fleet Recycling Association. Official website at http://www.afraassociation.org/ (accessed 03.13).
Allred, R.E., Gosau, J.M., Shoemaker, J.M., 2001. Recycling process for carbon/epoxy composites. In: SAMPE 2001 Symposium & Exhibition (SAMPE). Long Beach, CA, USA.
Alsop, S.H., 2009. Pyrolysis off-gas processing. In: SAMPE'09 Conference (SAMPE). Baltimore, MD, USA.
ASM, Inc — Aerospace Specification Metals. Properties of Aluminium 2024-T4 at http://asm.matweb.com/search/SpecificMaterial.asp?bassnum=MA2024T4 (accessed 03.13).
Carbon Ceramics Ltd. Official website in http://www.carbonceramics.com/ (accessed 03.13).
CFK Valley Stade Recycling GmbH & Co. KG. Official website at http://www.cfk-recycling.com/ (accessed 03.13).
Cornacchia, G., Galvagno, S., Portofino, S., Caretto, F., Giovanni, Casciaro, Matera, D., Donatelli, A., Iovane, P., Martino, M., Civita, R., Coriana, S., 2009. Carbon fiber recovery from waste composites: an integrated approach for a commercially successful recycling operation. In: SAMPE'09 Conference (SAMPE). Baltimore, MD, USA.
Conroy, A., Halliwell, S., Reynolds, T., 2006. Composite recycling in the construction industry. Composites Part A 37, 1216—1222.

Duflou, J.R., Deng, Y., Van Acker, K., Dewulf, W., 2012. Do fiber-reinforced polymer composites provide environmentally benign alternatives? A life-cycle-assessment-based study. Materials Research Society Bulletin 37, 374–382.

ECRC; European Composites Recycling Services Company. Official website at http://www.ecrc-greenlabel.org/ (accessed 03.13).

ELG Carbon Fibre—Recycled Carbon Fibre, Ltd. Official website at http://www.clgcf.com/ (accessed 03.13).

Council Directive 1999/31/EC of April 26, 1999 on the landfill of waste. The council of the European Union, Official Journal of the European Communities L 182(42), 1–19.

EU 2000/53/EC, Directive 2000/53/EC of the European Parliament and of the Council of 18 September 2000 on end-of-life vehicles. The Council of the European Union, Official Journal of the European Communities L 269(43), 34–42.

Fibrecycle Project. Official website at http://www.fibrecycleproject.org.uk/ (accessed 03.13).

George, P.E., 2009. End user perspective: perspective on carbon fibre recycling from a major end user. In: Carbon Fibre Recycling and Reuse 2009 Conference (IntertechPira). Hamburg, Germany.

George, P.E., Carberry, W.L., Connor, M.L., Allen, B.E., 2007. Recycled carbon fiber performance in epoxy and polycarbonate matrices. In: Composites Innovation 2007 – Improved Sustainability and Environmental Performance (NetComposites). Barcelona, Spain.

Heil, J.P., Hall, M.J., Litzenberger, D.R., et al., 2009. A comparison of chemical, morphological and mechanical properties of various recycled carbon fibers. In: SAMPE'09 Conference (SAMPE). Baltimore, MD, USA.

Heil, J.P., Litzenberger, D.R., Cuomo, J.J., 2010. A comparison of chemical, morphological and mechanical properties of carbon fibers recovered from commercial recycling facilities. In: SAMPE 2010 Conference (SAMPE). Seattle, WA, USA.

Hitachi, Ltd, 2010. Hitachi's involvement in materials resource recycling. Hitachi Review 59 (4).

Janney, M.A., Newell, W.L., Geiger, E., et al., 2009. Manufacturing complex geometry composites with recycled carbon fiber. In: SAMPE'09 Conference (SAMPE). Baltimore, MD, USA.

Jiang, G., Pickering, S.J., Walker, G.S., et al., 2008. Surface characterisation of carbon fibre recycled using fluidised bed. Applied Surface Science 254, 2588–2593.

Jiang, G., Pickering, S.J., Lester, E.H., et al., 2009. Characterisation of carbon fibres recycled from carbon fibre/epoxy resin composites using supercritical n-propanol. Composites Science and Technology 69, 192–198.

Job, S., 2010. Composite Recycling – Summary of Recent Research and Development. Report from Knowledge Transfer Network.

Karborek S.p.A. Official website at http://www.karborekrcf.it/ (last accessed in March 2013).

Kouparitsas, C.E., Kartalis, C.N., Varelidis, P.C., et al., 2002. Recycling of the fibrous fraction of reinforced thermoset composites. Polymers and Composites 23, 682–689.

Menzolit SMC 1800, preliminary data sheet, 2008. Available online at: http://www.menzolit.com/templates/rhuk_solarflare_ii/pdf/list_smc/SMC_1800.pdf (accessed 03.13).

Meredith, J., 2009. The role of recycled carbon fibre composites in motorsport applications. In: Carbon Fibre Recycling and Reuse 2009 Conference (IntertechPira). Hamburg, Germany.

Meredith, J., Cozien-Cazu, S., Collings, E., et al., 2012. Recycled carbon fibre for high performance energy absorption. Composites Science and Technology 72, 688–695.

Meyer, L.O., Schulte, K., Grove-Nielsen, E., 2009. CFRP-recycling following a pyrolysis route: process optimization and potentials. Journal of Composite Materials 43, 1121–1132.

MIT – Materials Innovation Technologies, LLC. Official website at http://www.emergingmit.com/ (accessed 03.13).

MIT-RCF; Reengineered Carbon Fiber, LLC. Official website at http://mitrcf.com/ (accessed 03.13).

Nakagawa, M., Shibata, K., Kuriya, H., 2009. Characterization of CFRP using recovered carbon fibers from waste CFRP. In: Second International Symposium on Fiber Recycling (The Fiber Recycling 2009 Organizing Committee). Atlanta, Georgia, USA.

Palmer, J., Savage, L., Ghita, O.R., Evans, K.E., 2010. Sheet moulding compound (SMC) from carbon fibre recyclate. Composites Part A 41, 1232–1237.

Pang, E.J.X., Pickering, S.J., Chan, A., Wong, K.H., 2012. Use of recycled carbon fibre as a heating element. Journal of Composite Materials first published on July 20, 2012.

Pickering, S.J., 2006. Recycling technologies for thermoset composite materials – current status. Composites Part A 37, 1206–1215.

Pimenta, S., Pinho, S.T., 2011. Recycling carbon fibre reinforced polymers for structural applications: technology review and market outlook. Waste Management 31, 378–392.

Pimenta, S., Pinho, S.T., 2012. The effect of recycling on the mechanical response of carbon fibres and their composites. Composite Structures 94, 3669–3684.

Pimenta, S., Pinho, S.T., 2013. The influence of micromechanical properties and reinforcement architecture on the mechanical response of recycled composites. Composites Part A 56, 213–225.

Pimenta, S., Pinho, S.T., Robinson, P., et al., 2010. Mechanical analysis and toughening mechanisms of a multiphase recycled CFRP. Composites Science and Technology 70, 1713–1725.

Pinero-Hernanz, R., Garcia-Serna, J., Dodds, C., et al., 2008. Chemical recycling of carbon fibre composites using alcohols under subcritical and supercritical conditions. Journal of Supercritical Fluids 46, 83–92.

RCF's/ELG Carbon Fibre's website http://www.elgcf.com/

Roberts, T., 2011. The Carbon Fibre Industry Worldwide 2011–2020: An Evaluation of Current Markets and Future Supply and Demand. Materials Technology Publications, ISBN 1 871677 64 5.

Scelsi, L., Bonner, M., Hodzic, A., et al. Potential Emissions Savings of Lightweight Composite Aircraft Components Evaluated through Life Cycle Assessment. eXPRESS Polymer Letters 5 (3), 209–217.

Song, Y.S., Youn, J.R., Gutowski, T.G., 2009. Life cycle energy analysis of fiber-reinforced composites. Composites Part A 40, 1257–1265 eXPRESS Polymer Letters 5, 209–217.

Szpieg, M., Wysocki, M., Asp, L.E., 2009. Reuse of polymer materials and carbon fibres in novel engineering composite materials. Plastics, Rubber and Composites 38, 419–425.

Technical Fibre Products Ltd. Official website at http://www.tfpglobal.com (accessed 03.13).

Turner, T.A., Pickering, S.J., Warrior, N.A., 2009. Development of high value, composite materials using recycled carbon fibre. In: SAMPE'09 Conference (SAMPE). Baltimore, MD, USA.

Turner, T.A., Pickering, S.J., Warrior, N.A., 2011. Development of recycled carbon fibre moulding compounds—preparation of waste composites. Composites Part B 42, 517–525.

Warrior, N.A., Turner, T.A., Pickering, S.J., 2009. AFRECAR and HIRECAR project results. In: Carbon Fibre Recycling and Reuse 2009 Conference (IntertechPira). Hamburg, Germany.

Witik, R.A., Teuscher, R., Michaud, V., et al., 2013. Carbon fibre reinforced composite waste: an environmental assessment of recycling, energy recovery and landfilling. Composites Part A 49, 89–99.

Wong, K.H., Pickering, S.J., Turner, T.A., Warrior, N.A., 2007a. Preliminary feasibility study of reinforcing potential of recycled carbon fibre for flame-retardant grade epoxy composite. In: Composites Innovation 2007 – Improved Sustainability and Environmental Performance (NetComposites). Barcelona, Spain.

Wong, K.H., Pickering, S.J., Brooks, R., 2007b. Recycled carbon fibre reinforced polypropylene composites: effect of coupling agents on mechanical properties. In: Composites Innovation 2007 – Improved Sustainability and Environmental Performance (NetComposites). Barcelona, Spain.

Wong, K.H., Pickering, S.J., Rudd, C.D., 2010. Recycled carbon fibre reinforced polymer composite for electromagnetic interference shielding. Composites Part A, 41 (6), 693–702.

Wong, K.H., Pickering, S.J., Turner, T.A., Warrior, N.A., 2009a. Compression moulding of a recycled carbon fibre reinforced epoxy composite. In: SAMPE'09 Conference (SAMPE). Baltimore, MD, USA.

Wong, K.H., Turner, T.A., Pickering, S.J., Warrior, N.A., 2009b. The potential for fibre alignment in the manufacture of polymer composites from recycled carbon fibre. In: SAE AeroTech Congress and Exhibition (SAE International). Seattle, Washington, USA.

Yip, H.L.H., Pickering, S.J., Rudd, C.D., 2002. Characterisation of carbon fibres recycled from scrap composites using fluidised bed process. Plastics, Rubber and Composites 31, 278–282.

CHAPTER 20

Recycling of Construction and Demolition Wastes

Vivian W.Y. Tam
School of Computing, Engineering and Mathematics, University of Western Sydney, Penrith, NSW, Australia

20.1 INTRODUCTION

"Housing for all" is invariably proclaimed as a national priority by all major political parties and adopted as a goal by the Government of India in the National Housing and Habitat Policy document (Laskar and Murthy, 2001; Jain, 2005). Integrated housing development not only satisfies the basic human needs but also facilitates holistic development within parameters of a planned welfare economy. Safe, secure, and affordable housings can increase employment and educational opportunities for individuals and enrich communities leading to a better civil society and better quality of life. From that, good social network of which lies clean environment, hygienic living, and quality housing can also be achieved (Tiwari, 2004; Jain, 2005).

India, with its large population, still witnesses an acute shortage of dwelling units. Despite the sharp increase in the usable housing stock from about 70 million units in 1961 to about 170 million units in 2001. The shortfall in 2001 was estimated at about 19 million dwelling units (Census of India, 2008). This has occurred because of the high population growth, especially in urban areas.

Previous studies revealed that populations in five most populous cities of India, namely Mumbai, Kolkata, New Delhi, Chennai, and Hyderabad, are set to increase at a scorching pace of >50% between 1995 and 2010 and more than double by 2025 (Census of India, 2008). The 2001 census statistics revealed that decadal population growth in the urban areas is one and a half times higher than the national average. All these statistics highlight to a high level of migration of population from rural and semiurban areas to a more urbanized form of settlement. The percentages of population staying in urban areas have steadily climbed from about 23.34% in 1981 to almost 28% in 2001 (Census of India, 2008).

Coupled with the demand for dwelling units, another major factor that has contributed to the buoyancy of housing activity is the affordability of properties. This, in turn, has been the result of a combined effect of stabilized property prices, high level of incomes, and low cost of borrowings. In fact, the boom witnessed by the housing finance sector can be heavily attributed to these factors (National Building Organization, 2003).

Further, if rural housing can be achieved through cost-effective means, such as usage of concrete blocks, the nation would be able to attain a fast sustainable growth. Development of Indian economy through a cost-effective route will be an engine of growth across the sectors (National Housing Bank, 2005).

This chapter aims to consolidate information and advice on the use of low-cost housing technologies over the traditional construction methods with a cost-effective analysis. Two case studies in India are used for the investigation. Construction costs for foundation, walling, roofing, and lintel for the traditional construction methods are examined and the low-cost housing technologies compared.

20.2 THE EXISTING LOW-COST HOUSING TECHNOLOGIES

Low-cost housing can be considered affordable for low- and moderate-income earners if the household can acquire a housing unit (owned or rented) for an amount up to 30% of its household income (Miles, 2000). In developing countries, such as India, only 20% of the populations are high-income earners, who are able to afford normal housing units. The low-income groups in developing countries are generally unable to access the housing market. Cost-effective housing is a relative concept and has more to do with budgeting and seeks to reduce construction cost through better management, appropriate use of local materials, skills, and technologies but without sacrificing the performance and structure life (Tiwari et al., 1999). It should be noted that low-cost housings are not houses that are constructed by cheap building materials of substandard quality. A low-cost house is designed and constructed as any other house with regard to foundation, structure, and strength. The reduction in cost is achieved through effective utilization of locally available building materials and techniques that are durable, economical, accepted by users, and not requiring costly maintenance (Miles, 2000).

Economy is also achieved by postponing finishing and implementing low-cost housing technologies in phases. High efficiency of workers, minimizing waste in design, and applying good management practices can also be achieved.

The Government of India and State Governments have been promising research in the fields of housing and construction activities (Government of India, 2008). This has led to a number of new alternative building materials and techniques aimed at reducing construction costs and improving the performance of conventional building materials and techniques. There are five common cost-effective building materials used for low-cost housing technologies (Gooding and Thomas, 1995; Building Materials and Technology Promotion Council, 2003):

1. Fly ash utilization: Bricks, Portland pozzolana cement, sintered aggregates, tiles, and lightweight aggregate solid and hollow blocks can be produced using fly ash as a raw material. The properties of the materials made from fly ash have been found to be quite comparable with the conventional materials for construction works. Fly ash can also be utilized for backfilling, lining irrigation canals, agriculture, and filling in road construction.
2. Red mud utilization: Red mud as a solid waste is generated during extraction of non-ferrous metals, such as aluminum and copper. Red mud can be used as a binder, as a cellular concrete additive, in making floor and wall tiles, colored composition of concrete, heavy clay products, red mud bricks, corrugated roofing sheets, and composite panels for door shutters. It is also used in the manufacture of aggregate and for making construction blocks.
3. Precast concrete frame: It is high quality, long lasting, and durable, and has a high rate of production and minimal requirement of raw materials. It is also fire-proof, water-proof, is produced utilizing unskilled labor, save on the cost of lintels, and its limited

mechanization results in high per capita output and the possibility of using industrial wastes, such as fly ash and blast furnace slag.
4. Ferro cement roofing components: Ferro cement is a highly versatile composite material consisting of cement mortar, chicken wire mesh, and welded mesh. It is increasingly being accepted as an appropriate cost-effective construction technique for different applications in housing and building. Ferro cement roofing components have a high strength to weight ratio, about 20% savings on materials and cost. It is also suitable for precasting and flexible in cutting, drilling, and jointing.
5. Bamboo mat corrugated roofing sheet: The sheets possess excellent physical–mechanical properties and are based on renewable resources requiring low energy. It may also be used as value-added products in the areas as an esthetically pleasing material. These sheets are not only highly water and weather resistant but are also resistant to decay, termites, and insects.

Low-cost housing is an innovative concept that deals with effective budgeting and following techniques that help reduce construction cost through the use of locally available materials along with improved skills and technologies without sacrificing the strength, performance, and life of the structure (Kumar, 1999; Civil Engineering Portal, 2008). Low-cost housing technologies aim to cut down construction costs by using alternatives to the conventional methods and inputs. It is about the usage of local and indigenous building materials, local skills, energy savings, and environment-friendly options. There are four common low-cost housing technologies used in the industry.

20.3 EARTH/MUD BUILDING

Mud brick, also referred to by the Spanish name of "Adobe", which means mud or puddle earth, generally refers to the technique of building with sun-dried mud blocks in either load bearing or nonload bearing construction (Kerali, 2001; Sustainable Earth Technologies, 2008). Mud bricks are increasingly becoming commercially available in a range of stabilized and nonstabilized bricks.

Mud bricks are typically 250 mm wide, 125 mm high, and 375 mm long and are normally made from earth with a clay content of about 50–80% with the remainder comprising a grading of sand, silt or gravel (Kerali, 2001; Sustainable Earth Technologies, 2008). Kaolin clays are the preferred clay types because of their nonexpansion characteristics. Stabilizing the mud brick with straw or other fibers is sometimes used where the soil mix displays excessive shrinkage behavior.

Mud brick has several advantages over the conventional fired clay or concrete masonry. The advantages include (Kerali, 2001; Sustainable Earth Technologies, 2008): (1) low in embodied energy; (2) utilization of natural resources and minimal use of manufactured products; (3) good sound absorption characteristics; (4) high thermal mass; (5) a claimed ability to "breadth"; (6) suited to a wide range of soils; (7) easily manufactured and worked; (8) flexibility in design/color/surface finishes; and (9) insulation properties similar to concrete or brickwork.

However, the specific architecture of earth houses usually leads to nonrighted, round-shaped walls, which can cause problems concerning the interior decoration, especially regarding furniture and large paintings. However, these problems can be anticipated during the conceptual design of an earth house.

20.4 PREFABRICATION METHOD

Prefabrication is the manufacture of an entire building or component cast in a factory so that they can be easily and rapidly erected on site

(Adlakha and Associates, 2003). Prefabrication is an easy and fast installation for any structure, such as houses, homes, storages, cabins, and garages. It is becoming popular to construct any building structure as it is cheap, fast to build, and durable.

Prefabrication can bring: (1) mass production of units; (2) reduction of cost and construction time on site; (3) the use of semi-skilled labor; (4) effective use of formwork; (5) improved quality of units; (6) form special shapes and surface finishes; (7) easy to demount and re-erect structures; (8) casting of units before the site becomes available; (9) built-in services and insulation; (10) accelerated curing techniques; and (11) the solution to the problem of lack of local resources and labor (Tam et al., 2003, 2004, 2005, 2006; Tam and Tam, 2006).

However, careful handling of prefabricated components, such as concrete panels or steel and glass panels is required. Attention has to be paid to the strength and corrosion-resistance of the joining of prefabricated sections to avoid failure of the joint. Similarly, leaks can form at joints in prefabricated components. Transportation costs may be higher for voluminous prefabricated sections than for the materials of which they are made, which can often be packed more compactly. Large prefabricated sections require heavy-duty cranes and precision measurement and handling to place in position.

20.5 LIGHTWEIGHT FOAMED OR CELLULAR CONCRETE TECHNOLOGY

It is widely used in the manufacture of single-skin lightweight concrete wall panels and uses tilt-up construction. This is an ideal situation for the manufacture of light commercial structures and factories as well as residential buildings. The multitude of applications for lightweight concrete includes (Pan Pacific Group of Companies, 2008): (1) aerated lightweight concrete blocks and lightweight tilt-up panels; (2) foam concrete floor screeds; (3) sound and thermal insulation; and (4) geotechnical and ornamental concrete applications.

Lightweight concrete technology has the following advantages (Building Materials and Technology Promotion Council, 2003): (1) rapid and relatively simple construction; (2) tremendous weight reduction, results in structural frames, footings or piles; (3) good thermal insulation properties; (4) low building costs; (5) reduction in handling and cartage costs; (6) economical in transportation; (7) low crane capacity required; (8) reduction in manpower; (9) substantial material saving; and (10) easy and fast production.

20.6 STABILIZED EARTH BRICK TECHNOLOGY

Stabilized earth brick technology offers cost-effective and environment-sound masonry system. The products have wide applications in construction for walling, roofing, arched openings, and corbels. Stabilized earth bricks are manufactured by compacting raw earth mixed with a stabilizer, such as cement or lime, under a pressure of about $20-40$ kg/cm^2 using a manual or mechanized soil press (Building Materials and Technology Promotion Council, 2003). The compressed earth block building system can be used in a variety of ways to construct buildings that are esthetic, efficient, and easy to build.

The housing technology is made very simple by using an interlocking clay brick system that is treated with ionic clay stabilizer formula (Road-Packer Group Limited, 1997). This improves its engineering properties, including compaction, density, bearing strength, and safety. This provides a low cost, durable product that can meet the needs of millions of low-cost housing units required annually around the world (Montgomery and Thomas, 2001).

20.7 CASE STUDY

To investigate the cost effectiveness of using low-cost housing technologies, two case studies in India are conducted. One is using the traditional construction methods and the other using low-cost housing technologies. Indepth interview discussions with project managers, quantity surveyors, engineers, foremen, and frontline workers for the two case studies are conducted. The costs involved at different stages of construction are compared and analyzed.

20.7.1 The Traditional Construction Methods

The traditional construction methods are used in the case study. The detail procedures of each step used for the case study are as follow:

- Foundation: The foundation is the lowest part of the structure that is provided to distribute loads to the soil, thus providing a base for the superstructure. Excavation work is first carried out, and then earth-work is filled with available earth and ends with watering and compaction in a 6-in thick layer.
- Cement concrete: Plain cement concrete is used to form a leveled surface on the excavated soil. The volumetric concrete mix proportion of 1:4:8 (cement: sand: aggregate), with a 6-in thick layer for masonry foundation and column footings are used. Plain cement concrete is finished on the excavated soil strata and mixed by manual process.
- Wall construction: Size stone masonry for the foundation is constructed for the outer walls and burnt brick masonry of a 9-in thick layer for the main walls and a 4.5-in thick layer for all internal walls. Good quality table-molded bricks are used for the construction.
- Reinforced cement concrete slab and beam: The normal procedure to cast reinforced cement concrete slabs is to make shuttering and provide reinforcement and concreting. Good steel or plywood formwork is used, with proper cover blocks between bars. Both the aggregate and sand used are clean, with the aggregate being three-fourths of an inch graded. After the concrete is poured, it is properly consolidated.
- Plastering: Plastering is used for the ceiling and the inside and outside walls. Joints are raked before plastering and proper curing is ensured.
- Flooring: For the flooring purpose, the earth is properly filled and consolidated in the ratio of 1:4:8 (cement: sand: aggregate) concrete.
- Plumbing: Good quality plumbing materials are used and have passed a hydraulic test before using it.
- Painting and finishing: Before the painting process, the surface is prepared with putty and primer and a ready made paint is used.

20.7.2 Low-Cost Construction Technologies

Cost-effective and alternative construction technologies, which apart from reducing construction cost by the reduction of quantity of building materials through improved and innovative techniques, can play a great role in providing better housing methods and protecting the environment. It should be noted that cost-effective construction technologies do not compromise with safety and security of the buildings and mostly follow the prevailing building codes. The detailed procedures of each step used for the case study are as follow:

- Foundation: Arch foundation is used in which walls are supported on the brick or stone masonry (Figure 20.1). For the construction of the foundation, the use of available materials, such as brick or concrete blocks, can be made to resist lateral force buttresses at the corner.

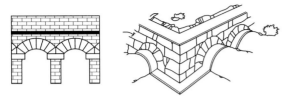

FIGURE 20.1 Arch foundation.

- Walling: Rat-trap bond technology is used (Figure 20.2) in the case study. It is an alternative brick bonding system for English and Flemish Bond. The reduced number of joints can reduce mortar consumption. No plastering of the outside face is required and the wall usually is esthetically pleasing and air gaps created within the wall help make the house thermally comfortable. In summer, the temperature inside the house is usually at least five degrees lower that the outside ambient temperature and vice versa in winter.
- Roofing: A filler slab roofing system is used, which is based on the principle that for roofs that are simply supported, the upper part of the slab is subjected to compressive forces and the lower part of the slab experiences tensile forces (Figure 20.3). Concrete is very good in withstanding compressive forces and steel bears the load because of tensile forces.

FIGURE 20.2 Rat-trap bond walling.

FIGURE 20.3 Reinforced cement concrete filler slab.

TABLE 20.1 Comparison between the Traditional Construction Methods and the Low-Cost Housing Technologies

Stages	Traditional Methods	Low-Cost Housing Technologies
Foundation	Stepped foundation Traditional foundations of rammed earth and stone or brick in mud or lime need high depth and width.	Arch foundation Arch foundation is supported on the brick or stone masonry, which saves materials such as cement, coarse rubble stone, and sand.
Walls	English bonds A large amount of brick and mortar are required, which brings a high investment cost.	Rat-trap bonds Less consumption of brick and sand reduces construction cost.
Lintel	Reinforced cement concrete lintels Use of concrete and steel in reinforced cement concrete lintels, which increases construction cost.	Brick arch lintels Cost is reduced by using brick for lintels and arches for span openings.
Roofing	Reinforced cement concrete roof As it uses a large amount of steel and concrete, it becomes costly. It also provides poor heat insulation.	Filler slab roof Filler materials replace redundant concrete, thus, self-weight of the slab is reduced and cost is lowered. It also reduces the heat inside the building because of the insulating cavity in the tiles.

Thus, the low tensile region of the slab does not need any concrete except for holding steel reinforcements together.

- Flooring: Flooring is generally made of terracotta tiles or color oxides. Bedding is made out of broken brick bats. Various patterns and designs are used, depending on the shape, size of tiles, span of flooring, and the client's personal preference.
- Plastering: Plastering can be avoided on the walls, frequent expenditure on finishes, and its maintenance is avoided. Properly protected brick wall will never lose its color or finish.
- Doors and windows: As door and window frames are responsible for almost half the cost of timber used, avoiding frames can considerably reduce timber costs. Door planks are screwed together with strap iron hinges to form doors, and this can be carried by "holdfast" carried into the wall. The simplest and most cost-effective door can be made of vertical planks held together with horizontal or diagonal battens. A simplest frameless window consists of a vertical plank of about 9 in wide set into two holes, one at the top and one at the bottom. This forms a simple pivotal window. Wide span windows can be partially framed and fixed to walls or can have rows of pivotal planks.

Table 20.1 summarizes the traditional construction methods and the low-cost housing technologies used in the case study.

20.8 COST-EFFECTIVENESS OF USING LOW-COST HOUSING TECHNOLOGIES

The construction methods of walling and roofing are selected for the detail cost analysis based on available resources from the interviews. Tables 20.2 and 20.3 summarize the cost analysis of the traditional construction methods and the low-cost housing technologies in the case studies for walling and roofing, respectively. It is found that about 26.11 and 22.68% of the construction cost,

TABLE 20.2 Cost Analysis of the Traditional Construction Methods and the Low-Cost Housing Technologies Used in the Case Studies for 1 m³ of Walling (Works Department, 2002)

No.	Item	Unit	Rate (US$)	Conventional Brickwork Quantity	Conventional Brickwork Amount (US$)	Rat-trap Bonded Brickwork Quantity	Rat-trap Bonded Brickwork Amount (US$)
MATERIALS							
1	Bricks	No	0.02	350.00	7.00	284.00	5.68
2	Sand	m³	0.32	0.28	0.09	0.17	0.05
3	Cement (10 kg/bag)	No	6.17	0.67	4.13	0.40	2.47
LABOR							
1	Mason (highly skilled)	No	1.70	0.35	0.60	0.35	0.60
2	Mason (2nd class)	No	1.49	1.05	1.56	0.80	1.19
3	Unskilled labor	No	1.06	2.96	3.14	1.96	2.08
	Add 2% tools and plant charges				0.34		0.25
	Add for scaffolding-superstructure: 0.42/m³				0.42		0.42
	Total (per m³)				17.71		13.08
	Savings						26.11%

TABLE 20.3 Cost Analysis of the Traditional Construction Methods and the Low-Cost Housing Technologies Used in the Case Studies for 1 m³ of Roofing (Works Department, 2002)

No.	Item	Unit	Rate (US$)	Conventional Slab Quantity	Conventional Slab Amount (US$)	Filler Slab Quantity	Filler Slab Amount (US$)
MATERIALS							
1	Concrete, including labor	m³	38.6	1.00	38.6	0.80	30.88
2	Reinforcement	Ton	36.12	0.80	28.89	0.38	13.72
3	Steel cutting, bending	Ton	3.87	0.80	3.09	0.38	1.47
4	Mangalore tiles	No	0.06	N/A	N/A	65.00	4.14
LABOR							
1	Mason (2nd class)	No	1.49	N/A	N/A	0.20	0.30
2	Unskilled labor	No	1.06	N/A	N/A	0.80	0.85
	Add 2% tools and plant charges				0.11		0.11
	Total (per m³)				84.32		65.20
	Savings						22.68%

20.8 COST-EFFECTIVENESS OF USING LOW-COST HOUSING TECHNOLOGIES

TABLE 20.4 Summary on the Experience on Recycling Technology and Practice

Construction and Demolition Waste	Recycling Technology	Recycled Product
Asphalt	• Cold recycling • Heat generation • Minnesota process • Parallel drum process • Elongated drum • Microwave asphalt recycling system • Finfalt • Surface regeneration	• Recycled asphalt • Asphalt aggregate
Brick	• Burn to ash • Crush into aggregate	• Slime burnt ash • Filling material • Hardcore
Concrete	• Crush into aggregate	• Recycled aggregate • Cement replacement • Protection of levee • Backfilling • Filler
Ferrous metal	• Melt • Reuse directly	• Recycled steel scrap
Glass	• Reuse directly • Grind to powder • Polishing • Crush into aggregate • Burn to ash	• Recycled window unit • Glass fiber • Filling material • Tile • Paving block • Asphalt • Recycled aggregate • Cement replacement • Man-made soil
Masonry	• Crush into aggregate • Heat to about 900 °C to ash	• Thermal insulating concrete • Traditional clay brick • Sodium silicate brick

TABLE 20.4 Summary on the Experience on Recycling Technology and Practice (cont'd)

Construction and Demolition Waste	Recycling Technology	Recycled Product
Nonferrous metal	• Melt	• Recycled metal
Paper and cardboard	• Purification	• Recycled paper
Plastic	• Convert to powder by cryogenic milling • Clipping • Crush into aggregate • Burn to ash	• Panel • Recycled plastic • Plastic lumber • Recycled aggregate • Landfill drainage • Asphalt • Man-made soil
Timber	• Reuse directly • Cut into aggregate • Blast furnace deoxidization • Gasification or pyrolysis • Chipping • Molding by pressurizing timber chip under steam and water	• Whole timber • Furniture and kitchen utensils • Lightweight recycled aggregate • Source of energy • Chemical production • Wood-based panel • Plastic lumber • Geofibre • Insulation board

including material and labor costs, can be saved by using the low-cost housing technologies in comparison with the traditional construction methods for walling and roofing, respectively.

Suggestions for reducing construction cost in this article are of a general nature and varies depending upon the nature of the building to be constructed and the budget of the owner. However, it is necessary that good planning and design methods shall be adopted by utilizing the services of an experienced engineer or an architect for supervising the work, thereby achieving overall cost-effectiveness.

20.9 RECYCLING TECHNOLOGIES AND PRACTICE

From a recent review study, it is found that there are well-established recycling technologies for construction and demolition waste in developed countries. Recycling technologies of 10 typical construction and demolition wastes have been investigated; the results are summarized in Table 20.4.

20.10 CONCLUSION

The dream of owning a house, particularly for low-income and middle-income families, is becoming a difficult reality. It is necessary to adopt cost-effective, innovative, and environment-friendly housing technologies for the construction. This chapter examined the cost-effectiveness of using low-cost housing technologies in comparison with the traditional construction methods. Two case studies in India were conducted. It was found that about 26.11 and 22.68% of the construction cost, including material and labor costs, can be saved by using the low-cost housing technologies in comparison with the traditional construction methods for walling and roofing, respectively. This was proven by the benefits and the trends for implementing low-cost housing technologies in the industry. Recycling technologies and practice of 10 major construction and demolition wastes were also suggested.

Acknowledgments

This chapter is an adaptation of: Tam W.Y. Vivian "2011" (Cost effectiveness of using low cost housing technologies in construction, *Procedia Engineering, The Twelfth East Asia—Pacific Conference on Structural Engineering and Construction*, 14, 156—160); Tam W.Y. Vivian and Tam C.M. "2008" (*Re-use of construction and demolition waste in housing development*, Nova Science Publishers, Inc., United States); Tam W.Y. Vivian and Tam C.M. "2006" (A review on the viable technology for construction waste recycling, *Resources, Conservation and Recycling*, 47(3), 209—221).

References

Adlakha and Associates, 2003. Study on Low Cost Incremental Housing Scheme for UP State. Building Materials and Technology Promotion Council, India.

Building Materials and Technology Promotion Council, 2003. Environmental Friendly Materials and Technologies. Building Materials and Technology Promotion Council, India.

Census of India, 2008. India Population Statistics. Census of India, India.

Civil Engineering Portal, 2008. Cuore Concrete — Nano Silica. Civil Engineering Portal.

Gooding, D.E.M., Thomas, T., 1995. The Potential of Cement — Stabilized Building Blocks as an Urban Building Material in Developing Countries. ODA Report, United Kingdom.

Government of India, 2008. Union Budget and Economic Survey 2008—2009. Ministry of Finance, Government of India.

Jain, S.C., 2005. Housing and GDP. National Real Estate Development Council, Ministry of Housing and Urban Poverty Alleviation, Government of India.

Kerali, A.G., 2001. Durability of Compressed and Cement-Stabilized Building Block. University of Warwick, United Kingdom.

Kumar, A., 1999. Sustainable building technology for mass application. Development Alternatives Newsletter 9 (11), 1—4.

Laskar, A., Murthy, C.V.R., 2001. Challenges before Construction Industry in India. Department of Civil Engineering, Indian Institute of Technology Kanpur, India.

Miles, M.E., 2000. Real Estate Development, Principles and Processes. Urban Land Institute, Washington D.C.

Montgomery, D.E., Thomas, T.H., 2001. Minimizing the Cement Requirement of Stabilized Soil Block Walling. University of Warwick, United Kingdom.

National Building Organization, 2003. Handbook of Housing Statistics. National Building Organization, Ministry of Housing and Urban Poverty Alleviation, Government of India.

National Housing Bank, 2005. Report on Trend and Progress of Housing in India. National Housing Bank, Reserve Bank of India.

Pan Pacific Group of Companies, 2008. Lightweight, Foamed or Cellular Concrete Technology. Pan Pacific Group of Companies, Australia.

RoadPacker Group Limited, 1997. Low Cost Housing Technical Information and Product Manual. RoadPacker Group Limited, Canada.

Sustainable Earth Technologies, 2008. Earth Building and Energy Efficient Construction. Sustainable Earth Technologies.

Tam, C.M., Tam, W.Y.V., Chan, K.W.H., Ng, C.Y.W., 2005. Use of prefabrication to minimize construction waste—a case study approach. International Journal of Construction Management 5 (1), 91–101.

Tam, W.Y.V., Tam, C.M., 2006. Evaluations of existing waste recycling methods: a Hong Kong study. Building and Environment 41 (12), 1649–1660.

Tam, W.Y.V., Tam, C.M., Chan, W.W.J., Ng, C.Y.W., 2006. Cutting construction wastes by prefabrication. International Journal of Construction Management 6 (1), 15–25.

Tam, W.Y.V., Tam, C.M., Shen, L.Y., 2004. Comparing material wastage levels between conventional in-situ and prefabrication construction in Hong Kong. Journal of Harbin Institute of Technology 11 (5), 548–551.

Tam, W.Y.V., Tam, C.M., Shen, L.Y., Lo, K.K., 2003. Wastage generation on conventional in-situ and prefabrication construction methods. In: CRIOCM 2003 International Research Symposium on Advancement of Construction Management and Real Estate.

Tiwari, P., 2004. A policy mechanism for housing construction activity to achieve social and environmental goals: a case of India. International Transaction in Operational Research 11 (6), 645–665.

Tiwari, P., Parikh, K., Parikh, J., 1999. Structural design considerations in house builder construction model: a multi-objective optimization technique. Journal of Infrastructure System 5 (3), 75–90.

Works Department, 2002. Revised Schedules of Rates 2000 Works Department and Analysis of Rates. Works Department, India.

CHAPTER 21

Recycling of Packaging

Ernst Worrell

Copernicus Institute of Sustainable Development, Utrecht University, Utrecht, The Netherlands

21.1 INTRODUCTION

To manage the environmental impact and increasing scarcity of resources, there is a strong need to change the way society produces and consumes materials. Reducing primary material production and consumption will save not only resources but also the energy that is required for production and waste processing. In our present society, the most important material flows consist of material chains in the build environment, transportation, and packaging. Packaging materials have a relatively high environmental impact. Approximately 40% of the municipal solid waste in Western Europe can be ascribed to packaging materials, along with at least 33% of all solid waste in the United States. As income levels increase and lifestyles change, packaging waste is likely to increase in transitioning and developing countries as well. For example, 2010 annual packaging waste generation in Bulgaria was estimated at around 43 kg/capita, compared to 196 kg/capita in Germany. Similarly, the composition of waste will vary between different countries.

In this chapter, the developments in packaging waste are discussed, focusing on the volume generated as well as the composition.

This will be followed by a discussion of different options to recover and collect packaging waste. Because there is a wide variety of collection schemes, we will discuss typical characteristics and a few selected examples. The chapter finishes with some general conclusions.

21.2 PACKAGING WASTE

The increasing use of packaging and the subsequent increasing generation of packaging waste can be illustrated on the basis of developments in the European Union; Eurostat has reported packaging waste figures since 2000. Figure 21.1 depicts the total volume of packaging waste in the European Union. It clearly shows the importance of the original EU member states (EU-15), as well as the small but growing contribution of the new member states (EU-12). Differences can be caused by many factors, including population, economic development, distribution of household sizes, and national traditions in packaging. Policies will affect the total volume of packaging waste. A large variety of policy instruments have been introduced in many places to manage packaging waste, varying from voluntary approaches to

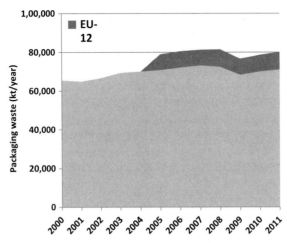

FIGURE 21.1 Development of packaging waste in the European Union from 2000 until 2011 (expressed in kt/year). The figure distinguishes the original 15 member states (EU-15) and the new member states (EU-12). Data for new member states is only available since 2005. *Based on data from Eurostat.*

taxation and bans of specific packaging applications or materials. The effectiveness of policy may vary between instruments and over time (see e.g. Rouw and Worrell, 2011; Dikgang et al., 2012).

To understand the underlying trends in packaging material use and waste generation, the concept of the environmental Kuznets curve (EKC) can be used. The EKC concept has been applied to several material studies that examined the relationship between material use and economic development. The studies show that there is no linear growth of the intensity of material use over time and outline the factors that influence material consumption development over time. Intensity of use is expressed as the demand for material in kilogram per unit of income (measured as gross domestic product) (see Eqn (21.1)):

$$IU = X_i/GDP \qquad (21.1)$$

where IU is the intensity of material used, X_i is the consumption of a specific material during year i, and GDP is the gross domestic product (the total output of an economy).

The EKC concept argues that material intensity first increases and starts decreasing after a certain level of development has been reached while income may continue to grow, resulting in a reduced material intensity. This process is called dematerialization or decoupling. Dematerialization refers to the absolute or relative reduction in the quantity of materials used and/or the quantity of waste generated in the production of a unit of economic output. Decoupling means that economic growth and material use is unlocked. The observed trend of the dematerialization process may be "distorted" by material substitution. Due to (technological) innovations and changing demands, materials replace one another over time. In structural applications, for example, the use of wood is successively replaced by iron, steel, aluminum, and plastic. Although the applied EKC depicts the intensity of material use (kg/GDP) over time, it does not necessarily explain the development of absolute material use over time. In fact, when the intensity of use declines, absolute material use can still increase, albeit to a smaller extent than GDP growth. For this reason, De Bruyn and Opschoor

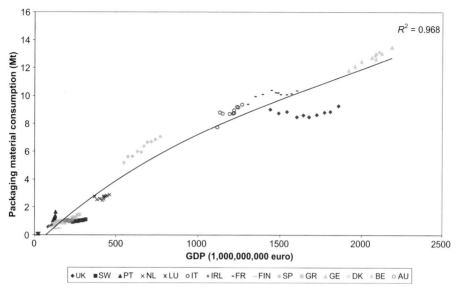

FIGURE 21.2 The relationship of packaging material consumption and national income for EU-15 countries in the period 1997–2006. *Rouw and Worrell (2011).*

(1997) distinguished weak dematerialization, which only implies a decline in the intensity of use, from strong dematerialization, which also includes a decline in absolute material use.

Income indeed has a strong correlation with packaging consumption in Europe. However, as Figures 21.2 and 21.3 depict, population seems to be a statistically slightly more robust

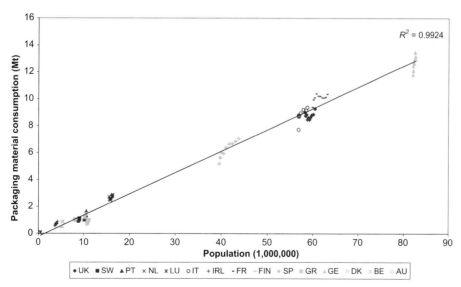

FIGURE 21.3 The relationship of packaging material consumption (tonnes) and population for EU-15 countries in the period 1997–2006. *Rouw and Worrell (2011).*

variable, given the comparable income levels of the EU-15 member states. The relationship between population and packaging consumption is linear for the EU-15 member states over the studied period. The relationship between GDP and packaging consumption shows leveling off over time—in other words, it demonstrates a certain degree of decoupling. This is most likely classified as weak dematerialization, as the total volume of packaging material is still increasing, even in the EU-15 countries. Other analyzed variables, such as GDP/capita, households, number of single households, and distribution of household sizes, show no statistically significant relationship with packaging consumption or packaging consumption per capita. Therefore, given the available data from the EU, there is no indication that single households consume more packaging materials per person than households with multiple people.

Despite the continuous growth in the use of material for packaging, large reductions are potentially feasible in the volume of packaging material (see e.g. Hekkert et al., 2000; Wever, 2009). These may even result in further cost savings along the total supply chain (Wever, 2009). Yet, little of the potential is currently realized, and typically only relatively minor opportunities are realized due to a large variety of factors (van Sluisveld and Worrell, 2013).

21.3 COMPOSITION

As consumption patterns change over time with economic developments, the composition of packaging waste will also vary. In recent decades, plastics have taken an increasingly larger share of the packaging market due to their versatility. Plastics are relatively easy to shape and can be more-or-less tailored to various packaging functions due to the use of barrier layers, composites, and additives. For example, more and more liquids are packed in plastic bottles (mainly polyethylene terephthalate (PET) and high-density polyethylene (HDPE)). In industrialized countries, plastics are forming an increasing part of packaging waste, despite its low weight. Moreover, national traditions in packaging and technology changes will affect the material composition of packaging and packaging waste. An analysis of the use of beverage containers in The Netherlands over time by Chappin et al. (2005) demonstrated the impact of changes in packaging on the material choice for this particular type of packaging, while statistical data on packaging waste composition collected in Europe demonstrate the overall changes in the volume and composition of total packaging waste. Traditions in packaging affect the composition as well. In contrast to most other countries, beverage cans in The Netherlands are mainly made of steel, not aluminum. This also explains why there is no separate collection system for aluminum cans in this country, while that is found in many other countries to retrieve the valuable metal. Also, national legislation affects the composition of packaging materials through the ban of certain materials due to environmental or health concerns.

Table 21.1 provides a rough breakdown of the typical composition of packaging waste as found in the EU and the United States. There are little data on the composition of

TABLE 21.1 Approximate Composition of Packaging Waste in the United States (2004) and the European Union (2010)

Material	United States (2004)	EU (2010)
Wood	4%	15%
Paper and cardboard	50%	40%
Plastics	26%	19%
Glass	16%	20%
Metals	4%	6%

Note that differences in the definition of packaging (e.g. inclusion of transport packaging like pallets) waste may affect the distribution.

packaging waste in developing countries, despite the growing use of packaging in these countries.

Plastics in packaging consist of multiple types that need to be separated before recycling. The key plastics used in packaging are PET (mainly for bottles, trays, and containers), HDPE (e.g. nonfood bottles), low-density polyethylene and linear low-density polyethylene (e.g. plastic bags, films), polypropylene (e.g. dairy and frozen foods packaging), and polystyrene (e.g. drinks, foams). PVC is still used in packaging, but its share is declining. The specific characteristics of PVC are only needed in niche applications, so the current use of PVC in packaging is generally driven by economics. Other plastics, including biobased plastics, are also used in lesser quantities. Note that even when the main constituent is a particular plastic type, the use of additives may make them different for recycling purposes. For example, PET containers (e.g. for fruit) should not be mixed with bottle-grade PET due to the use of additives in containers. Moreover, color may also necessitate the further separation of materials before recycling, such as with plastics and glass. This has resulted in more and more places in recovery systems that distinguish glass by color.

21.4 RECOVERY AND RECYCLING

Packaging waste is increasingly recovered for recycling. It is important to distinguish recovery and recycling when comparing the current achievements in packaging waste processing. Recovery is defined as the amount of material retrieved or recovered from the waste stream (both through separate collection and in waste separation plants) to replace (other) materials. Recycling is determined by the actual volume of material that is recycled to replace material (i.e. excluding the volume of material that is of too low quality to be recycled and is converted to refuse-derived fuel, incinerated, or landfilled). The point of measurement is also important in the latter because it can be measured at the entrance of the recycling facility or at the exit. Because incoming material may not meet quality demands or quality standards, it will be rejected in the process. Depending on the quality of the incoming material mix, the point of metering may hence considerably affect the reported recycling rate. Definitions may vary from one country to another, making comparisons sometimes difficult. Moreover, good measurements of the volume of packaging brought to market hardly exist, resulting in uncertainties in any recovery or recycling rates reported.

In the European Union, the recycling rate of packaging has increased from about 42% in 1997 to close to 63% in 2010 (measured by total weight), with the remainder being incinerated and landfilled. This makes the European Union one of the leaders in packaging recycling. Note that there is a large variation between the different member states and also between materials. Historically, metals, paper, and glass have achieved high recycling rates, building on long-term experience in many countries. In many Western European countries, the recovery and recycling rates of paper, glass, and steel are typically in the range of 70—80% (or even higher in selected cases). Aluminum recycling in a few selected countries (and states in the United States) approximates these high recycling rates, but is lower (typically 50% or less) in most countries. In contrast, recovery and recycling of plastics is in most countries still very limited. A few countries have a longer experience with plastic recycling (e.g. Germany) and they typically achieve higher recovery (over 80%) and recycling rates (around 50%), compared to an average recycling rate of 35% in the EU-27 (all figures are for 2011). In the United States, the U.S. Environmental Protection Agency

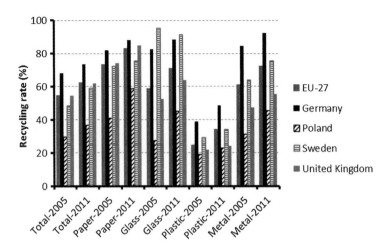

FIGURE 21.4 Recycling rate of packaging materials, plastic packaging, paper and board packaging, glass packaging, and metals packaging in the EU-27 and selected member states in 2005 and 2011. The member states are selected to illustrate the differences between countries. *Based on data from Eurostat.*

estimates the PET and HDPE bottle recycling rate at 29%, against 55% for aluminum and 70% for steel cans (all figures for 2011). Figure 21.4 depicts the recycling rates reported for the EU-27 and a few selected member states to illustrate the differences between countries and materials over time. Figure 21.4 clearly demonstrates the differences between countries for the recycling of different materials due to differences in history, markets, and policy success.

In many developing countries, there are little or no formal recycling schemes for packaging; however, there is a large informal sector of waste scavengers that collect, reuse, recover, and recycle packaging and packaging materials. The informal waste recovery sector often provides important income for low-income households. The large role of the informal sector makes it difficult to establish recovery and recycling rates for developing countries. Developing countries have taken initiatives to ban certain packaging to reduce the amount of waste in cities. For example, a growing number of developing countries now ban free plastic shopping bags in order to reduce plastic litter.

21.5 RECOVERY AND COLLECTION SCHEMES

Due to the relatively large share of packaging waste in municipal solid waste (MSW), packaging has always received a lot of attention in attempts to reduce waste. Policies and instruments have been introduced at all levels of government, varying from local initiatives by cities and towns, to national governments, and the organizations like the European Union. In contrast, the United States has no federal policy on packaging. Consequently, large differences in policies are found throughout the various states. However, at the same time that governments have actively pursued options to reduce packaging waste, the packaging industry introduces thousands of new packages a year, of which the majority without a lot of attention of the environmental consequences of the packaging in the waste stage (see e.g. van Sluisveld and Worrell, 2013). A difficulty in realizing opportunities for waste reduction and material efficiency is the relationship between the packaging and the packed product. If a change in packaging might result in damage to the product, the environmental impact may be higher.

However, little empirical research has studied the role of packaging, product damage, and user behavior in depth.

To attempt to decrease packaging waste, many governments in industrialized countries have reverted to extended producer responsibility (EPR) schemes, albeit with varying success. Packaging waste policy typically approaches the problem from a waste management perspective. This is slowly changing toward a more holistic system that includes a waste hierarchy. The hierarchy starts with prevention of waste or material efficiency (see also Chapter 30 and 33), followed by product reuse, recycling, incineration with energy recovery, and incineration/landfilling. In contrast to the hierarchy, we see that in many EPR schemes, prevention and reuse receive little or no attention. Most EPR schemes today seem to be focused on cost reduction in waste management through collective collection and recovery schemes. Table 21.2 summarizes selected examples of recovery and recycling schemes as found in many industrialized countries.

Reuse. Reuse of the packaging itself may result in higher environmental benefits, as the packaging does not need to be produced; instead, it is cleaned, rinsed, or refurbished to be used again. Collection cost and product design requirements for reuse may partially offset these gains. Typically, a refund system is used to encourage recovery. The best known example in many countries is the deposit system for bottles and crates. The bottles are collected at the point of sale, transported to the manufacturer (or industry), cleaned, and refilled. There are small variations in the system. For example, in The Netherlands, most common beer bottles are of the same design, allowing the bottles to be reused by most breweries. The recovery rates of deposit systems are generally very high (over 90%) and limit the contamination with other packaging materials. An analysis of beverage containers in The Netherlands over the period 1984—1999 has shown that reuse, next to material recycling, had the most impact on reducing waste (Chappin et al., 2005). A less visible system is the pooling of pallets by transport companies in Europe. The wooden pallets have a uniform design and are pooled by all participants in the system. They can be refurbished when damaged. Pallet and transportation companies are part of the pooling system for storage, refurbishing, and replacement. In other sectors (e.g. moving companies, business-to-business) crates and boxes are also reused, sometimes combined with a deposit system.

Separate collection. Of the recovery systems for recycling, separate collection can guarantee a higher quality of the collected material, limiting contamination with other materials. Historically, separate collection has been used for paper and glass for decades in industrialized countries, which has resulted in high recovery and recycling rates. Separate collection systems can be classified in different ways. Systems can be

TABLE 21.2 Selected Packaging Recovery and Recycling Schemes in Industrialized Countries

Type	Mechanism	Examples
Reuse	Deposit fee	Bottles (e.g. Germany, The Netherlands) Pallets (e.g. Europe) Boxes and crates (various)
Separate collection for recycling	Refund/deposit fee	Bottles and cans (e.g. California and other states in the United States) PET bottles (The Netherlands)
	Voluntary	Glass, paper (The Netherlands) PET bottles (France) Plastic bottles (Belgium)
Co-mingled/mixed collection for recycling	Fee-based	DSD system (Germany)
	Voluntary	UK (varies by city)

based on a refund or deposit fee system or based on voluntary action of the user. Voluntary systems can be separated into curbside collection and drop-off systems, where the user brings it to a given location for collection.

In packaging, *refund or deposit fees* have been used for a long time for reuse systems (see above) but also for collecting recyclables. For years, a large number of states in the United States have charged a small fee per bottle or can (typically between 3 and 5 US cents). The fee is returned when the container is returned at specific collection points. The refund system in the state of Michigan has been monitored for over 20 years and has shown on average collection rates of over 93%. The fees collected for remaining containers that were not returned covered the administrative costs of the system. Similar systems are found in other states of the United States, as well as in European countries. The Duales System Deutschland (DSD) system in Germany and some other systems charge fees based on the packaging material of manufacturers, but this fee is not refundable.

Voluntary systems build on the behavior of users and consumers to separate the recyclable materials from other waste. *Curbside systems* do collect the materials or packages at home at a regular schedule. Households have special containers or bags to separate and store the materials until collection. The containers can separate different materials or packages, or combine various materials in one. The latter is co-mingled or mixed packaging waste and is discussed below. In some countries and communities, part of the packaging waste is collected this way. For example, in The Netherlands, typically paper and paperboard packaging is collected separately with other paper, while some communities collect plastic packaging in special bags from homes. The alternative for a curbside system is a *drop-off system*. For example, for selected packaging, including glass bottles and plastics (and in some communities, metal cans) are collected in containers, distributed through neighborhoods and near shopping centers. In The Netherlands, glass is separated by color (i.e. clear, amber, and green) to allow for optimal recycling. Similar combinations of systems can be found in other countries. As evidenced by the high recovery rates of some of these materials, separate collection systems can result in effective and efficient recovery and recycling systems, generating relatively pure and clean recycled material streams. In the case of glass and plastics, separation of the various types can help to improve the recyclability of the collected material.

Mixed Collection. In some countries and/or urban communities, packaging waste is collected as a mix. The different systems vary, depending on the types of packaging and materials collected. In Germany, all packaging material consumption is part of the EPR-based DSD system. For the different types of materials and the volumes used, the packagers and producers pay a fee to the DSD system (recognizable by a green dot) for packaging sold in Germany. In the German system, paper is collected separately, but other packaging is collected as a mix in a separate container. The mix is treated in postcollection recovery centers to separate the different materials (including key plastics). In the UK, several communities collect all recyclables combined, including paper. While this potentially reduces the costs of collection and recovery, it can result in reduced quality of the recovered material. In particular, recovered paper is sensitive to contamination from food wastes, resulting in lower quality recyclable paper and fibers (Miranda et al., 2013). This will result in a lower recycling rate, despite a high recovery rate. Hence, a mixed collection system needs to be carefully designed to maximize recovery *and* recycling.

The economics of packaging recycling are not different than those of recycling in general (see Chapter 20). Due to the importance of the volume of packaging in waste, there have been a lot of systems in place in different countries to collect and recycle packaging materials. The economics are influenced by the costs and the benefits. Note that these may vary between different stakeholders, dependent on the financial organization of a recycling scheme (see e.g. Ferreira da Cruz et al., 2012 for an analysis of the packaging waste collection system in Portugal). The overall costs are also affected by the volume and quality of the recyclables collected. The costs are affected by the operation of the collection system, transport, separation (if necessary), and depreciation of capital equipment for each of the steps. The benefits are the results of reduced collection costs of MSW, avoided waste treatment (i.e. incineration, landfilling), and the value of the recyclable material (which is affected by markets and quality of the recyclable material).

21.6 CONCLUDING REMARKS

Packaging is the key factor in the volume and composition of municipal solid waste in most in most industrialized countries. Packaging material consumption is still increasing in industrialized countries, and even rapidly increasing in developing countries, despite policy attempts to limit the volume of packaging material and ban specific materials or specific types of packaging. Packaging policy is still driven by traditional waste management policy concerns, and only limited attempts have been taken to come to a more holistic approach to manage packaging material use, recycling, and waste. Policies in most countries have focused on increasing recycling of packaging material, which has been successful for various types of packaging and materials. The overall environmental and economic impacts of this focus are generally positive and may vary due to a variety of factors. In selected cases, however, the focus may have shifted away from more environmentally friendly packaging or product concepts.

Future developments will also affect the volume, composition, and recyclability of packaging, as they have done in the past. An important development that may have an impact on paper and cardboard use is the emergence and increasing importance of Internet shopping by consumers. Simultaneously, we witness emerging initiatives for more holistic packaging policies in selected countries, and increasing attempts to reduce the volume of packaging waste in the environment, in industrialized and developing countries alike.

References

Chappin, M.M.H., Hekkert, M.P., van Duin, R., 2005. Decomposition analysis of Dutch beverage packaging waste: an analysis of material efficient innovations. Resources, Conservation & Recycling 43, 209–229.

De Bruyn, S.M., Opschoor, J.B., 1997. Developments in the throughput-income relationship: theoretical and empirical observations. Ecological Economics 20, 255–268.

Dikgang, J., Leiman, A., Visser, M., 2012. Analysis of the plastic-bag levy in South Africa. Resources, Conservation & Recycling 66, 59–65.

Ferreira da Cruz, N., Simões, P., Marques, R.C., 2012. Economic cost recovery in the recycling of packaging waste: the case of Portugal. Journal of Cleaner Production 37, 8–18.

Hekkert, M.P., Joosten, L.A.J., Worrell, E., Turkenburg, W.C., 2000. Reduction of CO_2 emissions by improved management of material and product use: the case of primary packaging. Resources, Conservation & Recycling 29, 33–64.

Miranda, R., Monte, M.C., Blanco, A., 2013. Analysis of the quality of the recovered paper from commingled collection systems. Resources, Conservation & Recycling 72, 60–66.

Rouw, M., Worrell, E., 2011. Evaluating the impacts of packaging policy in The Netherlands. Resources, Conservation & Recycling 55, 483–492.

van Sluisveld, M.A.E., Worrell, E., 2013. The paradox of packaging optimization — a characterization of packaging source reduction in The Netherlands. Resources, Conservation & Recycling 73, 133–142.

Wever, R., 2009. Thinking About the Box: A Holistic Design Engineering Approach of Packaging for Durable Consumer Goods (Ph.D. thesis). Delft University of Technology, Delft, The Netherlands.

Further Reading

European Environment Agency Report, 2005. Effectiveness of Packaging Waste Management Systems in Selected Countries: An EEA Pilot Study. Copenhagen, Denmark.

Jedlička, W., 2009. Packaging Sustainability: Tools, Systems and Strategies for Innovative Package Design. Wiley, Hoboken, New Jersey, USA.

Lopez-Delgado, A., Pena, C., Lopez en, V., Lopez, F.A., 2003. Quality of ferrous scrap from MSW incinerators: a case study of Spain. Resources, Conservation & Recycling 40, 39—51.

CHAPTER 22

Material-Centric (Aluminum and Copper) and Product-Centric (Cars, WEEE, TV, Lamps, Batteries, Catalysts) Recycling and DfR Rules

Antoinette van Schaik[1], Markus A. Reuter[2,3]

[1]MARAS—Material Recycling and Sustainability, The Hague, The Netherlands; [2]Outotec Oyj, Espoo, Finland; [3]Aalto University, Espoo, Finland

22.1 INTRODUCTION

This chapter discusses recycling of various metals both in a material- and product-centric context, as this provides a good basis to discuss the various intricacies in detail and in context. As aluminum and copper are integral part of numerous products, their recycling will be discussed in a product centric context rather than the usual material-centric context, which is well documented in various texts. Therefore only a brief material-centric context is discussed with suitable references to other texts in this regard. Further on, the importance and effect of copper and aluminum on recycling in general are discussed in a product-centric context of complex sustainability enabling and consumer products.

22.2 MATERIAL-CENTRIC RECYCLING: ALUMINUM AND COPPER

22.2.1 Aluminum Recycling

Secondary recovery (recycling) is a critical component of the aluminum industry based on its favorable economic contribution to production and the reduced environmental impact compared to primary aluminum production. The aluminum industry is doing much to maximize the recycling of the various aluminum-containing metal alloys and other materials recovered from various EoL (End-of-Life) products (e.g. packaging, building, and automotive products).

Aluminum, like as any metal, can be recycled repeatedly without loss of properties, unless obviously the alloying components and impurities that are inevitably picked up and dissolved during its many cycles are maintained within the limits of the different alloy types. The high value of aluminum scrap is a key incentive for recycling. Such recycling benefits present and future generations by conserving energy and other natural resources. It is well documented that recycling of aluminum saves about 95% of the energy required as compared to primary aluminum production, avoiding corresponding emissions including greenhouse gases. Industry continues to recycle, without subsidy, all of the aluminum collected as scrap or from fabrication and manufacturing processes (http://www.world-aluminium.org/publications). The world average of recycling aluminum is 27%; with the highest figure for the United Kingdom (57.3%) (Figure 22.1). However, when considering only aluminum beverage cans, Brazil has been the leader for 10 consecutive years (Figure 22.2).

Aluminum cans are important in recycling because they have a high volume and a much shorter life-cycle than other aluminum products. In 2010, with a recycling rate of 97.6%, Brazil again beat the world record with 239,100 t of scrap cans (17.7×10^9 units, 48,500,000 cans/day, 2 million/hour). Currently, it takes about 30 days to buy, use, collect, recycle, remanufacture, refill, and return an aluminum beverage can to the shelves.

High-quality and pure scrap is remelted to produce wrought aluminum qualities. Refining under a salt (NaCl—KCl) slag is used for poorer scrap qualities. Furnace types include flux-free melting furnaces, tiltable and fixed-axle rotary as well associated salt slag processing when salt slag is used for the more unclean scrap types. Clean scrap of a defined quality that is massive, little oxidized and can contain a volatile component such as organic materials, is usually processed by a remelter under flux-free conditions to be once again fed into wrought aluminum production. Depending on oxidation

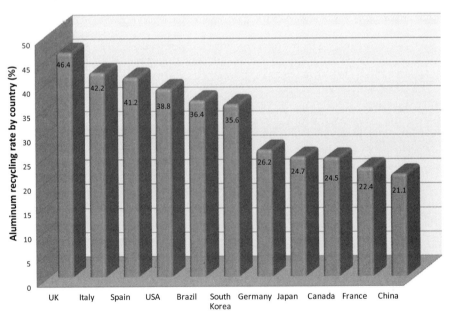

FIGURE 22.1 Ratio of scrap recovered and domestic consumption (%), 2009 (the Aluminum Association; www.abal.org.br).

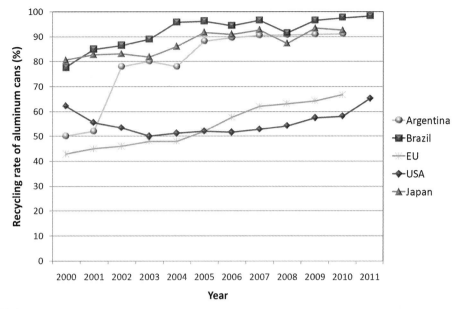

FIGURE 22.2 Recycling rates for beverage cans. *Sources: Brazilian Aluminum Association, www.abal.org.br, which collected data from the Brazilian Association of Highly Recyclable Cans, the Japan Aluminum Can Recycling Association, Camara Argentina de la Industria del Aluminio y Metales Afines, the Aluminum Association, and European Aluminium Association.*

conditions as well as surface area turnings, mixed scrap oils, coatings and the like, processing either flux free or refining under salt slag is required with subsequent alloying to achieve mostly suitable cast alloy types. If well defined, dross and other oxidic material finds its way back to the steel, cement and other industries, as discussed by Boin and Bertram (2005) and Bertram et al. (2009). Please also consult these two references for a good overview of aluminum recycling, as well as texts such as *Ullmann's Encyclopedia of Industrial Chemistry* (Wiley).

As mentioned, medium to very oxidized and dirty aluminum scrap are processed often in rotary melting furnaces, where the scraps are heated by the combustion natural gas, melted and refined under a top salt slag layer. As far as the energy and salt consumption, metal recovery and environmental impact are concerned, the melting efficiency of the aluminum scrap is a critical issue in the secondary aluminum industry (Figure 22.3).

Even with simple aluminum scrap, intimate connections to other materials can have a marked effect on recycling. The right side of Figure 22.4 shows smelting results from two scraps, A and B (on left). Metal recovery from scrap A is 84.3% by weight, but that from scrap B is 95.3%. Surface area is a major factor that determines metal recovery, but organic and other coatings and materials attached to the aluminum can create compounds, such as sulfides (e.g. Al_2S_3), phosphides (e.g. AlP), hydrides (e.g. AlH_3) or carbides (e.g. Al_4C_3), which result in losses and other issues. The losses result from the fact that the compounds are not aluminum and collect in the salt-slag phase; other issues are when these compounds react with, e.g. water or moisture in the air to form $H_2S(g)$, $PH_3(g)$, $H_2(g)$ and $CH_4(g)$, respectively, and Al_2O_3 (also lost), which are toxic, explosive or combustible. Hence, recovery is much affected by the purity and morphology of scrap. Figure 22.4 shows this effect of scrap cleanliness and surface area on the recovery and

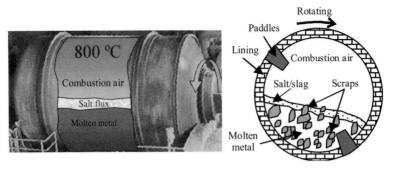

FIGURE 22.3 Illustration of the rotary melting furnace and the various main phases and phenomena inside the furnace.

FIGURE 22.4 The effect of scrap surface area (left) on recovery (right) of two scrap types. Top: AlSi9 alloy and the flux/metal after, which creates a clean melt and salt slag. Bottom: Melting recovery is held back by inactive alumina coating scrap, so scrap with a higher surface area/volume ratio has a lower recovery rate, creating under static melting conditions various droplets in slag (Xiao and Reuter, 2002).

losses. In addition, losses occur due to metals and materials linked to the aluminum, which cannot be removed and hence require after-refining dilution with high-purity aluminum, as these dissolve and/or form intermetallic compounds in the aluminum.

Even pure aluminum remelting as shown at the top of Figure 22.4 has oxidative losses due to oxidation and hence lowers the recovery and recycling rate of aluminum. These issues will be discussed in more detail later in the chapter (UNEP, 2013).

22.2.2 Copper Recycling

An integrated copper smelter can accept a variety of copper scrap types and recover various contained metals economically. A large fraction of relatively "pure" copper scrap can be accommodated in copper converters and anode furnaces (Figure 22.5), while waste electrical and electronic equipment (WEEE) and lower-grade scrap can be processed in a Kaldo or a top-submerged lance (TSL; www.outotec.com) furnace as shown in Figure 22.7 (UNEP, 2013). The amount of refined copper originating from secondary copper sources is given in (Figure 22.6).

UNEP (2013) shows various flowsheets that are well suited for recovering the many elements associated with copper-containing scrap and residue materials, and this in an economically feasible and environmentally friendly way, while still recovering many of the minor elements associated with copper-input materials. Usually a combination of pyro- and hydrometallurgy can extend the limits of recycling, thus reflecting the concept of carrier metal metallurgy, i.e. a segment of the Metal Wheel shown in Figure 22.18.

Usually pyrometallurgy distributes the contained metals and compounds into various phases such as copper, matter, speiss, slag, flue dust, and offgas, while the recovery of energy is possible. Hydrometallurgical processes can extract various valuable elements, such as bismuth, gold, silver, platinum group metals (PGMs) (iridium, rhodium, ruthenium, osmium, palladium and platinum), nickel and cobalt through a variety of processes including leaching, solvent extraction, electrorefining, electrowinning, precipitation and so forth. In summary, WEEE, PGM and metal-containing catalysts as well as other complex recycled materials require copper metallurgy and its deep knowledge.

Various texts exist, such as *Ullmann's Encyclopedia of Industrial Chemistry*, where more detail can be obtained. In the next section the intricacies and the relationship between product design, collection, physical separation and extractive metallurgy are discussed in a product-centric manner.

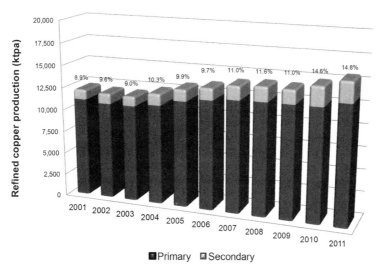

FIGURE 22.5 Refined copper production from primary smelting with the proportion of secondary copper superimposed on the primary production (International Copper Study Group; www.icsg.org).

	2007	2008	2009	2010	2011
World Mine Production	15,516	15,569	15,943	16,053	16,076
World Mine Capacity	17,900	18,551	19,254	19,560	19,824
Mine Capacity Utilization (%)	86.7	83.9	82.8	82.1	81.1
Primary Refined Production	15,165	15,391	15,407	15,732	16,126
Secondary Refined Production	2,738	2,823	2,841	3,250	3,470
World Refined Production (Secondary+Primary)	17,903	18,214	18,248	18,981	19,596
Seasonally Adjusted Refined Production - monthly [1]					
World Refinery Capacity	21,787	22,588	23,457	23,839	24,385
Refineries Capacity Utilization (%)	82.2	80.6	77.8	79.6	80.4
Secondary Refined as % in Total Refined Prod.	15.3	15.5	15.6	17.1	17.7
World Refined Usage [2]	18,196	18,053	18,070	19,346	19,830

FIGURE 22.6 Summary of primary and secondary refined production (note that Figure 22.5 shows the secondary production within the primary refined production data) (International Copper Study Group; www.icsg.org).

FIGURE 22.7 Top-submerged lance (TSL) and Kaldo technology for the recycling of copper and also e-waste and similar copper-containing materials (www.outotec.com).

22.3 PRODUCT-CENTRIC RECYCLING: COMPLEX SUSTAINABILITY ENABLING AND CONSUMER PRODUCTS

Products, Quantities and (Critical) Materials in EoL Consumer Products

This section gives a brief overview of products, product types and some of the compositions of various EoL consumer products, i.e. the "mineralogies" of these products. The dynamics of EoL developments are discussed, as well as how varying properties such as weight, composition and construction are affecting recycling performance and material recovery potential of the so-called "urban mine" over time. Knowledge of the time-varying mineralogy of the products in the urban mine, being the input of the recycling system, is vital to capture, understand and improve recycling and resource efficiency of these products and components and the contained materials, compounds and (minor/scarce) elements.

22.3.1 Dynamics of EoL Developments and the Urban Mine

The availability, weight, number and type of products (end-of-life vehicles (ELV), WEEE, batteries, catalysts, etc.) and the material composition from these products constitute the urban mine that is a source for resource recovery. The technological and economic potential to recover the resources from this modern mine, consisting of a wide variety of different and time-changing products and materials, is hence determined by the end-of-life vehicle and WEEE developments and their weight, composition and component content (e.g. printed wire boards containing critical resources, batteries, catalysts, etc.) for different products within these product groups. These will all change over time as a function of factors such as:

- Distribution of EoL products over time as a function of lifetime (usage) distributions of products (see Figure 22.8)
- Changing product design and product weight (see Figure 22.9)
- Product composition (e.g. as a function of legislative restrictions such as the ban on the use of chlorofluorocarbon coolants in refrigerators, lead-free soldering, changing technology, consumer demands, lightweight designs, energy demands in design, new vehicle concepts such as the electric vehicle, etc.) (see Figure 22.10 for an example on the inert content of WEEE products)
- Consumer behavior
- Disposal behavior and stocks (van Schaik and Reuter, 2004)

All of these factors determine the dynamics of material availability in the urban mine and hence the dynamic potential to recover materials from it. Recycling and recovery for the different materials will hence change as a function of time. Predicting the material recovery from EoL developments (see Figure 22.11) demands insight on all of the above-listed points and requires a dynamic modeling approach capable of predicting EoL dynamic distributions of product quantities, weights and composition (and construction) in relation to the dynamic (time-varying) material recovery from it. Such mathematical and recycling physics dynamic modeling is at the core of driving changes and controlling and optimizing resource efficiency over time. It will be topic of this chapter to discuss these models that can predict the product- and time-varying recycling performance for different EoL systems and mixtures of products, recovery of (precious/scarce) materials and leakage for minor elements for different (changing) plant configurations (including dismantling), shredder settings, as well changing designs and recycling trends into the future.

Predicting and understanding the potential of resource efficiency from the urban mine not only requires detailed knowledge on the

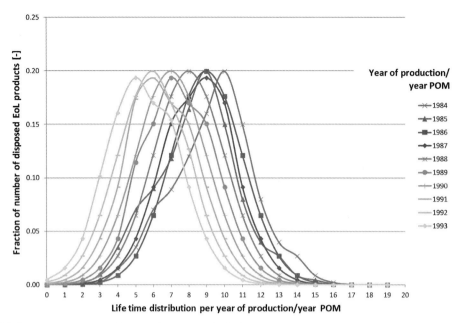

FIGURE 22.8 Age distribution of WEEE fitted as Weibull function (2D and 3D representation) (values and years are indicative and defined for illustrative purposes). (POM, put on market).

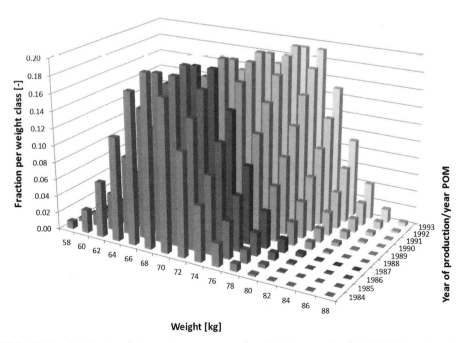

FIGURE 22.9 Weight distribution of electronic consumer product (for the example of an LHHA) as a function of year put on market. It must be noted that years and values for content are indicative for illustrative purposes. Sound data collection should be carried out in such way that these distribution functions can be defined. (LHHA, Large Household Appliances).

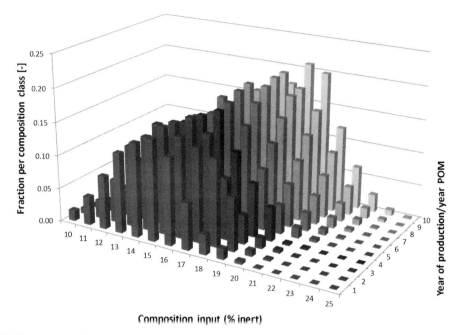

FIGURE 22.10 Example of time-varying distribution of composition of a WEEE product stream for the example of inert content (these graphs equally apply for any EoL product and material composition). It must be noted that years and values for content are indicative for illustrative purposes. Sound data collection should be carried out in such way that these distribution functions can be defined.

total flow of materials and recyclates (van Schaik and Reuter, 2004) but also requires these data as (statistical) standard-deviation and average-distribution values for each flow. The shape of this distribution (Normal, Weibull, or even multimodal) must be known as well and how these change over time. In addition, as the materials/metals contained in recyclates are the economic target, their average and standard-deviation distributions in the input recyclates should be known as well. A rigorous mathematical basis on how to combine these different distribution functions of quantities and weights of products with changing composition distribution function is required to estimate the urban mine and its material content over time, as illustrated by van Schaik and Reuter (2004) for ELVs; the methodology applied is similarly applicable for any other EoL product flow and its content. This depth and detail is not reached by many such studies, even if they are applied and accepted widely. Hence care should be taken when using derived data from these types of studies, which do not reflect and contain the depth of knowledge on the recycling system to justify its usage for prediction of resource recovery potential over time.

22.3.2 Requirements for Data Collection of EoL Streams, Products and the "Urban Mine"

Such understanding of data as described in the previous section (see also van Schaik and

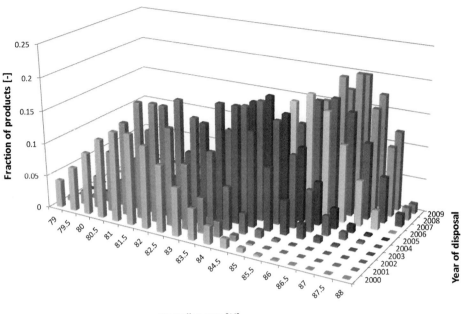

FIGURE 22.11 Example of dynamic recycling performance calculations of an e-waste product as a function of various years, also showing the distribution of recycling rates due to various distributed properties such as material liberation, complex interlinked materials in products due to functionality, distribution of quality of recyclates, distribution of analyses (all of a mean and standard deviation), range of products and changing designs, to name a few. (Values are indicative but can be derived by the dynamic simulation models based on detailed data collection on the urban mine products and EoL developments.)

Reuter, 2004) within a context of physical and metallurgical recycling technology and its economics, permits the calibration of detailed mass-balance models (van Schaik and Reuter, 2004, 2010) to estimate recycling rates as a function of time-changing EoL distribution. It also facilitates the construction of detailed Design for Resource Efficiency models for maximizing resource efficiency. Such theoretical knowledge is common practice in standard mineral processing and extractive metallurgy, used daily for optimizing metal recovery at all levels of the business. While the theoretical basis was developed for dealing with physical recycling, it is not generally applied. To advance recycling, the theoretical basis of physical recycling will have to be applied in industry, in order to match the sophistication of classic mineral processing that now is on a par with metallurgical processing.

23.3.3 Quantifying Use and Materials in EoL Automotive Applications

Continuous product development affects its design, components and weight. This, in turn, changes its composition and volume for waste-recycling streams, as discussed above and illustrated for end-of-life vehicles (ELVs) by Table 22.1. This increased use of many compounds, complexly bound together for functional reasons, dictates the physical and chemical properties of the materials available for recycling and hence the recycling rate. In

TABLE 22.1 The Average Composition of the Car (Nemry et al., 2008) and Changing Aluminum Content (International Aluminium Institute, 2007) to Mitigate the Weight Gain of Vehicles

Materials (kg)	Petrol	Diesel	Materials (kg)	Petrol	Diesel
Total content of ferrous and nonferrous metals	819	1040	Carbon black	2	2
Steel BOF	500	633	Steel	1	1
Steel EAF	242	326	Textile	0.4	0.4
Total content of iron and steel	742	959	Zinc oxide	0.1	0.1
Aluminum primary	42	43	Sulfur	0.1	0.1
Aluminum secondary	26	29	Additives	1	1
Total content of aluminum	68	72	Subtotal (4 units)	31	31
Cu	9	9	Battery		
Mg	0.5	0.5	Lead	9	9
Pt	0.001	0.001	PP	0.7	0.7
Pd	0.0003	0.0003	Sulfuric acid	4	4
Rh	0.0002	0.0002	PVC	0.3	0.3
Glass	40	40	Subtotal (4 units)	14	14
Paint	36	36	Fluids		
Total contents of plastics			Transmission fluid	7	7
PP	114	114	Engine coolant	12	12
PE	37	37	Engine oil	3	3
PU	30	30	Petrol/diesel	23	25
ABS	9	9	Brake fluid	1	1
PA	6	6	Refrigerant	0.9	0.9
PET	4	4	Water	2	2
Other	27	27	Windscreen cleaning agent	0.5	0.5
Miscellaneous (textile, etc.)	23	23	Subtotal	50	50
Tyres			Total weight	1240	1463
Rubber	4	4			

ABS = Acrylonitrile butadiene styrene; BOF = Basic Oxygen Furnace; EAF = Electric Arc Furnace; PP = Polypropylene; PE = Polyethylene; PU = Polyurethane; PA = Polyamide; PET = Polyethylene terephthalate; PVC = Polyvinyl chloride.

addition, the combining of such different materials also changes from simple rivets or nuts and bolts, to welding, gluing and molecular deposition, dictated by functionality and production cost affects the resource recovery from these products.

The average weight of a vehicle in the European Union has increased substantially over the years, as reflected by Table 22.1 (International Aluminium Institute, 2007) and as documented by van Schaik and Reuter (2004), from 856 kg in 1981 to 1207 kg in 2001. According to the International Aluminium Institute (2007), for example, the average weight of a VW Golf has increased from around 760 kg (Mk1) in 1974 to almost 1350 kg in 2002 (Mk5), for a Toyota Corolla from around 800 kg in 1970 to over 1300 kg in 2002 and for an Astra (Mk1) from around 800 kg in 1979 to almost 1150 kg (Mk5) in 2004. The same trend applies to the different models of the other car brands as well. Table 22.1 provides detail on the composition of an average vehicle and the changing aluminum composition of the car over time (International Aluminium Institute, 2007).

The recycling of ELVs already plays an important role in closing resource cycles for a variety of (critical and commodity) materials. Yet the modern car is ever more complex, with increasing electronics and other gadgets that render it an extremely complex product. This will become even more true for an electric car—a range of "critical" materials including rare earths elements (REEs) in EV batteries, hybrid electric motors and generators, LCD screens, glass (UV cut glass and glass and mirror polishing powders), catalytic convertors and diesel fuel additives and platinum group metals (PGMs) in fuel cells are planned to be needed in vehicles (Lynas Corporation, 2011).

22.3.4 Quantifying Products and Materials in Electronics

The recycling of e-waste involves a wide spectrum products (see Figure 22.12) and

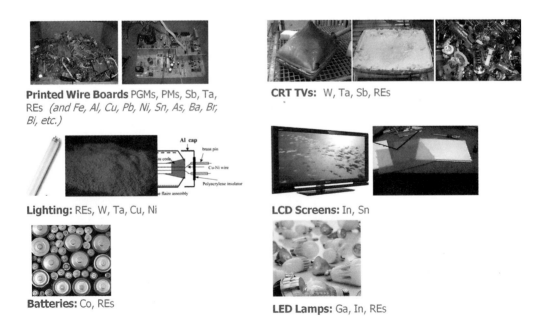

Printed Wire Boards PGMs, PMs, Sb, Ta, REs *(and Fe, Al, Cu, Pb, Ni, Sn, As, Ba, Br, Bi, etc.)*

CRT TVs: W, Ta, Sb, REs

Lighting: REs, W, Ta, Cu, Ni

LCD Screens: In, Sn

Batteries: Co, REs

LED Lamps: Ga, In, REs

FIGURE 22.12 Different types of WEEE with major critical material content.

materials, ranging from commodity metals and different plastic types (e.g. containing flame retardants), to precious metals as well as harmful minor elements. Crucial in all this is the control and prediction of recyclate quality as a function of (time-varying) product type and product design affecting the efficiency of the recycling processes applied (both physical and metallurgical/thermal) to recover the material and/or energy content of the product. van Schaik (2011) has determined, based on (publicly) available literature and data sources, the band width of the average critical material content present in a range of WEEE products. This is summarized in Table 22.2. van Schaik (2011) and UNEP (2013) give an extended overview of (critical) material content for the listed products from Table 22.2.

22.3.5 Material Composition and Quantities of Various Lighting Applications

Lamps and light emitting diodes consist of a variety of materials, including elements such as rare earth oxides (yttrium, lanthanum, cerium, europium, terbium and gadolinium) (USGS, 2002, 2010), mainly present in the fluorescent/phosphorus powders, as well as other substances such as indium, gallium and tungsten (Van den Hoek et al., 2010), which have among others been pinpointed as scarce/critical materials for Europe in a document produced by the European Union (EU, 2010). Fluorescent/phosphorus powders are mostly applied in fluorescent lamps (including TL) and LED lighting. Germanium can occur in the fluorescent powder used in high-pressure mercury lamps. Indium can be applied as coating in (low-pressure) sodium lamps. In addition, tungsten and tantalum can be present in lamps. In summary, the following materials and compounds may be present (Van Hoek et al., 2010; Franz et al., 2010; McGill, 2010):

- Metals (W including K, ThO_2, Re) (for incandescent filament and electrodes); Mo, Fe, Ni, Cr, Cu, Nb, Na, Hg, Zr, Al, Ba, Ni, Ta, Al and Zn (for seals, caps, getters, construction).
- Envelope materials, which include glasses, fused silica and translucent polycrystalline alumina (SiO_2, Na_2O, K_2O, B_2O_3, Al_2O_3, MgO, CaO, BaO, PbO, SrO).
- The fluorescent powders (also called phosphors) generally consist of a host lattice (e.g. $Ca_5(PO_4)_3(Cl,F)$, $BaMgAl_{10}O_{17}$, Y_2O_5, $LaPO_4$, $Y_3Al_5O_{12}$, YVO_5, etc.) doped by a few percent of an activator, which imparts the luminescent properties. The activators can be the metal ions (e.g. $Pb^{2+}/Mn^{2+}/Sb^{3+}$) or rare earth (e.g. $Eu^{2+}/Tb^{3+}/Ce^{3+}$).
- LEDs contain AlGaInP; InGaN; InGaN + (YAG:Ce); and InGaN + phosphor.
- Special materials such as getters (including W), amalgams, emitters, paints, cements and marking materials.

Various glass types are used in lamps. For fluorescent tubes, normal soda lime glass is sufficient, but for higher wall loading special soda, lead and barium glass are used. The difference in glass types, in combination with the different material mixtures applied in different lighting types, demands a well-considered sorting and/or mixing of lamps types in order to avoid cross-contamination of different glass types (lowering the quality of the glass fraction and limiting the applications afterward) and other materials. This will be discussed in more detail in the next section. While the above refers more generally to lamps, this chapter focuses on the recycling of fluorescent lamps and light emitting diodes (LEDs).

22.3.5.1 Fluorescent Lamps

The general construction of fluorescent lamps is generally a glass tube of between 30 and

TABLE 22.2 Average Amounts of "Critical" Materials in Different e-Waste Products Presented with Bandwidths (*Based on various literature sources*) (gram/ton equipment).[1] (van Schaik, 2011)

Quantites (gram/ton Equipment)	PMs			PGMs		Rare Earths (Oxides)					Other			
	Ag	Au	Pd	Pt	Y	Eu	Other REs	Sb	Co	In	Ga	W	Ta
Washing machine	0.59–0.64	0.14–0.15	0.07–0.08										
LHHA (excl. Fridges)	0.00–0.54	0.00–0.13	0.00–0.07										
Video recorder	67–94	3.1–4.3	1.0–1.4				*	*					*
DVD player	70–113	10–16	2.1–3.4				*	*					*
Hi-fi unit	54–71	2.5–3.3	0.8–1.1					*					*
Radio set	104–107	13.6–13.9	1.6–1.7				*						*
CRT TV	8.4–155	0.51–11	0.3–4.0		16–19	1.3–2.0	*	216	8				*
Mobile telephone	786–2440	81–800	63–610	15–36	*			*	19,289–45,509		*		*
Fluorescent lamps					1514–16,245							*	
LED					*					*	*	*	
LCD screen	*	*	*							+			
Battery (NiMH)					~80,000				~30,000				

[1] Not intended to be comprehensive for all "critical" materials due to partially lacking or limited availability of data in public sources.
(*): possibly present, not (enough) quantitative data available.
(+): 0.06–5.6 g in/m².
PM = precious metals; LHHA = Large Household Appliances.

120 cm in length and a diameter between approximately 22 and 30 mm. There are various metal, alloys and compounds that can be recovered—e.g. the white powder is the phosphor that contains the RE(O)s, tungsten filament, brass pins, Cu-Ni wires, polyacrylene insulator, aluminum cap, the glass tube, etc. Fluorescent lamps occur in many different shapes and sizes (e.g. as energy-saving fluorescent light bulbs), which will however often be processed in separate recycling routes due to differing compositions and complexity. Mercury content requires well-controlled suction through totally enclosed systems.

The various fluorescent phosphor powders can be recovered from the inside of the glass of the fluorescent tube, which could, depending on the lamp recyclate collection, be a mixture of the various compounds as summarized above and in Table 22.8. Also given above are the compositions of LEDs, which should be collected separately. This is also advisable for the phosphors to ensure optimal recovery of valuable materials.

22.3.5.2 Light Emitting Diodes

Light emitting diodes (LEDs) are basically p-n junctions acting as diodes. Electrons flowing into the p-type region and filling holes release energy in terms of photons. Materials for this are presently InGaN (blue light) and AlGaInP. To produce white light a combination of the blue LED with the popular $Y_3Al_5O_{12}:Ce^{3+}$ (YAG:Ce) phosphor produces white light, while red or red/orange phosphors $(Ca,Sr)AlSiN_3:Eu^{2+}$ produces color in the 2700–4000 K temperature range (Van Hoek et al., 2010). LED technology is applied in e.g. retrofit lighting applications. These lamps consist in addition to the LED emitting materials of constructional and functional materials such as aluminum applied the heat sink (also plastic-based retrofits are applied), PCBAs (printed circuit board assemblies) for both the LED and driver PCBA, and a range of other materials applied in the housing, socket, lens, for potting material, and so forth.

22.3.5.3 Quantities of Lamps and Application of (Critical) Materials in Lighting

van Schaik (2011) made an estimate of the range (bandwidth) of the amount of critical materials (in this case REOs) as applied and discarded at EoL stage in lighting based on available literature data on material composition of different lamp types, different composition of fluorescent powders, weights and numbers of different lighting applications, etc. Tables 22.6 and 22.7 give an overview of the estimated (calculated) amount of REOs in discarded lighting applications for Europe and The Netherlands in comparison to world demand, production and application of REOs (see Tables 22.3–22.9 for applied lighting and compositional data).

22.3.6 Batteries

Batteries convert chemical energy to electrical energy for powering a wide variety of cordless electrical products. The high variety of applications, from flashlights to vehicles, demands different battery chemistries and materials to able to meet a wide range of power, energy, size, weight, safety and cost requirements. The

TABLE 22.3 Summary Lighting Data (Literature Data) on Calculated REO Quantity per Lamp (van Schaik, 2011)

Literature and Calculated Data per Lamp		
Average-weight lamp (see Table 22.7)	(kg)	0.07–0.2 kg
% Fluorescent powder in lamps (see Table 22.6)	(%)	0.3–2.9%
Calculated average weight of REOs/lamp	(g)	0.2–2.3 g

TABLE 22.4 Calculated Quantities of REO in EoL Lighting in the Netherlands and Europe (Only Given for Fluorescent Lamps in Category 5 Lighting). (Based on Calculated Data of Weight of REO/Lamp; see Table 22.3)

Quantity REOs in EoL Population Lighting in the Netherlands and Europe (2010) (cat. 5 Lighting, Presented for Lamps with Fluorescent Powders)		
Discarded weight lighting, The Netherlands (fluorescent)	1,528,273 (Wecycle, 2013)[a]	(kg)
Total kg REOs in EoL lamps, The Netherlands (calculated)	2314–24,827	(kg)
Discarded weight lighting EU27 2010	44,489,000 (Chemconserve, 2011 http://www.chemconserve.com/)– 77,000,000 (UNU, 2007)	(kg)
Total kg REOs EU27 2010 (calculated)[b]	67,374–1,250,899	(kg)
EoL lighting REOs, The Netherlands versus Europe (2010) (calculated)	2.0–3.4%	(%)

[a] Corresponding to 11,273,167 items (#).
[b] Calculated based on average weight REO/collected kg lighting cat. 5; the assumption has been made that EoL population (presence of different lamp types) in EU is comparable to the Netherlands (real values could differ from this due to a different composition of EoL lighting population).

TABLE 22.5 Production Demand and Application REOs in 2010 Relative to the Quantity REO in Dutch EoL Lamps (2010)

Production, Demand and Application REOs 2010	EU27	World	
Items (number of lamps)	776,000,000[a] (UNU, 2007)– 988,000,000 (ELC, 2009)	6,512,000,000 (Van Hoek et al., 2010)	#
Average kg REO in produced lamps 2010 (calculated)	934,170–1,189,381	7,839,324[b]	(kg)
Production REO 2010		114,330,000 (Lynas, 2011)	(kg)
Demand REO 2010		127,500,000	(kg)
Application REOs in fluorescent powders (% of demand)		6%	(%)
Kg REO applied in fluorescent powders		7,650,000[b]	(kg)
% REOs in Dutch EoL lamps 2010 relative to world production 2010 (calculated)	0.03–0.32%		(%)

[a] Data 2007.
[b] The calculated averages of REO production in lamps (based on calculated gram REO/lamp) and literature data are corresponding.

TABLE 22.6 Material Composition of Lamps According to Various Literature Sources (%)

	UNU (2007)	Rabah (2004)	Rabah (2008)					ELC (2009)				
	Mix of lamps	Fluorescent lamps	Fluorescent lamps	Fluorescent lamps	Circular Fluorescent lamps	GLS	Halogen	Halogen integral	Fluorescent lamps	CFL-integral	CFL-non integral	HID	LED retrofit
Example lamp type			120 cm	60 cm	30 cm	60 W	35 W	35 W	36 W	11 W	13 W	MHL 400 W / HPS 150 W	7 W
Fe/steel	1.6											17.5 / 29.7	40.3
Cu	1.8					13	20	12.7	2.5	3.3	5.5		
Al	5.6	0.9	1.4	2.6	2.1								23.4
SS													
Brass			0.2	0.3	0.3								
Plastics	1.7					—	—	35.3	—	20.8	18.2	— / —	
Rubber													
Hg	0.0[a]		0	0	0								
Fluorescent powder	1.6	0.3	2.3	2.2	2.9								
Glass	85.9	97.6	95	93.2	93.2	78.3	79.8	25.3	95.8	54.2	72.7	81.3 / 70	12.1
Ceramics	0.4												
PWB (low grade)	0.9					—	—	24.1	—	20.8	—	— / —	21.8
Ni/Cu wire		1.1	0.1	0.2	0.2								
W		0.1	0.4	0.8	0.7								
Others	0.5	0.1	0.6	0.8	0.7	8.7	0.2	2.5	1.7	0.8	3.6	1.3 / 0.3	2.4
Total	100	100	100	100	100	100	100	100	100	100	100	100 / 100	100

[a] 45 ppm H g for 2006, estimation 2011 22 ppm H g.
Others = lamp cover and materials such as electrodes, paste and ceramic parts.
GLS = general lighting source; CFL-integral = compact fluorescent Lamp; HID = high-density discharge lamp.

TABLE 22.7 Weight of Different Lamp Types

Lamp type	Specification	Weight (gram)	Source
Mix of lamps (2006)	TL, TL luxurious, CFL, HID	144	UNU (2007)
Mix of lamps (estimation 2011)	TL, TL luxurious, CFL, HID	134	UNU (2007)
Fluorescent lamp	120 cm	192	Eurometaux's (2010)
Fluorescent lamp	60 cm	101	Eurometaux's (2010)
Circular	30 cm	117	Eurometaux's (2010)
GLS	60 W	23	Eurometaux's (2010)
Halogen	35 W	2.5–29	ELC (2009), Welz et al. (2011)
Halogen integral	30 W	79	ELC (2009)
Fluorescent lamp	36 W	120–226	ELC (2009), Welz et al. (2011)
CFL-integral	11 W	111–120	ELC (2009), Welz et al. (2011)
CFL-nonintegral	13 W	55	ELC (2009)
HID	MHL 400 W	240	ELC (2009)
	HPS 150 W	150	ELC (2009)
LED retrofit	7 W	124	ELC (2009)
Fluorescent lamp (TL)	<60 cm	70	Wecycle (2013)
Fluorescent lamp (TL)	>60 cm	200	Wecycle (2013)
Energy-saving retrofit	—	150	Wecycle (2013)
Energy-saving nonretrofit	—	70	Wecycle (2013)
High-pressure lamp	—	180	Wecycle (2013)
Sodium lamp	—	450	Wecycle (2013)
LED lamp	—	30	Wecycle (2013)
Other discharge lamps	—	180	Wecycle (2013)

GLS = General Lighting Service, CFL = Compact Fluorescent Lamp.

2010 global battery market size was reported at $60 × 10^9$ with a breakdown of 70% secondary batteries (rechargeable) and 30% primary batteries (nonrechargeable), as reported by UNEP (2013). The dominant primary battery types are alkaline and zinc, while lead/acid dominate secondary batteries (50% of total). Other secondary batteries include portable batteries, dominated by Li-ion (12%) and NiMH (3%); while HEV batteries (mostly NiMH) compose a small but growing share of the market at 3%. Inside a metal or plastic case, a battery consists of a positively charged cathode and a negatively charged anode, which are kept from touching by an inert separator membrane that allows ions to pass. The electrolyte is a liquid, gel or

TABLE 22.8 Composition of Fluorescent Powders in Lighting (%)

Material (%)	Rabah (2004) Fluorescent lamps	Rabah (2008) Fluorescent 120 cm	Fluorescent 60 cm	Circular 30 cm	Average
Ca sulfate	34.5	65.5	63.0	62.0	64.4
Ca (ortho) phosphate	61.5	32.0	34.5	35.5	33.1
Europiumoxide (Eu_2O_3)	1.6	1.5	1.5	1.5	1.5
Yttriumoxide (Y_2O_3)	1.7	1.0	1.0	1.0	1.0
Others (metal contaminations)	0.7	—	—	—	—
Total	100.0	100.0	100.0	100.0	100.0

TABLE 22.9 Additional Data for the Chemical Composition of the Most Frequently Used Powders (Schüler et al., 2011)

Phosphorous Powders without REs	Chemical Composition
Halophosphate (white, blue)	CaO, P_2O_5, MnO, Sb_2O_3, F, Cl
Phosphorous powders with REs	
Yttrium europium oxide (YOE)	~95% Y_2O_3, ~5% Eu_2O_3 (red)
Barium magnesium aluminate (BAM)	Al_2O_3, BaO, MgO, ~2% Eu_2O_3 (blue)
Cerium magnesium aluminate (CAT)	Al_2O_3, ~11% Ce_2O_3, ~8% Tb_2O_3, MgO (green)
Lanthanum phosphate (LAP)	~40% La_2O_3, ~16% Ce_2O_3, ~11% Tb_2O_3, P_2O_5 (green)

Table shows the REs as applied as activator. REs are also applied in the "host lattice" (yttrium, cerium, lanthanum).

powder that conducts ions from anode to cathode producing an electric current. The metal fraction of batteries is dominant, but varies greatly both in amount and in elements, with metals potentially contained in the anode, cathode, electrolyte and case/can, and being both valuable (Co, REE, etc.) and hazardous (Pb, Cd, etc.). Material compositions by weight of some batteries are shown in Table 22.10.

The presence of critical materials in rechargeable batteries makes these interesting for recycling. Eurometaux (2010) gives an overview of the past exponential growth of different battery technologies (UNEP, 2013).

22.3.6.1 Zinc/Alkaline Batteries

These primary batteries use a zinc anode and manganese in the cathode. Zinc batteries can be divided into Zn and Zn alkaline batteries. Bernardes et al. (2004) and Xu et al. (2008) give the ratio of 39% Zn and 51% alkaline within the EU.

22.3.6.2 Rechargeable Batteries

Li-Ion: Primary batteries use metallic lithium as cathode and contain no toxic metals; however, there is the possibility of fire if metallic lithium is exposed to moisture while the cells

TABLE 22.10 Material Compositions of Some Battery Types by Weight (UNEP, 2013)

| | Primary Batteries | | Secondary Batteries | | | | | |
| | | | Ni/MH | | | | | |
	Zinc/Alkaline	Ni/Cd	Button	Cylindrical	Prismatic	Toyota Prius II	Li-Ion	Li-Polymer
Fe	5–30%	40–45%	31–47%	22–25%	6–9%	36%	24.5%	1%
Ni		18–22%	29–39%	36–42%	38–40%	23%		2%
Zn	15–30%							
Mn	10–25%							
Cd		16–18%						
Co			2–3%	3–4%	2–3%	4%	27.5% (LiCoO$_2$)	35% (LiCoO$_2$)
Li								
REE			6–8%	8–10%	7–8%	7%		
Cu							14.5%	16%
Al								15%
K			1–2%	1–2%	3–4%			
Metals, unspecified						2%		
Graphite/carbon			2–3%	<1%	<1%		16%	15%
Plastics/polymer			1–2%	3–4%	16–19%	18%	14%	3%
H$_2$O			8–10%	15–17%	16–18%		3.5%	
Electrolyte						9%		
Other			2–3%	2–3%	3–4%	1%		23%
Source	UNEP (2013)	Müller and Friedrich (2006)				Öko-Inst. (2010)	Xu et al. (2008)	Meskers et al. (2009)

Various data can also be found in UNEP (2013), Bernardes et al. (2004), Rombach et al. (2008), Umicore; Xu et al. (2008).

are corroding. Lithium-ion secondary rechargeable batteries (LIBs), on the other hand, do not contain metallic lithium. Data for usage of Li-ion batteries are given in UNEP (2013).

Table 22.11 gives an overview of the composition of different types of NiMH batteries (including the content of the different REs). This table also gives a comparison between the composition of these types of batteries and different RE concentrates (Pietrelli et al., 2002). Table 22.12 gives the composition of electrodes of different types of NiMH batteries and the average ratio of different REs (Ce, La, Nd, Pr) present (Bertuol et al., 2006).

22.3.6.3 Lead Batteries

Used predominantly for conventional car batteries, such batteries have lead electrodes and a sulfuric acid electrolyte solution. Collection rates are typically high (>90% in many developed countries, though significantly lower globally) due to lead's characterization as a hazardous substance and the existence of well-established recycling systems.

22.3.6.4 Batteries in Electric Vehicles

The possible increase in the use of electric vehicles poses the question of whether there

TABLE 22.11 (a) Composition of NiMH Batteries, also in Comparison to (b) Typical Concentrations of REs in Different RE Concentrates

(a)

Element	Button Cell (%)	Cylindrical Cell (%)	Prismatic Cell (%)
Nickel	29–39	36–42	38–40
Iron	31–47	22–25	6–9
Cobalt	2–3	3–4	2–3
La, Ce, Nd, Pr	6–8	8–10	7–8
Graphite	2–3	<1	<1
Plastics	1–2	3–4	16–19
Potassium	1–2	1–2	3–4
Hydrogen, oxygen	8–10	15–17	16–18
Other	2–3	2–3	3–4

(b)

Element	Monazite Concentrate	Bastnasite Concentrate	Xenotime Concentrate	Turkish Ore Concentrate	Batteries Anode
La_2O_3	17	24.6	3.7	4.1	16.2
Ce_2O_3	29.9	47.1	6.8	3.5	12.6
Pr_2O_3	3.9	4.4	0.9	0.2	3.1
Nd_2O_3	11	12.6	2.7	0.5	6.6
RE total+Y_2O_3	64.8	89.9	45.6	8.4	

TABLE 22.12 Composition of Electrodes of Different Types of NiMH Batteries and the Average Ratio of Different REs (Ce, La, Nd, Pr) Present (Bertuol et al., 2006)

	Chemical analysis of the electrodes (wt.%)									
	Negative Electrodes					Positive Electrodes				
Metal	Battery 1	Battery 2	Battery 3	Battery 4	Battery 5	Battery 1	Battery 2	Battery 3	Battery 4	Battery 5
Ce	24.14	ND	25.04	23.06	ND	ND	ND	0.80	ND	ND
La	16.02	48.67	21.38	20.33	ND	ND	ND	0.84	0.78	ND
Nd	10.57	ND	9.36	9.09	ND	ND	ND	ND	ND	ND
Pr	1.63	4.35	1.14	0.96	ND	ND	ND	ND	ND	ND
Ni	ND	10.75	ND	10.60	34.80	31.28	29.31	32.96	32.56	29.07
Co	25.58	22.67	21.76	20.31	1.56	38.29	34.09	27.38	26.20	13.41
Mn	15.55	2.40	16.23	11.05	0.20	2.91	0.96	1.60	3.12	ND
K	4.58	6.90	2.32	2.14	ND	7.70	11.47	12.45	15.27	3.15
Fe	0.28	3.21	1.14	1.25	1.50	0.73	0.90	0.61	0.57	0.90
Zn	1.16	0.27	1.23	0.66	ND	18.40	21.45	22.30	19.60	0.18
Others	0.47	0.74	0.34	0.53	1.15	0.53	1.66	0.85	1.74	0.28
Cd	ND	ND	ND	ND	60.80	ND	ND	ND	ND	52.75
Total	99.98	99.96	99.94	99.98	100.01	99.84	99.84	99.79	99.84	99.74

ND, not detected
Ratio, Ce:La:Nd:Pr ~ 0.44:0.35:0.18:0.02.

are enough resources for these new products. Lithium is an important element in this regard due to the required battery technology. A recent study (EU, 2010) shows that if newly registered cars would globally consist of 50% electric vehicles, only about 20% of the currently known global lithium resources would have been used by 2050 (Schüler et al., 2011). This scenario considers the use of recycled materials and lithium demand for other applications. Even if market penetration would reach 85%, the resource would not be depleted by 2050, but the current resources that can be recovered by present technology will have been exhausted. Thus, developing recycling is important in addition to the development of technologies allowing the highest possible recovery of lithium from its natural deposits. Various projects have recently started in the EU to address the recycling of lithium-containing batteries for the automotive sector.

22.3.7 Catalyst Compositions

Catalysts come in a variety of forms for various applications, mostly situated on a substrate as summarized for various precious elements in Table 22.13. It is clear that the support type is usually some variation of (γ-) Al_2O_3, SiO_2, zeolites, honeycomb-type cordierite skeleton ($2MgO \cdot 2Al_2O_3 \cdot 5SiO_2$), SiC, (activated) carbon, etc. Molybdenum oxide (MoO_3) or sulfide (MoS_2) based catalysts also exist,

TABLE 22.13 Various Catalysts and Their Support Types and Their Loading of Precious Metals (Hagelüken, 2006)

Application	Catalyst Type Support	Precious Metals (PM)	PM-Loading	Catalyst Life (years)
OIL-REFINING				
• Reforming	Al_2O_3	Pt; Pt/Re, Pt/Ir	0.02–1%	1–12
• Isomerization	Al_2O_3 zeolites	Pt; Pt/Pd		
• Hydrocracking	SiO_2 zeolites	Pd; Pt		
• Gas to liquid (GTL)	Al_2O_3 (SiO_2 TiO_2)	Co+(Pt or Pd or Ru or Re)		
BULK & SPECIALTIES				
• Nitric acid	Gauzes	Pt/Rh; Pd	100%	0.5
• H_2O_2	Powder (black)	Pd	100%	1
• HCN	Al_2O_3 or gauzes	Pt; Pt/Rh	0.1%; 100%	0.2–1
• PTA	Carbon granules	Pd	0.5%	0.5–1
• VAM	Al_2O_3; SiO_2	Pd/Au	1–2%	4
• Ethylen oxide	Al_2O_3	Ag	10–15%	
• KAAP	Activated carbon	Ru		
HOMOGENEOUS				
• Oxo alcohols	Homogeneous	Rh	100–500 ppm*	1–5
• Acetic acid		Rh; Ir/Ru	"in process solution"	
FINE CHEMICALS				
• Hydrogenation	Activated carbon	Pd; Pt; Pd/Pt;	0.5–10%	0.1–0.5
• Oxidation	(powder)	Ru; Rh; Ir		
• Debenzylation				
• Etc				
AUTOMOTIVE				
• Catalysts	Cordierite monolith	Pt/Rh	0.1–0.5%	>10
	Ceramic pellets	Pt/Pd/Rh		
• Diesel particulate filter	Metallic monolith	Pt		
	SiC or cordierite	Pt/Pd		

including CoO, Bi_2O_3, P_2O_5, V_2O_5, TeO_2 and SnO_2 to catalyze a variety of chemical, organic and other reactions (Sebenik et al., 2012). Usually these catalysts are contaminated by carbon-containing residues originating from their various applications.

A typical car catalyst has the following approximate analysis (wt.%): Pt ~ 0.1, Rh ~ 0.01, Al ~ 23, Si ~ 15, Mg ~ 5.5, Ce ~ 0.61, Ba ~ 0.5, Ca ~ 0.12, C ~ 0.096, S ~ 0.04 (Rumpold and Antrekowitsch, 2012).

22.4 RECYCLING COMPLEX MULTIMATERIAL CONSUMER GOODS: A PRODUCT-CENTRIC APPROACH

This section will provide a brief overview of how EoL goods are recycled and how a product-centric approach based on the physics of design, technology and recycling is applied to understand and optimize resource efficiency through recycling.

22.4.1 Complex Product and Material Mixtures from EoL Sources

Cars (ELVs), electronics (WEEE) as well as batteries and catalysts applied in these and other products are examples of complex multimaterial products, in which the applied materials are complexly linked, connected and mixed by both constructive and chemical connections. This (functional) product complexity implies that metals and materials arriving at recycling operations of complex products are always mixed with other metals, alloys and compounds inclusive of plastics and other materials. In addition, most products are collected and/or treated in mixed streams, such as Large Household Appliances (LHHAs) (a mixture of washing machines, dryers, ovens, dish washers, etc.), small household appliances (SHHAs) (the exact definition differs per country, but these contain a wide range and variety of products such as mixers, vacuum cleaners, audio sets, coffee machines, tools, toasters, etc.) for WEEE. ELVs are sometimes processed in car shredder facilities together with the mix of WEEE LHHA and other steel-based appliances and contain more and more electronics as a consequence of design, safety, automation and functional developments in car design. This implies that EoL streams from both WEEE and ELV are covering a wide range of different products, for which the products are also different per brand/type/year of construction, etc. These EoL products have different (mechanical and constructive) properties and varying material content and contain a vast quantity and complexity of (physical and chemical) material mixtures and compounds.

These materials all have a (higher or lower) monetary value, which is maximized the purer the recyclates of recycling are. The interlinked nature and nonlinearity in recycling materials and energy from these type of products demands a product-centric approach to understand and optimize the recycling products such as ELVs, WEEE, batteries and catalysts on the basis of deep and fundamental knowledge and understanding of the physics and interconnections and dependencies in processing and recovery of the over 50+ elements and materials from these products.

22.4.2 First Principles of Recycling

Recycling of complex, multimaterial consumer products such as cars, electronics/WEEE and complex components such as batteries and catalysts demands an extended product-centric based network of different types of processes separating and producing many different (intermediate) recyclates (e.g. ferrous recyclate, copper recyclate, aluminum recyclate, PWB recyclate, plastic recyclate, etc.) and recycling product flows (such as metal, speiss, slag, matte, flue dust) in order to recover the wide range of materials present in the product and components. (See Figure 22.13; this will also be discussed in more detail in the appendices.)

The possibilities and limits of recycling are strongly influenced by the performance of the liberation and separation processes. Recycling of complex, multimaterial consumer products uses a network of different types of processes to first liberate and separate the different interconnected materials into a range of recyclates (steel recyclate, copper recyclate, aluminum recyclate, plastic recyclate, etc.), and then takes those as the input to the final treatment processes that work on the thermodynamics, such as metallurgical processing, as Figures 22.13 and 22.14 show (van Schaik and Reuter, 2012) for an overview and description of dry and wet sorting processes, UNEP. The selection and arrangement of processes, together with a thoughtful trade-off between collection, sorting and/or separate or combined processing of products for recycling, determines the ultimate quality of intermediates and recyclates and hence the material/energy recovery from it.

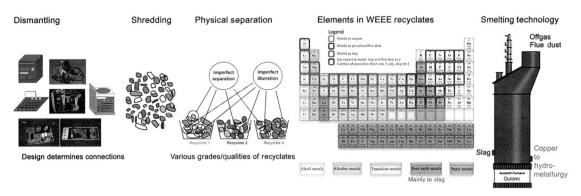

FIGURE 22.13 A key message concerning product-centric recycling: Product design must be linked to liberation, separation efficiency and quality of both recyclate and of final output streams, based on the thermodynamic limits of recycling (Reuter et al., 2005; van Schaik and Reuter, 2008, 2012a,b, 2013a,b).

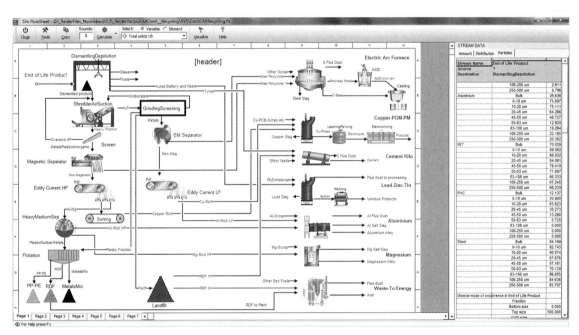

FIGURE 22.14 General flowsheet for recycling linking end-of-life products via physical separation to metallurgical recovery. Note the use of a simulator to simulate these complex flowsheets.

Maximizing resource efficiency requires a detailed understanding of the complete recycling chain, for recovering the maximum amount of metals in hydro- and pyrometallurgical processing facilities. This is only possible through a holistic systems approach, starting with product design and optimizing the technical and organizational setup of dismantling, shredding and separation technologies. Only then can the relevant material fractions be directed to the most appropriate metallurgical end-process. Figure 22.13 reflects this system, showing that any mismatch between the different steps produces unwanted residues, losses and ultimately leads to lower overall resource efficiency.

This also implies that policy should stress the building and maintenance of a suitable global metallurgical processing infrastructure and of well-channeled preliminary pathways, to ensure a maximum recovery of metals, minimize theft, and optimize manual sorting in the informal sector (Xinwen et al., 2011).

22.4.2.1 Product Design, Liberation, Recyclate Quality and Losses in Recycling

In the recycling processes for ELVs and WEEE, products and components are, after selective dismantling and/or depollution of components/fluids, usually broken/cut/shredded into small pieces and sorted into economic concentrates, in an attempt to separate out different mixed materials (van Schaik and Reuter, 2004, 2010). However, this is often only partially successful. This breaking up of EoL products and subunits often leaves different materials fixed to, or mixed with, each other due to partially inseparable (functional) connections (both constructive and chemically bound) between materials brought together in design for product functional/specification reasons. For example, materials that are purposefully attached to each other, as they would be in joints in a product, are likely to stay close together during shredding. Some parts of these products are on their own very complex mixtures of numerous materials, metals, alloys and compounds, joined in a functional, useful manner (e.g. batteries, catalysts, printed wire boards, electronic components, etc.). Metals used in small quantities next to each other on a printed wire board (PWB) or combined in electronic components on the PWB are also likely to stay together, even when the PWB is broken into pieces. When shredding circuit boards together with the product of which it is part, portions of the brittle components (e.g. ceramic capacitors) mounted on the surface could be broken to dust, or disperse during shredding to other fractions, which then, unless systematically collected, settles in the plant or on surfaces of other material streams or in the flue dust. Imperfection in separation and contamination of recyclates and material losses will also inevitably occur during mechanical sorting (sorting efficiency will never be 100%) due to the nature of sorting, physics and statistics in these type of processes (the classical grade/recovery relationship, where 100% recovery can never be achieved in combination with 100% grade) (Figure 22.15). Figure 22.16 gives some examples of the quality and composition of recyclate flows as a function of partial liberation and imperfect sorting. Therefore metals enter metallurgical processing as functionally mixed, rather than as single metal streams. These materials/compounds all interact in metallurgical processing, so in turn could increase the treatment charges incurred by custom smelters and processors of the recyclates obtained from physical recycling and/or dismantling and could result in unwanted losses of materials/metals and creation of residue streams in metallurgical processing due to prevailing chemistry and thermodynamics. This will be discussed in much more detail in the next sections.

Recent research based on extensive industrial and experimental data (van Schaik and Reuter, 2004, 2010; Reuter et al., 2005; Reuter and van Schaik, 2012a,b) has developed physics-based

FIGURE 22.15 Different levels of material connections and material combinations (both physical and chemical) of printed wire boards in e-waste.

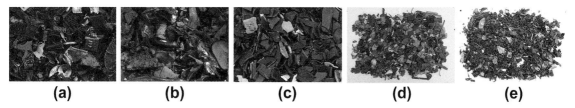

FIGURE 22.16 Recyclate quality from WEEE products due to inevitable imperfect separation and liberation. The distributions of particles contain (a) ferrous recyclate, (b) aluminum recyclate, (c) plastics recyclate, (d) printed wire board recyclate and (e) residue fraction.

models that simulate the effect of design on the liberation behavior and quality of recyclates from complex consumer products (Figure 22.17). Through this physics-based link between product design and resource efficiency through DfR (Design for Recycling), dynamic recycling simulation models for cars and e-waste/WEEE have been developed. Comminution-breakage laws as a function of material connections and related to particulate characteristics of recyclate flows are addressed and linked to sorting physics, chemistry and thermodynamics of high-temperature processing and resource recovery from recyclate streams.

22.4.2.2 Nonlinear Interconnected Nature of Design and Recycling of Complex Multimaterial Products

Metals are lost because the mix of metals brought together in design that cannot be sorted from each other (by liberation and mechanical sorting) entering metallurgical processes have physical and thermodynamic properties that do not allow the processes used to separate all

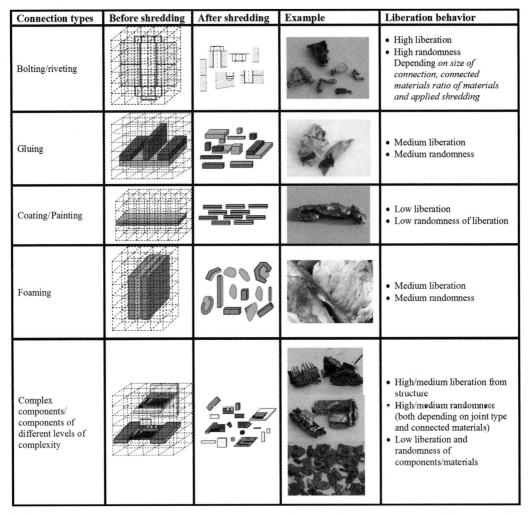

FIGURE 22.17 Characteristics of connection types related to their specific degree and nonrandomness of liberation behavior after a shredding operation (Reuter et al., 2005; van Schaik and Reuter, 2012).

of them from each other. The loss of metals when they cannot be separated during processing into thermodynamic and economic treatable fractions is often more common than generally assumed. In truth, it is never correct to believe—as is so often assumed in simpler material flow models that do not include chemical transitions between different phases—that "there is this much metal/element/compound in the waste product, and therefore this same amount of recycled metal can be produced." The valuable metals are often "lost" into process wastes in other recycled metals, residues and intermediate products due to economic reasons, among others.

Mixing of products and materials during end-of-life "waste" collection has to take place based on sound physics and thermodynamics understanding of the recycling chain, inclusive of the interactions between different materials/elements/compounds in different ratios in order to avoid unnecessary loss of metals. There are different metallurgical processes for the production of the most common commodity metals used in society—e.g. iron, aluminum, lead and copper—and these are also used for recycling and waste processing (see Figure 22.7), being implicitly part of the recycling systems of products (Figures 22.13 and 22.14).

Another reason for loss of these valuable metals is the lack of a systemic physics-based link between product design, material applications and combinations with final treatment process technology in the entire recycling system or chain. Much will be done if the often too simplified and linear approaches (e.g. MFA (Material Flow Analysis) type methods) are replaced by more rigorous approaches that embrace and recognize the complex interlinked and nonlinearity of recycling physics. Recycling modules in computer models often are relatively simple, as is still the general case for LCA (Life Cycle Analysis). Usually, they are based on rather simplistic models of systems and their design, often neglecting detailed physics. Recycling tools, models and derived guidelines have been developed and applied by Reuter (1999) and van Schaik and Reuter (2004, 2010) to try and correct this problem, embracing a more product-centric view to recycling. These models will be discussed in this chapter for the examples of ELVs, WEEE, and batteries and catalysts.

It must be realized that the potential recovery of materials from society is not fixed, but changes as a function of time, products, components, product mixture (input to recycling systems) and recycling routes. Wherever in the world recycling occurs, the infrastructures for all stages can be improved for achieving higher recycling rates.

Recycling takes place as a system, as illustrated in Figures 22.13 and 22.14. As each part of the system affects the success and viability of the others, the limits of recycling performance depend on the optimization of each of these related factors, recycling being a truly complex, multidimensional, dynamic problem. Hereafter we discuss the physical and organizational infrastructure toolbox for optimizing recycling rates.

Optimization and understanding of recycling requires deep systemic knowledge and understanding of the physics-based link between product design, material applications and combinations with the final treatment process technology (metallurgical processes) in the entire recycling system or chain. Deep metallurgical process knowledge is paramount as well as a thorough understanding of the process technology as well its economic drivers to maximize resource efficiency from complex products by a mindful balance of product mixing/sorting for processing, dismantling in combination with shredding/sorting and (pyro- and hydro-) metallurgical processing of recyclates from these products. The outcomes from recycling, i.e. the range, selection and purity of metals produced, are in turn a key determinant of the economic profitability—and so viability—of recycling in its entirety. Economy of scale recycling technologies and processes often already exist that can economically separate many metals into high-quality products (Figure 22.7), but whether the metals entering recycling are separated depends on the choice of processes used and also product functional design. Rapidly changing product design and product replacement (e.g. LCD display technology replaced by LED/OLED technology) poses challenges to physical recyclers to have recycling technology keeping pace with rapidly changing product properties and material usage, which poses different demands on sorting technology development and plant configuration in order to most optimally recover materials. It should be realized that a critical mass (over a long enough timespan) of EoL products with a more or less similar character is required to economically invest and build on technology development and plant design for shredding and sorting.

22.4.2.3 Metallurgical Processing Infrastructures, Process Thermodynamics and Recycling of Complex Products

The "metal wheel" in Figure 22.18 illustrates what generally happens to different metals in an end-of-life product entering the iron (Fe), aluminum (Al), copper (Cu), zinc (Zn) or lead (Pb) processes. Each circle indicates the destination of a metal found in mixed, end-of-life (electronic) product (waste), be it metal, slag, intermediate, flue dust, etc. The ideal situation for recycling would be a system that connects scrap and residues to all carrier (base) metals processes—a system that could be conceptualized as the center of the metal wheel—so that scrap and residues then flow, as determined by thermodynamics and economics, into the appropriate carrier metal technologies (or metal-wheel slices; an example is shown in Figure 22.18), with further links between each technology (or slice) for processing the residues from each technology. This is obviously utopian and cannot be totally achieved with existing technology. Nevertheless, the conceptualization can help understanding the intricate system and knowledge required for recycling and waste/residue processing, what is truly meant by a systemic view to recycling and what deep thermodynamic and process knowledge is required to "close" the loop. This suggests clearly that cradle-to-cradle thinking fails for even simple products due to the (limits imposed by the) laws of nature (thermodynamically) and illustrates that product resource efficiency has its limits.

The thermodynamic properties of each metal and their various compounds/alloys are particularly important. When metals and their compounds in end-of-life goods and recyclates have thermodynamic and physical properties that are compatible with a particular base metal's metallurgical infrastructure, the metallurgical processing technologies used by metals producers and refiners can usually separate and recover them economically from the various streams that are created such metal, matte, speiss, sludges, precipitates, slimes, flue dusts, fumes, slags, etc. minimizing the losses to streams that have a dumping/ponding cost attached to them. This is also the basis for design for recycling within the constraints of product functionality and performance demands.

The economics of recycling are affected by the degree to which metals can be separated, recovered and transformed into high-quality materials that can be applied, for example, in sustainability-enabling products. Where metals do not separate, they can either reduce the quality of the primary recycled metal product, because they cannot be recovered for use as separate metals themselves; and/or they tend to increase the energy needed during the recycling process. Elements within complex products, recyclate, or sludges, are not recycled individually. Instead, they pass through one of a wide range of combinations of processes; an example is shown in Figure 22.19. The choice of process is an economic and physics-based technological optimization puzzle for the recycling operation, one that is driven by the changing values of the metal and high-end alloy products. Secondary materials can be fed into the flowsheet, as shown by Figure 22.19, but this can only be done by experts who know the processes and suitable simulation software. The usual material flow analysis tools used by researchers active in recycling *do not* have the basis to do this and therefore cannot be used to advance and innovate in the recycling and waste-processing system (Reuter, 2011). This leads to the conclusion that a metal production infrastructure such as that shown in Figures 22.7, 22.14 and 22.19 must exist that allows flows of metals between carrier metal processes. This needs to be developed and nurtured over a long time, not least due to investment lead

22.4 RECYCLING COMPLEX MULTIMATERIAL CONSUMER GOODS: A PRODUCT-CENTRIC APPROACH

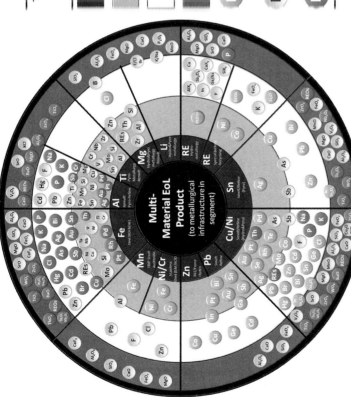

FIGURE 22.18 The "metal wheel," based on primary metallurgy, reflects the destination of different elements in commodity/base metal minerals (base or carrier) processed in hydro- and pyrometallurgy and electrorefining/winning infrastructure as a function of interlinked metallurgical process technology and different intermediates and products such as dross, sludges, slimes, precipitates, speiss, slag, metal, matte, flue dust, etc. Each slice represents the complete infrastructure for a carrier base metal refinery. This basic metallurgical know-how, based on the processing of concentrates originating from mineral ores (the elements are typical for the different ores), should be in place also for recycling in order to maximize resource efficiency (Reuter et al., 2005; Reuter and van Schaik, 2012a,b; Ullmann's Encyclopedia, 2012). As there are so many different combinations of materials, only physics-based modeling as discussed in this document can be used to provide the basis for good predictions in design for resource Efficiency models.

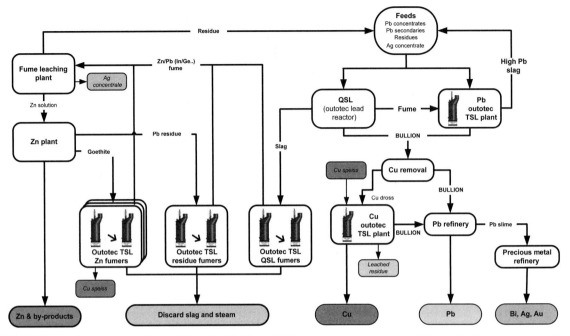

FIGURE 22.19 An example (Korea Zinc) of a Zn-Pb-Cu-Ni Segment in the metal wheel (Figure 22.18) that economically and optimally recovers the carrier metals Zn, Cu and Pb while treating electrolytes, sludges, Goethite and slags from one's own and other production facilities while economically recovering various minor elements of which a few are shown. Outotec Ausmelt TSL furnaces are used, highlighting the versatility of this type of technology (Huda et al., 2012).

times, significant capital costs, the need for cross-carrier metal expertise and because for each product type a different metal wheel exists.

22.4.2.4 Some Thermodynamic Detail is Required to Maximize Resource Efficiency

Economically it has been shown that it is wise first to treat ELV and WEEE streams, after suitable sorting and concentration, pyrometallurgically (a first rough separation at higher temperature that also permits energy recovery from plastics), followed by hydrometallurgy, producing the refined metals and materials of sufficiently high purity to return the EEE and automotive products. For residues, sludges, etc., treatment at high temperature is often preferred to capture elements such as iron in benign slags, as hydrometallurgy would then create further residues such as jarosite and Goethite. In some cases hydrometallurgy is preferred if materials are well sorted, especially for some high-value materials such as rare earths, but then contamination must be limited. At the heart of product-centric recycling is deep knowledge; a brief explanation of what is meant by this is given below:

- If aluminum (Al), for example, is not separated physically from the WEEE, it usually ends in the slag of pyrometallurgical processing as alumina (Al_2O_3) or various other compounds such as Al_2SiO_5, $CaAl_2SiO_6$, $CaAl_2O_4$, $FeAl_2O_4$ and $Fe_3Al_2Si_3O_{12}$, which can precipitate in the molten slag and create sometimes very disruptive operational issues due to viscosity, flow, foaming and separation problems. Slag chemistry also affects the distribution of elements into various phases and has to therefore be carefully managed. Figure 22.20 shows the effect of different conditions on the precipitation of some phases for the typical nonferrous smelting system as well as the final melting point when all possible phases are in the molten state. This would clearly indicate that care should be taken during physical separation to remove the aluminum before it is lost in the slag and in addition disrupts and/or limits operation and the recovery of elements from, for example, WEEE-sourced recyclates.

- Metals arise in numerous varieties due to, for example, their possible valancy states, which depend on what the conditions are within the process technology. This implies that metals can be recovered in the metal phase, can be volatilized into a flue dust and collected for further processing or into the slag or speiss phase, where it may go lost from the material cycle if the slag cannot be processed economically. To illustrate this, consider indium-tin-oxide (ITO) as well as As, Ce, Er, Ga, In, Nd, Sn, Ta, Tb and W-containing recyclate, processed under the temperature range as shown Figure 22.20. Figure 22.21 shows a selection of the possible metal, oxide and gaseous products that could arise at the prevailing partial oxygen pressure created by the addition of carbon, through waste plastics, coal or similar reductant types.

Figure 22.22 shows the recovery of metals into the metal phase (from which they can be recovered) in a TSL furnace as a function of temperature (T) and partial oxygen pressure (pO_2) as well as the solution thermodynamics of each phase. Note that a wide range is given, while in reality furnace operation, smelting campaigns, feed mixes, furnace type, etc. determine which oxygen partial pressure prevails in a narrow window. Thus it is incorrect to take a specific window that looks good for one metal only, one tries to operate in a window that maximizes recovery of many elements at the same time into the phases from which they are best recovered downstream during refining.

It is self-evident from Figures 22.20 to 22.22 that in order to recover metals from complex mixtures of recyclates, very careful control and understanding of the theory of metallurgical reactor technology is required to obtain metals that report to fractions from which these can

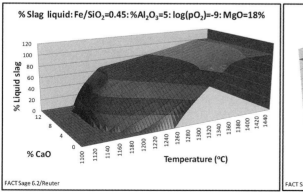
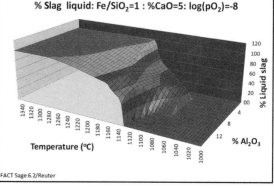

FIGURE 22.20 The effect of alumina on the slag chemistry, i.e. amount of solids in slag, melting temperature, etc. for different processing conditions, as simulated by FACT Sage (www.factsage.com).

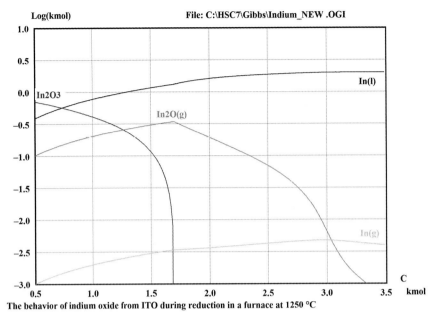

The behavior of indium oxide from ITO during reduction in a furnace at 1250 °C

FIGURE 22.21 The effect of operating conditions on metals and compound distribution of indium in ITO (Indium Tin Oxide) (90% In_2O_3 + 10%SnO_2) in a furnace at 1250 °C as a function of reductant (carbon) addition as simulated by HSC Chemistry (www.outotec.com/HSC).

be economically recovered. Note specifically what the simulations show:

- Rare earths, tungsten and tantalum report to slag, hence these should not be fed to nonferrous smelters as they go lost; but in steel smelters, with appropriate feed as metals these alloying elements impart important properties to steel.
- Indium, tin, arsenic, gallium and similar elements have various gaseous volatile species and report to flue dust—thus in appropriate smelting technologies these can be recovered to an extent.

Following a hydrometallurgical route also brings its own complexity, as many As-Sn-In-Ga cation and anion species appear in solution. All dissolved species have to be recovered from solution to produce high-value metal products by e.g. energy intensive electrowinning or precipitation and subsequent processing.

22.4.3 Developments in Recycling Technology for Current and Future Products

Due to rapidly changing material compositions and EoL products, it will be rather wise to invest much more time and effort in innovating, optimizing and maintaining the metallurgical processing infrastructure rather than (re)inventing new processes, as the basic structure in, for example the EU, can handle and recover many elements already. Also the CAPEX (Capital Expenditure) of metallurgical infrastructure is high, hence detailed understanding of this system is crucially important rather than simply (re-)inventing "new" process and wasting taxpayer R&D money on projects that are doomed to fail due to economy of scale, technological issues as well as lack of know-how. ELVs and e-waste (which may contain some of each of these metals, alongside many other

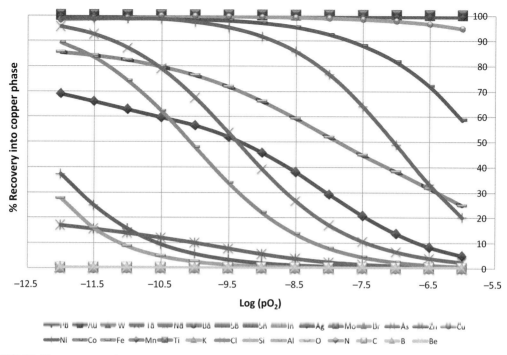

FIGURE 22.22 Recovery of a selection of the many elements present in the PCBA into the copper phase of PCBA smelting at 1300 °C, as is with copper present as solvent—a temperature too low to melt all oxide species, and thus the slag is under these conditions in some pO_2 ranges not totally molten with various spinels and other precipitates present. Note that many of the metals report to other phases and/or crowd close to the 0% recovery level and thus are only visible on a logarithmic scale. The solution models in FACT also do not fully cover the complete range of elements in the PCBA and the above is also to be considered only indicative. Filling the gaps in the solution models is the real challenge in DfR and *not* simplistic DfR rules.

minor and/or critical elements) can enter any of these processes Figure 22.22 however, in some cases materials are recovered maximally and in other cases lost maximally. It must be noted that GaAs, for example, may be present as a minute fraction of a percent in WEEE printed circuit boards (and cannot be economically recovered) and may well be locked up in slag due to thermodynamics, or end in flue dust and be environmentally benign as it is well ponded. Therefore recycling these elements in this case makes no (economic) sense. Of crucial importance is the existence of a system that can take care of all these mixes of elements while maximizing metal recovery and obviously profit and minimizing ecological damage, as reflected, for example, by the complex metallurgical flowsheet depicted by Figure 22.19 for Korea Zinc (Huda et al., 2012).

Each process in this flowsheet has different abilities to deal with mixes of sludges, materials and compounds. This example also excellently shows how in a smart system, residues and slags can be treated to maximally recover metals while producing benign residues that can be used for building materials, for example. This figure also suggests that establishing dedicated recycling facilities for *single* elements only (e.g. In, Ga, etc.), as is seemingly often championed even in the EU by academics and DfR people through studies, is economically and technologically challenging. It must happen in systems that can deal with the complexity and economic realities as suggested by Figure 22.19, i.e. the system must take care of all elements of end-of-life products at the same time.

22.4.4 Process and Recycling System Optimization: Optimization of Resource Efficiency and Environmental Performance of Complex Products

Rigorous economic and physics-based understanding of technology and systems for recycling make it possible to make informed decisions over the complete system boundary. Computer-based modeling and process simulation tools of the recycling technology and system performance of products have been developed and applied for the recycling of complex products by the authors to understand and capture the factors that improve design for resource efficiency. These models include expert rule-based modeling of the relation between design and liberation behavior, particulate quality, physics of sorting efficiency, recyclate quality, distribution of materials/compounds over all produced recyclates and recycling products and metallurgical process efficiency as a function of design and recyclate composition. This implies that sophisticated tools exist that enable the evaluation of resource recovery in metal and material processing systems, as shown by Figure 22.23 (Reuter and van Schaik, 2012a,b). These produce consistent mass and energy balances for all compounds and materials in a processing system, be it for a complete copper plant from rock to refined metal (Figure 22.23) or a complete recycling system for each liberated and unliberated mineral (physical material) and chemical element as a function of design, as shown in Figure 22.23 using the HSC Sim tool (www.outotec.com) linked to PE International's Gabi software (www.pe-international.com).

This differs fundamentally from the more simplified approaches such as LCA and MFA, which rely simply on total element and material flows, and hence are not providing the required depth to capture and predict the mass flows of recyclates and produced/created recycling products including the presence of contaminants and dispersion of (critical and/or toxic) materials during recycling and hence possible (toxic) residues for different designs and corresponding recycling solutions as the basis for LCA analysis or MFA predictions. These details are captured by the sophisticated tools as depicted by Figure 22.23. These simulations are typically based on deep process knowledge, thermodynamic simulation and slag chemistry as discussed in next sections in detail and shown in Figure 22.20, and economic and technological feasibility. Welz et al. (2011) used LCA methodology to investigate the impact of the use of lamps, but lacked the detail to fully integrate the impact of recycling.

Recycling is clearly a complex thermodynamic and resulting economic puzzle to solve, obviously with no one answer or set of design for recycling rules, and not solvable with beautiful credos such as cradle-to-cradle. It would also be self-evident that "mining" the urban mine, which has the sound of hype to it, will be rather a complex and even (economically) impossible task as reflected by the nonlinear effects (see Figure 22.20) and the complexity of the urban "ore body" and its "mineralogy" (Figure 22.15). It would also be clear that for each mixture and each condition this separation will be different, implying that general (MFA/LCA/DfR) methodologies that do not address this depth of mineralogy (i.e. compounds, alloys, transformations, etc.) will obviously and inevitably lead to false conclusions and uneconomic and unrealistic technological and policy recommendations. It is hence worrying to see that these methods are still well adopted and do not match the recycling industry and the sophistication of product design and its CAD (Computer Aided Design)

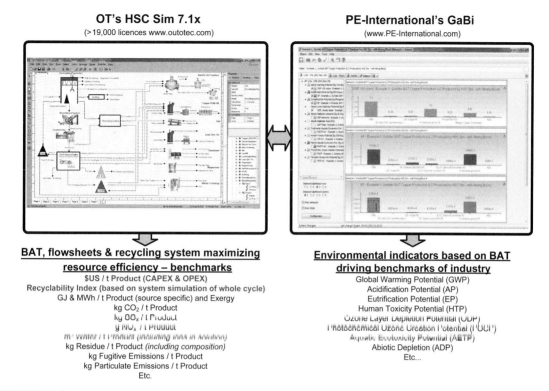

FIGURE 22.23 Linking rigorous process simulation tools with environmental impact assessment tools to optimize plants and complete systems. Linking particle-based simulation with molten metal and slags as well as aqueous solutions to capture the full complexity of recycling systems. (BAT, Best Available Technique).

and other tools. It also suggests clearly that all the figures in this chapter are snapshots and will be different for each new situation. This demands very rigorous modeling to reveal the opportunities and limits of recycling. Design for resource efficiency captures this detail, as discussed by Reuter and van Schaik (2012a,b, 2013a,b).

In the next sections, the recycling of cars, WEEE, batteries and catalysts will be elaborated upon, on the basis of the (dynamic) recycling optimization simulation models as developed by van Schaik/Reuter (2002–2013) to investigate existing and alternative processing routes for these products and/or product mixtures. Resource efficiency, environmental impact quantification of recycling as well as eco-labeling and design for recycling/resource efficiency will be discussed on this physics- and knowledge-based background by providing examples and discussions for the different products. Multidimensional flowsheets (as simply depicted by Figure 22.14) are given for the recycling routes of the various products and components showing the link of dismantling, shredding and physical separation with metallurgical processing represented by the HSC Sim process simulator (www.outotec.com). This product-centric approach, as captured by these flowsheets and models, provides the basis for discussing key aspects of recycling complex products. The presented flowsheets of the entire recycling system, network of processes and material flows provide a graphical and technological blueprint to predict and calculate the possibilities and limits of recycling and the optimization of resource efficiency through these recycling system models.

The physics-based systemic approach showing the various factors that optimize metal recycling, while embracing best available technology (BAT), has been explored and discussed extensively by Reuter/van Schaik (2002—2013) and UNEP (2013) and will be illustrated next with some specific examples and flowsheets for the recycling of ELVs and different WEEE products including the processing of batteries, catalysts and lighting. The approach and discussion holds equally for any other complex (consumer) product (including all other WEEE products/streams) that are not discussed specifically in this chapter.

22.5 AUTOMOTIVE RECYCLING/ RECYCLING OF ELVS INCLUDING AUTOMOTIVE BATTERY RECYCLING

Physics-based recycling simulation models have been defined and applied to calculate and optimize the recycling of cars by Reuter et al. (2006). Figure 22.24 gives a flowsheet and process simulation model (developed in HSC Sim process simulator, www.outotec.com) of recycling of ELVs, including the sorting and processing of batteries and catalysts.

These engineering- and industry-friendly simulation models define and provide the essential metrics to measure, control and improve recycling, determine recycling concepts as well as to enhance product designs facilitating high recycling and energy recovery rates while advising

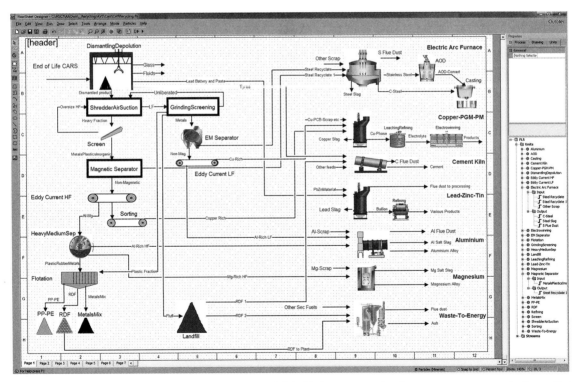

FIGURE 22.24 Flowsheet and process simulation model (HSC Sim process simulator, www.outotec.com) of recycling ELVs, including the sorting and processing of batteries and catalysts.

policy. Although similar models are discussed elsewhere in this document on WEEE recycling (van Schaik and Reuter, 2010), the following example for on ELVs illustrates some of the modeling that has been done to date. For the EU-funded sixth framework project SuperLight-Car managed by Volkswagen AG, the developed models have been applied to provide detailed and physics-based calculations on recycling/recovery of ELVs (including new design BIW concepts) and have also been applied to the determine optimal recycling concepts (combination and arrangement of processes) for a car body, while including the following aspects:

- Product design and liberation
- Separation physics of automated sorting
- Chemistry and thermodynamics of metal production and recycling systems
- Recyclate and recycling product quality (physical and chemical) as a function of product design choices and calorific values of the (intermediate) recycling streams
- Losses and emissions
- Optimization and selection of plant/flowsheet architecture with changing product design for different lightweight multimaterial car body concepts (see flowsheet for all processes) (Krinke et al., 2009).

22.5.1 Application and Results of Process Simulation/Recycling

The developed recycling simulation models have been applied for the following industrial applications:

- Calculation/prediction of the dynamically changing recycling and recovery rates of end-of-life vehicles and lightweight car design concepts and all individual materials in these products for different recycling scenarios, different recycling objectives (maximum total recycling/recovery, maximum recycling of metals, minimum production of waste, legislative constraints, etc.) as a function of distributed ELV population over time based on closed mass and energy balance calculations in the model (this approach has also been applied to different types of e-waste products); Table 22.14 shows that depending on the economic and other boundary conditions, the total recovery of materials and energy can change significantly. The various scenarios simulated by the multi-parameter optimization model included the following.

- **Scenario 1**: Optimization of ELV processing flowsheet without limitation of the thermal treatment stream
- **Scenario 2**: Optimization of ELV processing flowsheet with limitation of the

TABLE 22.14 Various Recovery Rates for End-of-Life Vehicles (ELVs) Depending on the Objective Function (Ignatenko et al., 2007)

Scenario	1	2	3	4	5
Total recovery	0.95	0.91	0.91	0.97	0.88
Metal recovery	0.74	0.74	0.72	0.63	0.73
Physical separation	0.73	0.74	0.72	0.63	0.72
Thermal processing	0.003	0.001	0	0	0.01
Material recovery	0.11	0.11	0.14	0.24	0.12
Plastics	0.02	0.03	0.05	0.02	0.04
Silica	0.004	0.05	0.05	0.03	0.05
From thermal processing	0.05	0.03	0.05	0.19	0.03
Energy recovery	0.1	0.07	0.05	0.1	0.04
Zn-rich dust (from thermal treatment)	0.003	0.002	0.002	0.05	0.002
Waste	0.06	0.09	0.09	0.03	0.12
Physical processing	0.01	0.03	0.05	0.002	0.06
Metallurgy plant	0.05	0.05	0.05	0.03	0.05
Thermal processing	0.001	0.001	0	0	0.01

thermal treatment stream restricted to 10% of the input (as required in the EU ELV Directive)
- **Scenario 3**: Maximum material and metal recovery scenarios with limitation of the thermal treatment stream restricted to 10% of the input (as required in the EU ELV Directive)
- **Scenario 4**: Minimum waste scenario without limitations for thermal treatment
- **Scenario 5**: Maximum Al recovery scenario with limitation of 10% of the input to thermal treatment

It should be self-evident that there can be no single recycling rate but only a distribution of recycling rates as different car models, different recycling scenarios, different recycling infrastructure and routes, etc. all create different recycling rates. It would be best therefore to focus on the economics of these processes to create a playing field in which recovery of materials and energy can be maximized.

This bullets below show various distributions and ranges of data to illustrate this:

- Prediction of grade (quality/composition) of all (intermediate) recycling streams (such as steel recyclate, copper recyclate, plastic recyclate, etc.) and recycling products (metal, matte, speiss, slag, flue dust, off-gas) (see Figure 22.25)
- Prediction of the dispersion, occurrence and appearance (chemical phase) of possible toxic/harmful/valuable elements in recycling products
- Provide insight into the limits and possibilities of recycling technology in view of the EU ELV directive on a physics basis (see Table 22.14)
- Define the most desired recycling concept for multimaterial designs such as the SLC concept (configurations of recycling plants/flowsheets) achieving optimal recycling results and/or minimize material losses.

These tools can start playing a significant role in benchmarking recycling efficiency through quantifying process performance with state-of-the-art environmental impact software. The HSC Sim and GaBi link in Figure 22.23 shows this connection, a basis to create benchmark BAT LCI data, complementing the present more averaged data for the industry. It is now even possible to perform exergy analyses of metallurgical and recycling plants directly by the use of simulators such as HSC Sim (Outotec, 1974—2013). These exergy data, in addition to LCA data, provide a basis to also perform an LCA analysis using, for example, GaBi, which

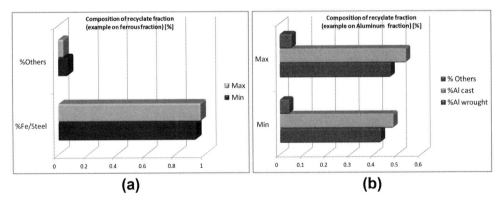

FIGURE 22.25 Quality of ferrous (a) and aluminum recyclate streams (b) including the quantification and specification of the composition/contamination of the recyclate streams (UNEP, 2013; SuperLightCar, 2005—2009).

maps the detailed simulation model in HSC Sim to GaBi to provide all the impacts as required to evaluate systems (www.outotec.com). Hence, all metallurgical process designs of whichever complexity, on numerous levels and all based on thermodynamic and reactor technology principles, can be benchmarked. In addition, as all compounds of each stream are known, toxicological detail can be investigated to a detail not possible within an LCA tool.

22.6 RECYCLING OF WASTE ELECTRICAL AND ELECTRONIC EQUIPMENT

Dynamic recycling simulation models have been developed and applied by van Schaik and Reuter for the recycling of waste electrical and electronic equipment (WEEE). Recent research based on extensive industrial and experimental data (van Schaik/Reuter, 2002—2013) has developed physics-based models that simulate the effect of design on the liberation behavior and quality of recyclates from complex consumer products. Through this physics-based link between product design and resource efficiency in DfR, the authors have developed dynamic recycling simulation models for cars as described above and e-waste/WEEE. Comminution-breakage laws as a function of material connections and related to particulate characteristics of recyclate flows are addressed and linked to sorting physics, chemistry and thermodynamics of high-temperature processing and resource recovery from recyclate streams as discussed in previous sections. Figures 22.26—22.28 give complete recycling flowsheets for different WEEE streams/products goods showing the link of dismantling, shredding and physical separation with metallurgical processing represented by the HSC Sim process simulator (www.outotec.com) for large household appliances (LHHAs) (Figure 22.26), small household appliances (SHHAs) (Figure 22.27) and LCD TVs/displays (Figure 22.28). The recycling flowsheet of SHHAs includes the sorting and processing of different types of batteries.

22.6.1 Application and Results of Process Simulation/Recycling

The recycling performance has been formulated for various existing and alternative recycling routes (including different recycling plants/flowsheet concepts, combination and arrangement/choice of processes and/or dismantling versus shredding and sorting, etc.) within statistical bandwidth of design, plant input and processing variations governed by physics. This all is required to cover the distributed and time-varying properties of EoL-product populations, changing designs and practical and industrial reality of recycling processes (Reuter, 2011). The models can be/ are used for informing OEMs, guiding consumer and policy on a first principles and environmental basis to make resource efficient decisions on rather daunting and complex problems. Some examples of derived results and application of these e-waste/WEEE dynamic recycling models, are given next.

22.6.1.1 (Dynamic) Product and Material Recycling Rate Predictions

Calculation of recycling and recovery rates and resource efficiency are performed based for:

- Different WEEE products for all different materials present (in both physical and chemical compounds, including recovery of metals/materials and critical metals, toxic materials, etc.)
- Different input mixtures of recycling plants (varying mixture/ratio of different products, such as the varying mixture of washing machines, dishwashers, washer-dryers, ovens, etc. in the LHHA flow)
- Different plants/settings/recycling objectives, etc.

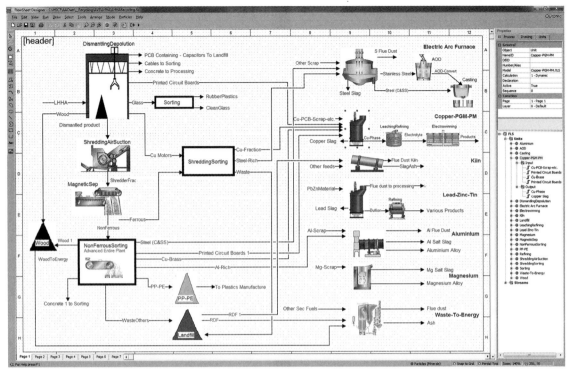

FIGURE 22.26 Flowsheet and process simulation model (HSC Sim process simulator, www.outotec.com) of recycling of large household appliances (LHHAs).

Figure 22.29 gives an example of recycling performance simulations showing the bandwidth of recycling rates as a function of the parameters of the recycling system (dependent on the objective of recycling and constraints). Figure 22.11 shows the predicted recycling rates for an e-waste product over time, being a function of changing EoL product populations (different types of products, from different production years) and changing product weights and material composition thereof. It depicts the capability of the models to capture the dynamic nature of product recycling and predict the input and process dependent distribution and bandwidth of recycling rates (and hence resource availability from it) in future, which is imperative to drive changes and improve recycling performance, apply design for recycling, etc. over time. Figure 22.30 underpins this, by giving an example of calculated recycling rates for a selected material present in the product.

22.6.1.2 Mass Flows of Produced Recyclates/Recycling Products (%/kg) and Distribution/Dispersion of (Critical) Materials Over Different Recyclate Streams

Key to the simulation/optimization models is that they produce closed mass balances for each liberated and unliberated mineral (physical material) and (chemical) element as a function of design as the basis for recycling/recovery rate predictions linked to product design choices. Figure 22.31 shows as an

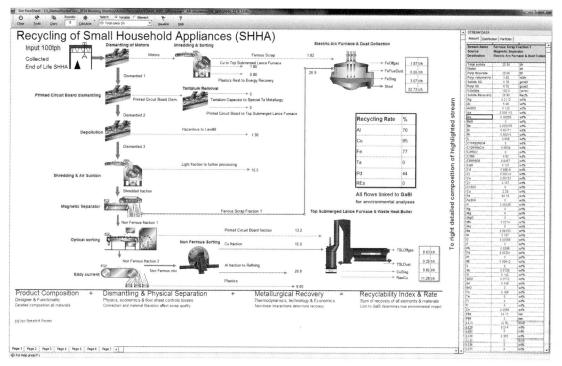

FIGURE 22.27 A complete recycling flowsheet for end-of-life small household appliances (SHHAs) including the sorting and processing of batteries, showing the link of physical separation with metallurgical processing represented by the HSC Sim process simulator (www.outotec.com).

example the calculation of the mass flow (relative to input/output of plant) of a selection of produced recyclate streams from recycling of e-waste streams. These predictions include a bandwidth of the results as a function of changing processing conditions and input composition (based on changing mixture of plant input, changing product weight, composition, etc.).

22.6.1.3 Calculation of Quality and Composition of Recyclate Streams and Recycling Products

The grade (quality/composition) of all (intermediate) particulate (liberated/unliberated/complex) recycling streams (such as steel recyclate, copper recyclate, plastic recyclate, etc.) and recycling products (metals, matte, slags, speiss, flue dusts) is calculated with the simulation models. Figure 22.32 gives an example of the calculated quality of an inert/waste stream recyclate. It reveals that the quality/composition (grade) of recyclates is not only calculated based on the major constituent but also includes the different other materials/contaminants present in the recyclate streams, due to imperfect separation and liberation (see Figure 22.16). The same applies for all streams in the recycling system (metals and nonmetal recyclates), as illustrated in the previous section for cars. On this basis, also the chemical composition of these recyclates is predicted (e.g. based on alloy composition of different metals, present minor elements, chemical compounds, etc.) in order

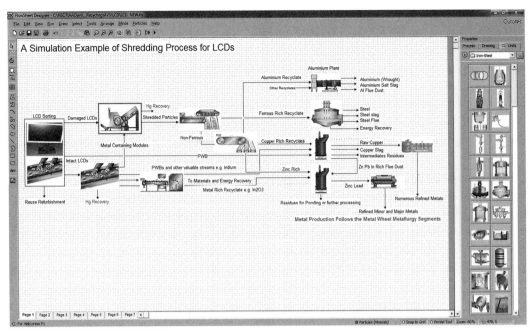

FIGURE 22.28 A recycling flowsheet for end-of-life LCD TVs/displays showing the link of physical separation with metallurgical processing represented by the HSC Sim process simulator (www.outotec.com).

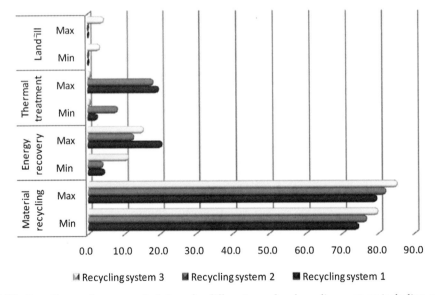

FIGURE 22.29 Recycling performance calculations for different recyclers/recycling systems including statistical bandwidths as a function of input variations, plant statistics, etc. The results are expressed per recycler/recycling system in terms of material recycling, energy recovery, thermal processing and landfill/produced waste. Material recycling rates are also calculated per metal/material (see Figure 22.30).

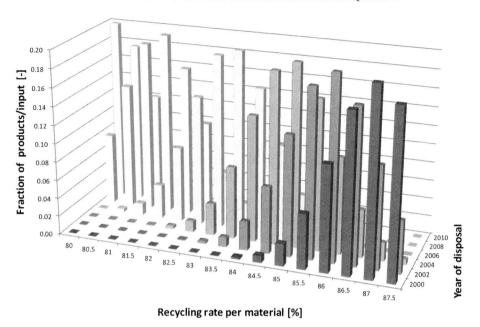

FIGURE 22.30 Dynamic recycling rates per material (example of total recycling rates given in Figure 22.11) as a function varying product weights, composition and plant input over time. Years and values are indicative for illustrative purposes.

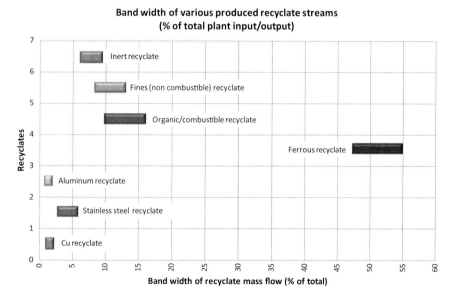

FIGURE 22.31 Bandwidth of various produced recyclate streams.

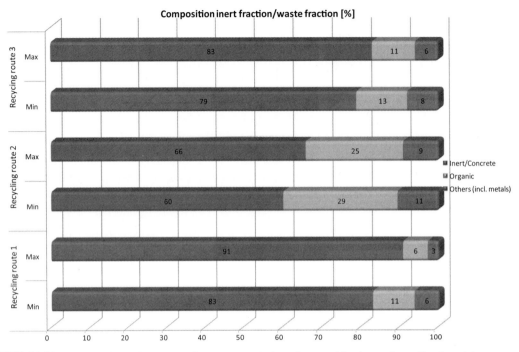

FIGURE 22.32 Quality of inert/waste recyclate stream including the quantification and specification of the composition of the recyclate stream.

to predict the quantities and composition (based on chemical composition/phases of the materials) of recycling products (metal, matte, speiss, slag, flue dust, off-gas). Figure 22.33 gives an example for the model-based calculation of the quality and composition of the recycling products from aluminum recycling for a particular WEEE stream and recycling system setup.

These quality predictions provide the basis for (1) evaluation of the marketability/economic value of the produced recyclates from dismantling and physical separation (e.g. smelter-charge for the streams that report to smelters and revenues in further processing, as with copper metallurgy); (2) a scientifically based estimate of the toxicity of each recyclate stream (based on the combination of materials in each particle); (3) assessment of applicability of the slag and/or requirements for further treatment;

(4) assessment of toxicity of the output streams (e.g. leaching behavior and/or landfill costs); and (5) product and recycling system layout-dependent environmental assessment (Figure 22.23).

22.6.1.4 Prediction the Impact of Different Operating Modes of Technology on the Total Recycling of Minor and Some Commodity Metals

The ability to predict the recovery of minor/scarce metals in addition to the recycling rates of the commodity metals is of great importance to enabling sustainable energy and other high-tech applications and making estimates on resource availability from the urban mine. A WEEE recycling model by Reuter/van Schaik (2010–2013), calibrated with industrial liberation data and experimental work, estimates

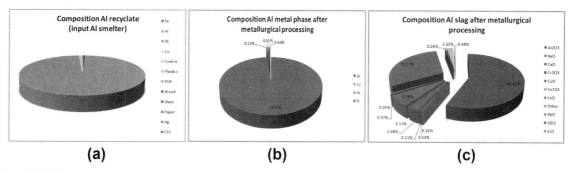

FIGURE 22.33 Metallurgical consequences of recyclate composition. Example of model-based calculation of the composition from recycling products after metallurgical processing shown for one value in the simulated recyclate composition bandwidth (a) composition Al recyclate (after physical processing); and recycling end-products of Al smelter (b) Al metal phase and (c) Al slag. For these calculations, the recyclate fractions of Al have been considered to be the only input to the smelter (something that will in practice not be the case—this can result in a different ratio between metal/slag/flue dust and composition when calculated for the normal industrial input of these processes).

that only about 15% of the gold is recovered by industrial preprocessing, though, depending on the processing method, it can also be close to 100%. van Schaik and Reuter (2010) have shown how different preprocessing methods change the balance of recycling rates between various metals.

22.6.1.4 Identification of the Dispersion, Occurrence and Appearance (Chemical Phases/Compounds) of Possible Toxic/Harmful Elements

Figure 22.34 gives an example of the prediction of the dispersion of PWB to various recyclate flows due to the low quantity and connected nature of PWBs in the design and the focus of the recycling of the illustrated product (LHHA) on the recycling of commodity metals in order to comply with the recycling/recovery quotas as imposed by the EU WEEE recycling directive. This figure immediately reveals that discussions on DfR of PWBs (also called PCBAs) can never just focus on redesigning the PWB assembly based on e.g. material compatibility tables as defined by van Schaik and Reuter (2013a,b); instead, the entire system should be known and detailed knowledge on the effect of liberation, shredding, dismantling and material mixtures, etc. on recyclate quality, dispersion and final destination of the PWB ending up in different carrier metals recyclates and processing routes (see Figure 22.18) should be incorporated, so that DfR makes sense from an industrial and product reality point of view. A singular component/material approach to DfR on these complex products and components will lead to ridicule and/or useless recommendations.

22.7 RECYCLING OF LIGHTING

Recycling of complex, multimaterial (consumer) products such as lamps (and WEEE in general) demands an extended and mindfully balanced network of different types of processes to recover the wide range of materials present as well as to ensure that (potentially) harmful materials such as mercury are captured. Figure 22.35 gives an example of a possible recycling flowsheet for fluorescent lamps as applied in industry (Indaver, 2013)

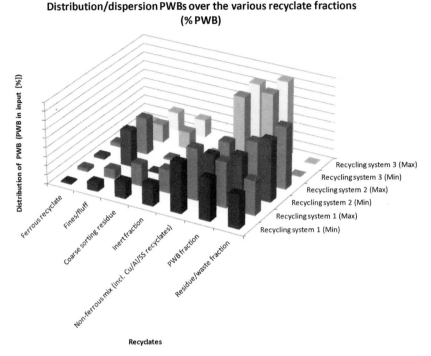

FIGURE 22.34 Example of calculated relative distribution/dispersion of PWB over different recycle streams.

and simulated by the physics-based process models discussed earlier in the chapter. This approach, which embraces best available technology (BAT), is discussed briefly in this chapter (for more details refer to various papers by UNEP (2013)), discussing the various factors and physical and metallurgical processing technologies required to optimize metal and material recycling from lamps. As discussed in detail by Reuter and van Schaik (2013a,b), it is self-evident that "classical" minerals processing and metallurgy play a key role in maximizing resource efficiency and ensuring that metals are recovered from EoL products for application in new products.

The recycling routes of lamps are determined by the type of lamps entering the EoL phase and how these are being collected and sorted for recycling and recovery (straight fluorescent tubes, fluorescent light bulbs and other formats, low-pressure sodium lamps, LED lamps, etc.). This chapter discusses in particular the recycling of different types of fluorescent lamps. Since LED lamps have only recently been introduced into the market, at this moment, the number of EoL LED lamps is still low. Setting up a recycling scheme for products requires a critical mass that justifies the investment and setup of new technology/separation plants adjusted to the particular properties of these types of EoL lamps. This section will in addition address the issues playing a role in the recycling of LED lamps and recovering materials from it. Special attention will be paid to the recovery of REs and critical materials such as In and Ga. Figure 22.35 illustrates the recycling of commodity metals and plastics in lighting.

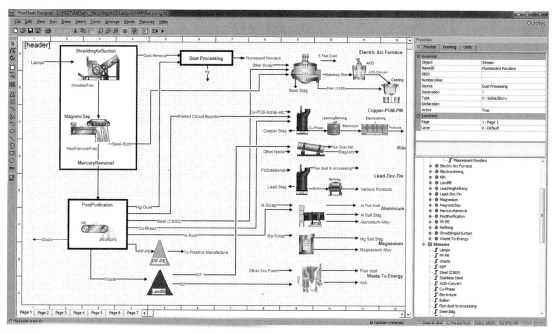

FIGURE 22.35 Flowsheet and process simulation model of recycling of lamps. Note the enclosed shredding facilities to ensure that mercury is captured.

22.7.1 Physical Recycling of Fluorescent Powders

It is evident from the section discussing the construction of fluorescent lamps that the phosphors from fluorescent lamps can be physically recovered relatively easily from the inside of the tubing (Rabah, 2004; Indaver, 2013; Alba, 2013). These powders should be able to find their way through normal RE processing routes, if these are available, to recover REs from them by preferably hydrometallurgical processing routes, which will be discussed in next sections.

22.7.1.1 Recovering Phosphors from Fluorescent Lamps

Rabah (2004) describes the physical recovery of the various metal components from fluorescent lighting. This should generally pose no problem, because the various metals can be recovered by the normal processing methods for recycling, as is the case for most currently recyclers. Fluorescent powders are separated from the lamps (see Figure 22.3) (Rabah, 2004; Indaver, 2013; Alba, 2013). By removing the powders from the tube/glass, the REOs are concentrated in the fluorescent powder fraction. Depending on the mixture of collected lamp types, the powder will be a mixture of the different (REO-containing) components, with a varying ratio, content and composition of the different components. Recovery of REOs demands technology that enables separation of the REOs from the other components in the powder and in which these can be recovered as REOs. This is possible by hydrometallurgical processing (McGill, 2010). This will be discussed in the next section. Further treatment and recovery of the fluorescent powders requires the availability of a suitable (hydro) metallurgical infrastructure (within Europe) and is depending on the economics of recycling and REO value. Within the EU only very limited amounts of REs are being produced, hence this infrastructure is very limited at present. In 2011, 97% of all REs were being produced in China (EU, 2010).

The major issue is containment of mercury; hence the physical separation should take place in an environment that ensures that no Hg escapes to the environment. This is a technical issue that is solved by BAT processors (Indaver, 2013; Alba, 2013) but increases the CAPEX and OPEX of the separation process.

In currently applied BAT physical recycling processes, cutting/shredding and recovery of the fluorescent powders is performed in a closed system (Figure 22.35). If done well, Hg should pose no problem. Care should be taken how this affects subsequent metallurgical recovery processes.

22.7.2 Recovery of LEDs

LEDs are in general a compact design of organic materials and phosphors. Hence it will be best to collect and recycle LEDs through high-temperature processing, which utilizes the contained calorific value, while recovering the valuable In and Ga from it. The REs will go lost to the slag of these high-temperature processes through, for example, copper pyrometallurgy, as depicted by the metal wheel in Figure 22.18; hence the present pyrometallurgical infrastructure in the EU has its problems to source these elements through these secondary copper/e-waste recycling routes. Contained aluminum will also be lost to slag (see Figure 22.20), but contained copper and precious metals would find their way into, for example, a copper melt if processed in a copper recycling route.

Perhaps crushing/grinding (cryogenic?) and subsequent physical concentration can separate organic from inorganic fractions to liberate materials for subsequent hydrometallurgical processing, but this may be uneconomic. The compactness of the LED design, liberation of materials/powders from the LED through shredding, and the occurrence of the fluorescent powders in combination with other materials, to name a few factors, will dictate whether such a processing route can produce pure enough recyclates for further treatment and hence its (economically) viability (van Schaik and Reuter, 2010). Therefore complete smelting could be the easiest while at least recovering the valuable In/Ga from it to an extent.

Retrofit LEDs contain a wider range of materials, which will be as far possible separated through shredding and sorting of the retrofit containing the LED. Limiting factors are the compactness of the design, use of connections that might not fully liberate during shredding (e.g. heat sink Al), the connection and material combinations of other materials to the LED and PCBA of the LED and driver, as well as the complexity and compounds applied in the different parts of the retrofit LED lamp (e.g. glass fiber and metallic fillers of plastics, etc.), which will complicate and limit the recovery of materials from the LED retrofit.

The LEDs (retrofits) have to be collected and treated separately from the other lamps, in order to maximize recovery of the valuable In/Ga from them. Mixing with fluorescent lamps should be avoided in order to prevent loss of valuable materials and mixing of different types of glass. This also puts demands on the possibility to identify LED retrofits from fluorescent lamps, which is difficult due to high similarity in appearance.

The recycling of the phosphor/fluorescent powders and LEDs will be the topic of the next sections in this chapter. In order to recover these materials from lamps, suitable physical and chemical separation techniques are required, which will be discussed briefly in this chapter.

23.7.3 Metallurgical Recovery of REO-Containing Fluorescent Powders from Lighting

23.7.3.1 Recovery of Rare Earths (Oxides)

Rare earths (REs) can be recovered as described by Jiang et al. (2005) in general terms from minerals. This technology can also be applied to concentrated REs from powders.

Separating the RE mixtures subsequent to leaching into individual rare earth compounds is a key issue. Rabah (2008) and McGill (2005) give an overview of the different methods used to separate REs into the various compounds and metals. Two principal types of process are used for the extraction of rare earth elements (McGill, 2005):

- Solid—liquid systems using fractional precipitation or crystallization, or ion exchange. Ion exchange processes are used in the production of small quantities of higher value heavy rare earth elements.
- Liquid—liquid systems using solvent extraction. This process is the most commonly used commercial process for the extraction of rare earth elements.

McGill (2005) discuss in detail different approaches to recover REs from various primary resources, discussing the efficiency of different extractants. McGill discusses the unique paths required to recover these rare earth elements, using an optimal mix of techniques that exploit the individual properties of the REs to separate them into high-purity oxides. These can subsequently be reduced by molten salt electrolysis or metallothermic reduction to metal and find their way into metal applications such as magnets.

The RE concentrates would contain a mix of Ca, Al, Si and Mg as part of the ore; the techniques to recover REs from concentrates take care of these. The phosphors contained in fluorescent lamps also contain these elements, and hence should not pose a problem in the recovery of the REs present in the phosphors.

It is evident from Figure 22.36 that the depicted RE oxides are relatively stable and cannot be recovered generally by normal reduction processes, e.g. by reduction with carbon (the depicted REs have rather negative delta G values). Therefore, generally speaking, RE metals are usually produced by molten salt electrolysis and metallothermic reduction (McGill, 2005).

The various techniques discussed by McGill (2005) have been applied to the recovery of Eu and Y from phosphors from fluorescent lamps. Rabah (2008) discusses the recovery of Y and Eu from a phosphor that contains 1.65% and 1.62% Y and Eu, respectively, using a sulfuric/nitric acid mixture to create $Y_2(SO_4)_3$ and $Eu_2(SO_4)_3$, which were subsequently separated into Y_2O_3 and Eu_2O_3. Under these conditions calcium sulfate is also created, but the RE separation techniques take care of this.

22.7.4 Recovery of Indium and Gallium from LEDs

Indium (In) and gallium (Ga), present in LEDs, can be recovered to an extent in normal pyrometallurgical smelting operations, e.g. during e-waste and copper scrap recycling (van Schaik and Reuter, 2010). This is clear as the carbon line in Figure 22.36 crosses the In_2O_3 and Ga_2O_3 lines at about 720 and 1000 °C, respectively, implying that with appropriate oxygen partial pressure in the pyrometallurgical reactor these elements can be distributed to phases from which they can be recovered. Indium distributes between copper and slag as a function of various variables, which include CaO in slag, slag composition, temperature, oxygen partial pressure, etc. as shown by Anindya et al. (2011). Also note that indium has gaseous species, e.g. InO(g), as does Ga, e.g. GaO(g), which will report to an extent to the flue dust and hence are recovered along this avenue to a lesser extent (Figure 22.21 and Figure 22.22).

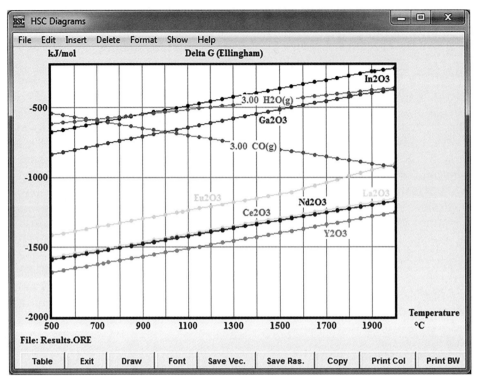

FIGURE 22.36 A general overview of the stability of a selected few RE oxides relative to indium and gallium oxide and CO(g) as a reducing agent (HSC 7, Outotec, 2013), from which is clear that these elements go lost in pyrometallurgical nonferrous smelters.

22.7.5 Recovery of Rare Earths from Optical Glass

A typical flowsheet that recovers REs from optical glass (which contains ~46% La_2O_3, 9% Y_2O_3 and 5% Gd_2O_3) includes, for example, alkaline leach followed by acid leach and conversion to $RECl_3$. These steps have usually been adapted but essentially are based on the techniques available for the recovery of REs from ores. Recycling of glass from lamps also includes the reuse of the separated and recovered glass fraction by the lamps producers and/or thermal treatments options in which the material content of the glass (or at least part of it) is a substitute for primary process inputs (Indaver, 2013). This processing route poses demands on the sorting of different glass types (e.g. by sorting TL from LED) from e.g. TL lamps and LED due to different composition of the glass used. Subsequently the RE mixture must be separated according to the various techniques summarized by McGill (2005) and also briefly discussed above.

22.7.6 Economics of Recovering Rare Earths from Lighting

The economy of scale, requirements for a critical mass of products and rapid changes in product and used technologies for lighting applications (fluorescent lamps, LED, OLED) would dictate that the phosphors and LEDs should be treated using existing technology. The key is therefore to produce suitable recyclates so that these can be channeled into the correct processing routes, if these are available. LEDs could find their way to normal copper scrap recycling facilities, of which there are a number in the EU at present. Fluorescent powders from lamps contain some REOs. Economic sorting and recovery could hence be considered if an economy of scale exists as well as the availability of an existing recovery infrastructure.

22.8 TECHNOLOGY FOR RECYCLING OF BATTERIES AND CATALYSTS

Various technologies exist for the recycling of batteries. These are discussed in detail in UNEP (2013). For the example of lead batteries, commercial technologies exist, e.g. TSL (www.outotec.com and www.recylex.fr). Short rotary kilns also represent state-of-the-art battery recycling (www.engitec.com). Specific commercial plants have been developed to recover metal content from the range of batteries listed in this chapter (www.recupyl.com, www.umicore.com, www.batrec.ch (part of Veolia group, www.veolia.com) and similar).

Important for all of these metallurgical plants is to receive well-sorted batteries in order to position the metallurgy in a sector of the metal wheel (Figure 22.18), in order to maximize recovery of metals and materials from them.

Most catalysts are processed pyrometallurgically, usually in plasma-type furnaces but also TSL (Outotec) furnaces (Rumpold and Antrekowitsch, 2012). The carrier material of catalysts makes this tricky because the alumina, silica, magnesia and others have a significant effect on the slag liquidus temperature and hence affect furnace operability as well as metal recovery (see Figures 22.20 and 22.37). In a nonferrous furnace, alumina can be handled to an extent

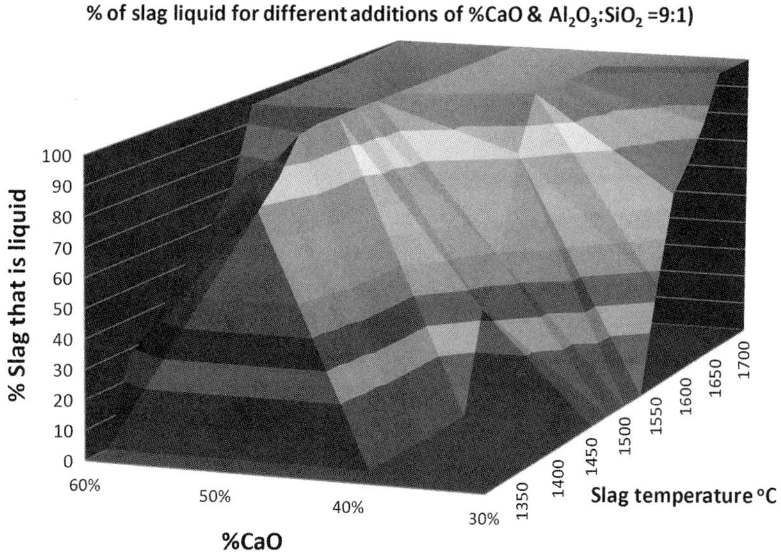

FIGURE 22.37 The effect of alumina, lime and magnesia on the slag liquidus temperature, showing that conditions must be controlled tightly to maintain a fluid slag at a relatively high temperature of above 1600 °C.

(Figure 22.20 for TSL furnace, www.outotec.com) if the temperature and partial oxygen pressure are controlled well. However, with little iron in the slag system, then rather high operating temperatures are required, shown by Figure 22.37 as achieved in plasma-type furnaces (www.tetronics.com). But once again, enough lime must be added, and this must be controlled tightly to ensure that the slag does not suddenly start to freeze or contain too much precipitated solids, which also increases the viscosity and can lead to possible foaming and catastrophic furnace failure.

Usually precious metals are then collected in metal phases, e.g. iron alloy, copper, and so on, and then further refined.

The high-temperature processing also takes care of all organic/carbon-based materials, using them as reductant/fuel, as the off-gas systems represent state-of-the-art processing and take care of dioxins, dust, etc.

22.9 DESIGN FOR RECYCLING AND RESOURCE EFFICIENCY

The functionality of products necessitates that materials are designed to interact in a complex manner, which creates by default very complex waste streams. To recover metals and materials from these requires a deep understanding of the physics of separation as well as the complete system. True DfR tools will incorporate the complex physics into the design tools to be able to show the inevitable losses that will appear due to the functional connections in products (Figure 22.15). Figure 22.18 shows the underlying thermodynamics and design issues that close the material cycle.

It must be noted that in various cases DfR makes no sense, as the metal value is so low and the functionality makes it impossible to design for recycling. Instead, time and money must be spent on designing the recycling systems for optimal resource recovery and resource efficiency. Therefore design for resource efficiency should be the driver for creating a recycling system that is built using BAT and is certified to meet this requirement.

Design for resource efficiency (DfRE), of which DfR is a subset, demands knowledge of liberation behavior, of the particulate quality of recyclates, of the separation efficiency linked to the compatibility and thus recovery and/or loss of material in metallurgical processing, and of the modeling thereof, all as a function of design choices, connection type and connected materials. Insight into this provides a technology- and industrial process-driven basis for DfR to optimize resource efficiency and closure of material cycles for both commodity and critical and scarce elements in complex consumer products, such as cars and e-waste/WEEE products.

Not all the metals can be extracted from EoL goods, for reasons explained in this chapter and shown by the metal wheel in Figure 22.18. The common, commodity metals like steel, magnesium and copper can be recovered relatively easily, as these are often used in relatively simple applications. However, the small amounts of precious/critical/valuable metals in, for example, WEEE can be harder to recover from often in excess of 50 elements.

A material compatibility matrix (Table 22.15) (van Schaik and Reuter, 2012, 2013) and the design for resource efficiency metal wheel (Figure 22.18) are developed as preliminary DfR tools for designers, which capture the physics of recycling. The simulation models and the wheel are based on system models that link product design to the complete recycling system, as shown in Figure 22.38. Table 22.15 considers some of these precious/critical metals and illustrates where they can be recovered using best current BAT practices. Although different appliances may all have similar suites of functional materials loosely called "mineralogies," i.e. contained elements and the functional connections, their recovery is not the same and hence their recycling rates are

TABLE 22.15 Compatibility Matrix as a Function of Metallurgical Recovery (Reuter and van Schaik, 2012a,b, 2013a,b, van Schaik, 2011). Developed from Experience and Estimations from Tools Such as Shown by Figures 22.18 and 22.23

Recoverability * (per equipment/application)	PMs		PGMs		Rare Earths (Oxides)			Other					
	Ag	Au	Pd	Pt	Y	Eu	Other REs	Sb	Co	In	Ga	W	Ta
Large Household Appliances (ex Fridge)													
Washing machine													
Small household appliances													
Video recorder													
DVD player													
Hifi unit													
Radio set													
TVs													
CRT TV													
LCD screens													
Lighting													
Fluorescent lamps													
LED													
Mobile telephone													
Batteries (NiMH)													

*Recovery is a function of processing route, product design etc. The table gives the recovery for the present most likely route, but could change if suitable technology exist.

Recovery possible	If separately recovered and/or if there is appropriate technology and recovery available.
limited recovery/recovery under certain conditions	If separately recovered. Partial or substantial losses during separation and/or processing/metallurgy. Recovery if appropriate systems exist.
No separate recovery	Pure recovery not possible. Lost in bulk recyclates during separation and/or during metallurgy into different non-valuable phases.
For a combination of colours	Depending on process route followed high recovery or high losses possible. Needs careful attention to design, infrastructure, legislation etc. This is especially possible for metals closely linked where one metal can be recovered while the other due to this selection of recovery then goes lost. This is driven by the thermodynamics, technology, design etc.

different. Table 22.15 shows that, depending on the product and the combinations of materials, the recovery of metals may be different due to chemistry, concentration and metallurgical processes being incompatible.

Therefore forcing recycling rates quotas for especially the minor metals would be a fallacy; rather, a focus should be to maximize recovery of the elements. Given the correct economic basis, BAT infrastructure and market-driven policy forcing, for example, zero landfill will help to maximize recovery and recycling rates. It is has always been the case in any metallurgical plant to recover valuable elements. If there is an economic incentive to do so, recovery will happen.

DfR should design products such that functional groupings in the product when dismantled produce the maximum recovery when recycled to a specific segment in the metal wheel. Obviously each functional group may have metals that are incompatible, but these will then be the loss from the system. Thus DfR should be guided by the possible metallurgical infrastructure that can recover metals economically. If product design can keep thermodynamically compatible materials close together, then metallurgical technology can deal with them. For example, the recycling of printed circuit boards is often carried out with copper metallurgy. This means that silver and gold can be easily recycled, but aluminum will be lost (see Figure 22.20). To reduce this loss, designers could avoid the presence of aluminum with copper, or could plan a design that facilitates the removal of aluminum during pre-processing. For example, heat dissipaters are often made of aluminum and could be designed for easy manual extraction (called "design for disassembly"). For other parts containing precious materials, the recovery ratio during dismantling or sorting could be increased by visually identifying such components.

It is not easy to make an LED or fluorescent lamp, car, electronic product, etc., any simpler, as the metals in close proximity create their functionality. This will always be at the heart of product design, but designers often have some leeway in the choice or arrangement of materials. Continuous innovation in materials can aim at delivering similar functionality with other, thermodynamically compatible, materials. In short, it is a truly complex challenge to design products for optimal recycling. However, what is good for DfR could, and most likely will, defeat functionality, especially that of a complex product.

Applying first principles, knowledge of material combinations in design linked to compatibility and hence recovery and losses of materials in metallurgical processing is critical in ensuring the supply and recovery of commodities and in particular critical resources from end-of-life products. Design for recycling thus needs a supportive infrastructure. Fortunately, the tools as discussed above (see Figure 22.38) now exist that provide designers with the information they need about recycling outcomes from different designs. These are based on the increasing and limiting factors for various combined materials in EEE on the bases of liberation, sorting and second law of thermodynamics as shown by the metal wheel, and allow the assessing of different designs. Where these are widely known and used, recycling outcomes can be improved. Examples of the application of these advanced tools for DfR/DfRE are discussed below for cars and WEEE.

The other part of a supportive infrastructure for design for resource efficiency is the creation of an economic or policy framework that motivates designers to consider recycling. This might be the case where consumers reward good design by preferring products they know to have high standards. More commonly, public policy can require, or influence, product manufacturers to take recycling into account, thus encouraging material innovation without constraining functionality. Eco-labeling, the topic in the next section, can support this and provide insight for consumers on the resource and environmental performance of products on a physics and industrial-reality basis. At the same time, such eco-labeling can help producers/OEMs to distinguish resource-efficient design from lesser products, hence providing a driver for design for resource efficiency.

Another aspect of design for recycling relates to supporting markets for recycled metal products. When recycling produces metals, or other compounds, that have the same purity (or grade) as primary metal, they can easily be sold at equivalent prices. This applies in most cases where state-of-the-art processes are used. Where recycling produces metal with impurities, this might fail to find markets, even when it has the physical or chemical properties needed for use in specific products. Manufacturers may not choose to use such recycled metals, either because they do not know the precise mix of elements and compounds on offer, or because they are unsure of the quality or consistency of the recycled metal on offer. These are not technical but marketing issues, which nevertheless influence the success of recycling. Designers can help to create markets by being willing to take up suitable recycled material.

22.9.1 Recyclability Index and Eco-Labeling of Products

The interactions in the recycling system are two directional along the chain from design, disposal, collection, sorting and (metallurgical) recycling (Reuter and van Schaik, 2012a,b). Each of the stages from design downwards impacts final recycling through the changes it creates in the physical properties of the material going for recycling. In the other direction, the realization of economic value from final recycled metals is the natural driver for collection activities, incentives for disposal and, potentially, DfRE. If products were all produced out of one metal, the system would be relatively simple, and the interactions linear, up and down the recycling "chain". In practice, as each product contains many metals as well as other substances imparting functionality, there are many recycling chains (from product to recycled metal) for each metal. These chains interact during design, collection and recycling. This creates a complex system, with a level of complexity that needs to be understood in a workable way by actors in the chain and policymakers, as reflected by Figure 22.38.

Figure 22.38 gives a general overview of various technologies applied during recycling and the processing of waste materials. If this is concretized more specifically for a product such as an SHHA, various specific technologies come into play, as reflected by the HSC (www.outotec.com) simulations (Figures 22.24, 22.26−22.28) of the interlinked system of technologies and activities. Rigorous simulation will provide a thermodynamic basis also for the environmental impact analysis that is linked to GaBi (www.pe-international.com). Through this link, it is possible to evaluate the environmental impact of different designs and scenarios based on actual environmental impact linked to the mass and material flows and the detailed compositions of each stream, including the design and recycling route-dependent recoveries, losses and the environmental footprint of created residues. This is not possible with currently applied LCA/environmental performance calculations based on general databases, which are therefore incapable of capturing the essential detail of recycling technology and design-dependent recyclability in view of recycling optimization and design for recycling.

Care should hence be taken by policymakers and OEMs when applying these simple linear calculations for recycling indexes, since these will be leading to false conclusions with respect to resource efficiency improvement and might lead to harmful decisions for the industry. Furthermore, using this basis can provide a rigorously based estimation of design rules, as shown in the next section. However, it must be noted that these are only loose estimations (much favored by the product designers who have no thermodynamic grounding); best would be to estimate recyclability of products with tools such as shown below run by true experts in the recycling field who have knowledge of metallurgy and process technological experience.

A result of this physics-based modeling approach of the system from design (CAD) to recycling is also a rigorously based estimation of the recyclability of products to determine an **eco-label** for a product, as these tools can be directly linked to design tools and their produced bill of materials, full material declaration and/or chemical content analysis. Figure 22.38 shows such a basis for the creation of a rigorous recyclability index based on BAT technology, moving away from the simplistic indexes still being developed and applied thoughtlessly by nonrecycling experts (e.g. OEMs and component designers), which do not represent the complexities and depth of recycling technology, its physics, chemistry and economics. Such a recycling knowledge-based **eco-Label** has the depth to really distinguish differences in product designs and to distinguish the more resource efficient design from the other. This might provide a driver for OEMs to design for resource efficiency. This approach differs fundamentally from many general DfR studies and applications as well as recycling indices as performed nowadays, relying on simplified one-dimensional DfR rules and recycling

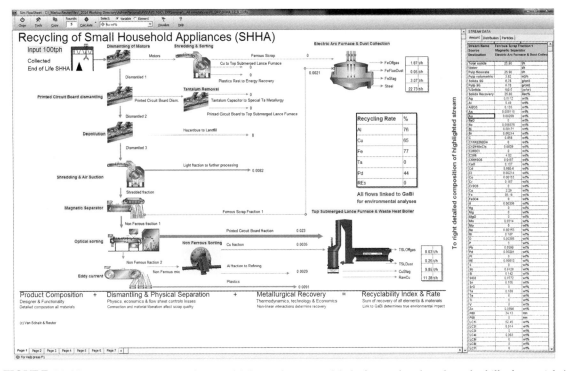

FIGURE 22.38 Rigorous estimation of a recyclability index or eco-label of a product based on the bill of materials/chemical material content through linking design tools with simulation and environmental impact tools as shown. The detailed composition, quality of recyclates and recycling products, dispersion and recovery, losses and fugitive emissions of all materials/elements (as illustrated here for the example of Au) as provided by the simulation models provide the in-depth knowledge and indispensable detail for rigorous recyclability index and eco-label (environmental assessment) calculations for a product.

calculations. These rules do not represent and reflect the realities of recycling, recyclate quality as a function of design, liberation and sorting efficiency and metallurgical process recoveries of different metals/materials as a function of recyclate quality and concentrations of materials in it as captured by the models as discussed here. The general approaches are not based on a proper process simulation basis that can include this required depth of nonlinear design to recycling (including slag chemistry and metallurgical processing efficiencies) to capture the real issues of resource efficiency. The discussed approach of Figure 22.38 (and other flowsheets given for ELVs and WEEE) applying a product-centric physics-based understanding of recycling allows by its depth for the evaluation of changes in design (for recycling) by including all effects, which one particular change or substitution of materials might have on the entire recycling system.

22.9.1.1 Some DfR Rules and Guidelines

Based on the expertise of the authors as discussed above and based on the developed recycling Process simulation models (van Schaik/Reuter, 2002–2013), a study was performed by the authors on DfR for NVMP/Wecycle (The Netherlands) to develop "product-centric simulation-based DfR rules" (van Schaik et al.,

2013), which take into consideration the possibilities and limits of recycling (UNEP, 2013). This includes recycling technology and physical limits due to functional linkages and combinations of materials in a product. Various recyclers have provided their input on design-related issues in recycling to develop these rules. van Schaik has investigated numerous recyclates to this end as well to understand the relationship between design and recyclate quality.

The following 10 DfR rules (5 fundamental and 5 derived) have been derived, addressing the technological and economical possibilities and limits in the entire recycling system from design to metallurgy in relation to material interactions, recovery, losses and emissions and resource efficiency:

1. DfR rules are product and recycling system specific, oversimplification of recycling by defining general DfR rules will not produce the intended goal of resource efficiency.
   a. Due to its functional and unique mix of materials, each product has a unique recyclability profile. This implies that every product has a unique set of DfR guidelines which are product and recycling system specific.
   b. These DfR guidelines are to be derived and iteratively refined for each product/product group separately by the application of recycling simulation models that can map any (BAT) recycling system and its opportunities and limits.
2. DfR needs model- and simulation-based quantification.
   a. DfR demands a tool (process simulation models) that is capable of quantifying the product's recycling profile and performance to pinpoint DfR issues of importance and to give priority within design adjustments to be implemented and insight in the effect of improved design on resource efficiency (recycling rate, toxicity, scarce material recovery/losses, environmental impact, etc.).
3. Design data should be accessible and available in a consistent format that is compatible with the detail required to optimize and quantify recycling performance of products for all metals, materials and compounds present.
   a. Detail and format of the product data on product material composition (including chemical compounds) and construction should have the resolution to quantify, identify and localize the commodity/critical/disturbing materials.
4. Economically viable technology infrastructure and rigorous tools must be in existence for realizing industrial DfR rules and methodology.
   a. Design must be based on a robust physical separation and sorting infrastructure that is minimally capable of producing economically valuable recyclates.
   b. A robust metallurgical infrastructure and system must be in place to ensure maximum recovery of all "critical" materials from complex recyclates and dismantled functional subunits of a product.
   c. Environmental footprint and eco-design should include the whole chain of processing to ensure that all materials, residues and fugitive emissions are tracked. This requires suitable global policy to be in place.
5. CAD, process and system design tools must be linked to recycling system process simulation tools to realize technology based, realistic and economically viable DfR.
   a. Linking of existing and industry applied process simulation tools to CAD/design tools is a necessary step forward to realize realistic and economically viable DfR. This is a rigorous basis for industrially useful DfR rules and methodology. The example in Figure 22.42 shows that based on rigorous recycling process simulation and

detailed product compositional data derived from CAD/design tools, recycling performance indicators and recycling indices including environmental analyses can be derived and provide the basis for DfR. Eco-labels for a product can be derived on this basis, which provide the depth to rigorously distinguish differences in product designs and identify the most resource efficient designs.

Various design for recycling guidelines have been derived by applying the above-listed fundamental DfR rules and principles. Recycling process simulation tools are used to define, validate and quantify the set of guidelines per product (of which the list below shows some possible guidelines). This physics-based approach also can set priorities between the different guidelines and quantify the necessity and potential result of DfR. Examples that follow the guidelines given below can be found in general DfR approaches; however, by the implementation of the fundamental rules given above and simulation as a basis to derive and refine them, unique sets of guidelines are derived per product as a function of material mix and (BAT) recycling systems, including a mindful consideration of product functional demands (whereas a fixed set of all possible guidelines will leave no room for the designer to design/construct a product). Input from recyclers was also used in defining examples for the certain guidelines. Some examples of these DfR guidelines are the following:

- Identify and minimize the use of materials that will cause losses and contaminations in recycling due to material characteristics and behavior in sorting.
  - Examples are the use and construction of concrete in washing machines; glass plates in refrigerators; application of isolation materials; and use of colored (other than brown or green) printed wire boards
- Identify components/clusters in a product that will cause problems and losses in recycling due to combined and applied materials.
  - A compatibility matrix (see Figure 22.44) can be derived per product based on the knowledge behind the metal wheel (Figure 22.18) and is useful for a quick first screening. It is important to realize that compatibility tables are not a DfR tool on their own, since they give *no* indication of resource efficiency, material recovery, losses and fugitive emissions.
- Design clusters or subunits in products that can be easily removed and that match with the final treatment recycling options (i.e. metal wheel), e.g.:
  - Examples are the use or removal of Al applied as heat sinks on a printed wire board; separate PWBs for different functions, e.g. power boards (high Fe content, which will be lost in PWB processing route), control boards; removal of Ta capacitors, etc.
- Labeling (including carefully considered standardization) of products/components based on recovery and/or incompatibility so that they can be easily identified from recyclates and waste streams. Thus design for waste stream sorting or design for (automated) dismantling/sorting are important.
  - Examples are (1) use of color- or identification-based labeling and easy to break connections for Ta capacitors (of crucial importance is metallurgical knowledge as captured in process simulation tools to understand what the quality requirements are for the existing industrial Ta production/recycling infrastructure); (2) labeling/identification-based waste stream sorting of CFL from LED lamps; (3) standardization of marking and identification (e.g. type of marking and position) of cooling liquid and gas and

marking of type of liquid and gas applied in fluid system and foam (these might be different); and (4) marking of the tapping point on compressors.
- Be mindful of liberation of materials in design (design for liberation)
  - Simulation and knowledge of liberation behavior in relation to design, particulate and recyclate quality and recycling (metallurgical) efficiency is crucial. Examples are avoiding bolts/rivets of dissimilar materials (e.g. Fe bolts to Al, PWB, plastics, etc.), as these generally produce a liberation problem and therefore creates cross-contamination of the different recyclate fractions (see Figure 22.39); and minimizing the use of nonreversible adhesives for incompatible and undesired material combinations, such as gluing of glass on the steel mask of a CRT TV, shrink films on tube lamps, sealed batteries, PUR foam glued to steel/aluminum/plastic, wood glued to plastic, etc.

22.9.2 Design for Recycling of Multimaterial Lightweight Automobiles

The models for cars and WEEE as discussed in previous sections (van Schaik/Reuter, 2002–2013) achieve a physics-based link between product design and resource efficiency through DfR and have simulated the effect of design considerations on the liberation behavior and quality of recyclates of complex consumer products based on extensive industrial and experimental data collection. In these models, such as breakage laws of comminution as a function of material connections, related to particulate characteristics of recyclate flows, are addressed and linked to sorting physics and chemistry and thermodynamics of

FIGURE 22.39 Particulate quality (determining recyclate quality and recovery/losses) as a function of design. An impression of unliberated and multimaterial particles as found in various WEEE recyclates.

high-temperature processing and recovery of resources from recyclate streams.

22.9.2.1 Example: Designing Car Bodies: The Super-Light Car Project

On this rigorous basis, van Schaik and Reuter have developed and applied fuzzy recycling models that have been applied for DfR in the EU sixth framework project SuperLightCar and provide real-time design-related recycling calculations (van Schaik and Reuter, 2007; Krinke et al., 2009). For the complex recycling and liberation process, a fuzzy set modeling approach has been developed with Matlab's Fuzzy Logic Toolbox® Matlab 1994-2014 The Mathworks www.mathworks.com), which is based on the knowledge provided by these physics-based recycling models. This approach is well suited for creating a link to product design through CAD software. The body-in-white (BIW) for the EU's SuperLightCar project is a novel construction of steel, light metals and polymers reducing the weight by 35% for a substantially lower manufacturing cost per kg (SuperLightCar, 2005–2009; Krinke et al., 2009). The ultimate objective was to reduce not only weight but also fuel consumption, creating the basis for future passenger-vehicle power

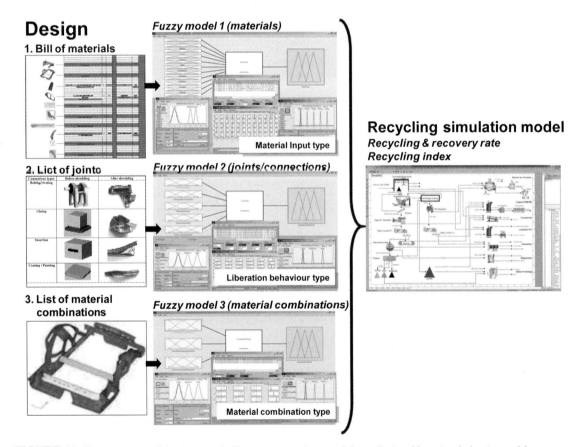

FIGURE 22.40 Overview of the fuzzy rule liberation recycling model, predicting liberation behavior and hence recycling/recovery rate as a function of design specifications (list of materials, joints and material combinations) (van Schaik and Reuter, 2007).

trains. The project created a tool that links CAD to a costing and LCA tool. Due to the EU's ELV legislation (95% material recycling and energy recovery by 2015), a major aspect of the project was to calculate the recyclability of the BIW. This was done with the ELV recycling model developed by Reuter et al. (2006), van Schaik and Reuter (2007) and van Schaik et al. (2002), linking the different tools between product design and recycling. The fuzzy set models were integrated with both CAD and LCA tools (Krinke et al., 2009; Reuter, 2011; van Schaik and Reuter, 2007) and have been applied to calculate recycling rates of this lightweight multimaterial design (including various design concepts) (see Figure 22.40). This also ensures that environmental models are provided with physics-based information on the EoL behavior of products and LCA can move away from general, too simplistic databases, which do not provide the dynamic and product design dependent recycling rates and creation of residues. Figure 22.41 illustrates examples of the recycling rate calculations performed by the fuzzy model for the SLC design concepts.

In addition to LCA data, this physics-based approach provides recycling rates and exergy data based on achieved recyclate quality and

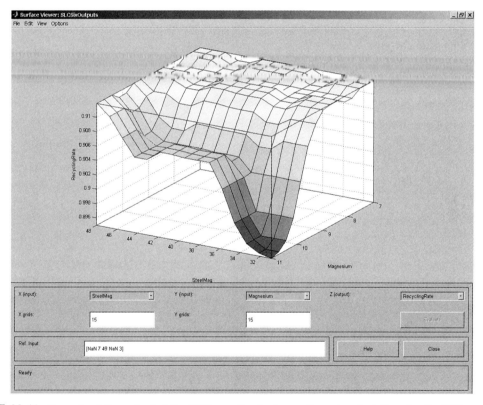

FIGURE 22.41 Example of the output of the SLC recycling tool: recycling-rate calculations for the SLC concept as a function of variations in design for steel and magnesium content. There is no single recycling value as it is a complex function of product mineralogy (van Schaik and Reuter, 2007). The figure shows why the simplistic one-dimensional recycling rate definitions by e.g. Reck, B.K. and Graedel, T.E. (2012) are only limiting cases as they do not consider complexity.

therefore design. This information helps OEMs in producing sustainable product designs, creating awareness of the costs and benefits of design, products and recycling systems for manufacturers, designers, legislators, recyclers, etc.. They also show the (un)feasibility and environmental (in)efficiency of imposed recycling/recovery targets (van Schaik/Reuter, 2002–2013). This DfR/DfS tool, incorporating the recycling model's input, provides environmental-impact data for different designs, while the costing tools provide the cost of each design.

22.9.3 Design for Recycling of Electrical and Electronic Equipment

The functionality and design of (waste) electrical and electronic equipment ((W)EEE), involving 50+ elements being applied in different compounds and phases, complicate recycling due to their ever-more complex structures producing partially unliberated, low-grade and complex WEEE recyclates. This complex interlinked structure of materials, compounds and recyclate flows demands that DfR needs to be addressed from a product-centric perspective and rigorously simulated to obtain a thorough understanding of recycling. Understanding the link between product design, particulate and interconnected nature of recyclates and industrial carrier metal processing infrastructures at a physics-based depth is at the core of DfR as illustrated by Figure 22.42 (van Schaik and Reuter, 2012).

WEEE recyclates can enter any of the different metallurgical processing infrastructures, which exist for the production of the most common commodity (carrier) metals used in society, e.g. iron, aluminum, lead and copper. The destination of the metal and metal mixtures to one of the different carrier metal routes is determined by the combination of design choices and design

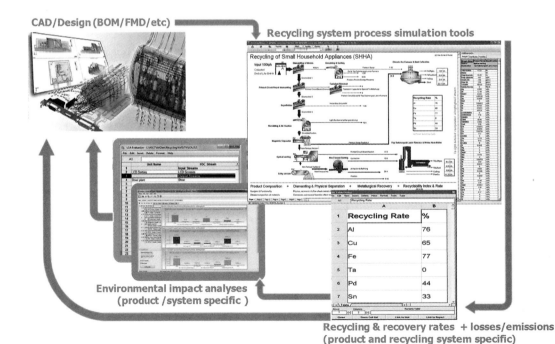

FIGURE 22.42 DfR for WEEE recyclates linking CAD with LCA and simulation tools.

characteristics (material selection, combination and connection types) (Reuter and van Schaik, 2012b, and van Schaik and Reuter, 2010), liberation and sorting efficiency, which create changes in the physical properties of recyclate grades (quality) going for recycling. The metal wheel succinctly expresses what design-determined linkages mean for processing of different metals during recycling and also shows the inevitable losses which represent the limitations of the system due to design requirements and related deficiencies in liberation and sorting. The computer-based modeling/simulation tools of the recycling performance of products in the recycling system as developed and applied by the authors and discussed in this chapter are being improved to help guide product design that facilitates more recycling. It includes expert rule-based modeling of liberation behavior and particulate composition of recyclate flows and is hence based on the realities of how products and their constituents break up and separate in likely BAT recycling processes. Figure 22.14 gives a simplified graphical representation of recycling flowsheets as simulated by these models for the WEEE recycling chain. Figures 22.24 and Figures 22.26–22.28 also illustrate how the various recyclate sorting routes (e.g. to produce ferrous, aluminum, copper, etc. recyclates) are connected to the various metal recovery routes of the metal wheel, and show the inevitable losses of metals if recyclates report to the incorrect metallurgical infrastructure (different segments of Figure 22.18 are superimposed on the flowsheets of Figures 22.24 and Figures 22.26–22.28). The metal wheel in Figure 22.43

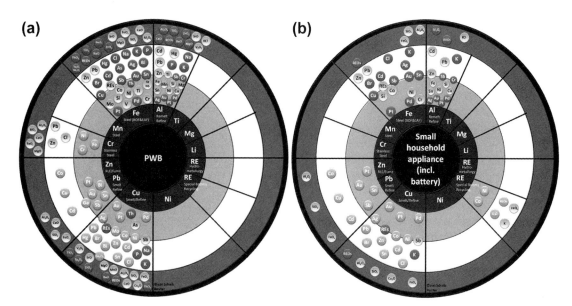

FIGURE 22.43 The metal wheel for recycling of (a) printed wire board (PWB) and (b) small household appliance with battery (an example product), showing the destination of elements in the product/component in the carrier metal processing infrastructures of each segment (recyclate flow). Note that this is an average picture, but requires the basis of extractive process metallurgy as well as the models depicted by Figure 22.24 and Figures 22.26–22.28 to fully determine to which recyclate flow and processing route elements go. Therefore, this figure is only an average representation and is to be used only if the thermodynamic, economic and technological complexity is understood.

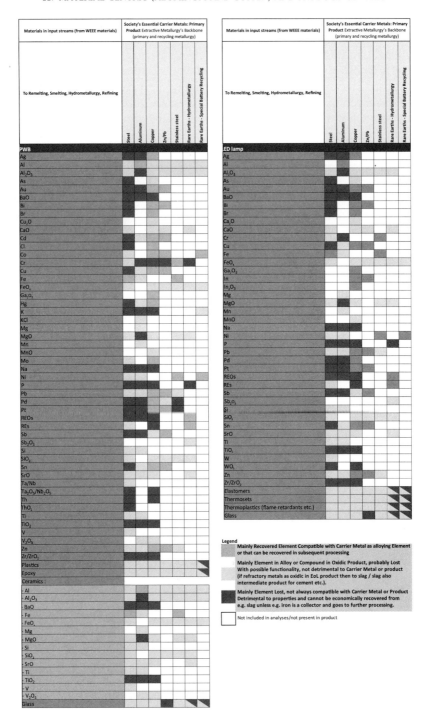

FIGURE 22.44 Design/material (in)compatibility matrices for a PWB and (LED) lamps, showing the (in)compatibility of elements in the product/component with the various carrier metal processing routes (Figure 22.18). Aggregated product data originating from different sources is presented. Product compositional data can also differ for different designs.

illustrates for the example of a PWB (printed wire board) and an example of a small household appliance (including battery) what happens to different metals entering the iron (Fe), aluminum (Al), copper (Cu), zinc (Zn) or lead (Pb) processing routes (Reuter and van Schaik, 2012a,b and Reuter et al., 2005)—where each pie of the wheel is understood to be a complete infrastructure. Design compatibility tables have been developed for a range of WEEE products on the basis of the metal wheels and on the background of earlier work by the authors on physics-based recycling simulation models for WEEE (as summarized in this chapter).

Figure 22.44 shows tables for PWBs from complex WEEE products and (LED) lamps (see also next section). Compatibility tables and metal wheels (see Figures 22.18 and 22.43) have been developed by the authors for a range of WEEE products, also specified for different components within the products in order to pinpoint design challenges in view of recycling on a detailed level and allow for the link to component level-driven (automated) disassembly developments. Innovating and optimizing product design in favor of recycling (if possible at all in view of functionality reasons) requires a deep understanding of physics of liberation and separation linked to thermodynamics, technology and metallurgy as depicted in the metal wheels and compatibility matrices. This recycling perspective, which takes into account the composition of recyclable streams and (metallurgical) processing infrastructures as applied for WEEE recyclate treatment, is at the core of the product-centric approach to recycling (UNEP, 2013). The compatibility matrices provide, being linked to recyclate quality as a function of design-related shredding and sorting efficiency technology, industry-relevant insights into technological and thermodynamic requirements for "creation" of economically valuable particles and hence recyclates by DfR and sorting. The tools are available to do this and cover the entire system from design to metallurgy in view of DfR for WEEE as illustrated by these examples.

22.9.4 Design for Recycling of Lighting

In the recycling system as depicted by Figure 22.35 for the example of fluorescent lamps, the products are broken to liberate different design-connected materials to allow for subsequent sorting of metals and materials into the different indicated recyclate fractions, including the removal of fluorescent powders and capturing Hg from the fractions (for fluorescent lamps). With respect to the shredding and sorting of the more material complex LED retrofit lamps, liberation and sorting of e.g. Al, In/Ga-containing fractions and PCB fractions from the other metals and materials is essential to allow for optimal recovery of the present materials in final treatment processes. The metal wheel in Figure 22.18 and the compatibility matrix for LED lamps in Figure 22.44 show consequences of current design on recoverability and hence options for improvement of design (by changing connections for liberation behavior, material combinations and applied materials) to improve recyclability.

Due to (functional) design considerations where materials are attached and combined in sometimes very small quantities in complex chemical compounds (e.g. In/Ga in LED, electronic components in PWBs, etc.) during shredding and sorting part of these materials are likely to stay connected and will end of up in mixtures of metal/materials in recyclate flows rather than as pure metal/material recyclate flows. Innovations in design such as the introduction of fracture lines in e.g. the relative robust Al heat sinks can be introduced to liberate these from the PWB and other materials.

22.9.4.1 Alternatives for the Use of RE(O)s in Fluorescent Powder in Lighting (and TVs)

Another option in DfR could be the substitution of problematic materials. This could also be a supportive solution for critical/scarce elements. Replacement of materials must, however, be possible from a product functional point of view (as well as economic). The USGS (U.S. Geological Survey, 2010) has indicated that yttrium in fluorescent powders cannot be replaced by other elements. Mn^{2+} is mentioned as a possible replacement for terbium (Jüstel, 2007) (lower quality). A potential decrease of 40% for the consumption of Tb was predicted (Lynas, 2011) as a result of technological developments in lighting. The U.S. Department of Energy (2010) indicates that there is no proven substitution for europium for fluorescent lamps and for the red phosphor in CRT TVs (although the latter is phasing out due to LCD and LED developments in TVs and displays). According to the same source, future OLEDs can be free from REs. LED technology can reduce the demand for lanthanum and terbium, although the use of and demand for cerium and europium will remain.

22.9.5 Requirements for the Application of Design for Recycling Tools and Rules: Physics-Based System Modeling of Recycling in Relation to Product Design

The complexity of products and recycling implies that compatibility tables and data from the metal wheels, examples of which are given by Figures 22.43 and 22.44, cannot simply be cut and paste into DfR tools. Useful design for recycling tools must be based on the rigorous methods discussed here. Simplistic linear methods without any physics basis will be of no use in guiding product design. The more complex the products are, the more rigorous the methods must become for capturing all nonlinear thermodynamic and physics interactions between materials, for a clear prediction of recyclability and resource efficiency. Recycling indices, based on physics-based understanding of the entire recycling system as discussed above, are a supportive tool to pinpoint problems in recycling and make the effect of DfR changes clear in the EoL phase. In summary, recyclability indicators will only be of use if they are based on rigorous physics, as otherwise they can mislead decision-makers and consumers.

22.9.6 Product Composition Data Requirements for Design for Resource Efficiency

The metal wheel and compatibility matrices in Figures 22.43 and 22.44 and underlying thermodynamics and technology dictate the level of depth to which product compositional data should become available from manufacturers and OEMs in order to make a sound design for recycling/resource efficiency assessment for complex multimaterial products. Not only should this data be provided on a general physical material level, in line with a simple material-centric approach, product data for complex multimaterial designs should also be generated and made available by manufacturing industries on a mineral and therefore product-centric basis. This implies that the mass composition of a product should become known on both a physical as well as chemical/compound level of detail for the entire product as well as for the different components. The latter is of importance to pinpoint design for recycling issues on a component level, rather than for the entire product. This will allow for much more detailed and industry-relevant design for recycling options and hence include the "mineralogy" of the urban mine, analogous to the minerals of geologically based mines. The same applies for data by sampling and analyses of the recyclate flows from sorting plants.

Recycling efficiency and recyclate quality are a function of design-connected materials. Therefore information on the "layout" (i.e. construction, connections, etc.) of the design is vital not only to capture and innovate material application in products but also to improve design (if possible!) from a material connection and combination perspective. The details of the metal wheels and design compatibility tables are hence subject to product data availability and rigor. The product-centric approach provides a theoretical basis for sampling, analyses and product design data generation. Figure 22.45 gives an example of the different levels and details required to capture the recycling profile of a product and pinpoint critical DfR issues.

Well-collected and formatted data, including information on their timelines as well as on standard deviation and average values, are important for evaluating recycling systems. Although some data are available, not many have a thermodynamic basis, and thus will be of little use for models based on that fundamental starting point. Many data are measured in such totally different ways that there is insufficient information for closing a mass balance, and certainly for producing statistically sound recycling-rate calculations, predictions, etc. While data compilations such as the Review of Directive (2002)/96 (UNU, 2007) give a snapshot of the "now", they are of little use for calibrating physics-based models that are needed for understanding and innovating recycling. Therefore, special attention should be paid in the future to collecting good data for a meaningful contribution to better resource efficiency. To perform dynamic modeling and simulation of recycling systems requires datasets with at least the following characteristics:

- Metal, alloy and compound details
- Identified connections between materials
- Standard deviation values of the composition of each metal, alloy, compound, and of total flow rates
- Thermodynamic and physical properties

Much can be gained by representing data, particles, liberation etc. in ways that are

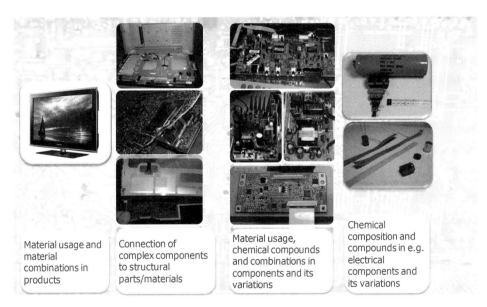

FIGURE 22.45 Example of different levels and details of data required (from BOM/FMD/chemical content analyses, etc.) to capture the recycling profile of a product and pinpoint critical DfR issues.

common to metals, minerals and materials processing. Standardization of data structures for design will help much to facilitate this.

References

Alba GmbH, 2013. Germany.
Anindya, A., Swinbourne, D., Reuter, M.A., Matusewicz, R., 2011. Indium distribution during smelting of WEEE with copper scrap. In: European Metal Conference, EMC 2011, GDMB (Goslar), Düsseldorf, Germany (June 26th–29th, 2011), vol. 1, pp. 3–14.
Bernardes, A.M., et al., 2004. Recycling of batteries: a review of current processes and technologies. Journal of Power Sources 130, 291–298.
Bertram, M., Martchek, K.J., Rombach, G., 2009. Material flow analysis in the aluminum industry. Journal of Industrial Ecology 13 (5), 650–654.
Bertuol, D.A., et al., 2006. Spent NiMH batteries: characterization and metal recovery through mechanical processing. Journal of Power Sources 160, 1465–1470.
Boin, U.M.J., Bertram, M., August 2005. Melting standardized aluminum scrap: a mass balance model for Europe. Journal of Metals, 26–33.
ELC (European Lamp Companies Federation), April 2009. Environmental Aspects of Lamps. Brussels, second ed. 17 pp.
EU, 2010. Critical Raw Materials for the EU (2010). Report of the Ad-hoc Working Group on Defining Critical Raw Materials. Enterprise and Industry Directorate General, 85 pp. http://ec.europa.eu/enterprise/policies/raw-materials/documents/index_en.htm.
Eurometaux's Proposals for the Raw Materials Initiative, Darmstadt/Brussels, 11th June 2010. http://www.oeko.de/oekodoc/1069/2010-115-en.pdf.
Franz, K.A., Kehr, W.G., Siggel, A., Wieczoreck, J., Adam, W., 2010. Luminescent Materials. Ullmann's Encyclopedia of Industrial Chemistry. Wiley-VCH Verlag GmbH & Co. KGaA, Weinheim, Germany, 41 pp.
GaBi, 2013. PE-International. www.pe-international.com.
Hagelüken, C., 2006. Precious metals process catalysts - material flows and recycling. Chemistry Today 24 (2), 14–17.
van den Hoek, W.J., Luuks, G.M.J.F., Hoelen, C.G.H., 2010. Lamps, Ullmann's Encyclopedia of Industrial Chemistry. Wiley-VCH Verlag GmbH & Co. KGaA, Weinheim, Germany, 53 pp.
HSC 7.1, Outotec Research 1974–2013, www.outotec.com.
Huda, H., Naser, J., Brooks, G., Reuter, M.A., Matusewicz, R.W., 2012. Computational fluid dynamic modeling of zinc slag fuming process in top-submerged Lance smelting furnace. Metallurgical Transactions B 43B (1), 39–55.

Ignatenko, O., van Schaik, A., Reuter, M.A., 2007. Recycling system flexibility: the fundamental solution to achieve high recycling rates. Journal of Cleaner Production 16 (4), 432–449.
Indaver, 2013. www.indaver.be.
International Aluminium Institute, 2007. Improving Sustainability in the Transport Sector Through Weight Reduction and the Application of Aluminium. http://www.world-aluminium.org/media/filer/2012/06/12/fl0000307.pdf (accessed date 07.12.12.).
Jiang, Y., Shibayama, A., Liu, K., Fujita, T., 2005. A hydrometallurgical process for extraction of lanthanum, yttrium and gadolinium from spent optical glass. Hydrometallurgy 76, 1–9.
Jüstel, T., 2007. Fluorescent Lamp Phosphors. Is there Still News? PGS, Seoul.
Krinke, S., van Schaik, A., Reuter, M.A., Stichling, J., 2009. Recycling and DfR of multi-material vehicles (as part of 'Life cycle assessment and recycling of innovative multi-material applications' by). In: Proceedings of the International Conference 'Innovative Developments for Lightweight Vehicle Structures', May, 26–27th 2009, Wolfsburg, Germany (Volkswagen Head Office), pp. 196–208. ISBN: 978-3-00-027891-4.
Lynas Corporation Ltd, May 2011. Rare Earths—We Touch Them Every Day. Investor Presentation. http://www.lynascorp.com/content/upload/files/Presentations/Investor_Presentation_May_2011.pdf.
McGill, I., 2005. Rare Earth Elements, Ullmann's Encyclopedia of Industrial Chemistry. Wiley-VCH Verlag GmbH & Co. KGaA, Weinheim, Germany, 46 pp.
Meskers, C.E.M., Hagelüken, C., et al., 2009. Impact of preprocessing routes on precious metal recovery from PCs. Proceedings of European Metallurgical Conference EMC.
Müller, T., Friedrich, B., 2006. Development of a recycling process for nickel-metal hydride batteries. Journal of Power Sources 158, 1498–1509.
Nemry, F., Leduc, G., Mongelli, I., Uihlein, A., 2008. Environmental Improvement of Passenger Cars (IMPRO-car). Joint Research Centre Institute for Prospective Technological Studies JRC 40598 EUR 23038 EN. ISBN: 978-92-79-07694-7, 216 pp.
Pietrelli, L., et al., 2002. Rare earths recovery from NiMH spent batteries. Hydrometallurgy 66, 135–139.
Rabah, M.A., 2004. Recovery of aluminium, nickel–copper alloys and salts from spent fluorescent lamps. Waste Management 24, 119–126.
Rabah, M.A., 2008. Recyclables recovery of europium and yttrium metals and some salts from spent fluorescent lamps. Waste Management 28, 318–325.
Reuter, M.A., van Schaik, A., 2013. 10 DESIGN FOR RECYCLING RULES, PRODUCT CENTRIC RECYCLING &

URBAN/LANDFILL MINING, Proceedings 2nd International Academic Symposium on Enhance Landfill Mining, 14–16 October 2013, Houthalen-Helchteren, Greenville, Belgium, pp. 103–117 (ISBN 978-90-8161-502-0).

Reuter, M.A., Van Schaik, A., 2013. Review on Metal Recycling. End-of-Life Products, Residues, Wastes, Slags, Design for Sustainability, Eco-Labelling. In: Malfliet, A., Yan, P., Pontikes, Y., Guo, M., Blanpain, B. (Eds.), proceeding of 3rd International Slag Valorisation Symposium. Leuven, 19-20th March 2013, pp. 115-134. (ISBN 978-94-6018-637-0).

Reck, B.K., Graedel, T.E., 2012. Challenges in metal recycling. Science 337 (6095), 690–695.

Reuter, M.A., 1999. The simulation of industrial ecosystems. Minerals Engineering 11 (10), 891–917.

Reuter, M.A., 2011. Limits of design for recycling and "sustainability": a review. Waste and Biomass Valorisation 2, 183–208.

Reuter, M.A., van Schaik, A., 2012a. Opportunities and limits of WEEE recycling—recommendations to product design from a recyclers perspective. In: Proceedings of Electronics Goes Green 2012+, 9–12 September 2012, Berlin, Germany, p. 8 (Paper A5-3).

Reuter, M.A., van Schaik, A., 2012b. Opportunities and limits of recycling: a dynamic-model based analysis. MRS Bulletin 37 (4), 339–347.

Reuter, M.A., van Schaik, A., 2013a. Review on metal recycling. End-of-life products, residues, wastes, slags, design for sustainability, eco-labelling. In: Malfliet, A., Yan, P., Pontikes, Y., Guo, M., Blanpain, B. (Eds.), Proceeding of 3rd International Slag Valorisation Symposium, Leuven, 19–20 March, 2013, pp. 115–134. ISBN: 978-94-6018-637-0.

Reuter, M.A., van Schaik, A., 2013b. Resource efficient metal and material recycling. In: Kvithyld, A., Meskers, C., Kirchain, R., Krumdick, G., Mishra, B., Reuter, M., Wang, C., Schlesinger, M., Gaustad, G., Lados, D., Spangenberger, J. (Eds.), REWAS 2013: Enabling Materials Resource Sustainability. TMS (The Minerals, Metals & Materials Society), pp. 332–340.

Reuter, M.A., Heiskanen, K., Boin, U., van Schaik, A., Verhoef, E., Yang, Y., 2005. The Metrics of Material and Metal Ecology, Harmonizing the Resource, Technology and Environmental Cycles. Elsevier BV, Amsterdam, 706 pp. (ISBN: 13 978-0-444-51137-9, ISBN: 10: 0-444-51137-7, ISSN: 0167-4528).

Reuter, M.A., van Schaik, A., Ignatenko, O., 2006. Fundamental limits for the recycling of end-of-life vehicles. Minerals Engineering 19 (5), 433–449.

Rombach, E., et al., 2008. Altbatterien als sekundare Rohstoffressource fur die Metallgewinnung. World of Metallurgy—Erzmetall 61 (3), 180–185.

Rumpold, R., Antrekowitsch, J., 2012. Recycling of platinum group metals from automotive catalysts by an acidic leaching process, Proceedings Platinum 2012. The Southern African Institute of Mining and Metallurgy, 695–714.

van Schaik, A., 2011. Kwantificering kritische grondstoffen in E-waste producten (in Commission of Vereniging NVMP), 51 p. www.producenten-verantwoordelijkheid.nl.

van Schaik, A., Reuter, M.A., 2004. The time-varying factors influencing the recycling rate of products. Resources, Conservation and Recycling 40 (4), 301–328.

van Schaik, A., Reuter, M.A., 2007. The use of fuzzy rule models to link automotive design to recycling rate calculation. Minerals Engineering 20, 875–890.

van Schaik, A., Reuter, M.A., 2010. Dynamic modelling of e-waste recycling system performance based on product design. Minerals Engineering 23, 192–210.

van Schaik, A., Reuter, M.A., 2008. Technology Based Design for Recycling and Eco-Design. In: Mishra, B., Ludwig, C., Das, S. (Eds.), Proceedings REWAS2008, Cancun, Mexico. The Minerals, Metals & Materials Society (TMS) Warrendale, Pennsylvania pp. 257–266 (ISBN Number 978-0-87339-726-1).

van Schaik, A., Reuter, M.A., 2012. Shredding, sorting and recovery of metals from WEEE: linking design to resource efficiency. In: Goodship, V. (Ed.), Waste Electrical and Electronic Equipment (WEEE) Handbook. University of Warwick, UK and A Stevels, Delft University of Technology, The Netherlands, pp. 163–211.

van Schaik, A., Reuter, M.A., Dalmijn, W.L., Boin, U., 2002. Dynamic modelling and optimisation of the resource cycle of passenger vehicles. Minerals Engineering 15 (11), 1001–1016.

van Schaik, A., Reuter, M.A., Wecycle/NVMP, 2013. 10 Design for Recycling Rules. Product Centric Simulation Based Design for Recycling (DfR). A Summary of the Study on Design for Recycling on E-waste Products, by MARAS in Commission of NVMP/Wecycle. Presented at the NVMP/Wecycle Conference "Grondstoffenterugwinning kritisch bekeken/A Critical View on Resource Recovery", August 29, 2013. Madurodam, The Hague, The Netherlands. www.producenten-verantwoordelijkheid.nl.

Schüler, D., et al., 2011. Study on Rare Earths and Their Recycling Final Report for the Greens/EFA Group in the European Parliament. Öko-institut e.V. Darmstad, Germany, 140 pp.

Sebenik, R.F., et al., 2012. Molybdenum and Molybdenum Compounds, Ullmann's Encyclopedia of Industrial Chemistry. Wiley-VCH Verlag GmbH & Co. KGaA, Weinheim, Germany, 46 pp.

SuperLightCar SLC, 6th Framework EU project. www.superlightcar.com. Project leader Volkswagen (2005–2009) project brochure http://www.superlightcar.com/public/docs/SuperLIGHTCar_project_brochure.pdf.

Ullmann's Encyclopedia of Industrial Chemistry, 2012. Various Metal and Other Relevant Chapters. Wiley-VCH Verlag GmbH & Co. KGaA, Weinheim, Germany.

Umicore 2013. www.umicore.com.

UNEP, 2011. Recycling Rates of Metals—a status report. In: Graedel, T.E., Alwood, J., Birat, J.-P., Reck, B.K., Sibley, S.F., Sonnemann, G., Buchert, M., Hagelüken, C. (Eds.), A report of the Working Group on the Global Metal Flows to the International Resource Panel.

UNEP (2013). A report of the Working Group on the Global Metal Flows to the International Resource Panel, Metal Recycling—Opportunities, Limits, Infrastructure. Reuter, M.A., Hudson, C., van Schaik, A., Heiskanen, K., Meskers, C., Hagelüken, C. 316 p.

United Nations University (UNU), 2007. 2008 Review of Directive 2002/96 on Waste. Electrical and Electronic Equipment—Study No. 07010401/2006/442493/ETU/G4, 347 pp.

U.S. Department of Energy, 2010. Critical Materials Strategy.

U.S. Geological Survey (USGS), 2002. Rare Earth Elements—Critical Resources for High Technology (USGS Fact Sheet 087-02).

U.S. Geological Survey (USGS), 2010. Mineral Commodity Summaries, Yttrium.

Wecycle, 2013. Nederland. www.wecycle.nl.

Welz, T., Hischier, R., Hilty, L.M., 2011. Environmental impacts of lighting technologies—life cycle assessment and sensitivity analysis. Environmental Impact Assessment Review 31, 334–343.

Xinwen, C., Streicher-Porte, M., Wang, M.Y.L., Reuter, M.A., 2011. Informal electronic waste recycling: a sector review with special focus on China. Waste Management 31 (4), 731–742.

Xu, J., et al., 2008. A review of processes and technologies for the recycling of lithium-ion secondary batteries. Journal of Power Sources 177, 512–527.

Xiao, Y., Reuter, M.A., 2002. Recycling of different aluminium scraps. Minerals Engineering 2002 15 (IIS1), 763–970.

CHAPTER 23

Separation of Large Municipal Solid Waste

Jan Thewissen, Sandor Karreman, Jorrian Dorlandt
Shanks/Van Vliet Groep, Nieuwegein, The Netherlands

23.1 INTRODUCTION

Large municipal solid waste (LMSW) is a waste stream that originates from households. It basically is the waste that is not collected frequently door to door because it either is too large to collect with the normal household waste or can be separated for recycling. In The Netherlands, as in many other countries, this waste stream is collected and treated separately. In this chapter we discuss the way LMSW can be separated and recycled into reusable materials.

LMSW is a mixed waste stream. It is waste generated by households and contains products of different composition, materials and sizes. Typical examples include old furniture, old mattresses, garden furniture and boxes. Usually the LMSW is generated once in a while, for example, during attic or basement cleaning. Frequently it also contains normal household waste that is taken along with the collection of LMSW. LMSW is a waste stream that has a fire risk due to the composition or the way it is collected; e.g. it may include small gas tanks or waste in bags that can cause overheating and may self-ignite, causing a fire.

LMSW is a waste stream that largely consists of materials and products which are relatively well suited for recycling because of the raw materials it contains. Examples are wood, plastics and metals. This waste stream often consists of products with a combination of these materials (for example, wood and glass, metal and plastics). This is a challenge for the sorting and recycling process. Table 23.1 provides an approximate breakdown of the materials found in LMSW.

23.2 THE CIRCULAR PROCESS FOR LARGE MUNICIPAL SOLID WASTE

Ideally products that are discarded as waste are reused in the same conditions (or with small adaptations). LMSW lends itself perfectly for this because of the composition. This is possible only if an extra step in the value chain

TABLE 23.1 Composition and Volume of Large Municipal Solid Waste Collected in The Netherlands in 2008

Waste Fractions	Volume (kt)
Mixed bulky MSW	672
Carpets and floor covering	13
Appliances	81
Bulky garden wastes	434
Furniture	40
Glass (windows)	9
Metals	83
Wood	384
Clean demolition wastes	444
Asbestos-containing wastes	13
Tires	3
Clean soil/dirt	111
Bituminous roofing	11
Others	58
Total LMSW	2356

Source: Corsten et al. (2013).

is added that focuses on reusing products after they are discarded by the previous owner. World-wide this is done in second-hand stores, which give a new life to these kinds of products. In The Netherlands also a separate system exists where municipalities or nongovernmental organizations have their own recycling agency.

At some stage in the life cycle of the products, these cannot or will not be used anymore. Disassembling into smaller components or materials for recycling is then the next best thing. Although initiatives exist, this is not yet economical and is confined to government-supported initiatives. This means the products are discarded as LMSW and sorting and recycling is the next best step in the process.

23.3 THE PRECONDITIONS FOR SORTING LARGE MUNICIPAL SOLID WASTE

To make it possible and economically feasible to sort LMSW, there are several base conditions. In short, these are as follows:

- There needs to be a collection system so LMSW can be collected separately.
- Sorting technologies, installations and knowledge need to be available.
- It needs to be economically feasible to sort LMSW. This means that the revenues from collecting and sorting LMSW must outweigh the costs.
- The government needs to create a playing field to stimulate sorting of LMSW. This is especially important in a situation were other (less sustainable) treatment options like incineration or landfilling have a more interesting cost structure.

23.4 COLLECTION SYSTEM OF LARGE MUNICIPAL SOLID WASTE

For the collection of LMSW there are two main systems in The Netherlands:

- In the first system the LMSW is collected by the municipal waste service at the source (the household). The LMSW is collected in bulk and then sorted at a sorting center or municipal waste depot.
- In the second system the households deliver the waste themselves to a waste depot of the municipality. The waste is separated into specific waste streams by the household themselves. All waste that is not, or cannot be, separated is then collected as the generic waste stream LMSW.

The waste collected at the source has a better quality due to the fact that the waste is often larger and better sortable (i.e. matrasses)

23.5 SORTING OF LARGE MUNICIPAL SOLID WASTE

Shanks in The Netherlands has much experience sorting LMSW, which will serve as the best practice on which the separation technology description in this chapter is based. This is done in a standardized recycling process, which consists of several process steps and includes sorting with cranes and specialized sorting installations. In general these include:

- pre-sorting of LMSW with cranes to separate the larger materials;
- shredding the waste into smaller fractions; and
- sorting the waste in the sorting line using several sorting methods.

The Shanks sorting lines were originally built to sort building and construction waste. LMSW can also be sorted by mixing this waste stream with building and construction waste. The reasons for this are as follows:

- LMSW is a relatively light waste streams and therefore needs a slow sorting process to maximize results. To make this economically feasible, it is mixed with a heavier waste stream so the throughput of the sorting line can be increased without compromising the sorting results.
- LMSW is a relatively dirty waste stream. Separate sorting of LMSW would create a dirty residual waste stream.
- In a traditional sorting line the contamination is sorted out of the waste. In modern sorting lines the product is sorted.

23.6 SORTING INSTALLATION

Shanks has two sorting installations capable of sorting LMSW. Both started sorting waste in the beginning of 2011 and since then have been modernized several times. The latest additions are a double optical divider installation, which enables us to separate plastics and paper from the waste and a new sieve. The newer of these two stands at our company, Van Vliet Groep in Nieuwegein. This installation and its sorting process are described in the picture in Figure 23.1.

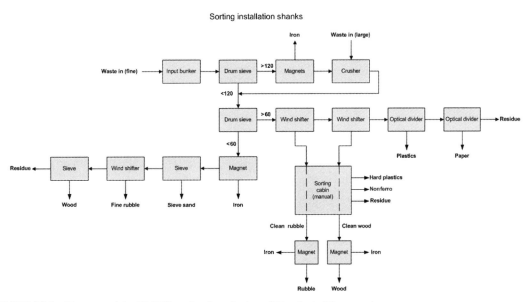

FIGURE 23.1 Diagram of the LMSW sorting installation of Shanks in Nieuwegein.

The installations use several standard sorting techniques like sieves, magnets and wind shifters to sort and separate waste. Basically a distinction is made between techniques using size (small/large) and weight (light/heavy). It is the order in which these are used that determines the effectiveness and efficiency of the sorting process. At Shanks this is the result of years of experience with sorting waste and experimentation. The sorting installations are capable of handling 25–30 t of mixed waste per hour. It is capable of handling more than 40 t of building and construction waste per hour.

23.7 SORTING PROCESS

The sorting process consists of several steps, which are described here.

23.7.1 First Steps

Before the LMSW is fed into the installation, a first sorting is done by cranes. These pick out large fractions like mattresses and large chunks of carpet. The LMSW is bulked and fed into the installation with the sole purpose to reduce the size of the waste. This step is done in the evening with a small crew. The reason for this step is that the LMSW needs to be reduced in size to make the complete sorting process more efficient (e.g. couches, chairs and bikes).

23.7.2 Into the Sorting Installation

The waste is first fed into a drum sieve. Here it is shaken so that it comes apart and breaks up. The smaller fraction (<120 mm) falls through the sieve holes and is transported to a second sieve. The larger fraction (>120 mm) leaves the sieve and passes a magnet where the first mono stream falls out of the process. The remainder is crushed and fed into the second sieve.

In the second sieve the process repeats itself. This sieve had smaller holes so the smaller fraction (<60 mm) falls out. This then passes several steps:

- a magnet to sort and remove metals;
- a sieve to sort and remove sand;
- a wind shifter to sort and remove small rubble; and
- a sieve to sort and remove wood.

What remains is residue that falls out of the process and will be transported to a waste incinerator where it is incinerated and converted into energy.

The larger fraction (>60 mm) in the second sieve leaves the sieve and passes two wind shifters. These blow out the lighter fractions consisting of a mix of materials like plastics. This fraction is then fed into two consecutive optical dividers to sort out plastics and paper. What remains is residue that is transported to either our Icopower energy pellet factory (to make refuse derived fuel (RDF)) or to a waste incinerator.

The fraction that has not been blown away by the wind shifter follows the process and enters the sorting cabin. Here a manual sorting is done to sort and separate recyclable fractions like hard plastics and nonferrous materials. The remaining fraction then passes magnets again to sort out any remaining iron. The result is a clean wood and rubble fraction.

During the process, waste falls out, for example, because it is too big to be wind shifted. This is fed into the sorting line again and processed again. This way the recycling efficiency can be increased.

23.8 RECYCLING EFFICIENCY

For LMSW the recycling efficiency is about 90%. The remainder is residue for which it is not economical to sort anymore. This is incinerated.

The LMSW also contains fractions that are sorted out but for which recycling is not always economical. Examples are textile (too polluted) and mattresses. These are also incinerated with energy recovery.

About 80% of the waste sorted in the Shanks installation in Nieuwegein (The Netherlands) is recycled into reusable materials.

In the European Union, waste incinerators are classified based on the degree of energy recovery, distinguishing the high-efficiency incinerators as R1. Since all waste incinerators in The Netherlands have a so-called R1 status, waste incineration is considered to be waste recovery. Our sorting installation therefore reaches 100% recovery. However, real recycling, of course, is a far more sustainable treatment then incineration.

23.9 THE FUTURE

Recycling has been done for ages. Currently more and more technologies become available to sort large municipal solid waste. The Shanks sorting installation proves that this is possible. We expect that in the future new and better technologies will be developed to further improve the efficiency of the sorting process. This will also have an effect on the economics. Sorting waste does cost effort and money so it must also be economically interesting to do. Legislation is an important factor that will determine if sorting LMSW will stay (economically) interesting. Currently we see some worrying developments in The Netherlands. Dutch legislation states that a waste stream should be treated in the best available way. For LMSW the minimum standard in The Netherlands is sorting and as such recycling. New legislation states that if municipal waste depots create facilities to collect and sort 18 different waste streams the remaining waste is not recyclable anymore and can go to incineration (with energy recovery). However, the experience with LMSW at Shanks (and other recycling companies) demonstrates that further recycling of LMSW and this fraction is possible and remains an environmentally preferable option than incineration.

Reference

Corsten, M., Worrell, E., Rouw, M., van Duin, A., 2013. The potential contribution of sustainable waste management to energy use and greenhouse gas emission reduction in the Netherlands. Resources, Conservation and Recycling 77, 13–21.

CHAPTER 24

Recovery of Construction and Demolition Wastes

Vivian W.Y. Tam

School of Computing, Engineering and Mathematics,
University of Western Sydney, Penrith, NSW, Australia

24.1 INTRODUCTION

Waste is defined as by-product material of human and industrial activities that have no residual values (Serpell and Alarcon, 1998). About 32.4 Mt of solid waste are generated in Australia annually (Productivity Commission — Australian Government, 2006), of which about 42% is from construction and demolition sectors. From that, about 7.8 Mt of material corresponding to about 57% of construction and demolition waste is recycled.

Among different types of construction and demolition waste, concrete waste constitutes the major proportions of about 81.8% of the total waste (Productivity Commission — Australian Government, 2006). From that, about 54% of the concrete waste is recycled. Moreover, the metal recycling rate is the highest at about 82%. The industry is highly motivated to recycle metal waste because it is profitable. Some demolition projects may even take risks in estimating the amount of metal waste collected on site and lowering the contract sum during tendering to improve the competitiveness.

This chapter aims to:

- Investigate construction and demolition waste generation;
- Examine the existing applications for recycled aggregate concrete;
- Explore mechanisms of concrete waste recycling technologies; and
- Explore a cost and benefit analysis using recycling concrete.

24.2 EXISTING RECYCLED AGGREGATE CONCRETE APPLICATIONS

Until recently, almost all demolished concrete has been dumped. Because concrete is such an essential, mass-produced material, like steel in the construction industry, much effort has been made to recycle and conserve it. Completed and repeated recycling can be theoretically suitable for concrete, as is the case for steel and aluminum (Noguchi and Tamura, 2001). Since concrete is composed only of cementitious materials, and the powders generated during the production of Recycled Aggregate (RA) can be reprocessed as

TABLE 24.1 Reuse of Demolished Concrete (Kawano, 1995)

Demolished Member	Artificial Reef, Paving Stone
Broken into 20–40 cm	Protection of levee
Crushed (−50 mm)	Sub-base, backfilling, foundation materials
Crushed and worn (−40 mm)	Concrete and asphalt concrete aggregate sub-base material, backfilling material
Powder (by-product of crushing)	Filler for asphalt concrete, soil stabilization materials

cement resources, this permits repeated recycling in a fully closed system. Recycling of concrete can be accomplished by reuse of concrete products, processing into secondary raw materials for uses as fill, road bases and sub bases, or aggregate for the production of new concrete (Howard Humphreys and Partners, 1994; Torring, 2000).

The most common way to recycle concrete rubble is as recycled aggregate (Coventry et al., 1999; Hendriks and Pietersen, 2000; Masters, 2001). Table 24.1 shows example of reusable concrete waste (Kawano, 1995).

Deformation properties of concrete made with secondary aggregate are less favorable than those of concrete made with natural gravel. There are two potential solutions to this problem (Hendriks and Pietersen, 2000): (1) substitute 100% gravel by secondary aggregate and increase dimensions of structure with about 10%; and (2) substitution of about 20% natural aggregate by mixed recycled aggregate, which does not reduce concrete quality with strength up to 65 MPa.

24.3 EXISTING CONCRETE RECYCLING METHODS

Basic equipment used to process virgin aggregate is similar to those used for crushing, sizing and stockpiling recycled aggregate. A recycling plant usually comprises crushers incorporating sieves, sorting devices and screens. The main recycling processes are crushing, sorting and screening, and thus to produce aggregate for use in civil engineering work, in landscaping and as a substitute for gravel in concrete products (Hansen, 1986).

Although the current concrete recycling method is used in many countries to produce recycled aggregate, the quality of the produced recycled aggregate is low, which limits their applications for low-grade activities such as road work, pavement and drainage (Jia et al., 1986; Poon et al., 2003; Tam, 2005; Tam et al., 2005; Commonwealth Scientific and Industrial Research Organization, 2006). Japan is now using advanced technologies to improve the quality of recycled aggregate so that it can be used for high-grade concrete applications (Kawano, 2003; Environmental Council of Concrete Organizations, 2006; Tanaka et al., 2006).

24.3.1 Heating and Grinding Method

The heating and grinding method makes hardened cement paste, which adheres to concrete waste softened by heating concrete waste to about 300 degree Celsius (Shima, et al., 2004). After that, parts of the hardened cement paste adhered to original aggregate in the concrete mass can then be separated by a grind process, resulting in clean original aggregate from the concrete waste.

24.3.2 Screw Grinding Method

The screw grinding method uses a shaft screw consisting of an intermediate part and an exhaust part with a warping cone to remove mortar adhered to the aggregate's surface (Matumura, 2005). Figure 24.1 shows the procedures of the screw grinding method.

24.3.3 Mechanical Grinding Method

The mechanical grinding method uses a drum body, which finely separates partition

FIGURE 24.1 Screw grinding equipment and grinded aggregates (Matumura, 2005).

boards with same-sized holes. The steel balls can move horizontally and vertically by rolling the drum. The quality of aggregate can then be improved in narrowing the inside space by using the partition boards (Kajima Corporation, 2006). Figure 24.2 shows the outline of the mechanical grinding equipment.

24.3.4 Gravity Concentration Method

After processing with a jaw crusher, an impact crusher and an improvement rod mill, aggregate of over 8 mm is divided into recycled coarse aggregate and mortar particles. Aggregate with sizes under 8 mm is divided into two types: recycled fine aggregate of 5 mm and 5–8 mm (Tatemastu et al., 2003). The wet gravity concentration machine is used to move (1) light-weight things such as mortar particle and wood waste upward; and (2) heavy-weight things such as aggregate grain downward. Figure 24.3 shows the equipment for the gravity concentration method.

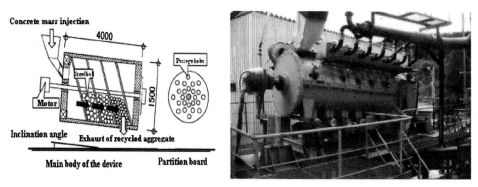

FIGURE 24.2 Outline of the mechanical grinding equipment (Kajima Corporation, 2006).

FIGURE 24.3 Gravity concentration method (Tatemastu et al., 2003).

24.4 COST AND BENEFIT ANALYSIS

To investigate the cost and benefit for concrete recycling, three construction and demolition companies, four recycling plants and two landfills in southeastern Queensland, Australia were visited. Site representatives were also interviewed. Currently, construction waste is dumped in landfills and new products are produced from the rocks and supplied to the site for new concrete production. It wastes energy to dump the waste and produce new materials for the production. Therefore, a concrete recycling method is proposed. The construction waste is sent to recycling plants for crushing as aggregate for new concrete production, which can save energy for dumping and producing new materials. This can also bring concrete materials into a closed-loop recycling process.

Detailed cost data for the current practice and the proposed concrete recycling method have been released by the Environmental Protection Agency (Environmental Protection Agency — Australia, 2007). However, some cost data are still not available from the agency, and in-depth interview discussions with site representatives from construction and demolition companies, recycling plants and landfills are thus conducted to estimate social and environmental cost required. Detailed interviews with companies are also conducted to validate the cost data and to ensure that consistent results are drawn. Detailed cost data are shown in Tables 24.2 and 24.3 for the current practice and the proposed concrete recycling method, respectively.

It should be noted that the estimated costs given in Tables 24.2 and 24.3 are confidential and could not be verified using different sources. The only source used to collect these cost data is in-depth interviews with company's onsite representatives. It is also believed that this practice is very common in the field of construction industry.

Based on the interview discussions, average construction waste generated from each site is about 115,200 t and a recycling plant has a production capacity of about 110,000 t/year. The expected life of the plant is estimated to be about 10 years. This information is used in the cost and benefit analysis.

24.4.1 The Current Method

In the current method, onsite construction waste is dumped in landfills and new products are produced from rocks and supplied on the construction site for new concrete production. Figure 24.4 shows the flow chart for the current method, and Table 24.2 summarizes the cost and benefit for the current method. Details are summarized as follows:

- *Construction waste* is the first stage to be considered in the cost and benefit analysis. In this stage, construction waste is sent to landfills and hence cost incurred is in the form of landfill charge (about $57/t), landfill space (about $220/t), transportation cost (about $5/t), air pollution (about 16.5% of landfill space charge), noise pollution (about 17.7% of landfill space charge), gas emission

TABLE 24.2 Cost and Benefit Analysis for the Current Method[1]

	Cost ($1000/year)	Benefits ($1000/year)
CONSTRUCTION WASTE		
Landfill dumping charge	6566.4[2]	
Landfill space saved by not dumping waste		18777.6[3]
Transportation	576.0[4]	
Air pollution	3136.3[5]	
Gas emission	3267.3[6]	
Energy consumption	4318.9[7]	
Noise pollution	3323.6[8]	
STRIPPING		
Equipment	145.0[9]	
Labor	45.0[10]	
Fuel	17.2	
Fixed overhead	40.6	
BLASTING		
Capital	137.8[11]	
Working capital	19.4[12]	
Equipment maintenance	30.1	
Labor	124.8[13]	
Fuel	15.9	
Fixed overhead	40.6	
Stockpiling	37.4[14]	
SORTING PROCESS		
Capital	168.4[15]	
Working capital	19.4[16]	
Equipment maintenance	35.2	
Labor	45.8[17]	

(Continued)

TABLE 24.2 Cost and Benefit Analysis for the Current Method[1] (cont'd)

	Cost ($1000/year)	Benefits ($1000/year)
Fuel	7.8	
Fixed overhead	40.6	
CRUSHING PROCESS		
Primary Crushing		
Equipment	165.1[18]	
Working capital	18.9[19]	
Equipment maintenance	30.1	10.2[20]
Labor	45.8[21]	
Fuel	9.8	
Fixed overhead	40.6	
Secondary Crushing		
Equipment	168.0[22]	
Working capital	19.3[23]	
Equipment maintenance	32.2	10.1[24]
Labor	45.8[25]	
Fuel	9.9	
Fixed overhead	40.6	
Shaper		
Equipment	90.0[26]	
Working capital	17.6[27]	
Equipment maintenance	22.3	
Fuel	8.9	
Fixed overhead	40.6	
Labor	45.8[28]	
WASHING, SCREENING OR AIR-SITTING		
Water	0.6[29]	
Fuel	7.8	

TABLE 24.2 Cost and Benefit Analysis for the Current Method[1] (cont'd)

	Cost ($1000/year)	Benefits ($1000/year)
FINISHED GRADED MATERIALS		
20 mm aggregate		550.0[30]
10 mm aggregate		1000.0[31]
7 mm aggregate		270.0[32]
75 mm aggregate		480.0[33]
Total	44,097.16	20.3

[1] Sources for cost data are from in-depth interview discussions with site representatives and information from Environmental Protection Agency (Environmental Protection Agency – Australia, 2007).
[2] About $57/t.
[3] About ($220–$57)/t.
[4] About $5/t.
[5] About 16.5% of landfill space charge.
[6] About 17.4% of landfill space charge.
[7] About 23% of landfill space charge.
[8] About 17.7% of landfill space charge.
[9] Bull dozer equipment cost is about $1,450,000.
[10] One person at about $45,760 per person per year.
[11] Blasting equipment cost is about $1,378,000.
[12] Fifteen percent variable operating cost of about $19390 per unit per year (pulverizer equipment).
[13] Two people at about $45,760 per person per year.
[14] One person at about $37,550 per person per year.
[15] Excavator equipment cost is about $1,684,000.
[16] Fifteen percent variable operating cost of about $19350 per unit per year (excavator equipment).
[17] One person at about $45,760 per person per year.
[18] Primary crusher equipment cost is about $1,651,000.
[19] Fifteen percent of variable operating cost of about $18,930 per unit per year (primary crusher).
[20] Different between the current method and the concrete recycling method.
[21] One person at about $45,760 per person per year.
[22] Secondary crusher equipment cost is about $1,680,000.
[23] Fifteen percent of variable operating cost of about $19,260 per unit per year (secondary crusher).
[24] Difference between the current method and the concrete recycling method.
[25] One person at about $45,760 per person per year.
[26] Shaper equipment cost is about $900,000.
[27] Fifteen percent of variable operating cost of about $17,630 per unit per year (shaper).
[28] One person at about $45,760 per person per year.
[29] About $0.005/t.
[30] About 23,000 t/year of $25/t.
[31] About 40,000 t/year of $25/t.
[32] About 18,000 t/year of $15/t.
[33] About 29,000 t/year of $16/t.

TABLE 24.3 Cost and Benefit Analysis for the Concrete Recycling Method[1]

	Cost ($1000/year)	Benefits ($1000/year)
CONSTRUCTION WASTE		
Dumping charge from recycling plants	2914.6[2]	
Landfill dumping charge		6566.4[3]
Landfill space saved by not dumping waste		18777.6[4]
Transportation		576.0[5]
Air pollution		3136.3[6]
Gas emission		3267.3[7]
Energy consumption		4318.9[8]
Noise pollution		3323.6[9]
STOCKPILING		
Labor	37.4[10]	
SORTING PROCESS		
Capital	168.4[11]	
Working capital	19.4[12]	
Equipment maintenance	35.2	
Labor	45.8[13]	
Fuel	7.8	
Fixed overhead	40.6	
EXCAVATION		
Equipment	156.2[14]	
Working capital	19.4[15]	
Equipment maintenance	34.9	
Labor	45.8[16]	
Fuel	7.8	
Fixed overhead	40.6	
CRUSHING PROCESS		
Primary Crushing		
Equipment	163.2[17]	
Working capital	20.5[18]	
Equipment maintenance	40.2	
Labor	45.8[19]	
Fuel	9.8	
Fixed overhead	40.6	
Magnetic Separation		
Equipment	120.8[20]	
Working capital	16.6[21]	
Equipment maintenance	15.9	
Labor	45.9[22]	
Fuel	8.7	
Fixed overhead	40.6	
Revenue from selling scrap (mainly steel)		187.2[23]
Secondary Crushing		
Equipment	166.6[24]	
Working capital	20.8[25]	
Equipment maintenance	42.3	
Labor	45.8[26]	
Fuel	9.9	
Fixed overhead	40.6	
Manual Removal of Remaining Contaminants		
Labor	37.4[27]	
Removal of large pieces of wood, paper, plastics etc to landfill	190.0[28]	
WASHING, SCREENING OR AIR-SITTING		
Water	0.6[29]	
Fuel	7.8	
Finished Graded Materials		
20 mm aggregate	506.0[30]	45[34]

(Continued)

TABLE 24.3 Cost and Benefit Analysis for the Concrete Recycling Method[1] (cont'd)

	Cost ($1000/year)	Benefits ($1000/year)
10 mm aggregate	800.0[31]	200[34]
7 mm aggregate	266.4[32]	3.6[34]
75 mm Rubble	462.0[33]	33.4[34]
Total	6738.06	37,654.61

[1] Sources for cost data are from in-depth interview discussions with site representatives and information from Environmental Protection Agency (Environmental Protection Agency – Australia, 2007).
[2] About $25.30/t.
[3] About $57/t.
[4] About ($220–$57)/t.
[5] About $5/t.
[6] About 16.5% of landfill space charge.
[7] About 17.4% of landfill space charge.
[8] About 23% of landfill space charge.
[9] About 17.7% of landfill space charge.
[10] One person at about $37,440 per year.
[11] Pulveriser equipment cost is about $1,684,000.
[12] Fifteen percent variable operating cost at about $19390 per unit per year (pulverizer equipment).
[13] One person at about $45,760 per year.
[14] Excavator equipment cost is about $1,562,000.
[15] Fifteen percent variable operating cost at about $19350 per unit per year (excavator equipment).
[16] One person at about $45,760 per year.
[17] Primary crusher equipment cost is about $1,632,000.
[18] Fifteen percent of variable operating cost of about $20,450 per unit per year (primary crusher).
[19] One person at about $45,760 per year.
[20] Magnetic Separator equipment cost is about $1,207,900.
[21] Fifteen percent of variable operating cost of about $16640 per unit per year (magnetic separator).
[22] One person at about $45,760 per year.
[23] 1872 t/year of about $100/t.
[24] Secondary crusher equipment cost is about $1,666,000.
[25] Fifteen percent of variable operating cost at about $20,780 per unit per year (secondary crusher).
[26] One person at about $45,760 per year.
[27] One person at about $37,440 per year.
[28] About 3328 t/year of about $57/t.
[29] About $0.005/t.
[30] About 23,000 t/year of $22/t.
[31] About 40,000 t/year of about $20/t.
[32] About 18,000 t/year of about $14.8/t.
[33] About 30,000 t/year of about $15.4/t.
[34] Difference between the current method and the recycling method.

(about 17.4% of landfill space charge) and energy consumption (about 23% of landfill space charge).

- *Stripping* is the stage where rocks are cleared and leveled. Equipment such as a bulldozer is required. Costs incurred in this stage include labor cost, fuel cost and fixed overhead cost.
- *Blasting* is the process where blasting equipment is used, and capital cost, equipment cost, working capital cost (about 15% of variable operating cost), operating cost including equipment maintenance cost, and labor cost, fuel cost and fixed overhead cost are estimated.
- *Stockpiling* is the stage where one labor is involved at the rate of about $18/h.
- *Sorting* is the stage where equipment such as excavator is used and capital cost, equipment cost, working capital cost (about 15% of variable operating cost), operating cost including equipment maintenance cost, labor cost, fuel cost and fixed overhead cost are estimated.
- *Crushing* includes primary crushing, magnetic separation and a secondary crushing process. It involves equipment such as a primary crusher, secondary crusher and shaper. In addition, capital cost, equipment cost, working capital cost (about 15% of variable operating cost), operating cost including equipment maintenance cost, labor cost, fuel cost and fixed overhead cost are estimated. In this process, the only benefit is the maintenance cost that can be saved compared to the recycling process because there is more wear and tear of the equipment blades. Hence, the difference between the maintenance cost for the equipment for the current method and the concrete recycling method is the benefit gained in this stage.
- *Washing, screening or air-sitting* is the stage that involves fuel and recycled water (about $0.005/t) to settle dust and all particles.

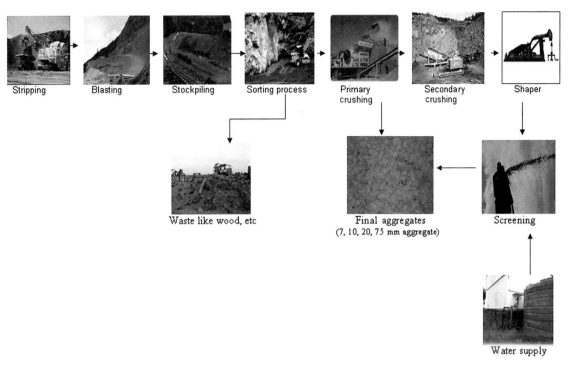

FIGURE 24.4 Flow chart of the current method.

- *Final product stage* is where the finished products of 7, 10, 20 and 75 mm aggregate are produced and sold at the rate of about $15, $25, $25 and $16/t, respectively.

24.4.2 Concrete Recycling Method

In the concrete recycling method, construction waste from the site is dumped to the recycling plant and new products are produced and supplied on the construction site. Figure 24.5 shows the flow chart of the concrete recycling method and Table 24.3 summarizes the cost and benefit for the proposed concrete recycling method. Details are summarized as follows:

- *Construction waste* is the first stage to be considered in cost and benefit analysis. In this stage, dumping charge is the only cost being incurred, at a rate of about $25.30/t. Dumping the construction waste in the recycling plant had many benefits such as avoided landfill charge (about $57/t), landfill space (about $220/t), transportation cost (about $5/t), air pollution (about 16.5% of landfill space charge), noise pollution (about 17.7% of landfill space charge), gas emission (about 17.4% of landfill space charge) and energy consumption (about 23% of landfill space charge).
- *Stockpiling* is the stage where one laborer is involved at the rate of about $18/hour.
- *Sorting* is the stage where equipment such as a pulverizer and an excavator are used and capital cost, equipment cost, working capital cost (about 15% of variable operating cost),

FIGURE 24.5 Flow chart of the concrete recycling method.

operating cost including equipment maintenance cost, labor cost, fuel cost and fixed overhead cost are estimated.

- *Crushing process* includes primary crushing, magnetic separation and secondary crushing

TABLE 24.4 Comparison of the Current Method and the Concrete Recycling Method

	The Current Method ($1000/year)	The Concrete Recycling Method ($1000/year)
Total cost	44,097.16	6738.06
Total benefit	20.30	37,654.61
Net benefit	−44,076.84	+30,916.55

process. It involves equipment such as a primary crusher, magnetic separator, and secondary crusher. Capital cost, equipment cost, working capital cost (about 15% of variable operating cost), operating cost including equipment maintenance cost, labor cost, fuel cost and fixed overhead cost are estimated. In this process, steel scrap is sorted out through a magnetic separation process, which adds to revenue and is sold at about $100/t.

- *Manual removal process* is the stage where the removal of pieces of wood, paper and plastics that are still in the crushed materials. For this stage, one laborer is involved at the rate of about $18/h.

- *Washing, screening or air-sitting process* is the stage that involves fuel and recycled water (about $0.005/t) to settle dust and all particles.
- *Final product stage*, where finished products of 7, 10, 20 and 75 mm aggregate are produced and sold at the rate of about $14.8, $20, $22 and $15.4/t, respectively. As these products are sold at a lower price compared to natural products in the market, it has a profit of about $45,000, $200,000, $3600, and $33,400 per year for 7, 10, 20 and 75 mm aggregate, respectively. At this stage, the benefit gained is the difference between the price of the same quantity produced by this method and the current method.

24.4.3 Comparison

Table 24.4 compares the results of the cost and benefit analysis of the current method and the concrete recycling method. It should be noted that the concrete recycling method is more beneficial than the current method. The concrete recycling method receives a positive net benefit of about $30,916,000 per year while the current method receives a negative net benefit of about −$44,076,000 per year.

There is no doubt that aggregate produced by the recycling methods is more economical in long term than using natural materials. But, one of the factors affecting the viability of aggregate recyclers is the availability of feed materials. If construction debris or other sources of feed are not consistently available or if there is some seasonality to the availability of local feed materials that limits the recycler's ability to operate at or near capacity, this can dramatically reduce operational profitability. The amount of material available for recycling is limited by sizes and changing conditions of the "urban deposit". Recycled material suppliers fail to meet demand for aggregate, so natural material production continues to be the primary source of aggregate in road construction in applications where they can substitute. At best, the contribution of recycled material will gradually grow until all available supply is consumed. Product pricing is often controlled by factors outside the direct control of the recyclers. The amount of material currently available from natural resources overshadows the amount of material available from recycling.

Product quality and uniformity can also pose a risk to the potential recyclers. Natural material producers continue to supply bulk materials for buildings and road construction because they are able to supply sufficient high-quality materials for a wide variety of high-grade applications. Unless the recyclers have established long-term contracts for consistent and high-quality feed material, it may be difficult for the recycler to maintain a predictable revenue stream because of uncertainty related to future feed availability and quality or market price fluctuations. This all affects the use of recycled materials in the industry.

24.5 CONCLUSION

This chapter focused on investigating construction and demolition waste generation, examining the existing applications for recycled aggregate concrete and exploring mechanisms of concrete recycling technologies. Concrete is found to be one of the most common types of construction and demolition waste. The use of concrete waste as recycled aggregate is limited to low-grade applications, including sub-grade and pavement. Four concrete recycling technologies help improve recycled aggregate quality and thus the use of recycled aggregate concrete as high-grade applications, including (1) heating and grinding method; (2) screw grinding method; (3) mechanical grinding method; and (4) gravity concentration method. The mechanisms of the methods were also discussed. A cost and benefit analysis using concrete recycling by a case study was also explored.

Acknowledgments

This chapter is an adaptation of: Tam, W.Y. Vivian "2009", (Review on concrete waste recycling technologies, *Concrete Materials: Properties, Performance and Applications*, Nova Science Publisher, Inc., United States, ISBN: 978-1-60741-250-2, 567-573.); Tam, W.Y. Vivian and Tam, C. M. "2008", (*Re-use of construction and demolition waste in housing development*, Nova Science Publishers, Inc., United States.) Tam, W.Y. Vivian "2008", (Economic comparison of concrete recycling: a case study approach, *Resources, Conservation and Recycling*, 52(5), 821-828.) Tam, W.Y. Vivian "2008", (Cost effectiveness of concrete recycling, *Progress in Waste Management Research*, Nova Science Publisher, Inc., United States, ISBN: 978-1-60456-235-4, 1-10. (Invited Article))

References

Commonwealth Scientific and Industrial Research Organization, 2006. Recycled Aggregate Applications as Subgrade and Pavement. Commonwealth Scientific and Industrial Research Organization, Australia.

Coventry, S., Wolveridge, C., et al., 1999. The Reclaimed and Recycled Construction Materials Handbook. Construction Industry Research and Information Association, London.

Environmental Council of Concrete Organizations, 2006. Recycled Concrete and Masonry. Environmental Council of Concrete Organizations, Japan.

Environmental Protection Agency – Australia, 2007. Construction and Demolition Waste, Waste Management and Resource Use Opportunities. Retrieved March 9, 2007, from: https://www.epa.qld.gov.au/publications?id=484.

Hansen, T.C., 1986. The second RILEM state of the art report on recycled aggregates and recycled aggregate concrete. Materials and Structures 1 (111), 201–246.

Hendriks, C.F., Pietersen, H.S., 2000. Sustainable Raw Materials: Construction and Demolition Waste. RILEM Publication, Cachan Cedex, France.

Howard Humphreys and Partners, 1994. Managing Demolition and Construction Wastes: Report of the Study on the Recycling of Demolition and Construction Wastes in the UK. HMSO, London.

Jia, W., Baoyuan, L., et al., 1986. Improvement of paste-aggregate interface by adding silica fume. In: Proceedings of the 8th International Congress on the Chemistry of Cement, Rio de Janeiro.

Kajima Corporation, 2006. Recycled aggregate concrete for within-site recycling. from: http://www.kajima.co.jp.

Kawano, H., 1995. The state of reuse of demolished concrete in Japan. In: Integrated Design and Environmental Issues in Concrete Technology: Proceedings of the International Workshop 'Rational Design of Concrete Structures under Severe Conditions' Hakodate, Japan. E & FN Spon, London.

Kawano, H., 2003. The state of using by-products in concrete in Japan and outline of JIS/TR on recycled concrete using recycled aggregate. In: Proceedings of the 1st FIB congress, Osaka, Japan.

Masters, N., 2001. Sustainable Use of New and Recycled Materials in Coastal and Fluvial Construction: A Guidance Manual. Thomas Telford, London.

Matumura, U., 2005. Concrete recycling technology. In: Proceeding of Annual Conference of Consultants, Hokkaido, Japan.

Noguchi, T., Tamura, M., 2001. Concrete design towards complete recycling. Structural Concrete 3 (2), 155–167.

Poon, C.S., Azhar, S., et al., 2003. Recycled aggregates for concrete applications. In: Materials Science and Technology in Engineering Conference – Now, New and Next, Hong Kong.

Productivity Commission – Australian Government, 2006. Waste Management: Productivity Commission Draft Report. Productivity Commission, Canberra, Australian Government.

Serpell, A., Alarcon, L.F., 1998. Construction process improvement methodology for construction projects. International Journal of Project Management 16 (4), 215–221.

Shima, H., Tateyashiki, H., et al., 2004. An advanced concrete recycling technology and its applicability assessment through input-output analysis. Journal of Advanced Concrete Technology 3 (1), 53–67.

Tam, W.Y.V., 2005. Recycled Aggregate from Concrete Waste for Higher Grades of Concrete Construction. Department of Building and Construction, City University of Hong Kong, Hong Kong, China.

Tam, W.Y.V., Gao, X.F., et al., 2005. Micro-structural analysis of recycled aggregate concrete produced from two-stage mixing approach. Cement and Concrete Research 35 (6), 1195–1203.

Tanaka, R., Miura, S., et al., 2006. Experimental Study on the Possibility of Using Permanently Recycled Concrete for Reinforced Concrete Structures. Centre National De Le Recherche Scientifique, France.

Tatemastu, K., Yamazaki, J., et al., 2003. Production of High Quality Recycled Aggregate by Gravity Concentration Method and Properties of Concrete Using its Recycled Aggregate from: http://www.concle.co.jp.

Torring, M., 2000. Management of Concrete Demolition Waste. Concrete Technology for a Sustainable Development in the 21st Century. E & FN Spon, London; New York, pp. 321–331.

CHAPTER

25

Waste Electrical and Electronic Equipment Management

Mathias Schluep
World Resources Forum Association (WRFA), St.Gallen, Switzerland

25.1 INTRODUCTION

The management of waste electrical and electronic equipment (WEEE, or e-waste) is a rather recent task on the agenda of organizations responsible for waste management. Experiences around the globe have shown that the usual waste manager, the municipality, is not adequately equipped to handle a complex waste stream, such as obsolete electrical and electronic equipment (EEE). Two new paradigms, however, started to change the management of e-waste: the closed loop economy and the extended producer responsibility (EPR). Some countries, mainly in Europe (e.g. Switzerland, Netherlands, and Belgium), started to experiment with new approaches in managing this high-technology waste stream more than 20 years ago. Soon, a new international framework took shape and the roles of important stakeholders along the end-of-life path of WEEE evolved: producers accepted the end-of-life responsibility for their products and initiated producer-responsible organizations to manage the material flows and the financing of unprofitable processing steps; the recycling industry went through a rapid evolution where specialists emerged amongst others for manual dismantling, mechanical processing, or final refining of secondary raw materials; the legislators carefully developed regulations defining responsibilities and promoting a constant improvement of the system's efficiency; and, last but not least, the consumer, from the big corporation to the small household, increasingly wanted a convenient and sustainable option to dispose of e-waste.

In developing countries and emerging economies, this new paradigm shift did not seem to take place until recently. Eye-opening reports from nongovernmental organizations (NGOs) on pollution from WEEE management in China, India, and African countries such as the Basel Action Network (Puckett et al., 2002, 2005), Toxics Link (Agarwal et al., 2003), and Greenpeace (Brigden et al., 2005, 2008) started to make their way to the mass media in the early 2000s. Poor people living in poverty in large in urban areas tried to recover valuables from the e-waste stream, putting themselves and their environment at risk (Sepúlveda et al., 2010). As a result, governments in these countries started to move WEEE up in their priority list of environmental issues that needed special legislative attention. Meanwhile, the e-waste topic was a high-priority area within various United Nations (UN) organizations and conventions. Between 2005 and 2010, various international cooperation projects were launched by multilateral UN organizations (e.g. in Africa (Secretariat of the Basel Convention, 2011; Magashi and Schluep, 2011; Wasswa and Schluep, 2008)), producers from the Information and Communication Technology (ICT) industry (e.g. Hewlett Packard (Schluep et al., 2008)), NGOs and governmental organizations (e.g. the Swiss e-Waste Programme (Widmer et al., 2005; Schluep et al., 2013)).

WEEE is often misunderstood as comprising only computers and related information technology equipment. However, according to the Organization for Economic Co-operation and Development (OECD), e-waste is "any appliance using an electric power supply that has reached its end of life". In this chapter, WEEE and e-waste are used as synonyms; they include all 10 categories as specified in the EU WEEE directive (European Union, 2003b), which has become the most widely accepted classification.

25.2 OBJECTIVES OF WEEE MANAGEMENT

The e-waste is usually regarded as a waste problem that can cause environmental damage if not dealt with in an appropriate way. However, the enormous resource impact of EEE is widely overlooked. EEE is a major driver for the development of demand and prices for a number of metals as shown in Table 25.1. Consequently, inappropriate disposal of e-waste not only leads to significant environmental problems but also to a systematic loss of secondary materials (Hagelüken and Meskers, 2008). Hence, the appropriate handling of e-waste can both prevent serious environmental damage and also recover valuable materials.

Besides the positive impact on resources, state-of-the-art recycling operations also contribute to reducing greenhouse gas emissions. Primary production (i.e. mining, concentrating, smelting and refining), especially of precious and special metals, is energy intensive and hence has a significant carbon dioxide (CO_2) impact. "Mining" our old computers to recover the contained metals—if done in an environmentally sound manner—needs only a fraction of this energy input (Hagelüken and Meskers, 2008).

Furthermore, the environmentally sound management of end-of-life refrigerators, air-conditioners, and similar equipment is significant in mitigating the climate change impact. The ozone depleting substances in these devices, such as chlorofluorocarbon and hydrochlorofluorocarbons (as well as fluorocarbons and hydrofluorocarbons), have a very high global warming potential.

Hence, the main services a comprehensive WEEE or e-waste management system has to deliver in order to ensure sustainability are (1) the collection of e-waste; (2) the recovery of valuables, such as secondary raw materials; and (3) the segregation and safe disposal of hazardous waste. Costs for unprofitable processes as well as administration, monitoring, and control to ensure quality have to be associated with all of these activities.

25.3 WEEE TAKE-BACK SCHEMES

The only tested formal e-waste management systems adhering to sustainability and extended producer responsibility principles are currently almost exclusively found in OECD countries (Sinha-Khetriwal et al., 2009, 2006; Ongondo et al., 2011). The European WEEE Directive

TABLE 25.1 Important Metals Used for Electrical and Electronic Equipment (EEE) (Schluep et al., 2009)

Metal	Primary Production[1] (t/y)	By-product from	Demand for EEE (t/y)	Demand/Production (%)	Main Applications
Ag	20,000	Pb, Zn	6000	30	Contacts, switches, solders
Au	2500	Cu	300	12	Bonding wire, contacts, integrated circuits
Pd	230	PGM	33	14	Multilayer capacitors, connectors
Pt	210	PGM	13	6	Hard disk, thermocouple, fuel cell
Ru	32	PGM	27	84	Hard disk, plasma displays
Cu	15,000,000		4,500,000	30	Cable, wire, connector
Sn	275,000		90,000	33	Solders
Sb	130,000		65,000	50	Flame retardant, CRT glass
Co	58,000	Ni, Cu	11,000	19	Rechargeable batteries
Bi	5600	Pb, W, Zn	900	16	Solders, capacitor, heat sink
Se	1400	Cu	240	17	Electro-optic, copier, solar cell
In	480	Zn, Pb	380	79	LCD glass, solder, semiconductor
Total			4,670,000		

[1] Based on demand in 2006. PGM = platinum group metals; CRT = cathode ray tube; LCD = liquid crystal display.

(European Union, 2003b), based on the concept of an extended producer responsibility (EPR) as an environmental policy, has set the global pace and standard in regulating e-waste management (Figure 25.1). A sister directive with the WEEE is the EU directive on the Restriction of the use of certain Hazardous Substances in Electrical and Electronic Equipment (RoHS) (European Union, 2003a), which aims at reducing the environmental impact of EEE, by forbidding certain quantities of specified hazardous material in certain products.

Although the WEEE Directive targets the end-of-pipe, the RoHS Directive clearly targets the beginning-of-pipe of the EEE lifecycle. Both had to be transposed to national laws, which increased the possibilities to accommodate national peculiarities but also increased the complexity for global producers to cope with dozens of deviations in the implemented procedures throughout the EU (Huisman et al., 2008). Several European countries had started to implement WEEE management policies before the EU WEEE Directive came into force. One of the oldest legislative frameworks is the Swiss Ordinance on the return, the taking back, and the disposal of electrical and electronic appliances (ORDEA) (Schweiz. Bundesrat, 2004). Its principles of defining a stakeholder's obligation is presented in Figure 25.2.

In the United States, 23 states have enacted some form of electronics recycling legislation. For example, some of these state laws established an electronics collection and recycling program and a mechanism for funding the cost of recycling under the EPR principle (GAO, 2010). Like in the USA, in Canada and Australia the implementation of EPR mechanisms is voluntary by the industries. It is more precisely known as extended product responsibility to emphasize

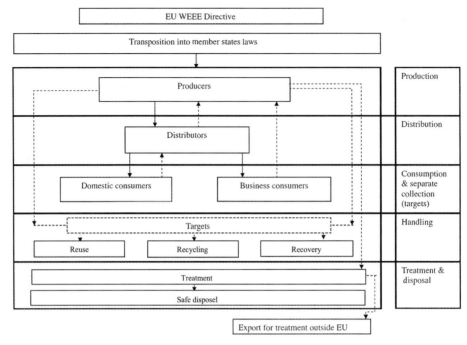

FIGURE 25.1 Simplified overview of the EU WEEE Directive (Ongondo et al., 2011).

that the responsibility is shared—the producer is not the only responsible party but also the packaging manufacturer, the consumer, and the retailer. It is thought that the biggest challenge for the United States, Canada, and Australia is legislation and compliance, as current take-back initiatives have either come in late or are not comprehensive enough (Ongondo et al., 2011). Probably the biggest issue is that the current take-back systems do not have a comprehensive product view and often only cover a few products.

Although the quantity of domestic e-waste per head is still relatively small in developing and transition countries, populous countries such as China and India are already huge producers of e-waste (Schluep et al., 2009). These countries also display the fastest growing markets for EEE, as well as the ones that are far from saturation (Yu et al., 2010). On top of WEEE generated out of domestic consumption, a considerable amount is—intentionally or unintentionally—imported via the trade of used EEE (Secretariat of the Basel Convention, 2011).

In these countries, a large, mostly urban, workforce of cheap and unskilled workers is abundantly available. This allows for the creation of many jobs in this partly profitable waste stream but also poses a considerable threat as they often miss the needed know-how and technologies for safe operation (Sepúlveda et al., 2010). Often, this is amplified by the lack of suitable laws and their enforcement, leading to a *laissez-faire* approach on all responsible sides resulting in, for instance, "cherry-picking" by the recyclers (i.e. get the best, dump the rest) (Sinha-Khetriwal et al., 2005). Most of the participants in this sector are not aware of associated risks, do not know better practices, or have no access to investment capital to finance even profitable improvements or implement safety measures.

Nevertheless, in recent years e-waste regulations have been established in a range of

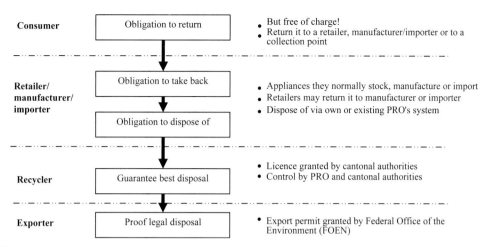

FIGURE 25.2 Schematic explaining the various stakeholders' obligations according to the Swiss WEEE legislation ORDEA (Widmer et al., 2008).

developing countries. The status of this development is currently changing constantly. Summaries of the recent development as of 2011 or earlier can be found as global overviews (Ongondo et al., 2011; Goodship and Stevels, 2012) or in different publications focusing on specific regions, such as for Africa (Secretariat of the Basel Convention, 2011; Schluep, 2012), Asia (Chi et al., 2011), and Latin America (Garcés and Silva, 2010). Most regulations are looking upstream as well as downstream, referring to sustainability and EPR principles similar to the European WEEE and RoHS directive. However, in most of these regulations, key elements and principles are still defined in very general terms. This might restrict their legal power, and the lack of subsequent implementation rules and measures makes enforcement currently difficult.

25.4 LONG-TERM TRENDS

It is likely that the quantities of discarded EEE will increase substantially in the foreseeable future as a result of fast innovation cycles and increased market penetration of cheap appliances—the latter being the main driver of e-waste volumes in developing countries and countries in transition (Schluep et al., 2013).

An important long-term trend that affects the waste flows is the ever-higher integration and miniaturization of digital electronics and, related to this, an increasing complexity of the material composition. The physical mass needed to provide capacities for storing, processing, and transmitting data is decreasing at a rapid pace (roughly along the lines of Moore's Law), which leads to a decrease of the average physical mass per device in use, despite increasing functionality. It has been observed using the example of mobile phones that this trend does not lead to a decrease in total mass flow, because at the same time the number of devices is increasing faster (Hilty et al., 2005). Historically, this miniaturization paradox can be explained by the general trend that processing capacity "is getting cheaper faster than it is getting smaller" (Hilty, 2008, p. 95). The miniaturization paradox has three effects on e-waste streams:

1. Total mass flow increases despite smaller and more lightweight devices.

2. More devices enter other waste streams because they are small and unremarkable; this trend increases with embedded electronics (Hilty, 2005; Kraeuchi et al., 2005; Koehler and Hilty, 2011).
3. Informal recycling becomes more difficult because of the higher integration density of the devices.

The e-waste management—the formal as well as the informal recycling industry, and even the definition of e-waste—will have to adapt to this general trend if the dissipation of valuable materials is to be slowed down in the long term.

References

Agarwal, R., Ranjan, R., Sarkar, P., 2003. Scrapping the Hi-tech Myth: Computer Waste in India. Toxics Link, New Delhi.
Brigden, K., et al., 2005. Recycling of Electronic Waste in China and India: Workplace & Environmental Contamination. Greenpeace International.
Brigden, K., et al., 2008. Chemical Contamination at e-Waste Recycling and Disposal Sites in Accra and Korforidua, Ghana. Greenpeace International, Amsterdam, The Netherlands. Available at: http://www.greenpeace.org/raw/content/international/press/reports/chemical-contamination-at-e-wa.pdf.
Bundesrat, S., 2004. Ordinance of 14 January 1998 on the Return, the Taking Back and the Disposal of Electrical and Electronic Equipment (ORDEE). Available at: http://www.umwelt-schweiz.ch/imperia/md/content/abfall/vreg_2004_e.pdf.
Chi, X., et al., 2011. Informal electronic waste recycling: a sector review with special focus on China. Waste Management 31 (4), 731–742.
European Union, 2003a. EU Directive 2002/95/EC of the European Parliament and of the Council of 27 January 2003 on the Restriction of the Use of Certain Hazardous Substances in Electrical and Electronic Equipment (RoHS).
European Union, 2003b. EU Directive 2002/96/EC of the European Parliament and of the Council of 27 January 2003 on Waste Electrical and Electronic Equipment (WEEE).
GAO, 2010. Electronic Waste Considerations for Promoting Environmentally Sound Reuse and Recycling. United States Government Accountability Office, Washington, DC. Available at: http://www.gao.gov/new.items/d10626.pdf.
Garcés, D., Silva, U., 2010. Guía de contenidos legales para la gestión de los residuos electrónicos. Centro de Derecho Ambiental, Facultad de Derecho, Universidad de Chile, Santiago de Chile. Available at: http://www.residuoselectronicos.net/?p=1789.
Goodship, V., Stevels, A., 2012. Waste Electrical and Electronic Equipment (WEEE) Handbook. Woodhead Publishing Limited, Cambridge, UK.
Hagelüken, C., Meskers, C., 2008. Mining our computers opportunities and challenges to recover scarce and valuable metals from end-of-Life electronic devices. In: Reichl, H., Nissen, N.F., et al. (Eds.), Electronics Goes Green 2008+. Fraunhofer IRB Verlag, Stuttgart, pp. 623–628.
Hilty, L.M., 2005. Electronic waste: an emerging risk? Environmental Impact Assessment Review 25, 431–435.
Hilty, L., 2008. Information Technology and Sustainability: Essays on the Relationship between ICT and Sustainable Development. Books on Demand, Norderstedt.
Hilty, L., et al., 2005. In: The Swiss Center for Technology Assessment (TA-SWISS), Bern, Switzerland (TA46e/2005), The Scientific Technology Options Assessment at the European Parliament (STOA 125 EN) (Eds.), The Precautionary Principle in the Information Society – Effects of Pervasive Computing on Health and Environment, second revised ed.
Huisman, J., et al., 2008. 2008 Review of Directive 2002/96 on Waste Electrical and Electronic Equipment (WEEE). Final Report. United Nations University.
Koehler, A., Hilty, L., 2011. Prospective impacts of electronic textiles on recycling and disposal. Journal of Industrial Ecology 15 (4), 496–511.
Kraeuchi, P., et al., 2005. End-of-life impacts of pervasive computing – are RFID tags a threat to waste management processes? IEEE Technology and Society Magazine 24, 45–53.
Magashi, A., Schluep, M., 2011. e-Waste Assessment Tanzania. Cleaner Production Centre of Tanzania & Empa Switzerland.
Ongondo, F.O., Williams, I.D., Cherrett, T.J., 2011. How are WEEE doing? A global review of the management of electrical and electronic wastes. Waste Management 31 (4), 714–730.
Puckett, J., et al., 2002. Exporting Harm, the High-tech Trashing of Asia. The Basel Action Network (BAN) Silicon Valley Toxics Coalition (SVTC), Seattle, WA, USA.
Puckett, J., et al., 2005. The Digital Dump, Exporting Re-use and Abuse to Africa. The Basel Action Network (BAN), Seattle, WA, USA.
Secretariat of the Basel Convention, 2011. Where are WEEE in Africa? Findings from the Basel Convention e-Waste Africa Programme. Geneva/Switzerland.
Schluep, M., 2012. WEEE management in Africa. In: Goodship, V., Stevels, A. (Eds.), Waste Electrical and

Electronic Equipment (WEEE) Handbook. Woodhead Publishing Limited, Cambridge, UK.

Schluep, M., et al., 2008. Assessing the e-waste situation in Africa. In: Electronics Goes Green 2008+.

Schluep, M., et al., 2009. Recycling — from e-Waste to Resources, Sustainable Innovation and Technology Transfer Industrial Sector Studies. United Nations Environment Programme (UNEP), Paris, France.

Schluep, M., et al., 2013. Insights from a decade of development cooperation in e-waste management. In: Proceedings of the First International Conference on Information and Communication Technologies for Sustainability. ICT4S 2013, February 14—16. ETH Zurich.

Sepúlveda, A., et al., 2010. A review of the environmental fate and effects of hazardous substances released from electrical and electronic equipments during recycling: examples from China and India. Environmental Impact Assessment Review 30, 28—41.

Sinha-Khetriwal, D., Kraeuchi, P., Schwaninger, M., 2005. A comparison of electronic waste recycling in Switzerland and in India. Environmental Impact Assessment Review 25, 492—504.

Sinha-Khetriwal, D., et al., 2006. Legislating e-waste management: progress from various countries. Elni Review 1+2/06, 27—36.

Sinha-Khetriwal, D., Kraeuchi, P., Widmer, R., 2009. Producer responsibility for e-waste management: key issues for consideration — learning from the Swiss experience. Journal of Environmental Management 90 (1), 153—165.

Wasswa, J., Schluep, M., 2008. e-Waste Assessment in Uganda: A Situational Analysis of e-Waste Management and Generation with Special Emphasis on Personal Computers. Uganda Cleaner Production Center, Empa, Kampala, Uganda; St. Gallen, Switzerland.

Widmer, R., et al., 2005. Global perspectives on e-waste. Environmental Impact Assessment Review 25, 436—458.

Widmer, R., Schluep, M., Denzler, S., 2008. The Swiss global e-waste programme. In: Waste Management Conference (WasteCon 2008). Durban, South Africa.

Yu, J., et al., 2010. Forecasting global generation of obsolete personal computers. Environmental Science & Technology 44 (9), 3232—3237.

CHAPTER 26

Developments in Collection of Municipal Solid Waste

Maarten Goorhuis
Senior Policy Advisor, Dutch Solid Waste Association, NVRD, Arnhem, The Netherlands

26.1 INTRODUCTION

Municipal solid waste (MSW) is a topic in which everybody is an expert, because everybody is a producer of MSW. Despite (or perhaps because of) the fact that we have so many experts on the subject, MSW is a difficult waste stream to manage. Its composition varies around the world and according to the season, plus there are about 7 billion producers of MSW. Proper management of MSW is essential for a society in order to provide for sanitation and to prevent environmental pollution. In more recent years, MSW management has gained interest as a new mine for our resources. A ton of old cellphones is now believed to contain more gold than a ton of gold ore. However, collecting and retrieving these materials is more complex than ever.

26.2 DEFINITION OF MUNICIPAL SOLID WASTE

The definition of MSW (also often referred to as municipal waste or urban/solid waste) varies from country to country.

The Organization for Economic Co-operation and Development (OECD) defines MSW as "waste collected and treated by or for municipalities. It covers waste from households, including bulky waste, similar waste from commerce and trade, office buildings, institutions and small businesses, yard and garden waste, street sweepings, the contents of litter containers, and market cleansing waste. The definition excludes waste from municipal sewage networks and treatment, as well as waste from construction and demolition activities" (OECD, 2011).

The European Landfill Directive defines MSW as "waste from households, as well as other waste which, because of its nature or composition, is similar to waste from households" (European Union, 1999).

Eurostat, which gathers the European data on waste management, uses a slightly different definition: "Municipal waste consists of waste collected by or on behalf of municipal authorities and disposed of through waste management systems. Municipal waste consists mainly of waste generated by households, although it also includes similar waste from sources such as shops, offices and public institutions" (Eurostat, n.d.).

In the United States, a definition can be found in the (United States Code, 2006), where it defines MSW as waste material that is:

1. generated by a household (including a single or multifamily residence); and
2. generated by a commercial, industrial or institutional entity, to the extent that the waste material:
   a. is essentially the same as waste normally generated by a household;
   b. is collected and disposed of with other MSW as part of normal MSW collection services; and
   c. contains a relative quantity of hazardous substances no greater than the relative quantity of hazardous substances contained in waste material generated by a typical single-family household.

The Pan American Health Organization defines MSW as: "solid or semi-solid waste produced through the general activities of a population center. Includes waste from households, commercial businesses, services and institutions, as well as common (non-hazardous) hospital waste, waste from industrial offices, waste collected through street sweeping and the trimmings of plants and trees along streets and in plazas and public green spaces." (Espinoza et al., 2011).

Although the definition of MSW varies around the world, some common characteristics can be established in the definitions of MSW:

- MSW is waste from private households.
- MSW includes similar waste from small enterprises, offices and other institutions.
- MSW is waste that is collected through the municipal waste collection system.

26.3 QUANTITIES OF MUNICIPAL SOLID WASTE

26.3.1 Global Municipal Solid Waste Generation

The amounts of MSW around the world vary depending on economic development and

TABLE 26.1 Waste Generation per Region in 2010 and Projection for 2025 (World Bank, 2012)

| | Current Available Data (2010 Estimation) | | | Projection for 2025 | | | |
| | | Urban Waste Generation | | Projected Population | | Projected Urban Waste | |
Region	Urban population (millions)	Per capita (kg/capita/day)	Total (t/day)	Total population (millions)	Urban population (millions)	Per capita (kg/capita/day)	Total (t/day)
Sub Saharan Africa	260	0.65	69,119	1152	518	0.85	441,840
East Asia and Pacific	777	0.95	738,958	2124	1229	1.5	1,865,379
Europe and Central Asia	227	1.1	254,389	339	239	1.5	354,810
Latin America and Caribbean	399	1.1	437,545	681	466	1.6	728,392
Middle East and North Africa	162	1.1	173,545	379	257	1.43	369,320
OECD	729	2.2	1,566,286	1031	842	2.1	1,742,417
South Asia	426	0.45	192,410	1938	734	0.77	567,545
Total	2980	1.2	3,532,252	7644	4285	1.4	6,069,703

income levels. Generally, the higher the economic development and income level, the higher the amount of MSW. According to research from the World Bank (2012), the amount of MSW is expected to nearly double between 2010 and 2025 from approximately 1.3 billion t/year to 2.2 billion t, representing an increase in per capita waste generation rates from 1.2 to 1.4 kg/person/day. This is illustrated in Table 26.1 and Figures 26.1 and 26.2.

Global data on MSW are often compromised due to inconsistencies in definitions and data collection methods. Especially in low- and middle-income countries, the reliability of MSW statistics is often poor due to unauthorized waste collection and treatment (e.g. waste dumping, local burning). The data in Tables 26.1 and 26.2 and Figures 26.1 and 26.2 are therefore an estimation.

Waste generation is influenced by income level. High-income countries produce significantly more waste per capita compared to low- and middle-income countries. Table 26.2 shows the distribution of waste generation levels according to country income level.

26.3.2 European Municipal Solid Waste Generation

In Europe, data on MSW generation are gathered and published by Eurostat and the

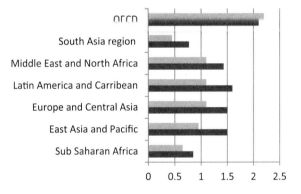

FIGURE 26.1 Waste generation per region (kg/capita/day) in 2010 and projection for 2025 (World Bank, 2012).

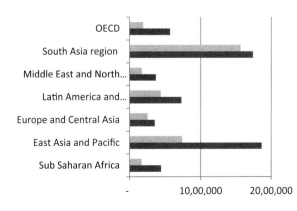

FIGURE 26.2 Total waste generation per region (t/day) in 2010 and projection for 2025 (World Bank, 2012).

TABLE 26.2 Waste Generation per Country Income Level in 2010 and Projection for 2025 (World Bank, 2012)

	Current Available Data (2010 Estimation)			Projection for 2025			
		Urban Waste Generation		Projected Population		Projected Urban Waste	
Region	Total urban population (millions)	Per capita (kg/capita/day)	Total (t/day)	Total population (millions)	Urban population (millions)	Per capita (kg/capita/day)	Total (t/day)
Lower income	343	0.60	204,802	1637	676	0.86	584,272
Lower middle income	1293	0.78	1,012,321	4010	2080	1.3	2,618,804
Upper middle income	572	1.16	665,586	888	619	1.6	987,039
High income	774	2.13	1,649,547	1112	912	2.1	1,879,590
Total	2982	1.19	3,532,256	7647	4287	1.4	6,069,705

European Environment Agency, based on data submitted by the different member states. In Figure 26.3, an overview is presented on the generation of MSW in 2010 in the different European countries.

26.4 QUALITY OF MUNICIPAL SOLID WASTE

MSW is composed roughly of organic waste and inorganic waste. Commonly, organic waste and paper waste make up more than half of MSW. The other half is plastic, glass and metal waste (mainly packaging waste) and other waste. Table 26.3 specifies the most relevant waste materials in MSW and their origin.

The waste composition is influenced by many factors, such as the level of economic development, geographical location and energy resources. When income levels rise, consumption also rises, which results in more waste generation, especially inorganic waste. In low-income countries, MSW composition is dominated by organic waste. Table 26.4 shows the waste composition per income level, whereas Figure 26.4 depicts the global composition of MSW.

26.5 MANAGEMENT OF MUNICIPAL SOLID WASTE

Responsibilities for the management of MSW are usually defined in national law. In most cases, municipalities are responsible for the management of MSW. This can either be a full responsibility for the collection and treatment of all MSW, or sometimes only for the management of household waste and waste from public spaces. In the latter case, the management of similar waste from small enterprises, offices and other institutions is a responsibility of the holder of the waste.

MSW is managed through (separate) collection and subsequent treatment of the waste. Originally, the management of MSW was driven by the need to provide for sanitation. In previous eras, uncontrolled dumping of MSW lead to the spread of vermin and pest diseases. In order to prevent this, it became necessary to

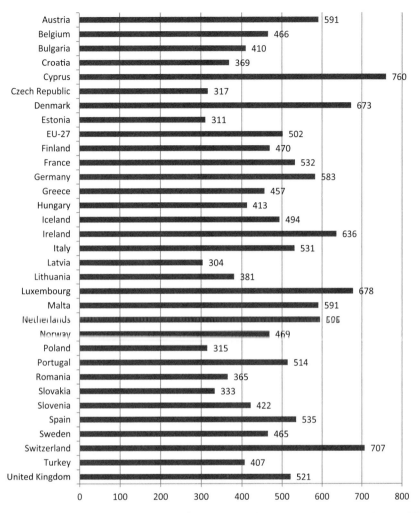

FIGURE 26.3 European waste generation (kg/capita/year) in 2010 (European Environment Agency, 2013).

collect the waste and take it out of the cities and villages. Until the industrial revolution, MSW consisted predominantly of organic waste, which could be spread on the land as a natural compost. With increased welfare and increased industrialization, the composition of the waste slowly shifted and became more inorganic. Due to an increase in consumption as well as in population, the amounts of MSW were growing. Instead of spreading MSW on the land, the waste was dumped in landfills. At the end of the nineteenth century, the first waste incineration plants were erected, mostly in or near large cities where the amounts of MSW were larger. However, landfill remained the dominant disposal method.

In the second half of the twentieth century, it became apparent that many landfills contained hazardous materials, which were leaking out of the landfill and causing environmental pollution. Also, waste incineration and flue gas cleaning techniques appeared to be insufficient for

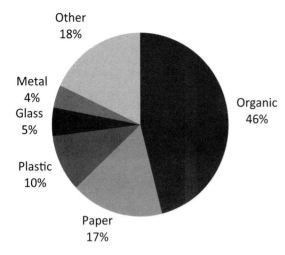

FIGURE 26.4 Global composition of MSW (World Bank, 2012).

TABLE 26.4 Composition of MSW (%) per Country Income Level, Estimate 2010 (World Bank, 2012)

Income Level	Organic	Paper	Plastic	Glass	Metal	Other
Low income	64	5	8	3	3	17
Lower middle income	59	9	12	3	2	15
Upper middle	54	14	11	5	3	13
High income	28	31	11	7	6	17

preventing the creation and spread of dioxins. MSW management had to be further developed in order to provide not only for sanitation but also for environmental protection. As a consequence, the concept of sanitary landfill, where waste is contained and monitored, was developed and waste incineration techniques and flue gas cleaning techniques were further improved.

Although the need for sanitation and the need to protect the environment from the negative effects of waste are still very valid reasons for waste management (certainly in low income countries), the current focus of modern waste management has shifted toward the recovery of resources from our waste. Modern management of MSW therefore combines the need for sanitation and environmental protection with the need to recover the resources that are contained in our waste.

The waste hierarchy has become a widely accepted guideline for waste management operations throughout the world, although the precise interpretation may vary from one place to another. The waste hierarchy stipulates a priority order in waste prevention and management legislation policy. Figure 26.5 shows the waste hierarchy according to the European Waste Framework Directive (directive 2008/98/EC).

Although the waste hierarchy seems to be quite clear in practice, several uncertainties

TABLE 26.3 Types of Waste and Their Sources

Type	Source
Organic	Biodegradable park and yard waste (leaves, grass, brush), food and kitchen waste from households, restaurants, caterers and retail
Paper and cardboard	Paper scraps, cardboard, newspapers, magazines, paper bags, boxes, wrapping paper, books, shredded paper (strictly speaking, paper is organic but usually not classified as organic waste)
Plastic	Bottles, packaging, containers, bags, lids, cups
Glass	Bottles, broken glassware, colored glass
Metal	Cans, foil, tins, empty aerosol cans, railings, bicycles
Other	Textiles, leather, rubber, multilaminates, electronic waste, appliances, ash, other inert materials

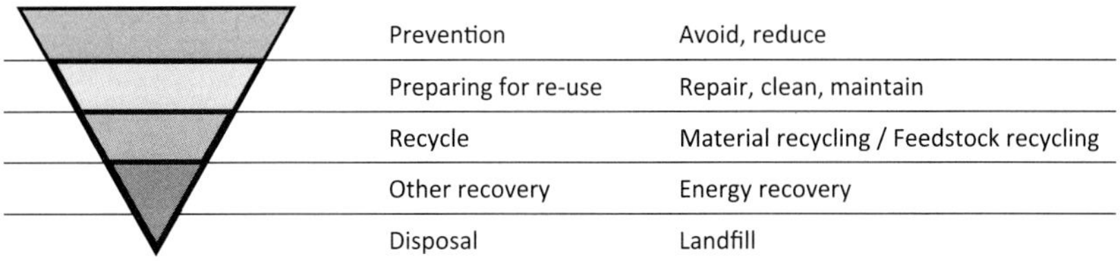

	Prevention	Avoid, reduce
	Preparing for re-use	Repair, clean, maintain
	Recycle	Material recycling / Feedstock recycling
	Other recovery	Energy recovery
	Disposal	Landfill

FIGURE 26.5 Waste hierarchy according to the EU directive 2008/98/EC. (For colour version of this figure, the reader is referred to the online version of this book.)

exist. On the one hand, the term *recycling* stands for a great variety of processes and methods that significantly differ in terms of environmental and economical benefits. The recycling of mixed plastic as opposed to material separated plastic, for instance, has a very different environmental and economic impact. Generally speaking, recycling operations that lead to materials with the same quality as the original have a better environmental performance than recycling operations that lead to materials with a lower quality. In the end, however, the optimal form of recycling should be decided based on a life cycle assessment.

26.5.1 Monitoring Municipal Solid Waste

To manage MSW, the monitoring of the quality and quantity of MSW is essential. Without information on these aspects, it is not possible to design and implement adequate solutions for the collection and treatment of MSW. Information on the waste composition is essential to evaluate the possibilities for collection, recycling and final treatment.

To determine the waste composition, sorting analysis can be carried out on a regular basis. To ensure the comparability of different sorting analyses, the process of the sorting analysis needs to be described. Different countries, such as New Zealand, already have a standard procedure for waste sorting analysis (New Zealand Ministry for the Environment, 2002).

Factors to take in account when performing a sorting analysis include the following:

- Research area: The area for which the composition needs to be determined. The different forms of waste collection that occur should be represented in the sample, as well as the different housing types that are present in the area (apartment buildings versus detached houses with gardens).
- Number and moments of sampling. The composition of the waste usually varies with the seasons. To get the best results, samples should be taken in different seasons.
- Sample size: The sample size has to be sufficient to get reliable data. In many cases, a minimum sample size of 750 kg is used. The larger the sample size, the more accurate the results.
- Collection and sampling procedures: In order to ensure the comparability of different sorting analyses, the process of collecting the samples, the sampling method and the sorting method needs to be carried out in accordance with set procedures and definitions.

Information on waste quantities is essential when evaluating logistics for collection of MSW and facilities for the recycling and treatment of MSW. Again, seasonal variations have to be taken into account. Preferably, collected waste is weighed and registered during the different phases of the waste management process: collection, recycling and final treatment.

26.5.2 Separate Collection and Waste Sorting

In order to reuse or recycle components of MSW, recyclables need to be isolated from the MSW stream. This can be achieved in different ways:

- Source separation of monostreams
- Source separation of mixed streams, followed by (mechanical) sorting
- Mechanical separation of mixed waste

The preferable methods very much depend on local circumstances. Source separation of monostreams usually offers the best quality; however, this also requires more space at the households to separate and store the different waste streams. Most commonly, one or more of the following waste fractions are collected separately:

- Organic waste
- Paper and cardboard
- (Plastic) packaging waste
- Glass
- Textiles
- Electrical appliances

Source-separated waste can be collected through various methods:

- Curbside collection systems
- Bring systems (drop-off sites)
- Recycling centers

Waste flow data are necessary to claim success or failure of a collection program, as was concluded by Berg (1993). He presented the following indicators for evaluating source-sorting systems:

1. Quantity of collected recyclables
2. Quality of recyclables (contamination rate)
3. Recycling rate (recovered material/potential recyclable amount)
4. Participation rate
5. Willingness to participate
6. Inhabitants' degree of satisfaction

Curbside collection systems usually lead to a higher participation rate and better quality of recyclables compared to drop-off systems (Dahlén, 2009). Curbside collection systems are especially useful in areas where people have enough space to store waste, such as in rural areas or in the outskirts of cities. In these cases, people can be provided with receptacles (e.g. wheelie bins) to store and offer separated waste streams for collection. Collection frequencies may vary depending on climate conditions and waste production. For citizens, a curbside collection system provides the highest service.

A drop-off site is more suitable in densely populated areas where there is little room for the storage of waste in or near the house. In this case, more effort is needed to offer separated waste for collection, which usually leads to a lower participation rate and lower yields compared to curbside collection systems.

Recycling centers are the backbone of any collection system, enabling citizens to offer separated and unseparated waste at their own convenience and/or to offer waste in larger quantities.

A literature review by Dahlén (2009) revealed 43 factors that were shown to affect the output of waste sorting programs, which they stratified into the following nine factors:

1. Property-close (curbside) or bring (drop-off) collection systems
2. Differences in the number and type of recycling materials collected separately
3. Mandatory or voluntary recycling programs
4. Use of economic incentives
5. Differences in information strategies
6. Residential structure (e.g. single or multifamily houses, urban or rural areas)
7. Social-economic differences
8. Households with private composting
9. Availability of alternative places of discharge

Halvorsen (2012) found many studies that have shown that monetary incentives and providing different types of recycling facilities matters. Recent experiences in the Netherlands have shown that increasing service for the collection of recyclables, by introducing recipients for different recyclables in a curbside collection system, combined with a decrease in the service for the collection of residual waste by replacing the recipient for a drop-off point, can further improve recycling rates and decrease the amount of residual waste (Goorhuis, 2012). This effect was even stronger when combined with a pay-as-you-throw system.

26.5.2.1 Source Separation of Mixed Dry Recyclables

Halvorsen (2012) concluded that increasing the supply of recycling services generally has a good effect on household recycling, and curbside collection and drop-off points are the two most effective collection schemes. He also concluded that if the recycling burden is already high (many materials to be separated and recycled), introducing services for new materials may reduce recycling overall. He suggested that this could be due to crowding-out effects.

To overcome these crowding-out effects and to lower the burden for citizens, source separation of mixed dry recyclables followed by industrial separation can be used as an alternative to the separate collection of monostreams. The separation of mixed dry recyclables can also be an alternative solution in those situations where there is not enough room in or near the house to store the desired amount of monostreams.

In many countries, a system of mixed dry recyclables is used for the collection of packaging waste where plastic packaging is collected together with metal packaging and beverage cartons. In Germany, this concept has been further developed into the so-called Wertstoftonne (value-bin), where the collection of mixed dry recyclables is not exclusive for packaging but for all plastic and metal, sometimes combined with other items such as small electrical appliances and compact disks.

A collection system for mixed dry recyclables requires a sorting facility where the mixed dry recyclables are separated into different monostreams. Such a sorting plant consists of a combination of different separation techniques such as sieves, windshifters, magnetic separation, eddy-current separation and near infrared. Depending on the precise composition of the input material and the desired quality of the output, handpicking can be used in the last phase of the sorting process.

The most important focus point in the sorting of mixed dry recyclables is the quality of the output material. This quality depends on the sorting techniques used and on the composition of the input material. Because some materials are easier to separate than others and some materials are more easily contaminated than others, the essence of a successful mixed dry recyclables scheme lies in the material composition of the mixed dry recyclables and the sorting techniques that are applied. Because sorting techniques are still developing, it can be expected that the application of mechanical sorting will grow in the future.

26.5.2.2 Mechanical Separation of Mixed Waste

The mechanical separation of mixed MSW has long been tried. In the second half of the twentieth century, installations for the mechanical sorting of mixed municipal waste were built. However, due to the large amount of wet organic waste in mixed residual waste, which contaminates all other waste, it is generally speaking not feasible to separate monostreams for material recycling after an integrated collection of mixed residual waste.

Today, the mechanical separation of mixed waste exists under the name mechanical-biological treatment (MBT). The MBT technique is usually focused on the separation and composting or digestion of the organic content in the waste. The remaining waste is landfilled or used for the production of energy pellets.

The European Landfill directive, which prohibits the landfill of untreated waste, has lead to an increase of the application of this technique in several EU countries. The compost that can be produced out of the organic content of the waste in an MBT operation will in most cases not meet quality standards for compost that can be used in agriculture. Instead, this compost can be used to cover landfills.

26.5.3 Assessment Framework for Municipal Solid Waste Management

To shape MSW management, municipalities have to make decisions on the collection and treatment of MSW. Apart from the environmental performance of the waste collection and treatment service, factors such as the total cost for waste management and the service provided to citizens are important to address. This assessment framework is also known as the waste triangle (NVRD, 2005). Figure 26.6.

The waste triangle has three areas of performance:

1. *Environmental performance.* The environmental performance describes to what extent environmental goals for MSW management are achieved. Indicators for environmental performance include (but are not limited to) the following:
   a. Total waste generation
   b. Amounts of recycled, recovered and disposed waste
   c. Waste separation rate and recovery rate
   d. Amount of separately collected waste per waste stream
2. *Cost performance.* The cost performance describes the cost effectiveness of the management of MSW. Indicators for cost performance include (but are not limited to) the following:
   a. Direct cost for (separate) collection per waste stream
   b. Direct cost for recycling, recovery and disposal per waste stream
   c. Indirect cost (overhead) for collection and treatment
3. *Service performance.* The service performance describes the amount of service delivered to the citizens for waste collection and treatment. Indicators for the service performance include (but are not limited to) the following:
   a. Collection frequency per waste stream
   b. Density of drop-off points
   c. Density of recycling yards
   d. Citizen satisfaction measurement

Ideally, the MSW management system combines a high service level with a high environmental performance at low cost. However, in practice, a high service level and high environmental performance often lead to higher costs. It is usually a political decision when determining which aspects of the waste triangle are more emphasized than others.

26.5.3.1 Treatment of Municipal Solid Waste

Processes of MSW treatment can be classified into the categories of landfill (disposal), incineration (thermal recovery) and recycling (including

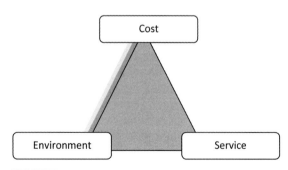

FIGURE 26.6 The waste triangle assessment framework.

TABLE 26.5 Classification of Advancement in Waste Treatment

Class	Borders	Remark
I	Landfill: >80%	Large room for improvement
II	Landfill: 40–80%	Considerable advance achieved; further improvement necessary
III	Landfill: 20–40%	European average
IV	Landfill: <20% Recycling: ≤60%	Far advanced in avoiding landfill, potential to increase recycling
V	Landfill: <20% Recycling: >60%	Current state-of-art
VI	Recycling: 100%	Optimal waste management, not possible in practice.

composting/digestion). Bartl (Goorhuis, 2011) classified the level of development of the waste management system based on the advancement of treatment methods according to the waste hierarchy (see Table 26.5).

Class I (i.e. landfilling >80%) is more or less the starting point of waste management. Even if open dumping has been replaced, at this stage a large amount of resources and energy are lost at the sanitary landfill.

In class II, the fraction of waste going to landfill is below 80% but still above 40%. The countries within this area have already achieved considerable advances in avoiding landfill. However, still more than 40% of the waste ends up in landfills; thus, these countries also have much room for improvement.

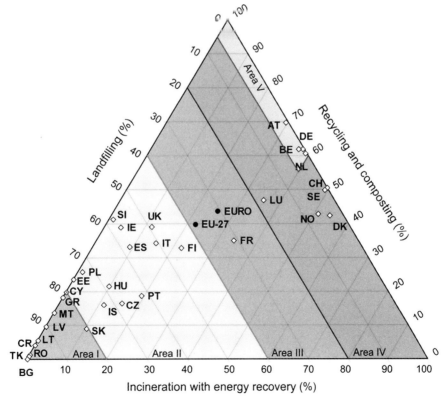

FIGURE 26.7 European countries moving toward recycling (Bartl, 2012). (For colour version of this figure, the reader is referred to the online version of this book.)

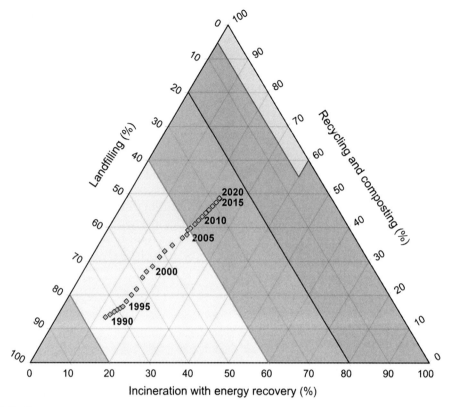

FIGURE 26.8 Europe moving toward recycling in the period between 1990 and 2020 (Bartl, 2012). (For colour version of this figure, the reader is referred to the online version of this book.)

If the rate of landfill is further decreased and lies within the range of 20 and 40%, class III is reached. The EU average for both EU-27 and the Euro countries can be found within this area.

Class IV is designated for countries that are more advanced in waste management and already have a landfill rate below 20%. However, it seems necessary to define an additional region, class V. On the one hand, it is characterized by a very low rate of landfill (<5%; i.e. near to zero). On the other hand, incineration has to be largely avoided (<40%). Only four European countries have reached class V to date. However, also the members of class V still have a considerable distance to the final goal of 100% recycling.

Theoretically, there is also a class VI, which represents optimal waste management where all waste is recycled. However, this situation is not achievable in practice because every recycling operation requires consumption of energy and input streams and also results in output streams (waste). In fact, every cycle in a recycling process means a certain decrease in quality (i.e. downcycling) and generation of waste (Bartl, 2012). A recycling rate of 100% is therefore a theoretical idea.

When plotting the data of individual countries in a triangle diagram (Figure 26.7), the advance of countries in terms of waste management becomes apparent. In most cases, the initial position is near the left corner (100% landfill). The final goal is to reach the top corner, where all waste is recycled and incineration as well as landfill is completely avoided. As an intermediate target, incineration can be found in the right corner.

Figure 26.8 sketches the significant improvements of European waste management from the past (1990) to the future (forecast until 2020). In 1990, European waste management depended heavily on landfill (74%). It had just crossed the border between class I and class II. Over the subsequent 15 years, the landfill rate had been drastically reduced, thus moving from class II into class III by about 2007. However, the speed of improvement will dramatically decrease in the next years and the progress in the period 2015 (estimated recycling rate: 47%) to 2020 (estimated recycling rate: 49%) will be more or less zero. It is obvious that Europe will not reach area IV within a midterm range and there is no indication that Europe will ever reach area V.

References

Bartl, A., 2012. Barriers and Limits for Recycling and Moving towards "Zero Waste".
Berg, P., 1993. Source Sorting. Theory, Method and Implementation. Institutionen för Vattenförsörjnings-och, Göteborg, Sweden.
Dahlén, L.E., 2009. Evaluation of recycling programmes in household waste collection systems. Waste Management & Research, 577–586.
Espinoza, P.T., et al., 2011. Regional Evaluation on Urban Solid Waste Management in Latin America and the Caribbean. Pan American Health Organization.
European Environment Agency, 2013. Managing Municipal Solid Waste. Publications Office of the European Union, Luxembourg.
European Union, 1999. Council Directive 1999/31/EC on the Landfill of Waste.
Eurostat. Glossary: municipal waste. http://epp.eurostat.ec.europa.eu/statistics_explained/index.php/Glossary:Municipal_waste (retrieved 23.06.13).
Goorhuis, M., 2011. Key issue paper. Waste prevention, waste minimization and resource management. ISWA.
Goorhuis, M., 2012. New developments in waste management in the Netherlands. Waste Management & Research, 67–77.
Halvorsen, B., 2012. Effects of norms and policy incentives on household recycling: an international comparison. Resources, Conservation and Recyling, 18–26.
New Zealand Ministry for the Environment, 2002. Solid Waste Analysis Protocol (New Zealand).
NVRD, 2005. Afwegingskader oud papier/karton, glas en textiel (Arnhem).
OECD, 2011. OECD Factbook 2011–2012: Economic, Environmental and Social Statiscs. OECD Publishing.
United States Code, 2006. United States Code. Supplement 4, Title 42-The Public Health and Welfare (United States of America).
World Bank, 2012. What a Waste—A Global Review of Solid Waste Management. World Bank, Washington DC.

PART III

STRATEGY AND POLICY

CHAPTER 27

From Recycling to Eco-design

Elisabeth Maris[1], Daniel Froelich[1], Améziane Aoussat[2], Emmanuel Naffrechoux[3]

[1]Laboratoire Conception Produit Innovation, Chambéry (LCPI), Institut Arts et Métiers ParisTech, Chambéry, Savoie Technolac, Le Bourget du Lac, France; [2]LCPI, Arts et Métiers ParisTech, Paris, France; [3]Laboratoire Chimie moléculaire Environnement (LCME), Université de Savoie, Le Bourget du Lac, France

27.1 INTRODUCTION

Regulatory requirements and the rising price of materials due to their scarcity urge product and system designers to integrate recycling and the reduction of environmental impacts throughout the life cycle as early as possible into the design process. In the 1990s, designers began to take into account the effects on the environment of product life cycle phases. Efforts must still be made, in particular in the manufacturing and end-of-life phases. One possible strategy—design for recycling—can be defined as designing a recyclable product and using recycled materials to replace virgin materials. Several factors slow down the design to make products recyclable, including technical barriers and barriers linked to the traceability of materials due to their potential contamination.

27.2 PRINCIPLE OF MATERIAL DESIGN FOR RECYCLING

Design for recycling is an eco-design strategy. Eco-design is a systematic approach allowing the design of more environmentally friendly products. For a company with a strategy to reduce its environmental impacts, the first stage is to review all the processes intervening in the design of a product and to find solutions to reduce the impacts on the product's life cycle. Several strategies can be adopted, as summarized in Figure 27.1. In the materials choice phase, one can choose less impacting materials; reduce the quantities of materials; improve process techniques, transport, and the usage phase; and optimize the life cycle and end-of-life of products.

It is widely accepted that the usage phase is highly impacting for energy-consuming products with long life cycles because these

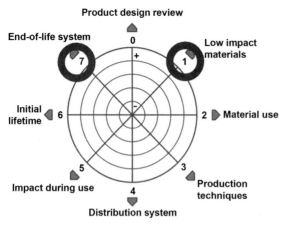

FIGURE 27.1 The wheel of eco-design strategies (Crul and Dieh, 2009).

impacts have a direct effect on climate change. Nevertheless, the impacts of the product manufacturing phase must not be forgotten because the effects on mining resources and nonrenewable energies are irreversible. These resources are limited and impact the extraction phase due to their increasingly low concentrations and therefore higher extraction costs. One solution is to reuse or recycle the product's component materials at their end of life.

27.3 ECO-DESIGN STRATEGIES FOR RECYCLING

A recyclable material must maintain its mechanical and chemical properties and be able to be sorted by recycling companies. Eco-designing recyclable materials means to make them transformable and sortable with an acceptable cost-to-performance ratio. On the wheel of eco-design, strategy 1 is to take into account the limits on resources, while strategy 7 is to take into account the limits of sorting processes. These two limits lead to the consideration of design constraints on the choice of materials and on their association.

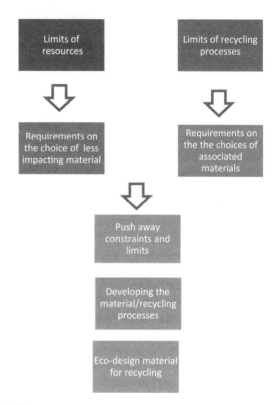

FIGURE 27.2 Principle of the eco design of materials.

To push these limits further means to develop the materials—sorting process combination, which is the definition of the eco-design approach for materials (Figure 27.2).

27.4 IS RECYCLING REALLY LESS IMPACTFUL ON THE ENVIRONMENT?

For polymers, it is important to remember that 95% of the energy required to produce 1 kg is due to the extraction and refining of the oil (Johnstone, 2005). When virgin polymers are replaced with recycled polymers at rates close to 100%, the recycling scenario is more advantageous than energy recovery. Furthermore, certain polymers can follow several recycling cycles (Assadi, 2002).

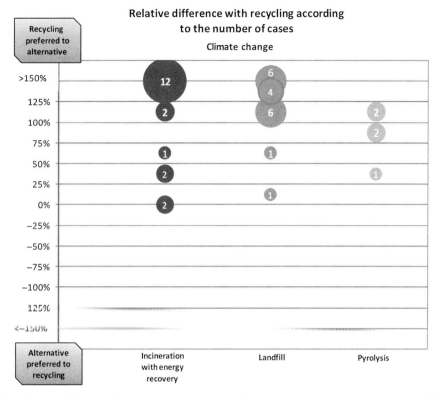

FIGURE 27.3 Relative differences between the various scenarios of polymer treatment at their end of life or recycling with respect to their effect on climate change. The bubble is proportional to the number of studies and the impact value of the same rank (WRAP, 2010).

A study conducted (WRAP, 2010) on the environmental life cycle analysis of the scenarios of different end-of-life polymer treatments shows that the number of studies in favor of the recycling scenario (Figure 27.3) is higher than the number favorable to incineration scenarios with energy recovery or burying in a landfill.

27.5 CURRENT LIMITS FOR ECO-DESIGN FOR RECYCLING STRATEGIES

In view of the medium-term shortages forecasted for certain resources, the efforts to improve material recyclability (Millet, 2003) have led material recyclers to improve sorting processes. They also encouraged designers to reduce the quantity of materials in products and to choose recyclable and associated materials so that they are potentially sortable. There are still several difficulties linked with recycling materials from end-of-life products. These products are collected and treated at their end of life (Figure 27.4), which requires sorting out complex mixtures after grinding (Reuter, 2006), despite eco-design efforts to recycle new products. The second difficulty is linked to the two most representative industrial sorting technologies (Table 27.1) in the recycling sector that do not make it possible to sort out materials using a physico-chemical sorting technique: density separation when the

FIGURE 27.4 End-of-life product scheme.

densities overlap or fast spectrometric sorting (near infrared) when polymers are dark-colored. In 2009, 24.3 Mt of polymer waste were generated in Europe, yet only 22.5% on average were recycled in all sectors combined (PlasticsEurope, 2010).

Other choices have been envisaged by designers to reduce environmental impacts, such as the use of biodegradable or biosourced materials. In the first case, the end-of-life impacts can be reduced to avoid the accumulation of waste in landfills.

The second case appears to reduce the use of nonrenewable resources, but in reality there is a transfer via another impact. Eco-profiles

TABLE 27.1 Current Limits of Detection Processes

Physical and chemical sorting	Densimetric sorting (Hwang, 1995; Altland, 1995)	☺ Low cost, industrial scale ☹ Not adapted for overlapped density $<0.12 \text{ g/cm}^3$
	Froth flotation sorting (Fraunholcz et al., 2004)	☺ Low cost ☹ Laboratory scale
	Triboelectric sorting (Hearn and Ballard, 2004)	☺ Low cost, industrial scale ☹ Sensitive to moisture and dust, does not sort complex mixed fractions, not able to sort certain polymers
Fast spectrometric sorting	Absorption near infrared (Huth-Fehre et al., 1998)	☺ Very fast sorting, industrial scale ☹ Does not detect dark polymers or polymers closed in formulation
	Absorption medium infrared (Florestan et al., 1994)	☺ Identification of white and dark polymers ☹ Sensitive to surface condition and reflectance signal very weak to identify C—H bonds, not compatible with fast detection
	X-ray fluorescence X (Kang and Schoenung, 2005; Biddle, 1999)	☺ Fast detection, industrial scale, industrial, detection of additives ☹ Does not detect polymers except poly(vinyl chloride)
	X-ray transmission: difference of atomic density analysis (Mesina et al., 2007)	☺ Fast detection, detection of metals at an industrial scale ☹ Possible for polymers but at a laboratory scale
	Laser-induced plasma spectroscopy (LIPS) (Solo-Gabriele et al., 2004; Anzano et al., 2006; Barbier et al., 2013)	☺ Identification of white polymers ☹ Weak signal, not compatible with fast detection
	Ultraviolet fluorescence (Pascoe, 2003)	☺ Many applications for food at the industrial scale ☹ No application for polymers

☺ Positive points for sorting, ☹ Negative points for sorting

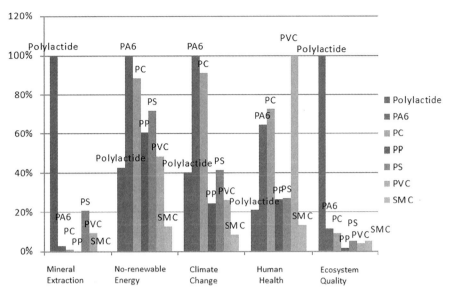

FIGURE 27.5 Comparison of environmental impacts per kilogram of polymers, Impact 2002 method.

(Figure 27.5) of various polymers from the mining extraction phase to the granule production phase were produced using the life cycle analysis modeling tool Simapro, the Impact 2002 method, and Ecoinvent data. The comparison of polylactic acids (PLAs; polymers originating from agricultural resources) with biodegradable and other polymers allowing the same type of application but originating from resins made with oil components shows that there is a transfer of the impacts. PLAs have a significant impact on mining resources due to the use of phosphate minerals as fertilizer for agriculture.

Nonrenewable resources are also significant due to the use of nitrogenous fertilizers originating from oil and the energy used by agricultural machinery.

In conclusion, the improvement of product design has driven the improvement of product recyclability, but limits still exist.

27.6 MARKET DEMAND

The recycling market is in full development (Johnstone, 2005), and the prices of recycled materials exist for metals and polymers. However, the market is still not sustainable and the prices of secondary raw materials are more volatile than those of primary raw materials.

Within the framework of a study carried out with the ADEME (French Environment and Energy Management Agency) and the club of French manufacturers CREER (Maris, 2007), a survey was conducted in 2008 with eight French groups concerning the acceptability of recycled materials, using interviews and questionnaires. The answers provided reliable data for seven companies consisting of manufacturers and original equipment manufacturers in the following sectors: electrical safety, automobiles, office furniture, and small electrical appliances, as well as waste collectors. This survey was completed with interviews with a recycled material producer and a consumers' association. The population targeted per company was classified by activity: the executive management, the buyers, the designers, the marketing department, and the quality and after-sales staff.

The survey results showed that all of the companies used recycled materials, including

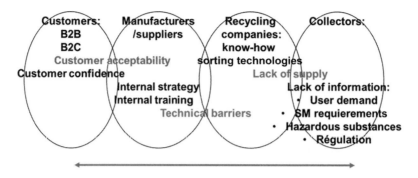

FIGURE 27.6 Barriers to the use of recycled materials.

for certain types of components whose technical performance requirements were fairly high, but they did not all integrate this usage into their strategy. The companies' motivations were both economic (linked to price increases or the scarcity of primary raw materials) and environmental. The main barriers expressed by the respondents (Figure 27.6) from the companies during the survey were:

- The quantity and quality of the recycled materials
- Technical barriers due to constraints on product applications (expressed by recyclers and manufacturers)
- Consumer acceptance

The consumer acceptance of recycled materials is a barrier linked to the traceability of materials due to their potential contamination.

According to the data published by PlasticsEurope (2009), the polymers the most often sold to transformers are polyethylene terephthalate (PET), polyethylene, and polypropylene (PP) for the following sectors: packaging, automotive, household appliances, and electrical products. PET applications are especially for colorless materials. PP is often colored with carbon black according to the application. These polymers represent the most important potential sources of recycled material.

27.7 CONCLUSION

Improving the environmental impact of materials means using recycled materials in new products and making this material recyclable when new products of different design and manufacturers' brands are treated together. To reach this objective, it is necessary to develop the materials—sorting process. Furthermore, while the demand for recycled materials already exists, it is necessary to find solutions to the problems linked to the risks due to lack of supply and the traceability of the recycled materials. Material traceability is a prerequisite to widespread acceptance by users and consumers.

References

Altland, B.L., Cox, D., Enick, R.M., Beckman, E.J., 1995. Optimization of the high-pressure, near-critical liquid-based microsortation of recyclable post-consumer plastics. Resources, Conservation & Recycling 15, 203–217.

References

Anzano, J., Casanova, M.-E., Bermudez, M.S., Lasheras, R.J., 2006. Rapid characterization of plastics using laser-induced plasma spectroscopy (LIPS). Polymer Testing 25, 623–627.

Assadi, R., 2002. Modifications structurales non réversibles lors du recyclage du poly (téréphtalate d'éthylène). ENSAM, Paris.

Barbier, S., Perrier, S., Freyermuth, P., Perrin, D., Gallard, B., Gilonstic, N., 2013. Identification based on molecular and elemental information from laser induced breakdown spectra: a comparison of plasma conditions in view of efficient sorting. Spectrochimica Acta Part B. http://dx.doi.org/10.1016/j.sab.2013.06.007.

Biddle, M.B., Dinger, P., Fisher, M.,M., 1999. An overview of recycling plastics from durable goods: challenges and opportunities. In: IdentiPlast II Conference.

Crul, M.R.R., Diehl, J.C., 2009. Design for sustainability: A Practical Approach for Developing Economies. UNEP, Division of Technology, Industry, and Economics, 2006 –124

Florestan, J., Lachambre, A., Mermilliod, N., Boulou, J.C., Marfisi, C., 1994. Recycling of plastics: automatic identification of polymers by spectroscopic methods. Resources, Conservation & Recycling 10 (1–2), 67–74

Fraunholcz, N., 2004. Separation of waste plastics by froth flotation—a review, part I. Original Research Article Minerals Engineering 17 (2), 261–268.

Hearn, G.L., Ballard, J.R., 2004. The use of electrostatic techniques for the identification and sorting of waste packing materials, UK. Resources, Conservation & Recycling 44, 91–98.

Huth-Fehre, T., Feldhoff, R., Kowol, F., Freitag, H., Kuttler, S., Lohwasser, B., Oleimeulen, M., 1998. Remote sensor systems for the automated identification of plastics. Journal of Near Infrared Spectroscopy 6, A7–A11.

Hwang, J.-Y., 1995. Separation of Normally Hydrophobic Plastic Materials by Froth Flotation, Patent US5377844.

Johnstone, N., 2005. Improving Recycling Market, Working Group on Waste Prevention and Recycling. OCDE.

Kang, H.Y., Schoenung, J.M., 2005. Electronic waste recycling: a review of U.S. infrastructure and technology options. Resources, Conservation & Recycling 45 (4), 368–400.

Maris, E., 2007. Incentives and Barriers for Improving the Using of Recycled Material, Technical Report, ADEME N°0602C0036, CREER.

Mesina, M.B., De Jong, T.P.R., Dalmijn, W.L., 2007. Automatic sorting of scrap metals with a combined electromagnetic and dual energy X-ray transmission sensor. International Journal of Mineral Processing 82, 222–232.

Millet, D., 2003. Intégration de l'environnement en conception: Entreprises et Développement Durable. édition Hermes Science Publishing, 230 p.

Pascoe, R.D., 2003. Sorting of plastics using physical separation techniques. Recycling and reuse of waste materials. Proceedings of the International Symposium, 173–188.

PlasticsEurope, 2010. The Compelling Facts About Plastics, an Analysis of Plastics Production, Demand and Recovery for 2009 in Europe. Ed. PlasticsEurope, (EuPC, EuPR, EPRO).

Reuter, M.A., van Schaik, A., Ignatenko, O., de Haan, G.J., 2006. Fundamental limits for the recycling of end-of-life vehicles. Minerals Engineering 19 (5), 433–449.

Solo-Gabriele, H.M., Townsend, T.G., Hahnc, D., Moskal, T., Hosein, N., Jambeck, J., Jacobi, G., 2004. Evaluation of XRF and LIBS technologies for on-line sorting of CCA-treated wood waste. Waste Management 24, 413–424.

WRAP, 2010. Environmental Benefits of Recycling. www.wrap.org.uk.

CHAPTER 28

Recycling and Labeling

Elisabeth Maris[1], Améziane Aoussat[2], Emmanuel Naffrechoux[3], Daniel Froelich[1]

[1]Laboratoire Conception de Produit et Innovation, Chambéry (LCPI), Institut ARTS et METIERS ParisTech Chambéry, Savoie Technolac, Le Bourget du Lac Cedex, France; [2]LCPI, ARTS et METIERS ParisTech, Paris, France; [3]Laboratoire Chimie moléculaire Environnement (LCME), Université de Savoie, Le Bourget du Lac, France

28.1 INTRODUCTION

It is difficult to identify and to sort out a certain fraction of plastic waste based on their intrinsic properties. One possibility is to add properties to the polymers to be recycled. Adding properties to a material in order to identify it by a signature is known as the labeling technique. This technique identifies a material by the signature of a tracer and not the material's intrinsic properties.

Based on a functional analysis to search for a solution for polymer recyclability, bibliographical research, and contacts with users and secondary material recyclers, a technical solution was validated for this technology. Polymer materials were chosen to test this sorting by labeling technology because they are currently recycled very little. Laboratory tests made it possible to determine the concentration limit for the tracers' incorporation into polymers.

28.2 FUNCTIONAL NEEDS ANALYSIS

An external functional analysis was performed according to International Organization for Standardization (ISO) standard NF X50-100 on the usage phase of the life cycle of the system in question to be able to deduce the main functions and constraints with which this new system will have to comply.

28.2.1 External Functional Analysis

The functional analysis (Figure 28.1) revealed two types of need: that of the users and that of the used product sorting chain operators. The constraints include the consumers of new products, the regulations and standards for new products, and the environmental and socioeconomic environment. Sorting must be fast, economical and produce good-quality materials.

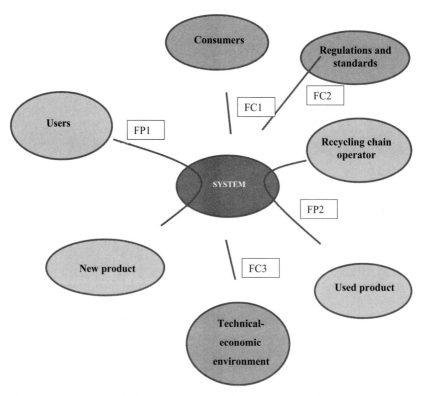

FIGURE 28.1 Representation of the main functions and constraints of the system in the octopus tool in the usage phase.

Users require high volumes of materials and long-lasting supply, with good traceability and a production cost lower than that of virgin materials. Polymer producers must guarantee the properties of the materials at the market price of the said materials.

28.2.2 Characterization of the Various Elements of the System's Environment

The users are new product designers and raw polymer material producers. The recycling chain operators include collectors, crushers, sorters and transformers. New products are products which have just been manufactured. A used product is a product that is disposed of by its last owner according to the European waste directive (Directive 2008/98/EC). A used product shifts from the status of waste to the status of a product again due to its recycling and also if the recycled materials meet certain requirements in order to be introduced on the market.

The technical-economic environment is composed of the market demand for secondary raw materials, and material prices. The consumer is a person who acquires goods and services for his own personal needs.

European environmental directives and standards aim to reduce the environmental impacts of products throughout their entire life cycle, from manufacturing to their end-of-life. The two principle functions are:

PF1: The system allows users to design new recyclable products.

PF2: The system allows recycling chain operators to recycle used products.

28.2.3 Main Constraint Functions

The main constraint functions are described here.

CF1: Consumer constraints:

- Material nontoxicity,
- Recycled material traceability

CF2: Constraints of the technical-economic environment

- Cost compatible with the prices of new materials
- High-speed automated industrial sorting of dark polymers at a speed of 3 m/s, with a 10 ms detection time, a distance between the detection system/source and the sample of 200 mm, and a sorting efficiency that must be 95%
- ISO properties of the recycled material with new materials for the targeted applications
- Because any material transformation is not 100% effective, the ultimate waste produced must comply with regulations and be accepted in incineration centers or landfills.

CF3: Regulatory constraints

The system respects the Registration, Evaluation, Authorisation and Restriction of Chemicals (REACH) and Restriction of Hazardous Substances (RoHS) regulations on the nontoxicity of the materials placed on the market, as well as compliance with product recyclability objectives, such as the regulations on product approval (RRR DIRECTIVE 2005/64/EC on the type-approval of motor vehicles with regard to their reusability, recyclability and recoverability) and European directives on waste treatment.

28.2.4 General Conclusions

This level of functional analysis made it possible to envisage several areas of research with regard to labeling for recycling technology:

- What types of polymers are necessary to label?
- What spectrometric sorting technology should be chosen?
- Which tracers should be used?

28.3 BIBLIOGRAPHICAL RESEARCH ON THE POLYMER LABELING PROCESSES

The concept of virgin polymer tracing system for sorting was developed by 1993. The first patent filed (British Petroleum Company, 1993) describes a method to identify polymers by detecting the fluorescence of certain tracers with rare earths in the near infrared (NIR) spectral domain, between 700 and 900 nm for tracer concentrations between 0.001 and 1 ppm. The source is a diode laser emitting in the NIR at 670 nm. The drawback of this method is the difficulty of detecting a signal in NIR when the matrix is dark colored. The carbon black used as a coloring agent absorbs all of the radiations in the NIR (Eisenreich et al., 1992).

A patent filed in 1994 (Bayer, 1994) describes two tracer systems with the same fluorescent emission wavelengths and different fluorescence durations. The identification principle allows a codification with four tracers. This method is currently used in the field of the biochemistry, and the experimental system includes a flash lamp and a programmable camera to delay the release of the camera by some nanoseconds after the excitation emission by the source. This system thus allows the identification of molecules that have the same fluorescence wavelength but not the same duration. This method encountered implementation difficulties for rapid automated sorting. In the case of industrial sorting, the samples are continuously lighted and thus it is impossible to differentiate between their durations.

In 1998, another study (Simmons et al., 1998; Ahmad, 2004) financed by a European program ended in a pilot for an application to sort plastic bottles for the packaging sector. Using a codification based on combinations of three tracers in concentrations between 0.5 and 20 ppm, the system made it possible to identify bottles made of high-density polyethylene. Patents were filed by the program partners (Lambert and Hachin, 2010). The experimental bench did not allow the identification of dark-colored polymers and no test was made on other types of polymer matrices.

In 2005, a patent (Eriez, 2005) was filed on the magnetic sorting of polymers containing magnetic tracers.

In 2007, an industrial project, TRITRACE, chaired by the Pôle National de Traçabilité, and financed by Région Rhône Alpes, has confirmed the industrial application of the use of the selected pigments by Tracing Technologies in the automotive and the electrical appliances industries. Results confirmed that no structural modification and no change in the appearance and/or glaze after an equivalent 13-year ultraviolet (UV) and weathering test. Polymers tested were black and white polypropylene and acrylonitrile butadiene styrene (Maris et al., 2011).

In 2007, a study on tracers validated automatic fast sorting for crushed dark polypropylene (Froelich et al., 2007). In the same study, a state-of-the-art review was conducted on the various labeling technologies. Two technologies were validated to sort black polypropylene: the detection of magnetic tracers, and X-ray fluorescent detection of tracers with rare earths. In 2008, a thesis was financed by the ADEME (French Environment and Energy Management Agency) to perform laboratory tests for the detection of X-ray fluorescence (Bezati et al., 2010). The results of these research projects show that detection by

TABLE 28.1 State-of-the-Art Tracer Technology for Polymer Sorting

Detection Principle	References	Tracer Type	Detection Wavelength	Source	Polymer Type	Development Scale
Fluorescence intensity	(British Petroleum Company, 1993)	Rare earths concentration between 0.001 and 1 ppm	Emission at 670 nm Detection between 700 and 900 nm	Diode	Transparent polymers	Laboratory
Fluorescence duration	(Bayer, 1994)	Fluorescence duration between 10^{-2} and 10^{-9} s	450–640 nm	Flash lamp	Low density colorless polyethylene	Laboratory
Fluorescence intensity	(Simmons et al., 1998) (Ahmad, 2004) (Lambert and Hachin, 2004).	Phosphor technology, Ltd 0.5–20 ppm	Ultraviolet excitation Detection visible light	Xenon lamp	Colorless packaging	Pilot Conveyor speed 3.5 m/s, purity rate 95%
Magnetic detection	(Eriez Magnetics, 2005)	Ferromagnetic particle : Polymag c = 1%	Electromagnetic or magnetic	Permanent magnet or metal detector	Production waste dark polyolefin and styrenic resin	Industrial
X-ray fluorescence intensity	(Froelich et al., 2007) (Bezati et al., 2010)	Rare earths oxides 0.1%	Excitation X-ray:10–5 kev; emission: 20–50 kev	X-ray generator	Dark polymers	Pilot

magnetic tracers is industrially viable but does not allow a codification with several tracers. The detection by X-ray fluorescence allows the detection of tracers with rare earth oxides in concentrations of 1000 ppm in black or painted polymers. The absorption of the X-ray fluorescence by the molecules of the ambient air does not allow reduction of the tracer concentrations below 100 ppm and detection times are still long with regard to the industrial constraints of fast sorting. The choice of the tracers is limited to rare earths.

In conclusion, these various studies (Table 28.1) all validated the polymer labeling technique. The technique chosen for this study is UV fluorescence because it allows the detection of low concentrations and meets the constraints of recyclers and end users. Certain aspects of the fluorescence of polymer tracer systems must be further explored, such as the possible signal quenching phenomena due to the interaction between the polymer tracers and their additives such as black carbon, the influence of polymer aging on the fluorescence, the choice of tracers that are less impacting on resources.

28.4 FIRST RESULTS OF DETECTION TESTS WITH POLYPROPYLENE SAMPLES

The objective is to calculate the detectable concentration limit to incorporate into a polymer in industrial sorting conditions for every tracer.

28.4.1 Experimental Conditions

Plates of 3-mm thickness, with and without tracer, were injected on a 750 t press by the French company Plastic Omnium at an injection temperature of 240 °C and an injection pressure of 105 bars. For every tracer, two dilution mixtures, 2% and 1%, were used. Ninety kilograms of mixture were used for every tracer. Plates were cut into 30×30 mm squares corresponding to 90 samples.

The choice of the fluorophores, which is very vast, was limited to inorganic molecules, particularly from two families—that of the complexes with lanthanides (T3 and T10) and that of barium magnesium aluminates (T1) (Figure 28.2). Works published in 2011 enabled us to determine the detection conditions for these polymer/tracer systems (Maris et al., 2011).

These samples were tested with a Fluoromax Jobin Yvon laboratory spectrometer (Figure 28.3) with an excitement wavelength of 325 nm. The detection of the fluorescence emission was in the visible spectrum, and the detection duration was 10 ms to satisfy the industrial constraints of industrial sorting machines (Figure 28.4).

28.4.2 Concentration Limits

The calculation of the detection limits makes it possible to define the tracer's concentration limit required for the incorporation into a polymer material so that it is detected in the scope defined by a measurement system with a defined confidence index.

The detection limit expressed as a quantity or a concentration of substance is the smallest measurement that can be detected by the blank with a defined confidence level and for a defined measurement procedure (Currie, 1995). The detection limit is the concentration in which we decide if an element is present or not and which allows us to distinguish it from the background noise signal or from the blank.

The fluorescence signal due to a substance overlaps with the signal due to the background (the polymer material in this case) and the noise due to the electronic instrumentation. Emissions due to the background are called the blank.

The detection limit (DL) and the concentration limit were calculated according to the

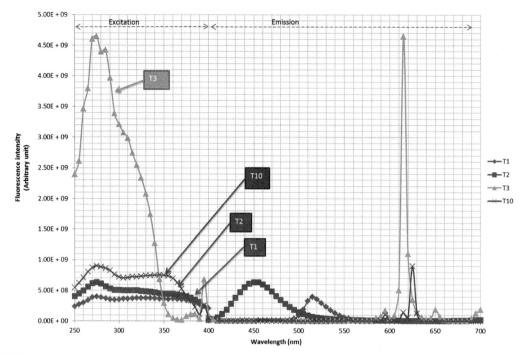

FIGURE 28.2 Spectrum of excitation and emission of the selected tracers T1, T3 and T10.

International Union of Pure and Applied Chemistry recommendation (Currie, 1995):

$$Y = AX + Y_0$$
$$\text{with } Y = B + S$$

where Y is the raw signal, B is the blank value or Y_0, S is the net signal, and X is the concentration of tracer and A the sensitivity.

$$S = AX$$

A is the sensitivity. Yield of the curve is $A = \dfrac{\partial y}{\partial x}$

$$X = (Y - Y_0)/A$$

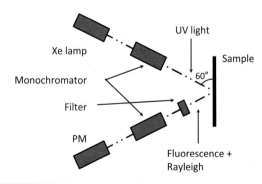

FIGURE 28.3 Diagram of the detection system.

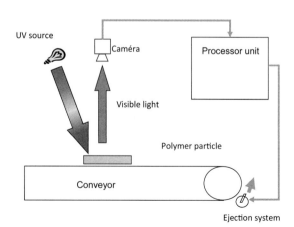

FIGURE 28.4 Diagram of the industrial sorting system.

The values of Y and Y_0 are evaluated by calculating the average of a sufficiently large number of measurements. The standard deviation on the measurement of the blank σ_0 and samples σ_Y is calculated. The calibration line is determined.

The DL is determined from the value of the standard deviation of the blank and that of the sample and the confidence index. The detection limit of the substance depends on the uncertainty due to the instrument, the uncertainty concerning the substance concentration, the uncertainty due to its dispersion in the sample, and the uncertainty due to the reproducibility of samples.

$$DL : Y_D = \overline{Y_0} + z_{1-\alpha}(\sigma_0 + \sigma_D)$$

where, Y_D is the detection limit in count per second, $\overline{Y_0}$ is the average of the measures of the blank in CPS (counts per second); $z_{1-\alpha}$ is the numerical factor depending on the confidence index, σ_0 is the standard deviation of the measurement of the blank (n measurement according to a defined protocol), and σ_D is the standard deviation of the measurement at the detection limit.

$$Y_D = AX_D + \overline{Y_0}$$

X_D is the tracer concentration at the detection limit. Thus:

$$X_D = z_{1-\alpha}(\sigma_0 + \sigma_D)/A$$

The precision of the measurement expressed by the relative standard deviation (rsd) decreases according to the concentration and the time of integration.

If the distribution of the population of the measurements follows a normal law (normal law table), for a risk of $2\alpha = 0.05$ and a reliable interval of 0.95, the coefficient $z = 1.96$; for a risk $2\alpha = 0.01$ and a reliable interval of 0.99, $z = 2.58$.

For $z = 1.96$ $Y_D = 1.96(\sigma_0 + \sigma_D)$
For $z = 2.58$ $Y_D = 2.58(\sigma_0 + \sigma_D)$

The sensitivity was calculated for a fluorescence intensity corresponding to the closest concentration to the concentration limit, at 100 ppm. The various tracers do not have the same quantum yield, which explains the differences in fluorescence. The values of the

TABLE 28.2 Values of the Fluorescence Measurements for Three Tracers with Concentrations of 0, 100, 500 and 1000 ppm

Tracers			Concentration (ppm)			
			0	100	500	1000
T1	$\overline{Y}$ (CPS)		6.36E+04	1.21E+05	1.84E+05	2.70E+05
	σ		2.39E+04	3.32E+04	4.29E+04	3.92E+04
	Rsd (%)		37.6%	27.5%	23.3%	14.5%
T3	$\overline{Y}$ (CPS)		7.52E+03	1.01E+05	6.23E+05	1.04E+06
	σ		2.49E+03	7.21E+03	3.34E+04	5.85E+04
	Rsd (%)		33.1%	7.1%	5.4%	5.6%
T10	$\overline{Y}$ (CPS)		7.52E+03	2.53E+04	9.22E+04	1.70E+05
	σ		2.49E+03	4.30E+03	8.62E+03	1.07E+04
	Rsd (%)		33.1%	17.0%	9.3%	6.3%

Rsd, relative standard deviation.

TABLE 28.3 Values of the Detection Limit Y_D and Concentration Limits X_D for a Confidence Interval 95%, $z = 1.96$, $2\alpha = 0.05$

	T1	T3	T10
Y_D (CPS)	1.75E + 05	2.65E + 04	2.96E + 04
A (CPS/ppm)	5,70E + 02	9,34E + 02	1,78E + 02
X_D (ppm)	196	20	75

fluorescence intensities for three tracers according to three concentrations are presented in Table 28.2. The calculation of the detection limit Y_D and the concentration limits X_D are expressed in Table 28.3. The concentration limits for T1, T3 and T10 are respectively 20, 196 and 75 ppm for a confidence interval of 95%.

28.5 CONCLUSION

Market demand for recycled materials is strong and unsatisfied. Recycling solutions are not yet sustainable due to technical barriers. Indeed, polymer fractions from end-of-life products are complex to sort out, particularly for black-colored polymers.

The chemical industry has endeavored to modify plastic materials to confer them with new properties (to resist aging, mechanical properties, etc.). Faced with these new challenges, the design of recyclable, easy-to-sort materials is a new approach compared to current product eco-design approaches. The principle is to add tracers to dark polymers to facilitate their sorting and recycling. These dark materials represent large volumes of nonrecycled waste (70% of the polymers from automobiles, 40% of the polymers from electrical and electronic products, and 30% from end-of-life packaging products are dark) and are particularly highly prized for the production of new goods.

At present, polymer recycling companies are small and specialized or very large, such as shredder companies or household waste collection companies. The big producers of consumer goods (automobiles, household electrical appliances, furniture) use more and more recycled materials to reduce the environmental footprint of their products and to reduce their production costs. Furthermore, recycling improves the social acceptability of polymers and is a satisfactory solution to the problems emerging across the planet: accumulations in the oceans, the destruction of marine species (birds, tortoises, dolphins) and numerous hygiene problems in developing countries. The producers of virgin polymers originating from the synthesis of polymers from oil have not yet taken a strong position in favor of recycling and consider it a rather marginal activity, which could represent competition. Nevertheless, the production of recycled polymers is complementary to the virgin polymers and helps to improve their acceptability and reduce their environmental impacts. The optimized recycling scenario is to date the most favorable treatment mode with respect to the environment and to the creation of jobs (WRAP, 2010).

Sorting and recycling methods are rapidly evolving and are increasingly effective and productive. The addition of tracers is a way to sort out the dark materials and also to ensure their traceability. The laboratory tests and the design of an industrial prototype made it possible to validate the concept of sorting dark plastics containing low concentrations of tracers (between 20 and 200 ppm). This new method will make it possible to "democratize" polymer recycling because it will not require a large investment for sorting.

Acknowledgments

We would like to thank the partners who actively contributed to this project named TRIPTIC, financed by the French National Research Agency: Tracing Technologies, who selected and supplied the tracers; Renault; POAE Company, who prepared the samples; Pellenc ST, who supplied the sorting system; RECORD (Cooperative Research Network on Waste and the Environment); and the two laboratories IMP, who supplied the master batches, and LIST.

References

Ahmad, S.R.A., 2004. New technology for automatic identification and sorting of plastics for recycling. Environmental Technology 25 (10), 1143—1149.

British Petroleum Company, Method of Identifying Polymer Materials, 1993, Patent GB2264558A.

Bayer, Method for Identification of Plastics, 1994, Patent US5329127A.

Bezati, F., Froelich, D., Massardier, V., Maris, E., 2010. Addition of tracers into the polypropylene in view of automatic sorting of plastic wastes using X-ray fluorescence spectrometry. Waste Management 30, 591—596.

Currie, Lloyd A., 1995. Nomenclature in evaluation of analytical methods including detection and quantification capabilities. Pure and Applied Chemistry Vol. 67 (No. 10), 1699—1723. IUPAC.

Eisenreich, N., Kull, H., Thinnes, E., 1992. Waste Management of Energetic Materials.

Eriez, Plastic material having enhanced magnetic susceptibility, method of making and method of separating Brevet US6920982, 2005.

Froelich, D., Maris, E., Massardier, V., 2007. Etat de l'art technico-économique sur les procédés et techniques d'incorporation de traceurs dans des matériaux polymères, en vue du tri automatisé des déchets plastiques des produits hors d'usage. Report No. 05—0907/1A. RECORD.

Lambert, C., Hachin, J.M., 2004. Method for authentication by chemical marking or tracing of an object or a substance, Patent WO2004040504.

Lambert, C., Hachin, J.M., 2010. Method for identifying a substance or object using a plurality of vectors, Patent US2010089804A1.

Maris, E., Aoussat, A., Naffrechoux, E., Froelich, D., 2011. Polymer tracer detection systems with UV fluorescence spectrometry to improve product recyclability. Minerals Engineering. http://dx.doi.org/10.1016/j.mineng.2011.09.016.

Simmons, B.A., Overton, B.W., Viriot, M.-L., Ahmad, S.R., Squires, D.K., Lambert, C., 1998. Fluorescent tracers enable automatic identification and sorting of waste plastics. British Plastics and Rubber. JUN, 4—12.

WRAP, 2010. Environmental Benefits of Recycling www.wrap.org.uk.

CHAPTER 29

Informal Waste Recycling in Developing Countries

Mathias Schluep
World Resources Forum Association (WRFA), St. Gallen, Switzerland

29.1 INTRODUCTION

Waste management systems are often associated with municipal collection infrastructure, mechanization, and the use of advanced recycling technology. However, especially in developing countries, waste management is mostly executed by a large urban workforce making a living by collecting, sorting, recycling, and selling valuable materials recovered from waste (Medina, 1997)—usually referred to as the "informal sector". In these countries, informal waste management fills the vacuum that is left behind by the municipalities, who are often lacking the legal framework, the capacity, and the resources for a formal waste collection and treatment system. Hence, besides earning their daily income, the informal sector contributes to public health, reduces costs related to municipal waste management, and recovers valuable materials that otherwise would be lost (Dias, 2012). High waste collection rates of up to 80% have been reported for various waste streams (Gunsilius and Gerdes, 2010; Secretariat of the Basel Convention, 2011; Wang et al., 2012; Schluep et al., 2009). Thus, it has been acknowledged that the informal sector related to waste management often reveals a great development potential (Nas and Jaffe, 2004; Medina, 2000; Dias, 2012).

However, discussions around the informal waste sector are also strongly related to adverse effects for humans and the environment. Due to their daily contact with garbage, people working in informal waste management are exposed to various health threats, including injuries, diseases, and acute and chronic effects. Especially when targeting waste streams containing hazardous substances, such as waste electrical and electronic equipment (WEEE, or e-waste), serious health effects are likely (Sepúlveda et al., 2010). In addition, based on the common association of waste being dirty, individuals working in informal waste management are often ascribed the lowest social status in society (Medina, 1997). At the absence of rules and regulations, only waste materials with a positive economic value get recovered, using readily available, low-technology, and low-cost practices. Other materials with no monetary incentive are either not collected, get dumped, or are burnt. These approaches lead to various adverse effects on the environment, including emissions from hazardous substances, which are constituents of the waste; auxiliary substances, used in recycling techniques; and byproducts, formed by the transformation of primary constituents (Sepúlveda et al., 2010). For example, the practice of recovering plastics through burning of cables has been identified to be a major source of dioxins (Secretariat of the Basel Convention, 2011).

Illustrated by two different stories from developing countries, this chapter intends to impart an understanding of how informal waste management is functioning. It discusses the environmental and socioeconomic effects, as well as failures and approaches on how to link informal and formal waste management activities.

29.2 DEFINING THE INFORMAL SECTOR

According to the International Labor Office (ILO, 2002a), the informal sector absorbs as much as 60% of the labor force in urban areas of developing countries. It should be noted that the same report also estimates that informal employment in developed countries makes up 15% of the urban workforce, indicating that informal economic activities is not a developing country issue only. Early estimations assume that up to 2% of the population in developing countries survives by recovering materials from waste (Bartone, 1988).

Despite informality having been subject to political and scientific discussions for decades (Chi et al., 2011), there is no clear definition of the informal sector. The reason might lay in differences of the nature of the informal sector in a specific geographical setup, depending on the cultural background and other socioeconomic factors. However, all definitions point toward similar elements and patterns, usually starting with the general description of the informal economy. For example, ILO (2002b) defines the informal economy to include all economic units that are not regulated by the state and all economically active persons who do not receive social protection through their work. This definition is more specified by the the Global Development Research Center (GDRC, 2013), focusing on the urban informal sector, which is the relevant demographic section for waste management. According to the GDRC, the urban informal sector is often defined to comprise the economic enterprises that employ less than 5—10 people and who satisfy one or more of the following criteria: (1) it operates in open spaces; (2) it is housed in a temporary or semipermanent structure; (3) it does not operate from spaces assigned by the government, municipality or private organizers of officially recognized marketplaces; (4) it operates from residences or backyards; and (5) it is not registered. In addition, other authors (McLaughlin, 1990) complete the definition by stating that informal enterprises rely on (1) the use of family and unpaid labor (apprentices) and reliance on manual labor rather than on sophisticated machinery and equipment; and (2) flexibility, allowing people to enter and exit economic activities in response to market demand.

It also has been acknowledged that there are many interdependencies between the informal and the formal sector (Flodman Becker, 2004). Market links exist through the trade of goods, raw materials, tools and equipment, and acquisitions of skills and know-how. The trade of raw materials obviously can refer to waste management, as the ultimate output of recycling is defined as (secondary) raw material.

29.3 INFORMAL SOLID WASTE MANAGEMENT

29.3.1 The Example of the Zabbaleen in Egypt

A well-documented example of informal solid waste management is the case of the Zabbaleen in Cairo, Egypt. The Zabbaleen (Arabic for garbage collectors) are a Coptic Christian minority who have been active in collecting, sorting, and recycling a substantial portion of solid waste in Cairo since the 1940s. Most of the following information have been taken from the local Zabbaleen community initiatives (Iskandar, 2003). It has been estimated that 3000 t of household waste have been collected each day by the Zabbaleen, which supports an estimated 40,000 people with a daily income. The recovery rate of material in this informal waste management system is estimated at 80%. All this is done with no payment from the Cairo City government. Over the years, the Zabbaleen have constantly developed themselves and their businesses by investing in their trade and building homes. They originally collected waste using donkey carts, but they later invested in small trucks to meet the municipality's requirements. They have also invested extensively in their workshops, buying machines for the recycling industries. As a result, their activities grew also beyond Greater Cairo and are now a large economic force of the nation's informal economy.

29.3.2 Waste as an Opportunity for Development

In addition to their core garbage business, the Zabbaleen have established community-based organizations and improved the infrastructure of Mokattam (their main settlement). This includes water and wastewater systems and paved roads. With the income of the waste business, the community also invested in new schools and, in the 1980s, began to enroll increasing numbers of their children—especially girls—in school. They have reduced their neonatal mortality rate through health programs and infrastructure, once the highest in Egypt, to the lowest in the nation. In short, by engaging informally in solid waste management, the Zabbaleen have developed themselves by expanding and professionalizing their businesses, establishing access to better sanitation, health and living infrastructure, as well as education and other services to the community.

29.3.3 Negative Effects by Formalization Approaches

The continuation of this beneficial relationship between informal waste management and community development has been jeopardized by the official privatization and formalization of municipal solid waste services through the city government (Fahmi and Sutton, 2010). Since 2003, the government of Egypt sold annual contracts reaching US$ 50 million to international companies to collect Cairo's solid waste (Fahmi and Sutton, 2006). The awarded international companies, however, were only required to recycle 20% of their collected waste, with the rest being dumped on landfill sites. In contrast, the Zabbaleen recycled 80% of their collected waste. In addition, citizens were not content with the new centralized garbage collection systems and preferred the door-to-door collection system offered by the Zabbaleen over the formalized approach. Still, the formalization had a negative effect on the Zabbaleen, making it harder for them to collect their waste material. In addition, in 2009 Egypt responded to the outbreak of the H1N1 virus (swine flu) by ordering the culling of its swine population. The act had a devastating effect on the Zabbaleen population, because raising pigs by feeding them organic waste was their main income. The cull actually interrupted the closed loop of municipal solid waste management as practiced by the Zabbaleen in Egypt. Since then, the Zabbaleen stopped collecting organic waste, leaving food piles to rot in the streets (Fahmi and Sutton, 2010).

29.4 INFORMAL E-WASTE RECYCLING

Developing countries and countries in transition are characterized by informal activities along the e-waste recycling chain (Schluep et al., 2009). Collection, manual dismantling, open burning to recover metals, and open dumping of residual fractions are normal practice in most countries. In smaller and less developed economies, these activities are usually performed by individuals, as volumes are too small to trigger the informal sector to specialize in e-waste recycling at large scale. Larger economies, especially countries in transition, such as India and China (Sepúlveda et al., 2010; Chi et al., 2011; Wang et al., 2012), as well as countries that are subject to the intense trade of secondhand equipment and illegal waste shipment, such as Ghana and Nigeria (Secretariat of the Basel Convention, 2011), reveal a large organized informal sector. The functioning of the informal sector can be grouped into the three main stages of the e-waste recycling chain—collection, preprocessing, and end processing. Approaches related to connecting informal recycling to formal systems is discussed with a description of the "Best-of-2-Worlds" philosophy.

29.4.1 Collection

In contrast to formalized take-back schemes, as found in Europe, where consumers pay (indirectly) for collection and recycling, in developing countries it is usually the waste collectors who pay consumers for their obsolete appliances and scrap material (UNEP, 2012). As a result, informal waste sectors are often organized in a network of individuals and small businesses of collectors, traders, and recyclers—each adding value, and creating jobs, at every point in the recycling chain (Sinha-Khetriwal et al., 2005). As many poor people rely on the small incomes generated in this chain, impressive collection rates of up to 95% of waste generated are achieved (Secretariat of the Basel Convention, 2011), which is far above what can be achieved by today's formalized take-back schemes (Huisman et al., 2008).

29.4.2 Preprocessing

Because labor costs are low in developing countries and countries in transition, informal and formal recyclers apply labor intensive preprocessing technologies, such as manual dismantling, as the primary treatment to separate the heterogeneous materials and components. A comparative study (Wang et al., 2012) of preprocessing scenarios revealed that material recovery efficiency improves along with the depth of manual dismantling. Purely mechanical treatment options, as typically applied in western countries with high labor costs, lead to major losses of precious metals, in particular, in dust and ferrous fractions (Chancerel et al., 2009; Meskers et al., 2009). Hence, manual recycling practices in developing countries have advantages, such as low investment costs, creation of jobs, and higher material recovery efficiency (Schluep et al., 2009).

29.4.3 End Processing

Subsequent to manual preprocessing practices, further "refining" techniques, such as the desoldering of printed wiring boards and subsequent leaching of gold, have been observed, especially in the informal sectors in India and China (Puckett et al., 2002; Rochat et al., 2008; Sepúlveda et al., 2010; Chi et al., 2011). There are indications that such processes are also applied in other larger developing countries, such as Nigeria (Ogungbuyi et al., 2012). In a pilot project in Bangalore, India, it has been demonstrated that besides being hazardous, informal end processing or refining practices are also inefficient. Improper sorting of printed wiring boards and subsequent wet chemical leaching processes for the recovery of gold, for example, revealed a combined yield of only 25%. In contrast, today's state-of-the-art integrated smelters, as used in most formalized recycling systems, achieve recovery efficiencies as high as 95% (Chancerel et al., 2009).

29.4.4 Connecting the Informal Sector to the Formal Recycling Industry

From these findings, it can be concluded that recovery efficiencies in informal recycling processes can differ considerably from those of formal recycling systems, even though there are individual strengths and weaknesses on both sides. Analyses have shown that the average material recovery yield over the entire recycling chain can be in a similar (low) range in informal and formal systems (Wang et al., 2012; UNEP, 2012). Taking this into account, an alternative business model for the informal sector has been piloted in Bangalore, which aims to combine "the best of both worlds" by transferring informal wet chemical processes to state-of-the-art recycling technologies (Rochat et al., 2008; Schluep et al., 2009). Through financial incentives and training, the informal sector was encouraged to concentrate on the preparation of the optimal fractions as input for the integrated smelter. A formal local cooperative was acting as an intermediate, buying the fractions from the small individual businesses in the informal sector on one hand and selling it to an integrated smelter on the international market on the other hand. Similar projects have been carried out by other initiatives and have been summarized as the "Best-of-2-Worlds" philosophy by the Solving the e-Waste Problem (StEP) initiative (Wang et al., 2012): "Under the observation of integrating best geographically distributed treatment options, the Best-of-2-Worlds philosophy helps to achieve the most sustainable solution for developing countries: to locally pre-process their domestically generated e-waste by manual dismantling; and to deliver critical fractions to state-of-the-art end-processing facilities in a global market."

This also highlights that the efficient and sustainable recovery of secondary resources from e-waste is a market opportunity for developing countries. This requires functioning reverse supply chains with adequate capabilities for recycling and refining as well as sufficient control supported over their material quality and environmental and social impacts of the related processes. Hence, the harmonization of international standards and the introduction of processes to distinguish "fair" secondary resources from materials recovered under substandard conditions (e.g. burning cables to recover copper) is instrumental to leverage these opportunities.

References

Bartone, C., 1988. The Value in Wastes. Decade Watch.
Chancerel, P., et al., 2009. Assessment of precious metal flows during preprocessing of waste electrical and electronic equipment. Journal of Industrial Ecology 13 (5), 791–810.
Chi, X., et al., 2011. Informal electronic waste recycling: a sector review with special focus on China. Waste Management 31 (4), 731–742.
Dias, S.M., 2012. Waste and development — perspectives from the ground. Field Actions Science Reports. Special Issue 6.
Fahmi, W.S., Sutton, K., 2010. Cairo's contested garbage: sustainable solid waste management and the Zabaleen's right to the city. Sustainability 2 (6), 1765–1783.
Fahmi, W.S., Sutton, K., 2006. Cairo's Zabaleen garbage recyclers: multi-nationals' takeover and state relocation plans. Habitat International 30 (4), 809–837.
Flodman Becker, K., 2004. The Informal Economy. Sida. Available at: www.sida.se/publications.
GDRC, 2013. Concept of Informal Sector. Available at: http://www.gdrc.org/informal/1-is_concept.html (accessed 19.08.13.).

Gunsilius, E., Gerdes, P., 2010. The Waste Experts: Enabling Conditions for Informal Sector Integration in Solid Waste Management — Lessons Learned from Brazil, Egypt and India. GTZ. Available at: http://gtz.de/de/themen/umwelt-infrastruktur/abfall/2841.htm.

Huisman, J., et al., 2008. 2008 Review of Directive 2002/96 on Waste Electrical and Electronic Equipment (WEEE). Final Report. United Nations University.

ILO, 2002a. Decent Work and the Informal Economy. International Labour Office, Geneva/Switzerland.

ILO, 2002b. Men and Women in the Informal Economy: A Statistical Picture. International Labour Office, Geneva/Switzerland.

Iskandar, L., 2003. Integrating local community-based waste management into international contracting. In: Proceedings of Solid Waste Collection that Benefits the Urban Poor, 9—14 March. The SKAT Foundation, Dar Es Salaam, Tanzania.

McLaughlin, S., 1990. Skill training for the informal sector: analysing the success and limitations of support programmes. In: Turnham, D., et al. (Eds.), The informal sector revisited. Seminar on the informal sector revisited. Paris 7—9 September, 1988. OECD, Paris.

Medina, M., 1997. Informal Recycling and Collection of Solid Wastes in Developing Countries: Issues and Opportunities. United Nations University/Institute of Advanced Studies, Tokyo, Japan.

Medina, M., 2000. Scavenger cooperatives in Asia and Latin America. Resources, Conservation and Recycling 31 (1), 51—69.

Meskers, C., et al., 2009. Impact of pre-processing routes on precious metal recovery from PCs. In: European Metallurgical Conference EMC.

Nas, P.J.M., Jaffe, R., 2004. Informal waste management. Environment, Development and Sustainability 6 (3), 337—353.

Ogungbuyi, O., et al., 2012. Nigeria e-Waste Country Assessment. Basel Convention Coordinating Centre for Africa (BCCC-Nigeria) and Swiss Federal Laboratories for Materials Science and Technology (Empa), Ibadan/Nigeria and St.Gallen/Switzerland.

Puckett, J., et al., 2002. Exporting Harm, the High-tech Trashing of Asia. The Basel Action Network (BAN) and Silicon Valley Toxics Coalition (SVTC), Seattle, WA, USA.

Rochat, D., Rodrigues, W., Gantenbein, A., 2008. India: including the existing informal sector in a clean e-waste channel. In: Waste Management Conference (WasteCon2008). Durban, South Africa.

Schluep, M., et al., 2009. Recycling — from e-Waste to Resources, Sustainable Innovation and Technology Transfer Industrial Sector Studies. United Nations Environment Programme (UNEP), Paris, France.

Secretariat of the Basel Convention, 2011. Where are WEee in Africa? Findings from the Basel Convention e-Waste Africa Programme, Geneva/Switzerland.

Sepúlveda, A., et al., 2010. A review of the environmental fate and effects of hazardous substances released from electrical and electronic equipments during recycling: examples from China and India. Environmental Impact Assessment Review 30, 28—41.

Sinha-Khetriwal, D., Kraeuchi, P., Schwaninger, M., 2005. A comparison of electronic waste recycling in Switzerland and in India. Environmental Impact Assessment Review 25, 492—504.

UNEP, 2012. Metal Recycling — Opportunities, Limits, Infrastructure. United Nations Environment Programme (UNEP), Paris/France.

Wang, F., et al., 2012. The Best-of-2-Worlds philosophy: developing local dismantling and global infrastructure network for sustainable e-waste treatment in emerging economies. Waste Management 32, 2134—2146.

CHAPTER 30

Squaring the Circular Economy: The Role of Recycling within a Hierarchy of Material Management Strategies

Julian M. Allwood

Department of Engineering, University of Cambridge, Cambridge, United Kingdom

30.1 IS A CIRCULAR ECONOMY POSSIBLE OR DESIRABLE?

Kenneth Boulding's famous 1966 essay "The Economics of the Coming Spaceship Earth" raised awareness of the contrast between an "open economy" with unlimited input resources and output sinks and a "closed economy", in which resources and sinks are bounded and remain forever a part of the concerns of the economy (Boulding, 1966). Boulding's essay is often cited as the origin of the phrase "circular economy", which is now widely used as an axiom in environmentally concerned discussions. At present, in parallel with the (failing) search for the miracle of unlimited renewable energy supplies and the (forlorn) creative approaches to hiding (sequestering) undesirable outputs underground, the dogma of today's pro-environmental discussions in politics and mass media reporting assumes that aspiring to a "circular economy" is one of the key technical fixes that will solve our environmental problems and allow the economy to keep on growing.

Governments, celebrities, nongovernmental organizations, and academics compete to be leading champions of the circular economy, but is that really correct? If global demand for steel has quadrupled in less than 50 years (USGS, 2012), largely to make high-rise buildings that are replaced on average every 40 years, is the pursuit of recycling really the most important option for reducing the impacts of your material production? If you are concerned about the criticality of elements used in modern electronics, is the best option to try to separate out all those elements from the complex mixed streams of waste electronics? Assuming he owned a car, should Kenneth Boulding's descendants today be driving a car made entirely from the atoms of his car built in the 1950s?

The key image created by the circular economy is of a fixed number of atoms currently formed into today's products that should be repeatedly reorganized into future products without requiring any further injection of new atoms. Is that really desirable? These rhetorical questions challenge the attractiveness of this target: if demand is growing, the circle cannot remain closed, and it may be a much more important priority to reduce the rate at which new material is required. If today's products require precise and complex mixing of atoms to create high-performance properties, the energy required to separate them from these products may be very much greater than the energy required to extract new material from ore or tailings. If technology is advancing so that the design requirements for today's products are unrelated to those of the past, it may not be possible to create tomorrow's products from today's stock of atoms in use.

Behind these challenges lies one primary question: how much energy is required to operate a circular economy? Almost all recycling processes operate by breaking down a solid waste stream into a liquid, which is then purified by some means. With today's technologies, some wastes cannot be broken down and some liquids cannot be purified, but in principle there is no atomic bond that could not in the future be separated under human control and with sufficient energy. However, the energy required to achieve this may be very much greater than that required to liberate an equivalent stream of atoms from other sources (e.g. ores or even mining tailings), and the process of doing so may lead to other environmentally harmful consequences such as the release of toxins or greenhouse gases.

The idea of a circular economy might be technically feasible if global demand for both the volume and compositions of products stabilized. However, for many materials, it would require significantly greater energy inputs than creating products by other means. If energy is produced from fossil fuels, as seems inevitable in the short- to medium-term future, the pursuit of this circular economy would therefore accelerate the release of greenhouse gases and consequent warming of the climate. Is this what its proponents intend?

Rather than having circularity as a goal, a more pragmatic vision for a material future would be to aim to meet human needs while minimizing the environmental impact of doing so. Because of the urgency of the challenge to curb global warming, the primary impact of concern is the emission of greenhouse gases, and in turn therefore the primary challenge is to deliver products with minimum energy. For some materials in some contexts, recycling (i.e. reducing old products to liquid, then reforming the liquid to new materials) may be the least energy-intensive solution, hence the importance of this book overall. However, a wider set of strategies with a greater potential impact across all material classes can be found under the umbrella term "material efficiency", which describes the aim to deliver material services with less input of material. Successful recycling can reduce the demand for new ore or biomass, but successful material efficiency reduces the total demand for material processing. This chapter aims to explore the means by which material efficiency can be achieved and to discuss how it might be applied and brought about for different major material groups.

However, before setting out on that exploration, we must return to Boulding's famous essay. The statement above suggested that a circular economy could be achieved "if global demand for both the volume and composition of products stabilized." That 12-word condition describes an environmental nirvana that virtually defies all imagination in current growth-driven economies, but prior to the industrial revolution was largely inevitable. Wrigley (2013) describes the operation of the United Kingdom prior to the exploitation of fossil fuel resources as an "organic economy" that was self-limiting: a householder with access to a certain amount of land could use the land for growing crops for food, for animal husbandry, for fiber (for clothing or structure), or for fuel (for comfort, cooking, or material production). However, with a finite amount of land per householder, these options must be traded, so the population could not grow in number or in material demand without sacrificing comfort or security. The availability of fossil fuels eliminated this constraint, thus allowing both forms of growth—without limit to date. Biologists and anthropologists report countless examples of species failing to self-regulate, and thus create their own extinction by exhausting their support systems; yet nearly 100 years ago, John Maynard Keynes anticipated that the goal of economic development was simply to reach a new equilibrium with a higher quality of life:

> "For many ages to come the old Adam will be so strong in us that everybody will need to do some work if he is to be contented. We shall do more things for ourselves than is usual with the rich today, only too glad to have small duties and tasks and routines. But beyond this, we shall endeavor to spread the bread thin on the butter to make what work there is still to be done to be as widely shared as possible. Three-hour shifts or a fifteen-hour week may put off the problem for a great while. For three hours a day is quite enough to satisfy the old Adam in most of us! ... Thus for the first time since his creation man will be faced with his real, his permanent problem-how to use his freedom from pressing economic cares, how to occupy the leisure, which science and compound interest will have won for him, to live wisely and agreeably and well. ... The strenuous purposeful money-makers may carry all of us along with them into the lap of economic abundance. But it will be those peoples, who can keep alive, and cultivate into a fuller perfection, the art of life itself and do not sell themselves for the means of life, who will be able to enjoy the abundance when it comes. "
> Keynes, 1930

Unfortunately—at least when assessing human impacts on the environment—Keynes was overoptimistic. Apart from specific (often religious) communities such as monastic communities or the Shakers, human demand for energy and material has proved insatiable. Figure 30.1 gives evidence of this showing per capita demand for energy, electronics, and cars growing apparently with no sign of saturation.

Growth—measured in particular by national gross domestic products (GDPs)—is now the primary goal of economic policy in virtually all nations; this both nurtures and feeds off the transient human desires for novelty and change that can be satisfied by material purchasing. A growing literature suggests that the pursuit of economic growth in this form does not lead to human well-being (e.g. Kasser, 2002; Layard, 2006), yet the reductionist view of mainstream economics that well-being (or welfare) arises from purchasing is both pervasive and supported by observations of purchasing behavior. Figure 30.2 (EPA, 2012) gives a compelling demonstration that, unless energy prices are very high, car purchasers in the United States prefer larger cars with greater acceleration: more (acceleration, volume, weight) with less (money) is an economic success arising from tremendous progress in techniques of production, but it apparently does not lead to increased human satisfaction, despite its consequence of increased material and energy production. This, and much other, evidence of individual demand for consumption, supported by government goals of GDP growth, creates the exact reverse of the 12-word condition required to allow circularity.

Boulding's proposed solution to this apparently insatiable human habit of demand for more goods was to move from economic

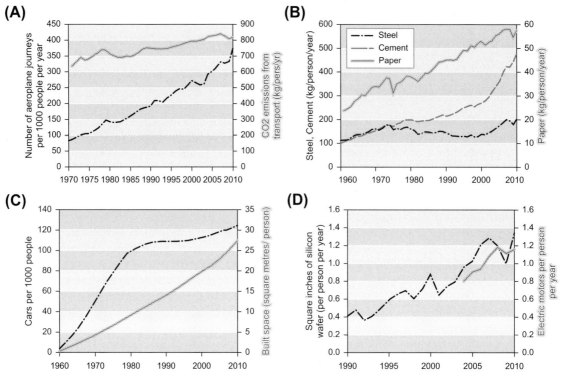

FIGURE 30.1 Trends of growth in global demand for material *(Adapted from Gutowski et al., 2013)*. The graphs have been normalized by population data from World Bank (2013) which correlates closely with UNDESA (2011). (A) Airplane passengers carried and transport CO_2 emissions from World Bank (2013); (B) Steel data from USGS (2012), cement data from USGS (2012b) and paper data from FAO (2012); (C) Global car production data from (USDoT, 2012, Table 1-23), converted to in-use stocks, assuming a 20-year car life with zero stock in 1960; Built space calculated from cement production USGS (2012b), assuming all cement used as concrete to create building floors lasting for 40 years. Cement usage estimated at 300 kg/m² based on bottom-up estimates for 200 mm floor slab in Goodchild (1993) and van Oss and Padovani (2003), and a top-down estimates of Chinese construction in Fu et al. (2013); (D) Silicon wafer production data from Winegarner (2011) and electric motor data from Zhou (2011).

measures based on annual flows (in particular, the GDP) to measures based on stocks. In contrast to the promotion of recycling inherent to the vision of a circular economy, such stock-based measures would promote conservation, maintenance, and renewal as the keys to delivering material well-being. An economy measured by its stocks would count increased material flows as a failure rather than an economic success. Demand for new cement within this vision would reflect a failure of conservation rather than a success in creating economic growth, where success would arise from the maintenance of lightweight, low-energy products over a long duration, with individual measures of success derived from quality of life, leisure, creativity, and other constructive values rather than from income alone.

Our vision of a future sustainable material economy is therefore not prescribed by the ambition to create a circular economy, but aims to minimize its total environmental impact. The key measures to promote this are frequently reported as hierarchies of options; one such

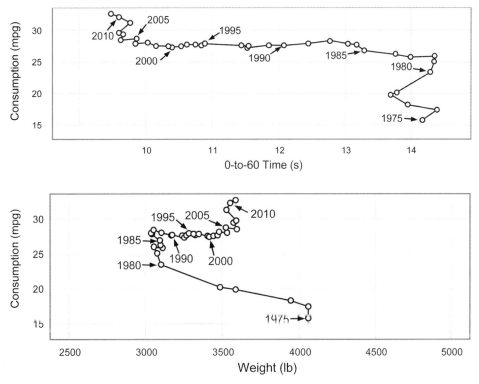

FIGURE 30.2 Evolution of car performance, consumption and weight in the U.S. 1975–2010 (EPA, 2012). Consumption is measured in a laboratory test, reflecting vehicle design parameters rather than driver behavior. The figures show that the energy crisis in the late 1970s drove regulation leading to improved consumption and reduced weight; once the crisis resolved, regulation was not tightened further, so manufacturers deployed technology developments to improve vehicle acceleration rates (dramatically) while also steadily increasing weights; this trend has shifted only recently, as the financial crisis of the late 2000s drove tighter regulation leading to slightly improved consumption, and no further increase in weight.

hierarchy is used to structure a wider search for material conservation options in the next section.

30.2 HIERARCHIES OF MATERIAL CONSERVATION

"Reduce, reuse, recycle" is a critical and intelligent mantra for the future of material management. However, as Figure 30.1 demonstrated, the preference in reality is to "redouble, replace, recycle-a-bit-if-it's-easy, reject." The British public is absolutely convinced of the benefits of recycling green glass bottles, so having done our bit for the environment, we can fly off to Spain for another short break because we must have our dose of winter sunshine or our human rights would be violated.

But "reduce, reuse, recycle" as promoted in the United Kingdom by the Department of the Environment, Transport and the Regions (DoETR, 1995) remains the correct ambition for reducing environmental impact. It is one statement among many versions (the Dutch prefer Lansink's ladder as described by Parto et al., 2007) of a hierarchy of material management options. (In some cases, this is referred to as a hierarchy of waste management. However, we prefer the term "material management hierarchy", as the label "waste" potentially disguises many other options for better managing materials to reduce the total impact of their production and processing.) The fact that recycling—the topic of this book—is the third choice is significant and correct: recycling is neither "green" nor free of impacts, so it is a less attractive option than reducing demand for material or reusing materials without energy-intensive processing. However, the challenge of such hierarchies is not to capture them in a short phrase but to turn the phrase into practice. Therefore, we will explore in this section the means by which such a hierarchy of material management could be brought about, and in the next section, will explore its applications to the major material classes.

30.2.1 Reduce

Demand for material production would be reduced if the overall demand for services created by materials reduced, such a reduction in service demand remains the "gold medal" of environmentally motivated materials management. Alternatively, but also with great potential for reducing the impacts of material production, the amount of material required to deliver each unit of service could be reduced by changes in design or production of goods. The strategy of "reduce" is at the top of any materials management hierarchy because it leads to a reduction in demand for primary materials production, which for bulk materials is generally the most energy-intensive process of material conversion.

30.2.1.1 Simplicity, Austerity, Poverty: Reducing Demand for Material Services

Our social status as humans is determined by an infinitely subtle comparison of ourselves with our peers. The competition is on many dimensions: I'm more influential than you, I'm more of an individual than you are, and, always among the metrics, I've got more stuff than you. Gravestones record families' key memories of their departed, with "much loved father/mother" being a common memory; yet during their lives—which may have been shortened as a result—those parents competed fiercely to get more stuff. Kasser (2002) surveyed a class of final-year students about to graduate and move to their first employment, and asked two questions. The first question was: "Next year, would you rather earn $100,000 knowing that your classmates were on average earning $200,000, or would you rather $50,000 knowing the others averaged $25,000." The answer was unequivocally the second—we measure our wealth relative to our peer group, not as an absolute. The second question was: "Next year, would you rather have four weeks holiday knowing your peers had eight, or would you rather have two weeks holiday knowing your peers had just one." In this case, the answer was clearly the first—we measure our leisure in absolute, not relative, terms. Yet, despite an absolute preference for increased leisure hours, government policy and business reward schemes focus on our instinctive competition for wealth, which, as Layard (2006) clearly explained, is one with no end point: If we measure our wealth relative to our peers, then the (brief) happiness created for an individual whose wealth increases is exactly balanced by a relative unhappiness in the peer group, while meanwhile the newly wealthy individual soon compares themselves against a new, slightly wealthier, peer group. Changes in ranking—our relative assessment of our earnings—can only be a zero-sum game.

The competition for status linked to wealth is an axiom of economic policy in most countries, the driving force for developing an economy fueled by money lending, and a widely assumed norm of recruitment and reward among employers. Yet, there is significant evidence that this does not actually tie up with individual preferences. When extra public holidays are announced, such as a Royal Wedding in the UK, most people do not use the extra day to take on extra work, but enjoy an expanded weekend. More broadly, even though working hours in the UK are higher than for most European countries (for the employed) with a current average of 43 h per week, personal aspirations for a better lifestyle says that the top priority of most employees is to have more leisure hours. (In Humphrey et al., 2011, in an extensive UK survey of attitudes of those who had thought about what they would want to do in later life, by far the most common aspiration [81% of responses] was to have more leisure.) With unemployment rates around 8% in the UK, we could apparently recalibrate our economy to eliminate unemployment if we all accepted an 8% cut in pay and working hours. Let's assume that a typical employed person in the UK spends 40 h at work, 10 h commuting, 80 h sleeping, washing and eating, and 20 h managing their domestic life, in which case they currently have 18 h for chosen leisure. Therefore, the potential reward for a 10% cut in pay and working hours is at least a 25% increase in chosen leisure time—which is what 81% of us aspire to "in later life". It seems like a tremendous payback, but apparently it is not one we wish to pursue. Instead, popular lifestyle magazines about houses and celebrities appear to demonstrate to us that it's only *after* you've earned more money that you can afford to have more leisure time. However, a conundrum in the glossy magazines of house decoration is that some of the most desirable internal decoration requires having very little stuff in the room: although it's difficult to accept that real people live in this way, the most upmarket properties have space around each beautiful item of furnishing or decoration so it can be seen well; just as in traditional Japanese tea ceremonies, very few but beautiful objects are used.

Simplicity seems then to be a viable aspiration for the rich, with decluttering of a lifestyle seen as a sign of health. In parallel with the requirement that to be thin/healthy in the US you must be rich, the lifestyle magazines suggest that if you are rich, you can afford to own fewer and more beautiful objects. This is clearly not borne out in practice—Druckman and Jackson (2008) provided clear evidence that income is the only predictor of personal carbon footprints in the UK—but to balance the apparently automatic urge to earn and spend more, we have a long history of cultural questioning about whether a simper lifestyle is in fact healthier or more liberated. This is strongly supported by the major religions, whether through Christ's suggestion to the rich man that to he give away all he owns, the voluntary simplicity of Buddhist monasteries, or the disciplines of Lent or Ramadan. Valuing the stock of what we already own, as Boulding proposed, as opposed to valuing the rate at which we increase or replace that stock, has a long cultural history despite being an opposite to standard economic policy aiming at growth in flows.

However, despite these hints, few people make the voluntary choice to earn less, and the idea of pursuing the health of simplicity is untouchable for politicians, for fear that they might be promoting austerity. In contrast to the chosen values of simplicity, austerity implies an enforced constraint on the freedom to earn or spend and has no place in any current political discussion. Only in times of war or crisis is austerity a politically possible choice: during the second world war, the population of the UK accepted rationing of many goods and voluntarily contributed excess material, such as iron railings, to support the cause of the war; following the 2011 Tsunami in Japan, which led to the shutdown of many of Japan's nuclear power stations, the population of Tokyo voluntarily reduced their summer electricity consumption by 25%. Without such a crisis, it is difficult to divert spending from high- to low-intensity activities to avoid rebound effects, because the recipient of our apparently low-intensity spending (e.g. in a service industry) may simply use the additional income to purchase high-intensity goods. In effect, the voluntary pursuit of demand reduction requires either that the population choose to pay more for less service or to earn less and therefore to reduce GDP. Thus, Boulding's alternative measures of economic health are central. Until we have metrics that demonstrate growth in quality of life while GDP declines, politicians will be unable to pursue policies other than those that aim to increase GDP.

Behind the enforced constraints of austerity lies the undesirable state of poverty, where basic needs for security, food, warmth, and shelter are not met. Clearly any discussion of strategies to reduce overall demand for the services provided by materials must recognize that in parallel with this overall demand reduction is a rebalancing between current rich and poor.

We seem to be swerving far from the theme of this book and must be careful not to stretch the editors' trust too far. However, if the goal of recycling is to reduce the environmental impact of delivering benefits from materials, we have little choice but to discuss the challenging fact that a greater reduction in impact will occur if in rich countries we can choose to want less new material to provide those benefits; without that choice, there is a danger that any other action will be only transient. Indeed, there is broad literature on rebound effects (e.g. Sorrell et al., 2009) that might be used to call into question whether any action other than reducing demand will truly be effective. If recycling saves energy and cost, where will the released energy and money be used instead? If material efficiency strategies lead to a reduction in demand for steel, how will the steel companies react in order to stimulate future demand for their products?

The discussion above has revealed hints that, despite a strong urge to consume more (to purchase more material goods), there may be real and important motivations to choose a different path with less material requirements in return for an improved quality of living. However, unfortunately we have not found significant evidence of individuals making that choice. Instead, the evidence of personal choices about diet makes depressing reading. Even though it is widely understood that obesity leads to greater chances of serious illness and reduced life expectancy, rates of obesity in rich countries are increasing rapidly, in part driven by preferences for highly processed prepared meals rather than healthier simpler ingredients prepared at home and at lower cost. If, as a population, we understand that obesity worsens our own life and yet we choose it (or at least choose not to take the actions that would avoid or counter it), what chance do we have that the population will choose to reduce their material consumption for the benefit of others, unseen, unknown, and even unborn?

Sadly, despite suggestions that those in richer countries may in some senses prefer a simpler life, current evidence of human preferences suggests that voluntary choice will be insufficient to lead to reduced demand for the services provided by materials. It appears that as humans, individually, even when presented with conclusive evidence of the self-harm created by our choices, we are incapable of constraining ourselves. However, collectively, it seems that we can develop sufficient shared trust in the evidence to agree to regulations that then constrain our individual choices—and that is ultimately the way that we must aim to reduce the impacts of our material production. Regulation that, for example, restricts cars to a mass under 300 kg per vehicle, guarantees that all buildings must be used for 500 years before being replaced, or prices the retirement of electronic goods above the cost of replacing them, is the key to reducing the environmental impacts of material provision. This far-reaching discussion is therefore central to a book on recycling. Without regulation that leads to a reduction in material output, we are greatly constraining the degree to which we can reduce the harmful impacts that are the cause of our interest in recycling.

30.2.1.2 Intensifying Use: Reducing Demand for Excess Capacity

The 60 million people living in the UK own 28 million cars ($\sim$5-seats each), which they use on average for 4 h per week with 1.6 people in each. If they could organize into groups of five at the right time and weren't too fixed on where they wanted to start or end up, they could provide the same amount of car-time with just 213,000 cars and thus reduce the environmental impacts of materials production for cars by 99.2%.

Our bedrooms are unoccupied for two thirds of each day, as are our nightclubs and restaurants, and our offices are used for 45 h out of 168. Therefore, if we slept, danced, and ate in our offices, we could reduce our need for built space in the UK to under a third of present levels. And in our streets at home, if we shared our lawnmowers, power-drills, washing machines, hair dryers, and coffee machines, we could comfortably save 95% of our material purchasing.

However, we do not do these things, nor do we want to. We own goods not just to use them, but to use them where and when we want them, without having to consult anyone else, and to show them to each other. I want to go to the shops now and not wait 20 min for the bus, and I want the freedom to decide on my own whether to mow the lawn in the morning or evening. "Convenience" is a watchword of growing prosperity, and perish the thought that I should inconvenience myself through shared use of the pillars of my liberty such as my lawnmower and washing machine.

The opportunity to reduce material demand through more intense use of a reduced stock of products has had very little attention to date, although if a real constraint were applied to material supply (e.g. in war), shared use would instantly become normal practice. However, over the past 20 years, several attempts have been made to share car ownership and increase the intensity of car use through "car pools" and sharing schemes. To date, take-up of these schemes has been low, and several reasons have been proposed for this: Prettenthaler and Steininger (1999) studied a car-sharing scheme in Austria; they found that while considering just the journeys completed, 70% of the households would save money by car sharing. However, in reality the car had other functions—as a meeting place, for storage, as a symbol; when these were taken into account, 9% of households would still benefit.

The rise of budget airlines has occurred because the capacity utilization of conventional airlines was relatively low, and new operators were able to reduce the capital costs per flight by increased use of their planes. In other examples (such as in high-performance computing, extreme scientific equipment, production equipment in continuous lines, and elements of road infrastructure), high-value equipment is used near to capacity to allow spreading of the capital costs. However, the reverse remains the case for domestic scale equipment and goods, and in particular for cars, and there is a rich opportunity to unlock more value from these largely idle assets. One of the ambitions of the research groups examining "product service systems" has been to identify the conditions under which companies might benefit from selling the service of using a product rather than selling the product itself. However, no clear strategy to persuade customers to prefer this option has yet emerged.

Shared use of goods appears to be a powerful weapon to reduce material demand, yet runs counter to the norms of recent Westernized development. Potentially, our increasing preference for urbanization might allow a re-exploration of shared ownership. Rates of car ownership in large cities with good public transport infrastructure are lower than those in rural areas, and with increasing land costs leading to smaller dwelling spaces, other shared services may become attractive to future city-dwellers.

30.2.1.3 Life Extension: Reducing Demand for Replacement

Approximately half of the world's emissions of CO_2 from industry arise from producing materials for construction, and most of the resulting built space is unoccupied for most of the time. But on top of this, buildings which are designed to last for 100 years or more are knocked-down and replaced on average every 40 years in the UK or nearer to every 20 years in China. Not only is the embodied energy of construction underused through excess capacity, it is also underexploited and could deliver its intended service over a much longer period.

Figure 30.3 provides a summary of reasons why owners choose to replace goods, based on whether their needs have changed or whether alternative offerings on the market today are more attractive than what's already owned. Exploration of the application of this simple grid to a range of steel-intensive products suggests that relatively few products—mainly those related to infrastructure (or in China, buildings that were constructed with poor quality control)—are replaced because they have worn out. Much more common is replacement because users' needs have changed or replacement because a new offering is more attractive. For nonresidential buildings, decisions to replace may arise because of changes in regulations (e.g. allowing a taller building or one for a different purpose on the same site, or requiring different safety features), changes in fashion (from multiple offices to open plan space) or technology developments (in heating ventilation and air conditioning (HVAC) or communications technology). As the costs of construction in developed economies are dominated by labor and therefore the costs of a complete building refurbishment are comparable to the costs of replacing the building, owners will often prefer replacement as a means to an uncompromised development.

What can be done to counter this preference for replacement over refurbishment? A striking contrast to the example of commercial buildings in the UK given above is found in the history of the world's steel rolling mills as virtually every rolling mill ever made continues to operate today. Figure 30.4 illustrates how the cost and the embodied energy (the energy required to make the materials) in both products builds up. Both for the commercial building and the rolling mill, the bulk of their embodied energy is in their structural frame, and this forms a relatively low fraction of their total cost. However, the highest value part of the rolling mill is the control system, which includes only a very small fraction of the embodied energy, is easily removed and replaced, and which due to technology development is most likely to lead to a need for product replacement. Therefore, both

	Degraded	Inferior
The performance of the product has declined ...	... relative to when it was bought	... relative to what is currently available
The desire for the product has changed ...	Unsuitable	Worthless
	... in the eyes of its current user	... in the eyes of all users

FIGURE 30.3 A framework for understanding product replacement decisions *(from Skelton, 2013)*. The two rows of the framework distinguish between failure that arises from a change in the state of the product, and failure that arises from a change in the desires of the user. The columns distinguish between changes that affect only the current individual product and user, and more systemic changes that come about through developments elsewhere in the market.

30.2 HIERARCHIES OF MATERIAL CONSERVATION

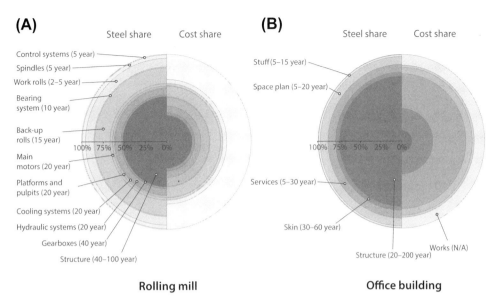

FIGURE 30.4 "Onion skin" models to illustrate the composition of steel and cost in (A) a rolling mill and (B) a representative office building. Developed with data from industrial case studies, and described in Cullen et al., 2012 and reproduced with permission.

for the owner of the mill and for the supplier of mill equipment, it makes commercial sense to upgrade the control system to allow the mill to last longer. In contrast, for the office building, the highest fraction of cost is the HVAC and communication services, which must be replaced long before the main structure but are difficult to access without also removing all other interior fittings in the building. It is the failure of these services (combined with changed requirements for building layout) that therefore currently drives most replacement decisions for commercial buildings. Countering this preference would require that buildings be designed with easily removed service systems, as well as excess capacity to allow the installation of novel service technology and layout adjustments in future.

Most products are a combination of subassemblies and components. As the onion skins of Figure 30.4 illustrate, it is rare for all subassemblies to fail after the same lifespan. A rapidly developing challenge to the ambition to create longer lasting products (particularly to maintain for longer their structural core, which dominates embodied energy) is that products are increasingly dependent on electronics. For example, Braess and Seiffert (2005) described how the average number of microelectronic control units in Volkswagen cars rose from around 10 in 1999 to over 40 by 2005. The number of electric motors and computer processing units (CPUs) in most goods is increasing, but the technical lifespan of these components remains short: most technology-related devices have been completely superseded within 3 years. The insight created by the onion skin diagrams in Figure 30.4 is of vital importance to ensure that the embodied energy of the structural core of products is allowed to continue in use: unless the motors and CPUs and their associated wiring, sensors, and interfaces can be replaced repeatedly, the embodied energy of structural components will be written off as rapidly as the electronics, leading to a rapid increase of demand for material production. This is a particular concern for car manufacturing, where the use of electronics is rapidly expanding and becoming more and more integrated through the whole car, as well as for "active" buildings, with sensors or small actuators built into structural components. Design for long life requires that the different layers of the onion skin model for the product can easily be separated and replaced in order to exploit the structural core for as long as possible. Without such design being integrated into the next generation of cars, we will see car lifetimes dropping from around 15 years at present to nearer to 5 years.

30.2.1.4 Lightweight Design: Reducing Excess Material Use

Buying material costs money, so surely no one would use more material than they needed to make their product—it would not make any sense, would it? It seems extremely unlikely that any designer or business would want to use more material than necessary in their product, but this is only the case if considering material costs in isolation. In reality, two other features of cost drive different decisions. The economies of scale favor the production of standardized components, which can be used in many applications over those that are optimized for a particular product. Also, labor costs, particularly in developed economies, are high relative to the costs of bulk materials, which are produced with extraordinary efficiency; therefore, if it is possible to reduce labor by increasing material use, that is often the choice taken to minimize costs. This section will look at the effect of these two cost drivers.

A small number of industries, particularly in aerospace, aim to produce the lightest possible product that will meet their customer's needs: every extra kilogram of material used to make a plane or rocket is a kilogram less of cargo that could be lifted and hence could generate revenue. Reducing the weight of planes and rockets is so important that the industry has always been ready to incur extra production costs to save material. However, this is not the case in most applications: the average car in the UK is already 20 times heavier than the average passenger. As we saw in Figure 30.2, customers prefer larger, heavier cars with more features; separately, the performance of buildings is unrelated to their weight. Buildings must be built with sufficient material to meet local safety regulations, but customers are largely unaware of how much material is used beyond this. Customers for cars and buildings are therefore largely indifferent to the mass of the product they purchase, while remaining highly sensitive to the cost of making their building.

Cars and buildings could be made lighter if their components were further optimized. The panels and plates that form the structure of the car could be made from ribbed sheets and have an internal frame of tailored lightweight members rather than heavy floorpans; the steel or reinforced-concrete beams that form the horizontal structure of commercial buildings could be given the "fish-belly" shape shown in Figure 30.5, with more material placed where the bending loads on the beam are greatest. Optimizing individual components for their specific load conditions typically saves around one third of the material currently used (as confirmed in a set of case studies by Carruth et al., 2011). Potentially, we could reduce demand for material production and therefore its harmful impacts by one third by efficient design. Surely that is rational? It is clearly an important strategy, but there is no commercial incentive for this approach at present because the savings would require a much more tailored approach to producing each component. If every beam in a steel-framed structure were optimized, each beam would be made individually and the economies of scale denied, as tooling and labor was used specifically in each case. Furthermore, downstream on the

FIGURE 30.5 Schematic illustrations of optimized 'fish-belly' beams with more material at the location of maximum bending moment, allowing an overall reduction in beam mass. *From Carruth and Allwood, 2012.*

construction site, effort would be required to manage and control a large supply of individual components, rather than having a stock of identical parts that can be swapped between each other. As a result, for many products, at least one third more material is used than required; this leads to cheaper products than if they were lighter.

However, even without this optimization, we still prefer excess material use if it allows us to save labor. A study of 24 commercial building projects in London has shown that the final buildings, as designed by the leading structural engineers in the UK, used on average double the amount of material that would have been required to meet the Eurocode safety standards. Interviews with the engineers behind these choices revealed that this is a consequence of cost-savings: the cost of a day of engineering design time is roughly equivalent to the cost of a ton of structural steel, so it is not economically sensible to design the building to meet the code, when less time can be used to create a design that definitely exceeds it (Moynihan and Allwood, 2013).

Counter to expectations, it makes good business sense to overspecify materials when doing so allows a greater saving in labor costs, and this is a difficult issue to overcome. Product certification may be helpful to raise customer awareness of the inefficiencies of design, and in due course, to create added brand value for products that are designed to avoid excess. In transport, lighter vehicles have the benefit of reduced fuel consumption, and rising fuel prices and new regulation may stimulate customers to pursue more efficient vehicles. (The world record performance is 15,000 mpg and the UK fleet average is 35 mpg, so there is certainly a lot of space for improvement. The key to creating fuel efficient cars is to make them smaller and lighter.) Lightweight design of some components (particularly in static products such as buildings) may reduce the loads on others, thus creating a virtuous circle of benefit. Finally, novel and more flexible production technologies, such as flexible rolling processes or computer-based layout systems for reinforcement bars, may reduce the cost penalty of component optimization, and equipment manufacturers may be able to find new market opportunities in delivering such equipment.

30.2.1.5 Improving Yield: Reducing Scrap Rates in Production

The energy-intensive materials industries produce products that no one actually wants. No one has a coil of strip steel, an aluminum bar, or a bag of cement powder at home to show their friends. These products are instead intermediate goods designed to trade off the benefits of the economies of scale against the particular requirements of a family of customers. However, because they are 'averaged' geometries, these intermediate products actually do not meet any specific requirements, so converting them to final components in products creates scrap. Scrap is the automatic consequence of the economies of scale in material production. Just as the materials industries

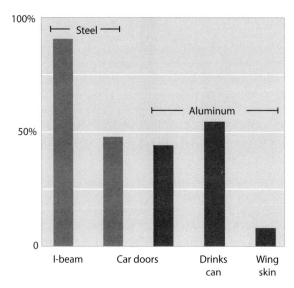

FIGURE 30.6 Yield losses in production of five representative components, illustrated as a percentage ratio of the mass of the final component to the mass of metal originally cast. *Data from industrial case studies as described in Milford et al., 2011 and image adapted from Allwood et al., 2012 with permission.*

have been extremely efficient in creating intermediate products, downstream manufacturing (supported by its tutors in the computer-controlled machining industry) have become experts in the efficient generation of scrap. The construction sector has some skill in scrap creation. The car industry is extremely good at it and confidently scraps about 30% of what it purchases, but the real experts at scrap generation are the aerospace industry. Typically, the manufacture of an airplane involves the conversion of 90—95% of all purchased materials into scrap, with the airplane left over as a minor byproduct. That's efficiency!

Figure 30.6 shows the ratio of the mass of liquid metal produced to the mass of various final components, and Figure 30.7 shows the effect of this scrap production on the embodied energy of a final component—a car door panel, in this case. Losing around half of the liquid metal, particularly due to the blanking and stamping processes required to form the door panel, causes the embodied energy in the part to double.

Once again, the choice to create all this scrap is based on good business logic. It minimizes costs, but at the cost of causing a significant excess of material production. Approximately 25% of all steel made in the world each year is converted into scrap (Cullen et al., 2012) and never features in a product; for aluminum, the figure is around 50% (Cullen and Allwood, 2013).

Two broad approaches can be used to reduce yield losses. Firstly, collaboration between designers of products and processes along complete supply chains is likely to reveal significant opportunities for material saving, such as by adjusting final component designs so that they can be made from tessellating blanks. Secondly, new technology developments may allow more efficient use of intermediate stock materials without generating so much scrap. For example, the clothing and

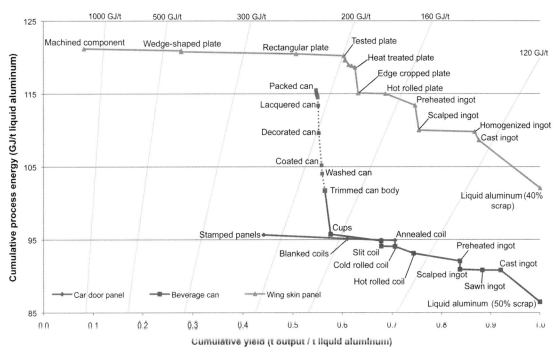

FIGURE 30.7 The process causes of yield losses for three aluminum components, with contours showing the effect of yield loss on the embodied energy of the resulting component. *From Milford et al., 2011.*

textiles industry is more efficient that the sheet metal industry at cutting blanks from fixed width strips, because it uses laser cutting that allows clever tessellation of different components in the same strip. As a separate example, Figure 30.8 shows to alternative routes to producing metal washers: a conventional route creating scrap between the outer circumference of the washer and scrap from the punched-out hole; a novel route invented by the Sakamura Press Company in Japan and now licensed to the San Shing Fastech Company in Taiwan, in which washers are made by cold-forging and then piercing a bar, with no scrap at all. This is a wonderful example of material efficiency in action, and it can operate at a sufficiently high production rate to compete effectively with the conventional process in the open market for metal washers.

The key to reducing yield losses is to design production processes that minimize cutting and trimming. Following years of technology development focused on the economies of scale producing standardized parts, there is great potential for invention and innovation in this area.

30.2.2 Reuse

The strategies of Section 30.2.1 under the banner "reduce" occur at the top of the hierarchy of material management options, because reducing demand for the production of materials eliminates all of its associated impacts. They also apply to all material goods, whether they are consumable (used up while providing a service) or durable (lasting to provide service

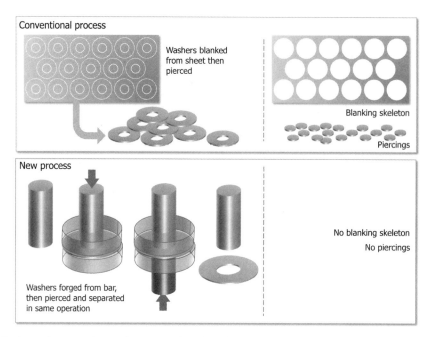

FIGURE 30.8 A novel process for producing metal washers from a solid bar, with no yield losses: the bar with diameter equal to the pierced hole in the washer is forged through a die onto a solid anvil, so the base of the rod spreads out to the thickness and outer diameter of the washer. A second tool set allows a punch of the same diameter as the bar to separate the newly spread washer from the bar.

more than once). Strategies leading to the reuse of materials follow in this section, applicable only to durable goods and where possible are preferable to recycling, because the impacts of production are primarily associated with the creation of materials rather than their conversion into products. Reusing material implies additional deconstruction and manufacturing steps, avoiding the energy intensity and long supply chains of recycling.

In developed economics today, the idea of reuse carries overtones of poverty and substandard goods. In other countries and former times, reuse has been entirely normal practice and in some cases even adds to the value of goods. For example, in the "organic economy" discussed above, where iron could be produced only by redirecting biomass for fuel from other competing uses, all iron goods would be reused because the material had so much intrinsic value. Figure 30.9, from the Rongo Town in Kenya, shows reuse in action today, including both production from scrap metal and component resale.

Four forms of reuse are considered here: reuse of products, components, and materials in turn, and the diversion of manufacturing scrap prior to recycling.

30.2.2.1 Reusing Products: Upgrade and Secondhand Sales

In most economies, secondhand markets are well established for many goods. Despite Akerlof's famous concern about "lemons" (Akerlof, 1970), the buyer cannot easily learn as much as the seller knows about the condition of the goods. Such markets emerge easily when goods are discarded not because they

FIGURE 30.9 Examples of material reuse in Rongo Town Kenya *(Courtesy of Chris Cleaver)*: (A) a scrap metal hawker; (B) small hoe blades formed from scrap metal; (C) cooking stoves and metal boxes made from scrap metal. *Images © Christopher Cleaver, 2013.*

have worn out but because the current owner's needs have changed.

The challenge of creating an effective secondhand market is to find a way for buyers and sellers to share information and to reduce the transaction costs of doing so to the point that it is attractive to trade. For housing, secondhand sale is the most common, and it is well regulated to deal with Akerlof's lemon. For cars, secondhand markets are also well developed, albeit not regulated; the transaction costs are low, a large market exists, and when values become too low for further trades within one country, cars can be shipped to other countries where their relative value is higher. For low-value goods in developed economies, such as most clothing, books, and crockery, for example, the transaction costs are generally too high for individuals to trade, so charity shops exist to provide an alternative to disposal. However, the relatively low value of used clothing is problematic even when supplied for free, as the costs of sorting may exceed the price potential buyers may offer. Donated clothing in the UK is largely baled and shipped to Eastern European countries, where labor costs are sufficiently low to justify sorting used clothes by type and quality. For goods with values between cars and clothes, such as white goods, furniture, and leisure equipment, the Internet has created a means to reduce transaction costs sufficiently to create new secondhand markets, and similar services in business-to-business trades create markets for industrial equipment and commercial furniture and fittings.

However, secondhand markets cannot prosper for goods that are rapidly superseded (such as electronics) or where regulation has changed so that a previous generation of goods cannot be operated legally. For example, single-skin oil tankers have been outlawed by the International Maritime Organization, so they must be scrapped while new double-skinned tankers are produced to take their place. In these cases, it is not possible to resell the entire product; however, it may be possible to disassemble it into modules, components, or individual materials that retain value greater than their scrap price.

30.2.2.2 Reusing Components: Modularity

As discussed in Section 30.2.1.3, the end of a product's life is generally determined by one subassembly, module, or component, while the remainder of the product has residual life. If this life cannot be extended for the whole of the original product, it may be possible to exploit the residual life of the modules through partial disassembly of the original product into modules, some of which can then be reused in other assemblies. Famous examples of this strategy include the much discussed reuse of modules in Xerox copying machines (Ayres et al., 1997), re-use of modules from car-breakers (Subramoniam et al., 2009), and reuse of disposable cameras (Bogue, 2007).

The costs of disassembly are generally greater than those for assembly due to the loss of economies of scale: assembly plants have been optimized since the days of Henry Ford to simplify repetitive tasks when the same operation is repeated many times. In an equivalent disassembly line, each succeeding operation would be different due to the variety of goods and their different condition at end of life. However, many strategies of design for disassembly have been developed to create joints that can be reversed with ease and to plan for the separation of components materials.

The real challenge of component reuse is not the logistics and management of the operation, but the design of long-lasting interface protocols that allow subassemblies to last through several generations of products. Famously, Michael Dell won the battle for the market for personal computers in the 1980s by developing a modular design strategy, in which the various key components of a computer (motherboard, disk drive, monitor, etc.) could evolve at different rates, allowing customers to choose the most up-to-date configurations. This was possible because the physical and software connections between the components evolved much more slowly than the components themselves. The Lego toy series has the same property; the geometry of the connectors has remained constant for many years, allowing a continuous development of new components, which can integrate into existing kits. However, for many goods, such standardized interfaces either do not exist or would be difficult to maintain, which inhibits a much broader application of component re-use. For example, steel-framed commercial buildings are generally bolted together, so they can be disassembled relatively easily at the end of life. However, each building is designed to maximize the floor space created on each individual plot of land. Therefore, without standardization of the spacing of columns in commercial buildings, it is difficult to develop a market for used structural beams. Similarly, the technology of cars develops sufficiently rapidly (e.g. the design of different fuel injection, air filtering, or electronic systems) so that the volume and geometry of shape required to accommodate each new generation of engines is different and has little leeway. Therefore, it is very difficult to reuse structural parts of the car, even when they can be disassembled and have a high residual life when the original car is discarded.

The development of standardized open-architecture and interfaces as a basis for new product development would greatly enhance the possibility that components and subassemblies could be reused to exploit their full lifespans. In addition, it potentially creates new business opportunities based on nurturing the value of subassemblies over longer periods. A few emerging examples of this approach, with the design of flat-pack buildings and modular hotel structures, show that some of this commercial potential is immediate. However, counter-examples, such as the regular redesign of power-supply connectors for Apple computer products or Hewlett—Packard printer cartridges, also demonstrate how module reuse can be prevented to maintain revenue streams based on replacement sales rather than component maintenance.

30.2.2.3 Reusing Material: Cutting Bits Out

If modules cannot be disassembled and reused, one last strategy for avoiding new material production is to re-use single-material components, following further manufacturing steps. The intermediate goods created by the materials processing industry are typically of a large geometry to maximize the number of components that can be cut from the stock downstream; most manufacturing therefore involves cutting smaller pieces out of bigger pieces of material. Although old material can retain perfect properties to allow further use and metals often retain sufficient ductility to allow some reshaping, the habit of manufacturing to cut small pieces from large ones creates a difficulty for material reuse. Unless the component can be reused directly, as discussed in the previous section, old material can only be used to make smaller components in future. Therefore, material reuse generally occurs only for large pieces of single material. Such reuse is common for wood, particularly for valuable hardwoods, which may be difficult to source from new forests under more cautious management than previously. However, in developed economies it is rare for other materials, except at a domestic scale.

In developing economies, as illustrated in Figure 30.9, material reuse is common. A large-scale example of this occurs with ship breaking in India. Asolekar (2006) has documented the flow of steel from the coast at Alang, where old ships are driven onto the beach, via manual disassembly, to hot rolling mills where the old ship plates are rolled into reinforcing bars for new construction. This example of material reuse is practiced at significant scale, potentially providing up to one sixth of India's current steel demand. However, it has limited potential. Even in this case, the reinforcing bar made from old ship plate is less strong than the new bar. Recycling by melting, with careful control of alloying, might allow sufficient material saving in construction to justify the additional energy cost of melting.

30.2.2.4 Diverting Scrap

A final option for reusing material without reducing it to liquid form in a recycling process occurs upstream, with the opportunity to divert manufacturing scrap to an alternative use rather than sending it for recycling. Allwood et al. (2012) described how Abbey Steel in Kettering in the UK has for 30 years bought blanking scrap from car body panel makers in the UK and cut it into regular blanks for manufacturers of smaller components. This strategy could be applied for the scrap material of all sheet materials. It reflects the material inefficiency of current practice, where each manufacturer cuts out their own blanks, creating the 20—30% scrap mentioned above. If suitable cutting technology exists (such as laser cutting of textiles), yield losses would be minimized by tessellating the largest possible number of geometries from the same sheet. Therefore, a parallel to diverting scrap would be to coordinate blanking requirements over a much wider range of customers.

A specific opportunity for diverting aluminum scrap from machining operations has received attention in research (albeit not yet in practice). Aluminum is a highly reactive metal that can be welded in solid-state at temperatures well below the melting temperature. Thus solid-state welding, for example by extruding machining chips (e.g. Tekkaya et al., 2009), could in future be used to divert aluminum scrap into new productive uses without melting.

30.2.3 Recycle

The rest of this volume provides the most up-to-date assessment of recycling, so it is included here to complete the hierarchy of waste management options, and allow a brief introduction to the limits to recycling that will guide the discussion on options by material in Section 30.3 of this chapter. Recycling is usually practiced when it is technically possible and when it saves energy (and emissions) compared to primary production. It may also be important if a resource is scarce, either due to an absolute shortage (concern about critical elements, most recently the rare earths, raises the specter of exhausting the world's total stock, but usually this is a transient problem while new mines are opened), or due to competing uses (e.g. the growing competition for biomass for food and fuel may in future compete with the use of biomass for paper and board).

However, recycling is not a universal panacea:

- For some materials, notably cement, ceramics, and composites, there is as yet no recycling route by which the material can be returned to its original structure and quality.
- For some materials, recycling generally involves a loss of quality. For many bulk metals, it is currently impossible control the alloy content of recycled material with the same precision applied to virgin metal; for paper, the processes of pulping used paper lead to shortening of fibers and a reduction in quality; the great variety of compositions used in plastics creates great difficulty in generating new high quality material from mixed waste streams. As a result, in many cases recycled material must be mixed with virgin material to produce acceptable products, thus reducing the net benefits of recycling.
- For many critical metals that are used in compounds (as alloys, or in electronics applications), the energy required to separate them as part of a recycling process may be significantly greater than the energy needed for virgin production.
- For some materials, notably glass, the energy required for recycling is similar to that required for virgin production as energy is dominated by the high temperature of melting, combined with the diseconomies of scale in used glass collection, sorting, and logistics. For other materials, such as paper, the emissions benefit of recycling may be less than the energy saving, if primary papermaking is powered by used biomass, where recycling is powered by electricity from the grid.

Notwithstanding these caveats, recycling remains an attractive option within the hierarchy of material management options for reducing the environmental impacts of some material use. However, the intention of this chapter is to set recycling in context: if the primary motivation for interest in recycling is to reduce the environmental impacts of materials production, it is only one of a portfolio of options, and not a universal goal to be sought and celebrated.

30.2.4 Downcycle, Decompose, Dispose

Even if the above strategies to reduce material demand are applied vigorously and recycling applied wherever possible, human activities will continue to lead to streams of waste material.

The limiting option for this waste is to bury it underground, which is a widely practiced management technique for difficult problems (see also nuclear waste, carbon sequestration, red mud, toxic dumping, and psychotherapy). This can be done with more or less care, but such waste streams may also have value as substitutes for lower value material, or some benefit may be had from their decomposition.

Downcycling refers to the practice of using unwanted material for an application of less value than its original purpose. Common examples include the use of mixed streams of high-purity aluminum alloys to create less pure casting alloys, and the use of crushed construction and demolition waste (and in the UK, crushed green glass) as a substitute for aggregates in road and other construction. This practice is fundamentally wasteful, as it would have been better to maintain the original higher value of the material. However, as an option of last resort, it at least may save some of the energy and resource consumption of the lower value of material that is replaced.

Two forms of decomposition in common use in waste management are composting biomass waste and incinerating municipal waste for energy. Anaerobic digestion is increasingly used to convert biodegradable waste and sewage sludge to methane and carbon dioxide, which are then combusted for heat and power or upgraded to biomethane. This option is near the bottom of the materials management hierarchy because the energy recovered is substantially less than that required to produce the biobased products being decomposed—and, for example, it would have been a better strategy to avoid wasting food rather than to have it available for anaerobic digestion. However, as landfill gases are released from biomaterials left to rot in the ground and for wood fibers, methane releases may have a higher global warming potential than the CO_2 absorbed during tree growth, so such digestion is a better option than straight disposal.

Energy recovery refers to the process of combusting waste materials for heat and eventually power generation. It includes the substitution of municipal waste for fossil fuels in cement kilns, and the incineration of wood, plastics, and other waste for power generation.

If downcycling or decomposition are not possible, the remaining, last-ditch, materials management option is to store the unwanted material, usually underground in a landfill. This process is much studied, and research over decades has driven regulation to better management of landfill sites to avoid soil and water contamination from leaching.

30.3 WHEN IS RECYCLING NOT THE ANSWER?

The opening section of this chapter questioned whether the circular economy is really a desirable goal, and the second section has shown how recycling is in fact one among many strategies for conserving the value of materials and reducing the total impact of their production and disposal. This section now examines how these strategies play out in practice for the major material groups discussed elsewhere in this book. For each material, we first explore whether recycling is possible now, or will be in future, and evaluate the potential for recycling as a long term material management strategy. This evaluation is then set against the other options for material management raised in Section 30.2, to provide practical guidance on the strategies that should be deployed for each material in order to have most effect in reducing the environmental impact of delivering the services they provide. The sections are ordered according to the greenhouse gas (GHG) emissions impact of current world production of each material group (as reported by IEA, 2008).

30.3.1 Iron and Steel

Global demand for steel production quadrupled between 1960 and 2010 and has now reached 1500 million ton/year, equivalent to ~200 kg/year for every person alive on the planet. Because it is ferromagnetic, steel is easily separated from other materials in mixed waste streams, so has one of the highest recycling rates, estimated to be up to 90% of arising waste (Graedel et al., 2011). The main losses of steel from this return loop occur with reinforcing bars inside reinforced concrete and steel used for subsurface applications, where extraction is too costly. Other steel that is not separated for recycling but dumped in landfill rapidly oxidizes into rust, which can be processed back into steel by conventional production routes. Manufacturing steel goods from scrap rather than ore leads to around half the total GHG emissions per product, and the cost benefit of this energy saving provides sufficient commercial motivation for recycling to need no further legislative stimulus. However, despite the high collection rate, recycled steel contributes only around one third of current demand because of the rate of demand growth, and this is expected to rise to around 50% by 2050.

Figure 30.10 illustrates global flows of steel in 2008 and demonstrates that steel recycling is closely coupled to primary production: scrap steel is used in primary production (from ore) to assist in cooling the melt, and virgin steel is used in secondary production (from scrap) to control metal composition. The flow of metal from secondary production into products demonstrates that secondary steel is widely used

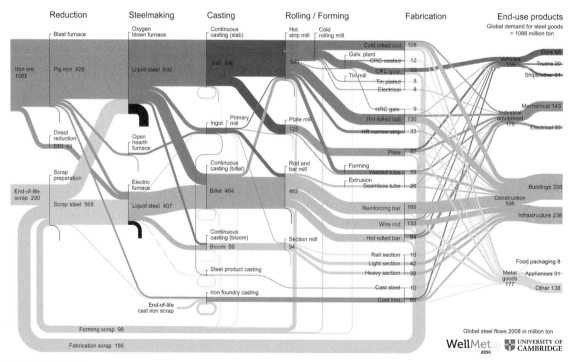

FIGURE 30.10 The annual global flow of steel in 2008 from production (from ore or scrap) to end-use, including scrap generation and return. *From Allwood et al., 2012.*

for producing reinforcing bars, and other bars and sections. However, the higher grade applications of steel—as rolled sheet or plate for subsequent forming into panel components—are entirely made from virgin metal: developments of steel metallurgy have allowed significant increases in strength through precisely controlled alloying with more diverse elements, but these have not been matched by equivalent developments in refining unwanted elements out of the molten metal created from scrap. Removing copper from liquid iron is particularly challenging, so manual sorting is used even in developed economies to remove possible sources of copper from scrap feeds. More generally, due to the economics of current steel-making, the main approach used at present to purify liquid steel is to bubble oxygen through the melt to remove the impurities that preferentially bond with oxygen. More advanced extraction methods, such as electrolysis, used in processing metals from waste electronics could be developed further to improve steel refining, but these are not yet applied.

The achievement of a circular economy for steel—the most recycled metal—is thus a distant dream for two key reasons: demand growth limits the relative availability of scrap and processes for refining molten scrap have not been developed to allow composition control comparable to what is achieved with virgin production.

However, the strategies of material efficiency could be applied to allow delivery of greater value from less steel, and the high rates of steel use at present are in part due to the extraordinary efficiency of its production, and hence its low price: a ton of steel in the UK costs around £800, which is comparable to the cost of a day's work by a structural engineer. Around a half of all steel is used in construction, as reinforcing bars, sections, and cladding, and total requirements could be reduced by avoiding overspecification, selection of ideal compositions (particularly for reinforcing bars in China and India, which are currently much less strong than those used in the EU and USA), and by maintaining buildings or building components for much greater periods. In the UK, designing buildings that met rather than exceeded the requirements of the EuroCode safety standards would half constructional steel demand, and if in addition buildings were maintained for 200 years rather than the current 40 before replacement, we could maintain the same building stock with one tenth of the new steel.

Typically 70% of the mass of a finished car is steel, but with yield losses of around 50%, the average 1.3-ton car in the UK has required production of 2 tons of steel; steel use would be cut immediately by a move toward lightweight vehicles. Fuel consumption in cars is strongly correlated with vehicle mass (see Figure 30.11), so the key to developing a future low-emitting car fleet is not to create electric cars (the current target of EU policy, despite there being no excess supply of "low-carbon" electricity available) but to create light cars. Small, light cars (say 300 kg steel per car, comparable to Volkswagen's L1 concept car, and in line with the 500-kg aluminum Lotus Caterham 7 and the 600-kg steel Tata Nano) made

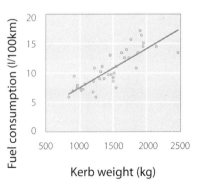

FIGURE 30.11 The effect of vehicle mass on fuel consumption for a representative range of cars on the road in the UK today, clearly demonstrating that fuel efficiency arises from mass reduction. *From Allwood et al., 2012, with permission.*

with low yield losses (say 10% losses) could provide the same level of car service with around one sixth of today's steel requirement. This would be halved again if cars were maintained for twice as long.

These two examples, from construction and cars, underline the potential for demand reduction as the key strategy to reduce the environmental impacts of steel production. Steel recycling is already virtually universal, but it cannot at present create the same quality of material as production from virgin ore, unless a different scrap stream is created and maintained with perfect cleanliness for each distinct composition. Reducing the number of compositions in use would simplify future recycling loops, and some opportunities to control steel properties by processing rather than alloying are under development, but there is strong commercial pressure against this.

30.3.2 Cement

Cement is not currently recycled because doing so would require energy inputs comparable to making new cement, and there is no shortage of resources to make new cement. Therefore, the recycled product would require as much energy and be of lower quality than new cement, so there are no commercial processes for cement recycling. However, significant confusion has been created by efforts within the industry to promote the (rather questionable) benefits of "concrete recycling". Any discussion of a future circular economy for cement must therefore begin by clarifying the distinction between cement and concrete. Cement is a binder and is never used on it its own, but it provides the adhesion of a composite material, generally either mortar (cement plus sand plus water) or concrete (cement plus sand plus water plus small stones, known as aggregate).

Most cement (Portland cement in particular) is supplied as a dry powder, to which water is added to trigger a hydration reaction. This reaction cannot easily be reversed, and a circular cement economy is not currently a possibility. The one possibility being explored in this area is that there is some evidence that following use, the hydration reaction may not be complete, so finely ground used cement can be used in small amounts as a partial substitute for new cement. Dosho (2008) explored this opportunity in great detail. In contrast to the ideal of a circular economy, cement is therefore a one-way material.

Faced with this knowledge, the cement industry has wisely looked for alternative strategies to reduce the impact of cement, and given the public popularity of the word *recycling*, has attached it not to cement, but to concrete: old concrete, which would otherwise be dumped in the ground (along with any embedded steel reinforcing bars, which are too difficult to extract) may instead be crushed, and resulting granular material can be used as a substitute for the aggregate required by new concrete. Estimates of the value of this approach vary, but a concern is that the range of particle sizes in crushed concrete is much broader than the range in new aggregate, and in turn this may drive up requirements for cement to make new concrete, and hence worsen the overall impacts. There is no evidence that recycling concrete in this form provides a substantial reduction in emissions.

In contrast to this problem with recycling, there is significant potential to reduce the requirement for new cement when making concrete through avoiding overuse and partial substitution by other materials, such as ground granulated blast furnace slag or pulverized fly ash, which are produced with less additional emissions than new cement. As described above, finely ground used cement can also be used as a substitute.

Demand for new cement can therefore be reduced by dilution of cement with other powders. However, in addition, as with the use of steel in construction, demand for concrete can be reduced by maintaining buildings and infrastructure over longer lifespans and by intelligent design, as illustrated in Figure 30.12.

As yet, the idea of modular design of reuseable concrete components has not developed, but interest in off-site fabrication of modules is growing rapidly as clients aim for shorter construction times. At present, such modules are typically joined "wet" on site so that at end of life, the building cannot be disassembled. If new techniques allow "dry" mechanical joints between prefabricated modules, then potentially those joints can at end-of-life be reversed and the modules reused. This potentially offers an interesting new business model to cement companies: rather than selling a low-margin powder in the greatest volume possible, they can sell or lease structural modules, whose value can be maintained over several generations of building.

Without any option for recycling, there can never be a circular economy for cement. The opportunity to create concrete blocks and components, which are themselves reused, has had little if any attention, but the fact that bricks (particularly from older houses) are commonly reused suggests that this may have some potential. However, without a credible return loop and given its very high environmental impact, the role of cement in the world's material economy requires re-examination: cement is abundant, cheap, strong, and convenient but harmful. An environmentally constrained economy would reframe cement as an exotic material to be used in niche applications. With no possibility of creating a circular economy for cement, reducing its environmental impacts depends entirely on reducing its production.

30.3.3 Plastics

Plastics are an extraordinarily versatile family of materials with two particularly attractive properties that have caused an explosion in their use: their composition can be adjusted easily to create an infinite variety of colors, textures, and mechanical properties that can be tailored to each application; final parts are produced directly from molten feedstocks, allowing precise optimization of every design with cheap molds in standard processes. These two properties, and plastics' low cost, have led to enormous diversity in the uses of plastics. Unfortunately this mitigates very strongly against the creation of a circular plastic economy: thermoplastics can easily be recycled (production scrap is often fed directly back into the machine that created it) but only if the material being recycled is of a consistent composition. As one of the key attractions of plastics is their variety in composition, this inhibits plastics recycling. Despite increasing

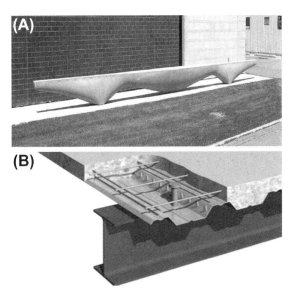

FIGURE 30.12 Lightweight design applied to concrete (A) through the use of fabric formwork to shape the beam to a 'fishbelly' as seen above for steel and (B) through composite design with shear connectors between steel and concrete allowing full use of concrete's compressive strength while avoiding tension.

sophistication in labeling different plastics, rates of recycling remain relatively low.

An interesting beacon of hope for a circular economy for plastics arose with the development of several processes claiming to turn "anything to oil" by thermal depolymerization. However, several practical implementations have failed, largely due to the use of intense acids in the process causing rapid degradation of the equipment. Claims about the energy yield of such processes suggests that something like one sixth of the energy created (as oil) was consumed by the plant, but the overall material yield remains fairly low, with around 25% of input mass being converted to oil. In comparison, using waste plastics as fuel for heat and power generation can occur with high efficiency, and significant commercial value: Subramanian (2000) reported energy densities for plastics around 45 MJ/kg, which is similar to that for petrol or diesel. Given the difficulty of managing the complex plastic waste stream, such power generation may well be the best use of plastics—far from a circular economy, when the material is best used as consumable fuel.

The strategies to use less materials that have such high potential for steel and cement, as discussed above, have less impact for plastics because, by their nature, they are already used to create optimized lightweight products. Therefore, the key to reducing the impacts of plastic production lies at the top of the hierarchy of options, in reducing overall demand. Several studies, such as Hekkert et al. (2000a,b) have shown how plastic packaging requirements can be cut, and the idea of reusing supermarket carrier bags seems to have extraordinary appeal to politicians. However, there are few opportunities for reusing plastics in other applications because they are used to make small complex components within larger assemblies, and they are difficult to separate.

Achieving a circular economy for plastics is technically easy if their composition is constrained, and there is an energy benefit in achieving this. The two key alternatives appear to be to use waste plastic as a fuel for energy recovery or to reduce overall demand for the type of products in which plastics are used, but as they permeate so widely, this is a difficult strategy to evaluate.

30.3.4 Paper

Paper is widely recycled. In the UK, paper recycling is now widely practiced and accepted through curbside separation of domestic and commercial waste. Interestingly, its benefits are less clear than for the metals. The energy requirements for recycling paper are lower than those for primary pulp making, but recycling degrades the fibers in the paper and the energy for primary processes is often supplied at least partially from the waste biomass of tree harvesting. So, the GHG emissions of paper recycling may be greater than those of primary papermaking, although recycling carries a co-benefit of reducing the demand for new biomass, which will face increasing competition in future as demand for food and fuels grows. Recycling (or combustion) also avoids the complications of paper decomposition in landfill, which leads to methane releases. A closed circular economy for paper is not technically possible, due to fiber shortening during the process, and because some paper applications (e.g. for sanitary use) have no return loop. However, public interest in paper recycling is high, so collection rates are good.

The services created by paper use could be delivered with less paper: office paper could be thinner, and print-on-demand services could avoid overproduction of books, magazines, and newspapers. There has been considerable discussion about whether "eReaders" will cause a reduction in paper use or whether, as with claims dating back to the 1970s about the "paperless office", the advent of computers will continue to increase paper consumption due to the wide availability of printers. This discussion remains unresolved until future figures on paper demand can be correlated against eReader use. The use of paper for archiving will remain attractive long into the future, due to the rapid turnover of computer technologies.

A form of circular economy for paper could be developed if, instead of recycling, the paper were cleaned for reuse without destroying its structure. This was normal practice in the use of ancient palimpsests, or indeed of chalkboards in schools. Leal-Ayala et al. (2012) have demonstrated that toner print on office papers could be removed by carefully calibrated lasers to allow reuse of paper. However, as with the steel, cement, and plastics, the astonishing efficiency of current production leads to very low prices for paper, so there is little motivation for either user or supplier to develop alternative material loops.

30.3.5 Aluminum

Aluminum production from ore is extremely energy intensive due to the electrolysis step required to extract alumina from bauxite. Because this process uses 20 times more energy than that required to melt an equivalent mass of solid aluminum, the aluminum industry has expended significant marketing efforts, making the claim that aluminum is a "green material" and that recycling uses only 5% of the energy of primary production. In reality, this statement is rarely true. Creating a product from recycled material normally involves a full supply chain with many processes, so without question recycling uses less energy than primary production, but usually the reduction is to around 30% rather than the much vaunted 5% figure.

Furthermore, aluminum recycling is inhibited much more than for steel by the difficulty of purifying the liquid metal. Figure 30.13 illustrates the consequences of this problem: nearly a half of all aluminum produced each year becomes internal scrap and is recycled within the factories in which it was created. End-of-life recycling makes only a small contribution to the overall flow. Apart from the contribution of closed-loop can recycling, all end-of-life recycling for aluminum involves downcycling from the high purity wrought alloy systems to the lower purity cast alloys.

A circular economy for aluminum is therefore not possible at present, because it would require perfect separation and cleaning of all alloys. Within individual sites this occurs, and much effort is spent on managing internal scrap by alloy type. However, it cannot in general be achieved for postuse waste streams, and the proliferation of alloys developed by the industry to meet specific user requirements further inhibits the possibility of future alloy separation. The one example where a circular economy is approached is in the recycling of drink cans, where melting the three components of the can (the body, the head, and the ring-pull, each made from different alloys) leads to a composition that, when diluted by around 5% virgin aluminum, can be used for making new can body stock. The energy cost of this route is around 30% that of making cans entirely from virgin aluminum, and rightly the industry would like to expand it, if collection rates could be increased. Unfortunately, the convenience of cans is that they can be used and discarded easily, so despite the relatively high value of the material in a used can (around 3–4 UK pence per can), collection rates are estimated to be around 50%, with the remainder being lost in landfill dumps.

The most striking feature of Figure 30.13 is the high yield losses of the aluminum supply chain. Better integration between companies combined with new process development could significantly reduce these. In general, this would reduce the size of the recycling loop, but would not reduce requirements for primary aluminum unless recycling leads to degrading wrought alloys to casting alloys, which must then be

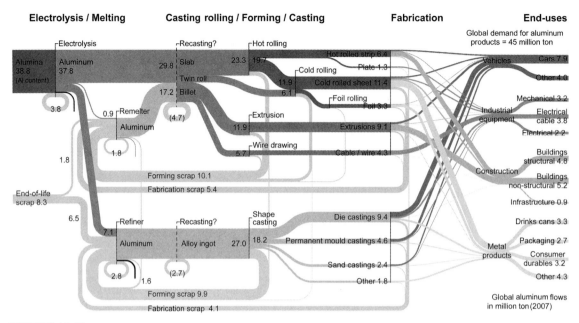

FIGURE 30.13 The global flow of aluminum in 2007, showing that production from ore dominates, in total only one half of the liquid aluminum made each year reaches final products, and that the (higher purity) wrought and (lower purity) casting alloys have some separate recycling loops, but recycling leads on average to a downgrading of ore purity. *From Cullen and Allwood, 2013.*

replaced by new virgin production. This happens in the aerospace industry; although it uses only 1—2% of the world's aluminum production, it turns 90—95% of what it purchases into scrap (machining chips) that are not separated and thus are downcycled.

Aluminum has properties that make it attractive in many applications, including its low density, high corrosion resistance, high conductivity, and high ductility. Using such a valuable material for disposable packaging much of which is lost through disposal seems extravagant, and the properties of the material would be better exploited in more advanced applications. However, as with plastics, aluminum is generally used in rather small components where its properties have particular value, and these are already optimized and difficult to reuse.

Two significant opportunities for reuse to reduce overall demand for aluminum occur in construction and power cables. Aluminum extrusions and roll-formed sheets are used as the purlins and cladding of light industrial buildings, which typically have short lifespans and in which the components are not damaged. They could therefore be reused, and already at least one UK supplier offers 20-year leases on such components to ensure they are reused at end of life. Aluminum extrusions are also widely used in curtain walls and window frames. Although developments in Europe are aiming to ensure that these materials are separated from old buildings prior to demolition to ensure the metal is recycled, they could also potentially be reused in future. Similarly, aluminum in electrical cables could potentially be reused at end of life.

Development of a circular economy for aluminum is currently not technically feasible due to the difficulty of alloy control. However, given the energy benefit of recycling, a priority

30.3.6 Clothing and Textiles

Clothing in developed economies is generally discarded before it is worn out, and most of the population own far more clothing than needed for comfort and function. The rise of 'fast fashion' with short fashion seasons and new outlets in supermarkets led to an increase of clothing purchasing in the UK such that by 2005, they were on average purchasing one third more garments per person than in 2000. As a result, several studies have attempted to estimate the amount of unused clothing in the nations wardrobes, claiming that only one third of all clothes in the UK are worn in each year.

Clothing in this sense is an icon of the problem of reducing the impacts of material production; for clothing in particular, recycling offers little benefit. There are no recycling routes for natural fibers. Although most of the growth in clothing purchased in the UK has been through manmade fibers, such as nylon and rayon, they are increasingly used in mixed-material composites, so they cannot easily be separated for end-of-life recycling. In fact, it is not clear that recycling the materials used to make clothing would offer much benefit even if widely practiced: the dominant environmental impacts of clothing occur in use (washing and drying are energy intensive), in coloring and texturing fibers (requiring intense use of potentially toxic chemicals), and in (fertilizer- and insecticide-intensive) cotton agriculture; only the last of these would be avoided by fiber recycling. The most widely quoted application of textiles recycling occurs in manmade carpets, with the Interface company apparently a leading practitioner.

Instead, the key to reducing the impacts of clothing production would be to reduce the rate of purchasing and hence production. However, in developed economies, proposing this is rather akin to suggesting that plants should not flower! Leasing, which is widely used for working uniforms and hotel bed-linen, could in principle be used more widely, and a few well-motivated groups have explored clothes sharing. Charity shops create some hierarchies of cascading, but these have limited effect on demand in developed economies and may not be entirely beneficial to developing economies, where early economic growth that might occur through domestic clothing manufacturing is stifled by a free supply of discards.

We could live well with smaller stocks and flows of clothing, and indeed with repair and upgrade could largely have a circular clothing economy, but prefer not to do so. Acting to change this preference voluntarily (it would be accepted without question in a crisis) requires a sociological shift of extraordinary scale; as yet, the mechanisms required to create this shift are unknown. However, the eco-fashion designer Kate Fletcher (Fletcher, 2008) has created a positive first hint of possibility by observing that fast fashion creates disposable clothing: easily purchased and easily discarded after one outing. In contrast, a shirt embroidered or repaired by a friend or relative for personal reasons converts the commodity to a part of a life-story, which takes on a significance far greater than its appearance.

30.3.7 Glass

The British public love putting their bottles in recycling bins. According to several studies, they consider this to be the most important action they should take to conserve the environment for future generations. The fact that we largely discard green glass, because we like drinks made in other countries, where we make mainly clear or brown glass because that is what we need for our own famous drinks, somehow does not register. Thank goodness we're busy making new roads and can use the crushed old glass within them!

Glass recycling is restricted by color and has a limited energy benefit: the most energy-intensive step of both primary and recycled glass production is melting, which is the same for both routes, so the benefit of recycling is to avoid the heat of reaction as the raw materials combine (Sardeshpande et al., 2007). This leads to a theoretical energy saving of around 25% for the recycling route, although this is partially offset by the additional energy costs of collection.

Glass is mainly used for flat applications such as windows (made by the float glass process) and containers (made by blowing). Glass bottles have become lighter with improved process control, but there is limited potential for further saving. Plate glass made by the Pilkington float glass process is already so well controlled that there is no opportunity for reducing thicknesses.

Glass degrades in use only very slowly. Glass is a super-cooled liquid, so the 2000-year-old Roman glass in the British Museum has become cloudy, but the 500-year-old Elizabethan glass remains clear. Therefore, it could in principle be reused many times. However, such reuse is rare: windows from old buildings are smashed as part of the demolition route, as the cost of deconstruction, storage, and resale would not be met by low secondhand prices. Glass bottles, which used to be the standard for drinks purchasing, have been replaced in many cases by plastic bottles; even when still used, the costs of collection, cleaning, and refilling are too great to allow commercially viable reuse.

A circular economy for glass would save little energy compared to the current largely one-way system, so the key to reducing impacts is to make less. This would occur if buildings lasted longer (so less window production were required) or if we reduced our total requirement for glass containers. New approaches to cleaning and returning glass containers—possibly based on refilling the same containers within supermarkets, rather than returning old containers for recycling and buying new ones—might be possible.

30.4 DISCUSSION

The phrase "circular economy" creates an attractive image of a beneficial loop of continuous material recycling without the harmful environmental impacts of new production. In fact, this chapter has shown that recycling is only one among the lower orders of a hierarchy of options for reducing those impacts, and the circular economy is rarely if ever the key to doing so. Reducing demand and reusing products, components, and materials are strategies with greater potential to reduce unwanted impacts. They can be applied across all material groups. These strategies have had relatively little attention to date, and possibly therefore have the most potential for development. In contrast, recycling is already well developed when it is commercially attractive, is impossible for some materials, and for most recycling systems it is only possible to maintain material quality by mixing recycled material with virgin material.

The survey of options applied to key material groups in the previous section has raised several themes that span across materials:

- Economists will often respond to engineer's reports of technical limits by denouncing their failure to think of substitutes. Surely if it is difficult to recycle cement, some clever chaps in the future will find a way to substitute it with something else. Substitution is obviously a warm comfort blanket for economists, but sadly it has much less potential than they realize: we use such vast quantities of steel and cement that the only two materials available in comparable quantity are wood and stone. We used to use wood and stone, and moved to steel and cement because they were stronger and more convenient, and the (at the time) free availability of fossil fuels made them readily available. Stone is in fact considerably less versatile than cement, because it cannot be poured, and apart from the problem of competing demands for biomass, wood production at present requires sufficient energy (for kiln drying) that its benefit as a substitute for steel is marginal. Ashby's book on ecomaterials selection provides a thorough basis for exploring substitution (Ashby, 2009), but demonstrates few if any easy options to reduce impacts.

- "What about innovation then? You technical people never have the imagination to see what might happen in future," cry the economists. Well, those same technical people have been thoroughly interested in looking for innovative materials for decades and have been searching the outer reaches of the periodic table to find them. As a result, innovative materials of recent discovery tend to have exotic properties (conductive polymers, superstrong graphene, carbon nanotubes), but these will be useful in highly technical applications, not as substitutes for the few bulk materials that dominate the environmental impacts of production. Steel has become stronger with new exotic alloying elements. However, because that comes at the cost of a loss of ductility, the value of further strength improvements is unclear. We do not make reinforcement bars from ceramics, because we want them to cope with some aspects of building failure. At present, a range of new formulations for cement are being promoted heavily (IEA ETSAP, 2010; provides a survey), but as yet without clear evidence of product properties or total process impacts. It is much more likely that innovation in materials will deliver new electrical properties than deliver a replacement for steel and cement.

- Can we reduce the total number of compositions in use for each class of material? For every recycling loop in current operation, a reduction in the number of compositions would simplify and improve the quality of recyclate. Yet this is absolutely counter to commercial developments and would be extremely difficult to enforce. Creating simpler waste streams with less variety is more likely to occur when materials have sufficient value that companies will collaborate to simplify their return.

In effect, all of the challenges to a circular economy raised in this chapter arise fundamentally from the low cost of bulk materials, which in turn arises from the extraordinary efficiency with which they are made and the relatively low cost of energy, particularly to the energy-intensive industries. Platinum is recycled perfectly because everyone involved with it understands the value of retaining it and using it again, so a massive increase in energy pricing which raised steel and cement to the value of platinum would bring about the instantaneous adoption of every strategy in the materials management hierarchy described above!

However, such an increase is unlikely to happen. Twenty years of international negotiation aiming at a cross-border carbon price have so far failed, and there is not even any political will to impose fuel tax on aviation fuel. Climate change presents too many urgent challenges to wait longer for such agreement, so instead we need to pursue other implementation mechanisms, of which the key appears to be regulation. At a simple level, glass bottles are reused in some European countries because of regulation enforcing a deposit to be paid against their return. More broadly, smog in cities, river water quality, and acid rain have been controlled in Europe by regulation, and globally, CFCs were eliminated by regulation. The primary goal of work around materials management and its environmental impact is therefore to nurture the conditions under which the public will be ready to vote in politicians to pursue regulation. This happens easily in war or crisis (Fukushima), and also where broad idea has become familiar. Therefore, a primary goal of material activism today is both to reduce material demand, and to prepare mindset for overall demand regulation.

Regulation is a negative end-point for this chapter, so instead we will end with a celebration of the cultural value of materials. Tanya Harrod's wonderful survey of the symbolism of materials in different cultures (Harrod, 2013) reminds us that materials create objects of beauty, that working with them allows us to express a part of our humanity, and our shared appreciation of them is part of our cultural identity. Ready access to fossil fuels and a narrow set of measures of economic success have in less than 100 years reduced our perception of materials as being the building blocks of our legacy to commodities of convenience. If we can take back some of the wonder of creativity and association offered by the best of material goods, may be the regulation we need could be at our common request and not a constraining imposition.

Acknowledgments

The author is supported by a Leadership Fellowship provided by the UK Engineering and Physical Sciences Research Council (EPSRC) reference EP/G007217/1.

References

Akerlof, G.A., 1970. The market for 'lemons': quality uncertainty and the market mechanism. The Quarterly Journal of Economics 84 (3), 488—500.

Allwood, J.M., Cullen, J.M., Carruth, M.A., Cooper, D.R., McBrien, M., Milford, R.L., Moynihan, M.C., Patel, A.C.H., 2012. Sustainable Materials: With Both Eyes Open. UIT Cambridge, Cambridge, UK.

Ashby, M.F., 2009. Materials and the Environment: Eco-informed Material Choice. Butterworth-Heinemann, Burlington, MA, USA.

Asolekar, S.R., 2006. Status of management of solid hazardous wastes generated during dismantling of obsolete ships in India. In: Proceedings of in the International Conference on "Dismantling of Obsolete Vessels".

Ayres, R.U., Ferrer, G., van Leynseele, T., 1997. Eco-efficiency, asset recovery and remanufacturing. European Management Journal 15 (5), 557—574.

Bogue, R., 2007. Design for disassembly: a critical twenty-first century discipline. Assembly Automation 27 (4), 285—289.

Boulding, K.E., 1966. The economics of the coming spaceship earth, originally published. In: Jarrett, H. (Ed.), Environmental Quality in a Growing Economy. Resources for the Future/Johns Hopkins University Press, Baltimore, MD, pp. 3—14.

Braess, H.H., Seiffert, U. (Eds.), 2005. Handbuch Kraftfahrzeugtechnik, fourth Aufl. Vieweg, Wiesbaden.

Carruth, M.A., Allwood, J.M., 2012. The development of a hot rolling process for variable cross-section I-beams. Journal of Materials Processing Technology 212 (8), 1640—1653.

Carruth, M.A., Allwood, J.M., Moynihan, M.C., 2011. The potential for reducing metal demand through lightweight product design. Resources, Conservation and Recycling 57, 48—60.

Cullen, J.M., Allwood, J.M., 2013. Mapping the global flow of aluminium: from liquid aluminium to end-use goods. Environmental Science and Technology 47, 3057—3064.

Cullen, J.M., Allwood, J.M., Bambach, M., 2012. Mapping the global flow of steel: from steelmaking to end- use goods. Environmental Science and Technology 46 (24), 13048—13055.

DoETR (Department of the Environment, Transport and the Regions), 1995. Making Waste Work; a Strategy for Sustainable Waste Management in England and Wales. HMSO, London.

Dosho, Y., 2008. Sustainable concrete waste recycling. Proceedings of the Institution of Civil Engineers Construction Materials 161 (CM2), 47—62.

Druckman, A., Jackson, T., 2008. Household energy consumption in the UK: a highly geographically and socio-economically disaggregated model. Energy Policy 36 (8), 3177—3192.

EPA, 2012. Light-Duty Automotive Technology, Carbon Dioxide Emissions, and Fuel Economy Trends: 1975 through 2011. U.S. EPA-420-R-12-001, Office of Transportation and Air Quality, March 2012.

FAO, 2012. FAOSTAT, Food and Agriculture Organization of the United Nations.

Fletcher, K., 2008. Sustainable Fashion and Textiles: Design Journeys. Earthscan, London.

Fu, F., Pan, L., Ma, L., Li, Z., 2013. A simplified method to estimate the energy-saving potentials of frequent construction and demolition process in China. Energy 49, 316—322.

Goodchild, C.H., 1993. A Report on the Comparative Costs of Concrete & Steel Framed Office Buildings. British Cement Association, Crowthorne, UK.

Graedel, T.E., Allwood, J.M., Birat, J.-P., Buchert, M., Hagelüken, C., Reck, B.K., Sibley, S.F., Sonnemann, G., 2011. What do we know about metal recycling rates? Journal of Industrial Ecology 15 (3), 355–366.

Gutowski, T.G., Allwood, J.M., Herrmann, C., Sahni, S., 2013. A global assessment of manufacturing: economic development, energy use, carbon emissions, and the potential for energy efficiency and materials recycling. Annual Review of Environment and Resources 38, 12.1–12.26.

Harrod, T., 2013. 'Visionary rather than practical': craft, art and material efficiency. Philosophical Transactions of the Royal Society A 371, 20110569.

Hekkert, M.P., Joosten, L.A.J., Worrell, E., Turkenburg, W.C., 2000a. Reduction of CO_2 emissions by improved management of material and product use: the case of primary packaging. Resources, Conservation and Recycling 29, 33–64.

Hekkert, M.P., Joosten, L.A.J., Worrell, E., 2000b. Reduction of CO2 emissions by improved management of material and product use: the case of transport packaging. Resources, Conservation and Recycling 30, 1–27.

Humphrey, A., Lee, L., Green, R., 2011. Aspirations for Later Life. UK Government Department of Work and Pensions Research Report no. 737. www.gov.uk/government/uploads/system/uploads/attachment_data/file/214511/rrep737.pdf (accessed September 2013).

IEA, 2008. Energy Technology Perspectives 2008: Scenarios & Strategies to 2050. IEA (International Energy Agency), Paris.

IEA ETSAP, June 2010. Cement Production, Technology Brief I03. www.etsap.org.

Kasser, T., 2002. The High Price of Materialism. MIT Press, Cambridge, MA.

Keynes, J.M., 1930. Economic possibilities for our grandchildren. In: Keynes, J.M. (Ed.), Essays in Persuasion. W.W.Norton & Co., New York, 1963, pp. 358–373.

Layard, R., 2006. Happiness: Lessons from a New Science. Penguin.

Leal-Ayala, D.R., Allwood, J.M., Schmidt, M., Alexeev, I., 2012. Toner-print removal from paper by long and ultrashort pulsed lasers. Proceedings of the Royal Society A 468 (2144), 2272–2293.

Milford, R.L., Allwood, J.M., Cullen, J.M., 2011. Assessing the potential of yield improvements, through process scrap reduction, for energy and CO_2 abatement in the steel and aluminium sectors. Resources, Conservation and Recycling 55 (12), 1185–1195.

Moynihan, M., Allwood, J.M., 2013. Utilisation Study – Forgotten the Exact Title, Currently under Review.

Parto, S., Loorbach, D., Lansink, A., Kemp, R., 2007. Transitions and institutional change: the case of the Dutch waste subsystem. In: Parto, S., Herbert-Copley, B. (Eds.), Industrial Innovation and Environmental Regulation: Developing Workable Solutions. United Nations University Press, Tokyo.

Prettenthaler, F.E., Steininger, K.W., 1999. From ownership to service use lifestyle: the potential of car sharing. Ecological Economics 28, 443–453.

Sardeshpande, V., Gaitonde, U.N., Banerjee, R., 2007. Model based energy benchmarking for glass furnace. Energy Conversion and Management 48, 2718–2738.

Skelton, A.C.H., 2013. The Motivations for Material Efficiency: Incentives and Trade-offs along the Steel Sector Supply Chain (Ph.D. thesis). University of Cambridge.

Sorrell, S., Dimitropoulos, J., Sommerville, M., 2009. Empirical estimates of the direct rebound effect: a review. Energy Policy 37, 1356–1371.

Subramanian, P.M., 2000. Plastics recycling and waste management in the US. Resources, Conservation and Recycling 28 (2000), 253–263.

Subramoniam, R., Huisingh, D., Chinnam, R.B., 2009. Remanufacturing for the automotive aftermarket-strategic factors: literature review and future research needs. Journal of Cleaner Production 17, 1163–1174.

Tekkaya, A.E., Schikorra, M., Becker, D., Biermann, D., Hammer, N., Pantke, K., 2009. Hot profile extrusion of AA-6060 aluminium chips. Journal of Materials Processing Technology 209, 3343–3350.

UNDESA, 2011. World Population Prospects, the 2010 Revision (Updated 28 June 2011). Department of Economic and Social Affairs, United Nations.

USDoT, 2012. Research and Innovative Technology Administration Bureau of Transportation Statistics National Transportation Statistics. U.S. Department of Transportation.

USGS, 2012. Iron and Steel Statistics and Information. United States Geological Survey (accessed 08.01.13).

USGS, 2012b. Cement Statistics and Information. United States Geological Survey (accessed 08.01.13).

van Oss, H.G., Padovani, A.C., 2003. Cement manufacture and the environment Part II: environmental challenges and opportunities. Journal of Industrial Ecology 7, 93–126.

Winegarner, R.M., 2011. Silicon Industry 2011. Sage Concepts. Available: http://www.sageconceptsonline.com/report1.htm (accessed 08.01.13).

WorldBank, 2013. The World Bank Data. Available: http://data.worldbank.org (accessed 08.01.13).

Wrigley, E.A., 2013. Energy and the English industrial revolution. Philosophical Transactions of the Royal Society A 371, 20110568. http://dx.doi.org/10.1098/rsta.2011.0568.

Zhou, V., 2011. Global Motor Market Study. International Copper Association Ltd. online at. http://industrial-energy.lbl.gov/ (accessed 08.01.13).

CHAPTER 31

The Economics of Recycling

Pieter van Beukering, Onno Kuik, Frans Oosterhuis
Institute for Environmental Studies (IVM), VU University, Amsterdam, The Netherlands

31.1 INTRODUCTION

Recycling is generally considered to be an important strategy for alleviating the pressures of society on the environment. Natural resources are saved, emissions are decreased, and the burden of solid waste is reduced. At the same time, recycling creates employment and attracts investments. In recent years, many countries have experienced large increases in recycling. Besides domestic causes, international trade has played an important role in the expansion of the global recycling sector. In recent decades, international trade of recyclable materials has increased significantly. This chapter aims to identify the main economic drivers of recycling, address the private and external costs of recycling-related activities, and demonstrate the effectiveness of the most important economic instruments to promote recycling.

31.2 ECONOMIC TRENDS AND DRIVERS

Waste is a byproduct of consumption and production activities. Considering the increasing scarcity of natural resources, waste is also seen as a potentially valuable input to production, substituting for virgin resources. Recycling is generally considered as an important strategy for alleviating the pressures of the economy on the environment by saving natural resources, decreasing the emissions of pollutants in the environment, and reducing the burden of solid waste. Recycling also contributes to economic activity, attracts investment, and creates employment. Globally, drivers for recycling vary between industrialized countries and developing countries. In the industrialized countries, drivers for recycling of waste result from increasing disposal costs, increased public concern about the health and environmental impacts of waste disposal, and the fear of future scarcity of certain natural resources. In developing countries, recycling is mainly driven by direct economic motives. International trade, including trade between industrialized and developing countries, has played an important role in the expansion of the global recycling sector. As is the case for any commodity, international trade in recyclable waste allows countries to exercise their comparative advantages to increase the efficiency of the allocation of resources of the global recycling industry. However, there is also fear that (hazardous) waste is shipped to foreign destinations to avoid costly regulations for disposal and storage.

31.2.1 Generation of Waste

The volume of waste that is generated each year in the world is vast and increasing. Large volumes of waste are generated by mining operations, construction and agriculture, but it is difficult to put a number on these volumes as they are often not recorded. It has been estimated that extraction waste, mostly from mining, in the United States alone amounted to 2,500,000,000 t per year in the 1990s—an order of magnitude larger than either industrial or municipal solid waste (Porter, 2002). In developed countries, most manufacturing firms have financial incentives to minimize waste disposal costs. Nevertheless, the Organization for Economic Cooperation and Development (OECD) reports a volume of manufacturing waste in OECD countries in 2010 of 458 Mt, excluding the United States and Canada (OECD, 2013a). Municipal solid waste supply in OECD countries was 661 Mt in 2010, a modest increase from 641 Mt in 2000 (OECD, 2013b).

Research has established that the supply of household solid waste grows proportionally with population, grows less than proportionally with household income, and decreases less than proportionally with increasing charges of waste collection services. In economic terms, this means that household demand for waste services has unitary elasticity with respect to population and is inelastic with respect to income and price (Johnstone and Labonne, 2004). It has been investigated if the supply of municipal solid waste would eventually start to decline with rising income, according to the so-called environmental Kuznets curve (EKC) hypothesis. On the one hand, higher-income families consume more and would therefore produce more waste; on the other hand, it has been suggested that they tend to have a consumption pattern that does not favor waste-intensive goods (Ferrara, 2008). The evidence on the EKC hypothesis is mixed: some researchers found a turning point where household waste supply starts to decline (Abrate and Ferraris, 2010) and others have not (yet) (Mazzanti and Zoboli, 2009; Nicolli et al., 2012).

31.2.2 Recycling

Given this vast amount of waste, there is much interest in reuse or recycling—if not for the potentially valuable materials and energy that can be extracted from the waste, then at least for the increasing costs of disposal and storage. In many countries, an increasing share of industrial and municipal waste is designated for recovery operations, including incineration with energy recovery, composting, and recycling. For example, in the United States, the share of municipal solid waste that is being recycled has increased from 14% in 1990 to 26% in 2010. Among all countries, Germany has the highest share of recycling of municipal solid waste (46%) (OECD, 2013b).

Recycling rates differ per material recycled. According to the industry itself, the global recycling rate of paper is currently 56% (ICFPA, 2013). Europe is the leader, with a paper recycling rate of 70%. Recycling rates for metals vary from very high (gold) to negligible for many specialty metals, such as lithium and tellurium. Recycling rates tend to be higher when the metals are used in large quantities in easily recoverable applications (e.g. lead in batteries, steel in automobiles) or when they have a high value. Increasingly however, small quantities of (rare) metals are used in complex products such mobile phones (Graedel et al., 2011). In this context, Porter (2002) distinguishes between *economies of scale* in recycling (unit costs of recycling go down when the supply of waste material increases) and *diseconomies of scope* (unit costs of recycling go up when the number of different recyclable materials and applications increases). Next to these techno-economic factors, key drivers for recycling are the scarcity, costs, and volatility of prices of substitute primary materials; the cost of waste disposal; public pressure; and government regulations.

31.2.3 International Trade

Recycling has become global business. International trade in recyclable resources is increasing at a fast rate. In 2007, international trade in waste exceeded 191 Mt, almost five times as much (in physical weight) as the international trade in passenger automobiles (Kellenberg, 2012). The most important incentive to international trade is the difference in recycling costs and benefits across countries. Van Beukering (2001) has shown that a number of trade theories predict trade flows of recyclable materials from the north to the south. The basic idea is that the north is abundantly endowed with capital and recyclable or secondary resources and the south with labor and primary resources. All other things equal, this would lead to net export of recyclable resources from the north to the south. This tendency would be enforced by income differences that cause southern consumers to demand lower quality and therefore cheaper materials and products. Finally, laxer environmental standards of waste handling and disposal in the south may also be an additional factor for trade. Empirical observations have supported these predictions. For example, Lyons et al. (2009) showed on the basis of detailed trade statistics that the United States is a large net-exporter of recyclable materials (such as iron and steel, paper, plastics, aluminum, copper, nickel, and zinc) to developing countries, especially Mexico and some Asian countries, including China.

An illuminating way to picture the patterns of international trade in recyclable materials is to make a distinction between the recycle recovery rate and the recycle utilization rate. The recycle recovery rate is the ratio between domestic collection of recyclable material and total consumption of the material (both of recycled and primary origin). The recycle utilization rate is the ratio between the consumption of the recycled material and the total production of the material. For net-exporters of recyclable materials (the north), the recovery rate exceeds the utilization rate. For net-importers (the south), the reverse is true (Grace et al., 1978). Van Beukering (2001) showed that this relationship holds for international trade in recycled paper and that the difference between recovery rate and utilization rate in the industrialized countries and developing countries also increased between 1970 and 1997. Hence, international trade in waste plays an important role in bridging domestic gaps between demand and supply of recyclable materials. However, there is also some evidence that a part of international trade in waste has less welfare-enhancing consequences, when it is in fact a cover-up of dumping waste in places with lax environmental regulations at the expense of human health and the environment. This is called the "waste haven" effect. Using bilateral waste trade data between 92 developed and developing countries, Kellenberg (2012) has indeed found robust statistical evidence of such a waste haven effect, even when controlling for relative productivities in recycling industries, effects of the Basel Convention on the Transboundary Movements of Hazardous Wastes and Their Disposal, and other potentially relevant covariates. Hence, while international trade offers a solution to the oversupply of recyclable material in the north and the lack thereof in the south, the present situation is still far away from the ideal of a "closed loop" industrial ecosystem (Lyons et al., 2009).

31.3 ENVIRONMENTAL AND SOCIAL COSTS AND BENEFITS

Recycling is widely assumed to be environmentally beneficial (Craighill and Powell, 1996). It slows down the exhaustion of scarce resources and limits the use of landfill space. Recycling, however, also generates significant environmental impacts through the collection, sorting, and processing of materials into new products. Therefore, it is unclear when recycling is to be preferred to the use of virgin goods or when waste recovery should replace landfilling or incineration. Due to differences in environmental effects, studies about the desirability of recycling sometimes lead to opposite conclusions (Leach et al., 1997; Ackerman, 1997; Bartelings et al., 2005; Eshet et al., 2006; Erikssonn and Baky, 2010).

31.3.1 Economic Valuation

In economics, environmental and social effects are generally defined as external effects. An external effect, or externality, is said to exist if an economic agent's decision has an influence on another agent's well-being or production possibilities and the former does not (properly) take these effects into account. Because of its unwanted nature, solid waste is often considered an externality. The extent to which solid waste actually is an externality depends on the method by which it is processed. Clearly, if waste is littered or illegally dumped, the externality will be substantially larger than if the waste is recycled or reused in a sustainable manner. For many other waste processing methods, however, this choice between recycling and alternative methods is less straightforward.

Policy makers generally use the level of externalities to determine the preferred ranking and mix of waste management options. Generally, landfilling is considered an environmentally less-favorable option than recycling. However, whether the level of externalities of landfilling always exceed those of recycling or incineration is unclear and requires the relevant external effects to be valued in economic terms. The main reason to express external effects in monetary values is that it allows for the comparison between private costs of various waste management options and the environmental and social costs and benefits related to these options. Economic valuation can express external effects in monetary units. This approach facilitates the comparison of multiple environmental impacts and allows for a trade-off between the benefits and costs of many kinds of environmental improvement.

Ideally, economic valuation forms an integral part of the overall environmental assessment of waste management. As shown in Figure 31.1, an example of such an integrated assessment commonly applied in the waste management sector is the impact pathway approach (COWI, 2000). This approach proceeds sequentially through the lifecycle or pathway of an economic process, linking impacts to burdens, and subsequently valuing these impacts economically. First, overall emission levels and other external effects are determined in physical terms. Then, the impacts of these effects on economic activities and human well-being are assessed. Next, these impacts are translated into monetary values.

An advantage of the impact pathway approach is that it enables the comparison of the benefits of some environmental improvement in the waste sector with the costs to realize such an improvement. Because many studies have applied valuation methods, standard values can be derived for most pollutants or impacts. These allow translating emissions directly into costs bypassing the elaborate impact pathway approach. The valuation of external effects encounters various problems. First, because external effects, by definition, occur outside the market, market values for these effects are generally absent. Therefore, special techniques are required for the estimation of external values. Second, external effects of recycling-related processes occur at various locations. When it is not possible to value these effects at each individual location, values estimated at one location need to be transferred to other locations. Therefore, a disadvantage of this approach is that the complexity of economic

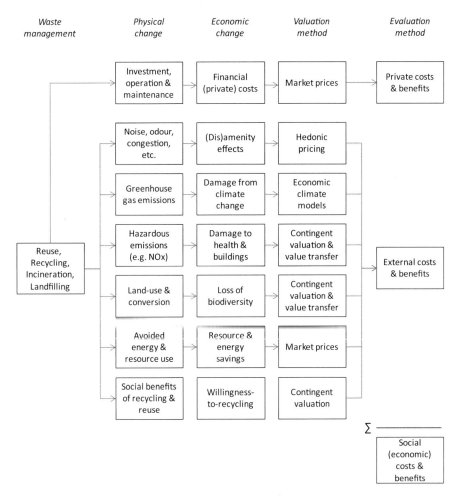

FIGURE 31.1 Economic valuation and the impact pathway approach.

31.3.2 External Costs and Benefits

As shown in Figure 31.1, various categories of external effects are associated with waste management. This section illustrates how some of these external effects affect welfare and thus can be economically valued. First, pollution from recycling-related activities can have many impacts on human health, ranging from short periods of coughing to premature death. The valuation of human health impacts remains one of the most controversial aspects of any valuation study. Many reactions to the monetary valuation of these impacts are partly caused by the unfortunate choice of terminology, such as the "value of a statistical life".

Second, climate change is an important impact category in the valuation of externalities related to waste management. Methane (CH_4)

emissions from landfills contribute approximately 16% to the world's total methane emissions (WRI, 1996). Depending on the efficiency of the incinerator and the composition of the burned materials, waste-to-energy practices may avoid carbon dioxide (CO_2) emissions through supplying electricity to networks. Recycling is usually less energy intensive than primary production processes and therefore may reduce the impact on global warming. Comprehensive climate models that are linked to economic models calculate the costs of climate change. They include agricultural damage, increased morbidity and mortality, damage caused by sea level rise and by extreme events, and loss of species.

Third, disamenity effects of waste-related processes are likely to make up a significant share of the externalities caused. Landfill sites or incinerators generate substantial social costs for their neighboring population. Disamenity effects may occur in different forms. The trucks that transport the waste to and from the sites may cause noise externalities as well as congestion. The landfill site may emit noxious odors and create visual pollution. Increased health risk, or at least an increased perception of higher health risk, is caused for the people living in the vicinity of an incinerator or landfill.

Fourth, all industrial processes involve the (direct or indirect) use of material resources. These resources may be renewable (e.g. planted forests) or nonrenewable (e.g. fossil fuels, ores). When valuing the use of resources, one may ask whether the actual market price (i.e. the internal costs of one unit of the resource) adequately reflects the real cost to society of using that resource unit. There are two ways in which the social cost of resource use may deviate from its market price. There may be external effects in mining and processing the resource. The environmental impact of (nonrenewable) resource extraction tends to increase as the rate of depletion grows. Ideally, these effects should be incorporated in the analysis and valued accordingly.

Moreover, the market price may lead to a higher rate of resource depletion than would be optimal, not taking into account the needs of future generations.

Fifth, recycling may also generate positive externalities as it caters people's willingness-to-recycle. Environmental consciousness plays a crucial role in recycling activities (Ackerman, 1997; Jones et al., 2010). Without the voluntary participation of consumers in waste recovery programs, the level of recycling would be significantly less. From the purely private perspective of the consumer, landfilling or incineration of household waste in a distant site or plant might be the most optimal way to eliminate waste residues. The increased awareness of pollution, however, can motivate consumers to play their part to alleviating the waste problem. Consumers may be interested in waste management for other reasons than purely money savings on their bills (Sterner and Bartelings, 1999). This willingness to recycle can be considered a positive externality.

31.3.3 Private and External Costs

The ultimate purpose of the impact pathway approach is to determine the real costs and benefits of recycling and other waste management options. The external costs of waste management have been the subject of a relatively large number of studies in recent years. Most of these studies focus only on one specific externality. Only a few studies attempt to aggregate a number of the most relevant environmental effects, thereby allowing for a fair comparison internationally (Ascari et al., 1995; COWI, 2000; Dijkgraaf and Vollebergh, 2004; Bartelings et al., 2005). By combined published sources, Kinnaman (2006) estimated the marginal cost of landfilling solid waste in Europe between $5.39 and $8.76 per ton (Kinnaman, 2006). Isely and Lowen (2007) estimated these costs at $5.26 per ton in the United States. Illustrating the composition of the social costs of waste

TABLE 31.1 Private and External Costs of Incineration and Landfilling (€/Ton of Waste)

Private and External Effects	Incineration	Landfill
• Gross private costs	125	40
• Displaced costs—energy	−21	−4
• Displaced costs—materials	−3	0
Net private costs	101	36
• Health	7.09	0.70
• Disamenity	9.09	3.50
• Transport-related	1.67	1.25
• Solid waste	0.11	0.00
• Climate change	0.11	4.21
• Other environmental pollution	0.13	0.52
• Displaced effects—energy	−7.63	−1.14
Total external effects	10.57	9.04
Total social costs	111.57	45.04

Sources: Bartelings et al. (2005) and Dijkgraaf and Vollebergh (2004).

management alternatives, Table 31.1 summarizes the various categories of the private and external costs of landfilling and incineration (Bartelings et al., 2005). On the one hand, private costs are estimated taking into account displacement benefits. On the other hand, externalities are estimated accounting for climate change, other environmental pollution, land use effects, health effects, and disamenity effects.

Both external and private costs are surrounded by large uncertainties. Kinnaman (2009), for example, reported the marginal cost of operating a municipal recycling program at roughly $120 per ton for the first ton recycled, yet these costs decrease with economies of scale by an estimated $2.13 per 1000 t recycled (Kinnaman, 2009). For external costs, the uncertainties are even higher, depending on contextual (e.g. location) and methodological assumptions (e.g. discount rate). The best way to deal with such uncertainties is to conduct an extensive sensitivity analysis to evaluate whether the ranking of preferred waste management options changes with different methodological assumptions (Erikssonn and Baky, 2010).

31.4 ECONOMIC INSTRUMENTS

For most products and materials, 100% recycling is neither feasible nor desirable from a social cost-benefit point of view. Although there may be disagreement on the optimum level of recycling, many citizens and policy makers feel that current levels are still below that optimum. For certain categories of products and waste, targets are set to increase the recycling rates. Some of these even have a binding character. For example, in the EU minimum recycling rates have been determined for packaging waste, batteries, cars, and electronic equipment. Member States are obliged to achieve these targets by specified dates.

One way of stimulating people and companies to recycle more is by making it financially attractive for them to do so. Policy instruments that aim to provide such a financial incentive are called economic (or market-based) instruments.[1]

---

[1] There is no clear distinction between the two terms and they are often used interchangeably (e.g. in EEA, 2005). Possibly the term *economic instrument* refers more to the financial incentive that the instrument conveys, whereas *market-based* emphasizes the role of the market mechanism in achieving the environmental objective. In that sense, instruments such as ecolabeling could also be called "market-based," although they do not provide financial incentives and therefore are not economic instruments.

They can be applied in many different ways and at many different points along the value chain from resource extraction to disposal. Basically, we can distinguish two types of economic instruments for recycling:

- Those that create a financial disincentive for nonrecycling behavior (e.g. dumping waste or sending it to landfills and incinerators);
- Those that create a financial incentive for recycling.

Taxes and charges[2] (but also penalties on littering) are examples of the first category; subsidies and public facilities belong to the second. In practice, mixtures of incentives and disincentives are also common. Examples include, for instance, tradable landfill permits, differentiated tax rates, and deposit-refund systems. The latter will be discussed separately below.

31.4.1 Taxes and Charges

"The polluter pays" is a key principle in environmental policy. In the area of waste prevention and recycling, it can be applied in various ways. Taxes at the end of the chain, on landfilling and incineration, improve the competitive position of recycling *vis-à-vis* these disposal options. Taxes (or charges) on (potentially) waste-generating products make it more attractive to use less of them, to use them for a longer time, and/or to switch to alternatives that generate less waste. A classic example is the tax on plastic bags in Ireland. After its introduction in 2002, the use of plastic bags dropped by 90% (Convery et al., 2007). Taxes can also be applied at the beginning of the chain (i.e. on natural resources), but such taxes appear to be less effective in terms of promoting recycling (see e.g. Söderholm, 2006).

Charges for waste collection and processing can also play a role in stimulating recycling. Many households still pay fixed charge rates for municipal waste collection. Making these rates variable (i.e. related to the amount of waste offered: "pay-as-you-throw") can incite them to display a less wasteful and more pro-recycling behavior. Differentiating the rates (e.g. by applying a reduced or zero rate to separated types of waste such as glass, biowaste, and paper) may further enhance this impact. Such differential and variable rate (DVR)-based schemes have been shown to be very effective (see e.g. OECD, 2006; Oosterhuis et al., 2009; Hogg et al., 2011), especially in the presence of a comprehensive system for the collection of segregated materials for recycling. In some countries, such as South Korea, DVR schemes are already applied on a nationwide scale.

31.4.2 Subsidies and Public Facilities

Although subsidies in general conflict with the "polluter pays" principle, they may still be useful policy instruments in situations where market imperfections (e.g. transaction and monitoring costs) preclude the use of "first best" tools, such as waste disposal fees. For example, Dinan (1993) and Eichner (2005) concluded that the optimum instrument package should include subsidies on reuse or recycling. Likewise, Fullerton and Wu (1998) showed that, if market signals cannot be corrected by means of disposal charges, welfare can be improved by subsidies to recycling or recyclability (in combination with product taxes).

In particular, positive incentives for recycling may play a role in cases where there is a high risk of illegal waste disposal practices, for instance, because the cost of enforcement would

---

[2] The distinction between a tax and a charge is related to the destination of the revenue. Tax revenues tend to accrue to the general public budget (although they may sometimes be earmarked for specific purposes), whereas the revenues from charges are spent on purposes that are directly related to the product or activity on which the charge is levied.

be prohibitively high. An example is ship waste: there are no policemen at sea who might fine for waste dumping. Port reception facilities can make it more attractive for captains to refrain from dumping. Likewise, households can be encouraged to separate their waste by providing a well-organized system of collection facilities (e.g. glass containers). Clearly, the incentive does not need to be a financial one: recycling behavior can also be made attractive by reducing the time and effort involved.

31.4.3 Deposit-Refund Systems

Deposit-refund schemes (DRS) are basically a combination of two instruments: a tax on the purchase of a certain product and a subsidy on the separate collection of the same product in its after-use stage. International experiences show that DRS can achieve very high return rates (Ten Brink et al., 2009; Ecorys, 2011; Van Beukering et al., 2009) and that they do lead to a reduction in litter (Hogg et al., 2011; Ecorys, 2011). On the other hand, the handling and administration costs can be substantial. DRS are widely applied in the area of drinks packaging, mainly on a voluntary basis. Some countries, however, apply mandatory DRS (e.g. Denmark, Germany, and a number of states in the USA).

In principle, there is no reason why DRS should be restricted to drinks packaging. At the theoretical level, several authors have studied the conditions under which DRS can be considered a useful policy instrument to attain maximum welfare and how they should be designed. For example, Aalbers and Vollebergh (2008), using a general equilibrium model, found that DRS can provide the optimal incentives to recycling, landfilling, and dumping, taking into account the possibility of waste mixing and the efforts needed to keep waste streams separated. Calcott and Walls (2005) argued that a deposit-refund should be applied to *all* products, combined with a disposal fee (waste tax).

31.4.4 The Importance of Instrument Mix Design

Economic instruments can play an important role in making recycling a more attractive waste management option and in reducing the attractiveness of other, environmentally less desirable options. Shaping a recycling policy, however, is not a matter of picking one or more instruments from a pre-existing toolbox. In reality, a tailor-made instrument mix will have to be designed that takes into account specific situation characteristics and conditions. These include, among others, the price-responsiveness of the producers and consumers involved (in economic terms: the price elasticity of supply and demand), as well as numerous social, cultural, political and institutional factors. Designing economic instruments for recycling is therefore an art that requires skills and creativity, but which can also benefit from theoretical insights and practical experiences gained by researchers and practitioners.

31.5 CONCLUSIONS AND DISCUSSION

In the first part of this chapter on the economics of recycling, we explained the main economic drivers and developments of recycling. It shows how recycling takes place in a volatile, dynamic, and globalizing world, which complicates the projections in future developments in the recycling sector.

Next, the chapter demonstrated which private and external costs of recycling-related activities need to be taken into account and how these effects can be valued in economic terms. Internalization of externalities leads to better decisions on which waste management policies to pursue. However, a proper valuation study incorporates all externalities across the (international) life cycle. Therefore, valuation remains difficult because of spatial and temporal variations and boundary issues.

Finally, the most important economic instruments to promote recycling are identified and explained. Economic instruments may have a larger role to play in waste policies of most countries in the world. Economic instruments are most effective in changing behavior by targeting the subject as directly as possible. Major immediate effects from economic instruments are limited because "low-hanging fruit" has already been harvested. Real-world conditions (e.g. trade effects, long-term contracts, high costs of recycling) may limit the effectiveness of economic instruments. DVR schemes (differential and variable rates in waste collection charges) are an essential precondition for the effectiveness of many other economic instruments.

References

Aalbers, R.F.T., Vollebergh, H.R.J., 2008. An economic analysis of mixing wastes. Environmental and Resource Economics 39 (3), 311–330.

Abrate, G., Ferraris, M., 2010. The environmental Kuznets curve in the municipal solid waste sector. (Working Paper n 1). Moncalieri, Italy: Hermes, Fondazione Collegio Carlo Alberto.

Ascari, S., Cernuschi, S., 1996. Integration of pollutant dispersion modelling and hedonic pricing techniques for the evaluation of external costs of waste disposal sites. In: Baranzini, A., Carlevaro, F. (Eds.), Econometrics of Environment and Transdisciplinarity, vol. I. Proceedings of the LIst International Conference of the Applied Econometrics Association (AEA), 1–12 April 1996, Lisbon, pp. 156–173.

Ackerman, F., 1997. Why Do We Recycle? Markets, Values and Public Policy. Island Press, Washington DC.

Bartelings, H., van Beukering, P.J.H., Kuik, O., Linderhof, V., Oosterhuis, F., 2005. Effectiveness of Landfill Taxation. IVM. Report R.-05/05. p. 154.

Calcott, P., Walls, M., 2005. Waste, recycling and "design for environment": roles for markets and policy instruments. Resource and Energy Economics 27 (4), 287–305.

Convery, F., McDonnell, S., Ferreira, S., 2007. The most popular tax in Europe? Lessons from the Irish plastic bags levy. Environmental and Resource Economics 38, 1–11.

COWI, 2000. A Study on the Economic Valuation of Environmental Externalities from Landfill Disposal and Incineration of Waste. European Commission DG Environment, Brussels, 43968–43974.

Craighill, A.L., Powell, J.C., 1996. Lifecycle assessment and economic evaluation of recycling: a case study. Resources, Conservation and Recycling 17, 75–96.

Dijkgraaf, E., Vollebergh, H., 2004. Burn or bury? A social cost comparison of final waste disposal methods. Ecological Economics 50, 233–247.

Dinan, T.M., 1993. Economic efficiency effects of alternative policies for reducing waste disposal. Journal of Environmental Economics and Management 25 (3), 242–256.

Ecorys, October 2011. The Role of Market-Based Instruments in Achieving a Resource Efficient Economy. Report for the European Commission, DG Environment. Ecorys, Cambridge Econometrics and COWI, Rotterdam.

EEA, 2005. Market-Based Instruments for Environmental Policy in Europe. Technical Report 8/2005. European Environment Agency, Copenhagen.

Eichner, Th, 2005. Imperfect competition in the recycling industry. Metroeconomica 56 (1), 1–24.

Eriksson, O., Baky, A., 2010. Identification and testing of potential key parameters in system analysis of municipal solid waste management, resources. Conservation and Recycling 54 (12), 1095–1099.

Eshet, T., Ayalon, O., Shechter, M., 2006. Valuation of externalities of selected waste management alternatives: a comparative review and analysis. Resources, Conservation and Recycling 46 (4), 335–364.

Ferrara, I., 2008. Waste generation and recycling. OECD Journal: General Papers 2008/2 (Paris: OECD).

Fullerton, D., Wu, W., 1998. Policies for green design. Journal of Environmental Economics and Management 36 (2), 131–148.

Grace, R., Turner, K., Walter, I., 1978. Secondary materials and international trade. Journal of Environmental Economics and Management 5 (2), 172–186.

Graedel, T.E., Allwood, J., Birat, J.-P., Reck, B.K., Sibley, S.F., Sonnemann, G., Buchert, M., Hagelüken, C., 2011. Recycling Rates of Metals—A Status Report. A Report of the Working Group on the Global Metal Flows to the International Resource Panel. UNEP, Nairobi.

Hogg, D., Sherrington, C., Vergunst, T., 2011. A comparative study on economic instruments promoting waste prevention. Final Report to Bruxelles Environnement. Eunomia, Bristol, November 8, 2011.

Isely, P., Lowen, A., 2007. Price and substitution in residential solid waste. Contemporary Economic Policy 25 (3), 433–443.

ICFPA, 2013. ICFPA Sustainability Progress Report. International Council of Forest & Paper Associations.

Johnstone, N., Labonne, J., 2004. Generation of household solid waste in OECD countries: an empirical analysis using macroeconomic data. Land Economics 80 (4), 529–538.

Jones, N., Evangelinos, K., Halvadakis, C.P., Iosifides, T., Sophoulis, C.M., 2010. Social factors influencing perceptions and willingness to pay for a market-based policy aiming on solid waste management. Resources, Conservation and Recycling 54 (9), 533–540.

Kinnaman, T.C., 2006. Examining the justification for residential recycling. The Journal of Economic Perspectives 20 (4), 219–232.

Kinnaman, T.C., 2009. The cost of municipal curbside recycling and waste collection. Working Paper. Department of Economics, Bucknell University.

Kellenberg, D., 2012. Trading wastes. Journal of Environmental Economics and Management 64 (1), 68–87.

Leach, M.A., Bauen, A., Lucas, D.J., 1997. A system approach to materials flows in sustainable cities: a case study of paper. Journal of Environmental Planning and Management 40 (6), 705–723.

Lyons, D., Rice, M., Wachal, R., 2009. Circuits of scrap: closed loop industrial ecosystems and the geography of US international recyclable material flows 1995–2005. The Geographical Journal 175 (4), 286–300.

Mazzanti, M., Zoboli, R., 2009. Municipal waste Kuznets curves: evidence on socio-economic drivers and policy effectiveness from the EU. Environmental and Resource Economics 44, 203–230.

Nicolli, F., Mazzanti, M., Lafolla, V., 2012. Waste dynamics, country heterogeneity and European environmental policy effectiveness. Journal of Environmental Policy & Planning 14 (4), 371–393.

OECD, 2006. Impacts of Unit-Based Waste Collection Charges. ENV/EPOC/WGWPR(2005)10/FINAL, May 15, 2006. Organisation for Economic Co-operation and Development, Paris.

OECD.Stat, 2013a. Generation of Primary Waste by Sector (Thousand Tonnes) [Online]. http://stats.oecd.org. accessed June 20, 2013.

OECD.Stat, 2013b. Municipal Waste—Generation and Treatment [Online]. http://stats.oecd.org. accessed June 20, 2013.

Oosterhuis, F.H., Bartelings, H., Linderhof, V.G.M., Beukering, P.J.H. van, 2009. Economic instruments and waste policies in the Netherlands. Inventory and options for extended use. IVM report (R-09/01). Institute for Environmental Studies, VU University, Amsterdam.

Porter, R.C., 2002. The Economics of Waste. Resources for the Future, Washington, DC.

Söderholm, P., 2006. Environmental taxation in the natural resource extraction sector: is it a good idea? European Environment 16 (4), 232–245.

Sterner, T., Bartelings, H., 1999. Household waste management in a Swedish municipality: determinants of waste disposal, recycling, and composting. Environmental and Resource Economics 13, 473–491.

Ten Brink, P., Lutchman, I., Bassi, S., Speck, S., Sheavly, S., Register, K., Woolaway, C., 2009. Guidelines on the Use of Market-Based Instruments to Address the Problem of Marine Litter. Institute for European Environmental Policy (IEEP), Brussels, Belgium.

Van Beukering, P.J.H., Linderhof, V., Bartelings, H., Oosterhuis, F., 2009. Effectiveness of unit-based pricing of waste in the Netherlands. Waste Management 29 (11), 2892–2901.

Van Beukering, P.J.H., 2001. Recycling, International Trade and the Environment: An Empirical Analysis. Kluwer Academic Publishers, Dordrecht/Boston/London.

WRI, 1996. World Resources 1996–97. World Resources Institute Oxford University Press, New York.

CHAPTER 32

Geopolitics of Resources and Recycling

Ernst Worrell

Copernicus Institute of Sustainable Development, Utrecht University, Utrecht, The Netherlands

32.1 INTRODUCTION

Traditionally, products and materials were reused and recycled in society. When mass production reduced the costs of materials and products, attention to recycling decreased as it became economically less viable or less important because of rising income. Since the 1980s, however, the increased environmental impacts associated with materials production and waste disposal, and increasing costs of raw materials, also have increased the attention to and role of recycling. Since the publication of the report "Limits to Growth" by the Club of Rome in 1972, increasing costs of raw materials resulting from rapidly increasing demand for materials and consumer products from a growing and more affluent global population also has increased the attention for (future) availability of materials and has led to concerns about supply disruptions. This is called scarcity. The fear of scarcity of resources has fueled political discussions as we have entered a new era of geopolitics. In recent years, this has become evident for the supplies and reserves of strategic and critical materials, such as rare earth metals and other specific metals. Rare earth metals increasingly are used in modern technology, ranging from efficient electric motors to computers and mobile phones, and there are few alternatives. Now, virtually all rare earth metals are mined and produced in China. China has flexed its "rare earth metal muscle" in a territorial conflict with Japan and has threatened buyers that it must produce more value-added products in China to guarantee access to the reserves. Until the 1980s, the United States was the key global supplier of rare earth metals.

Although physical scarcity of materials can be debated (Tilton, 2002), scarcity may be due to various reasons other than physical reserves in the Earth's crust. The rare earths case shows that scarcity may not necessarily mean that stocks are close to depletion but rather that other factors also affect scarcity, including concentration of production in a limited number of geographic locations, limits in mining capacity for so-called hitchhiker elements that mainly are found in combination with other minerals, environmental pollution from producing and refining the metals, and national politics. This issue has come to the forefront as rare earth metals play a key role in information and communications technology as well as in new technologies needed in the transformation toward a sustainable energy system. Renewable and new-generation technology (such as photovoltaics, wind generators, fuel cells), energy storage (e.g. batteries) and highly efficient technologies (e.g. electric motors) need the rare earth metals to function, making access to these materials essential for the future survival of the economy (AEA Technologies, 2010; European Commission, 2010). In summary, scarcity of materials may be the result of the following:

- Depletion of existing reserves, while no new reserves exist or are identified
- High investments in new mining technology may be limited by current (relatively low) prices of these materials
- Access to resources may be limited by natural occurrence in ore bodies of mined minerals (hitchhiker elements)
- Slow response of the mining sector to price dynamics of materials, as the regulatory approval and development of new infrastructure of a mining project may take many years
- Environmental impacts of new mining and production may restrict access or the production of the materials
- Producing countries may limit or tax exports as part of domestic economic strategies

These issues are not necessarily new; in 1981, the U.S. Environmental Protection Agency commissioned a study on critical materials scarcity for air pollution control equipment (e.g. platinum) (CRA, 1981). Technological progress in catalyst loading, platinum recycling and increased mining have helped to secure platinum availability despite strong growth in demand. New, however, is that the scarcity is now felt for many critical materials simultaneously and is combining with other challenges like climate change and water availability at times of unprecedented growth of the consumption of all materials and resources.

Response strategies consist of three elements:

- Resource efficiency. Using the scarce materials as efficiently as possible. For example, the platinum loading of flue gas catalysts has decreased consistently over time, and so too is the platinum need in fuel cell technology.
- Alternative materials. If physical characteristics allow so, alternative materials may be identified that can replace the scarce materials. In the past, this has happened in special circumstances such as war (e.g. replacement of oil by other fuels) or if raw materials have become too expensive (e.g. copper by aluminum in power transmission lines).
- Recovery and recycling of scarce materials. Recycling materials from wastes is a key option to maintain control of the materials and make them available for future applications. In the past, this has been driven mainly by economics and environmental policy, and high recycling rates exist for most metals. For many of the new critical materials, however, no or limited recycling infrastructures exist, or they may be recycled in regions with low labor costs, sometimes the same regions that produce the primary materials.

The new geopolitics of raw materials make reconsidering the role of resource efficiency, as well as the role of recycling in waste management, essential. These geopolitics also are changing waste management from an "end-of-pipe" policy and technique to a tool integrated in the life-cycle management of raw materials throughout the supply chain, next to the environmental and public health considerations of waste management. This means a complete rethinking of our approaches to waste management, which is only slowly taking hold in even the most advanced industrial countries. This chapter briefly discusses resource availability and scarcity, the role of recycling to mitigate the risks of geopolitics around resources and the future role of recycling in resource management.

32.2 RESOURCES, SCARCITY AND GEOPOLITICS

Mineral deposits are not distributed evenly. A *mineral deposit* is any accumulation of a mineral or a group of minerals that may be economically recoverable. The value of a deposit depends on the availability, access and economics. Mineral deposits occur at those locations where geological processes have concentrated specific minerals in sufficient quantities to be mined (often in so-called basins). A *mineral reserve* is the part of the resource that has been evaluated fully geologically and is commercially and legally mineable. Reserves may be regarded as "working inventories", which are revised continually as factors, such as mining and processing technology, change. The *reserve base* includes the mineral reserve plus those parts of the resources that have the potential to become economically viable beyond current technology and economics (European Commission, 2010). Beyond these estimates, less certain and speculative estimates of hypothetical and speculative resources are based on assumptions of known mineral bodies. These resources may be reasonably expected to exist under similar geological conditions or deposits may exist in geological settings where mineral discoveries have not (yet) been made or in types of deposits as yet unrecognized for their economic potential.

The uncertainties associated with resource estimates are large. Reserve estimates have been replenished constantly. Over the past 50 years, the extractive industries sector has succeeded in meeting global demand, and the calculated lifetime of reserves and resources has been extended continually further into the future. For example, the total amount of copper extracted from the Earth is estimated at 320 Mt. This remains small compared with the estimated global reserves of 690—3000 Mt. Reserve estimates over the past 20 years have not changed much. Technological progress in exploring, mining and processing minerals has enabled supply to keep up with demand. Current economically mined copper reserves have a much lower ore grade than ones that were mined previously. Improved mining technology (and higher copper prices) has allowed for exploitation of these reserves and an increase in the reserve estimate. Moreover, the metal is highly recyclable, with recycling rates well over 50% and the annual contribution of secondary input to primary copper production of around 35% — a ratio that has remained fairly constant in the past 10 years despite a huge increase in primary production. Hence, published reserve figures do not reflect the total amount of mineral potentially available, and compilation of global reserve figures are not reliable indicators of long-term availability. Still, these figures often are used to estimate future availability of (scarce) materials. Various studies have documented a large variety of estimates of future availability, which often are driven by factors such as (current) reserve estimates, current and future demand patterns, and other uncertainties. Despite the large uncertainties, these estimates can signal future supply problems.

The key causes for the current debate on scarcity therefore are not found in the global availability of the materials (especially metals), but rather have to be sought in the regional distribution of the material deposits, access to these deposits and national development strategies in developing and emerging economies that aim to reserve their resource base for their exclusive use. This has become increasingly apparent during the past decade with the mushrooming of a variety of government measures, such as export taxes, quotas, subsidies, price-fixing or restrictive investment rules (European Commission, 2010).

32.3 RECYCLING IN THE GEOPOLITICAL CONTEXT

The attention to scarcity has shifted recycling to the foreground as a strategy to become less vulnerable for supply disruptions and increasing prices. Combined with the need to reduce the overall environmental impact of mankind on the world, feeding a material-hungry society has raised the need to change the current linear industrial system to a circular system. In this world, the stocks of materials in our buildings, equipment and infrastructure (also called the technosphere) represents a large stock of materials, next to stocks found in the Earth. This is reflected in such terms as "landfill mining", the "urban mine" (e.g. materials stored in the urban infrastructure and buildings) or the "urban forest" (i.e. paper stored in society). It takes time before the urban reserves can be mined or harvested, however, as buildings are demolished and cars and appliances are scrapped. Furthermore, most stocks are held in the infrastructure of industrial countries, while most demand for the materials is in the developing countries investing in infrastructure. Hence, in the context of scarcity and geopolitics, the two different aspects to recycling are as follows:

- Current or potential future shortages of critical materials, in which recycling represents a key strategy to maintain control of the future supply for a growing need
- Future unbalance of supply and demand of recoverable and recyclable materials

Various studies have explored the need for recycling to be able to supply future needs of critical materials. For example, Roelich et al. (2012) explored different scenarios for the 2035 demand of neodymium (needed for future wind turbines and premium-efficiency motors). The study demonstrated that recycling is essential to meet future demand in scenarios in which the energy system transitions to meet global greenhouse gas emission levels. Wanger (2011) explored the future demand for lithium (e.g. in batteries), potential environmental impacts and restrictions on future production, as well as the need for increased recycling. Current recycling strategies of lithium batteries often focus on cobalt and nickel, and not on the lithium itself, reducing the environmental impact already by half compared with producing primary materials (DeWulf et al., 2010). The likely future prices of lithium will make lithium recovery interesting. In the case of lithium, however, recycling might not be sufficient to meet a rapidly increasing demand over the next decades (Wanger, 2011), making a combined strategy of recycling, resource efficiency and searching for alternatives necessary. Hoenderdaal et al. (2012) showed similar results for dysprosium (e.g. used in permanent magnets and wind turbines). These studies highlighted that currently a small fraction of the critical materials are recycled. Hence, the need is great for early planning and investment in an infrastructure for collection, recovery and recycling of the materials to ensure sufficient time for developing new resources and to increase resource efficiency, even though recycling currently may not be economically attractive. This will require active policy to ensure that an effective and efficient recycling infrastructure is developed (Alonso, 2010; Schüler et al., 2011).

Analyses of stocks and flows of metals have been done for a number of metals, including iron. Iron (and steel) is the most abundantly used metal found in almost every application. Stocks can be broken down into natural stocks and anthropogenic (or in-use) stocks. Natural stocks are deposited in the Earth through geological processes. Anthropogenic or in-use stocks are defined as the total amount of a material contained within commodities used in human society with a potential for future recovery. These include waste rock found in tailings, metal held in strategic stockpiles by governments or companies, and in-use applications — such as buildings, infrastructure, transport and appliances—as well as in landfills. Gerst and Graedel (2008) summarized the literature to estimate in-use stocks and found that although the first articles were written on the subject in the 1950s and 1960s, 70% of all publications have been written since 2000. Global in-use steel stocks reached around 12,000—13,000 Mt in 2005, a figure twice that of just 25 years before (Hatayama et al., 2010; Müller et al., 2011). Of the world total, construction and buildings account for around 60% of in-use stock due to their longevity. Per capita in-use stocks vary among different regions. In Asia per capita in-use stocks are estimated at 1.5 t, and estimated at 5 t/capita in North America, Commonwealth of Independent States, Europe and Oceania. The most rapid growth of steel demand and production, however, will take place in Asia and other developing regions. The world's largest steelmaker, China, is importing large quantities of steel scrap and is expected to meet its own scrap demand only between 2035 and 2050. Steel scrap is traded internationally as any other commodity, but as future costs of iron ore increase, scrap may become a critical material to maintain a competitive national steel industry, and this may lead to export restrictions or taxes. Because in-use stocks are not distributed evenly, this may lead to new future geopolitical conflicts.

References

AEA Technologies, 2010. Review of the Future Resource Risks Faced by UK Business and an Assessment of Future Viability. DEFRA, London, UK.

Alonso, E., 2010. Material Scarcity from the Perspective of Manufacturing Firms: Case Studies of Platinum and Cobalt. Massachusetts Institute of Technology, Cambridge, MA.

Charles River Associates (CRA), 1981. Scarcity, Recycling and Substitution of Potentially Critical Materials Used for Vehicular Emissions Control. Prepared for the US Environmental Protection Agency, Ann Arbor, MI.

DeWulf, J., van der Vorst, G., Denturck, K., 2010. Recycling rechargeable lithium-ion batteries: critical analysis of natural resource savings. Resources Conservation & Recycling 54, 229—234.

European Commission, 2010. Critical Raw Materials for the European Union: Report of the Ad-hoc Working Group on Defining Critical Raw Materials. European Commission, Brussels, Belgium.

Gerst, M., Graedel, T.E., 2008. In-use stocks of metals: status and implications. Environmental Science & Technology 42, 7038—7045.

Hatayama, H., Daigo, I., Matsuno, Y., Adachi, Y., 2010. Outlook of the world steel cycle based on the stock and flow dynamics. Environmental Science & Technology 44, 6457—6463.

Hoenderdaal, S., Espinoza, L.C., Marscheider-Weidemann, F., Graus, W., 2012. Can a dysprosium shortage threaten green energy technologies? Energy 49, 344—355.

Müller, D.B., Wang, T., Duval, B., 2011. Patterns of iron use in societal evolution. Environmental Science & Technology 45, 182—188.

Roelich, K., Kemp Benedict, E., Chadwick, M., Dawkins, E., 2012. Resource Scarcity, Climate Change and the Low Carbon Economy. Stockholm Environmental Institute (SEI), York, UK.

Schüler, D., Buchert, M., Liu, R., Dittrich, S., Merz, C., 2011. Study on Rare Earths and Their Recycling. Öko-Institut, Darmstadt, Germany.

Tilton, J.E., 2002. On Borrowed Time? Assessing the Threat of Mineral Depletion. Resources for the Future, Washington, DC.

Wanger, T.C., 2011. The lithium future — resources, recycling and the environment. Conservation Letters 4, 202—206.

CHAPTER 33

Recycling in Waste Management Policy

Ernst Worrell

Copernicus Institute of Sustainable Development, Utrecht University, Utrecht, The Netherlands

33.1 INTRODUCTION

Traditionally, waste management policy has been directed at the sanitary disposal of waste at the lowest direct costs for the stakeholders. This has, for example, resulted over the past decades in the shift from landfills to sanitary landfills and waste incineration. However, in pre-industrial societies, when materials were relatively expensive, earlier waste management strategies build upon formal and informal routes of recycling recovered most of the recyclables. This more traditional organization of waste management is still found in developing countries where scavengers or waste-pickers are an important factor in the recycling infrastructure of those countries, albeit often part of an informal economy. Also in times of stress (e.g. economic crisis, war) there is an increased focus on recycling. In fact, during World War II, recycling became patriotic, and countries used massive campaigns to urge more efficient use of resources and recycle. Evidence exist of recycling through re-melting of glass and bronze in societies as early as 400 BC. In more modern times even money was impossible without recycling, as many banknotes consisted of cotton from recycled clothing. Recycling was then economically more attractive than production from primary resources, also due to the allocation of costs and the negligence of external costs of materials production and waste management. Only, when low-cost mass production techniques reduced the costs of materials and products since the industrial revolution, we see a decreasing attention for recycling, as it became economically less attractive.

In this period, waste management strategies focused mainly on sanitation and public health since the growth of cities in the industrial revolution. Since the 1970s the increased environmental impacts associated with materials production and waste disposal, and rising energy costs, have shifted the focus of waste management policy toward an increased role for recycling. More recently, increasing costs of raw materials due to rapidly increased demand from a growing and more affluent global population, and the resulting geopolitics (see Chapter 32) has shifted recycling to the foreground in waste management. Combined with the need to reduce the overall environmental impact of mankind on the world and "feeding a material-hungry" society, this has raised the need to change the current linear industrial system to a circular system. In such a world, the stocks of materials in our buildings, equipment and infrastructure (the "technosphere") represents a large stock of materials, next to stocks found in the earth, and aptly named the "urban mine". This stock can be managed in a similar way as primary resources, opening ways to change from a linear system of material and waste management toward a circular economy, mimicking natural ecosystems. This chapter discusses the changing role of recycling as part of (solid) waste management, starting with a brief history, followed by a discussion of recycling as the key to and an integral part of modern waste management. This chapter excludes special or hazardous waste management.

33.2 A BRIEF HISTORY OF WASTE MANAGEMENT

As society industrialized its production and simultaneously saw an increasing share of the population moving into urban areas (as labor became more and more centralized), consumption of materials and resources increased dramatically. As over 90% of the consumed materials and resources end up as waste, this massive increase in materials production and use (see Chapter 1) also generated an increasing waste stream. Due to the increased concentration of consumers, this led to local waste disposal problems, which threatened public health and the local environment. In this era, regulation and policy was hence driven mainly by concerns over pollution impacting human health, with little if any attention paid to resource efficiency, and also (initially) without consideration of broader environmental impacts. Hence, waste management was framed as a technical, end-of-pipe solution to collect and dispose of waste, primarily by landfilling. This was later expanded to include incineration, with the first incinerator build in 1874 in Nottingham (United Kingdom).

In the 1960s, increasing concerns on environmental damage due to our industrial society, led to a change in emphasis toward sanitary landfilling to reduce the impact of landfilling through leaching of hazardous pollutants and other negative impacts (e.g. odor). Simultaneously, as space was becoming more expensive in large urban areas and costs of waste transport and disposal increased, incineration became an economically viable technique to reduce the volume of waste and hence the costs of final disposal.

As energy costs increased in the early 1970s, energy recovery from incineration gained more attention, while in this era also recycling gained further attention, as recycling reduced the energy costs of producing materials while simultaneously reducing the costs of waste management. Waste-to-energy facilities were especially developed in those countries with high costs of landfilling (e.g. dense populations in industrialized countries) and energy, and the capacity of this technique is still growing around the world as urbanization and consumption increases.

This has led to including more environmental and sustainability concerns in designing waste management strategies. Both in Europe and North-America this has led to introducing a waste management hierarchy. The European Union formalized this in 1975 in a Waste Framework Directive. While the hierarchy has been named differently in various countries, e.g. the "3 R's", the "waste ladder", these are essentially similar around the world:

- *Reduce*: preventing the generation of waste through product design
- *Reuse*: reuse of products (e.g. refillable containers)

- *Recycle*: reuse of materials contained in products
- *Recover energy*: incineration in waste-to-energy or industrial (e.g. clinker kilns) facilities
- *Disposal*: sanitary landfilling

This hierarchy forms the basis for modern waste management policy, although in practice the hierarchy is often not observed in the design and implementation of (local) waste management programs. Around the world, progress is slow, as by 2010 still 38% of all municipal solid waste was landfilled. In 2010, European municipal waste landfilled (38%), incinerated (22%), recycled (25%) and composted (15%). Within the EU recycling rates still vary dramatically, from a high in Germany (45% of waste treated) to lows in the single digits (e.g. 1% in Romania, and 4% in Slovakia). In the United States, about 54% of municipal solid waste was landfilled, 12% incinerated, 26% recycled and 8% composted (2010).

For recycling, a distinction can be made between high- and low(er)-quality recycling. It is assumed that high-quality recycling results in replacing (part of) the primary product or material by reused or recycled materials, i.e. the more of the primary product or material is replaced, the higher the quality of recycling. Alternatively, the recycled material may replace a different material, which is considered low-quality recycling. In the latter case, the characteristics of the replaced materials should be used to determine the environmental impact of recycling. This type of recycling is generally referred to as downcycling, as the use of the material or product is often of a lower quality and functionality than when the original primary material is replaced. However, in current statistics no information is available on the nature of the recycling process and the degree to which virgin or primary materials are substituted by the recovered and recycled material.

33.3 INTEGRATING RECYCLING IN WASTE MANAGEMENT POLICY DESIGN

Modern integrated waste management involves a wide array of suitable processes and programs including source reduction (prevention), recycling, transformation (e.g. incineration, composting) and final disposal (e.g. landfilling), to meet multiple objectives including public and environmental health, and (increasingly) a more sustainable and resource efficient society. The European Commission has recognized the importance of resource efficiency and made it one of the seven flagship initiatives that are part of the Europe 2020 strategy that aims to deliver sustainable, smart and inclusive growth. The focus on resource efficiency should help to achieve the European Union's targets on reducing greenhouse gas (GHG) emissions, improving the security of supply of raw materials, and make the European economy more resilient to price increases of energy and commodities (European Commission, 2011). Various ways to use resources more efficiently include prevention, reuse and recycling, while incineration and anaerobic digestion may recover part of the embodied energy of materials. The strategic goals put an increasing focus on the first three steps of the waste treatment hierarchy.

For example, The Netherlands is, as the rest of the world, at the beginning of a transition to a more sustainable production-, consumption- and energy system. Waste management is a key element in achieving sustainable resource management. The Netherlands has a long history in (research on) waste-to-energy and saving resources, and has been successful in the past to recover materials from waste. Waste and resources management was also central to the development of the second National Waste Management Plan (Landelijk Afvalbeheerplan, LAP2) in the Netherlands (VROM, 2010), which is implemented for the period 2009–2021. The LAP2 aims to:

- Limit the total waste volume to 68 Mt in 2015 and 73 Mt in 2021 (60 Mt in 2006);
- Increase waste recovery to 85% in 2015;
- Increase municipal waste recovery from 51% in 2006 to 60% in 2015;
- Maintain recovery of construction and demolition waste at the 2006 level of 95%.

The LAP2 should also contribute to the reduction of GHG emissions as set out in national policy, and will try to achieve this by focusing on recycling, anaerobic digestion and incineration. A model calculation by Corsten et al. (2013) explores various scenarios to evaluate the impact of increased recycling efforts beyond the current policy goals. The explorative analysis of waste processing shows that current waste management in the Netherlands already contributes a lot (in 2010, 33% of municipal solid waste was recycled) to recover materials, saving energy and reducing CO_2 emissions, compared to a situation where waste is not recycled or incinerated. While incineration is important in The Netherlands the largest contribution in GHG emissions mitigation actually comes from recycling. In the 2008 reference situation, waste processing contributes to a substantial reduction in energy use by over 106 PJ and a reduction of 4.5 $MtCO_2$ emissions per year. About 70% of the energy savings are due to current recycling processes and 30% due to incineration with energy recovery. The currently mitigated CO_2 emissions are solely due to recycling and use of wastes as refuse derived fuel (RDF). The incineration of waste creates additional CO_2 emissions, despite the avoided emissions in electricity generation, which is a result of the large volume of plastics in the waste.

When recycling of selected materials from the waste stream is increased, this can result in additional emission reductions of 2.3 $MtCO_2$ annually, compared to the reference situation in 2008. This is equivalent to a potential improvement of more than 45%. Figure 33.1

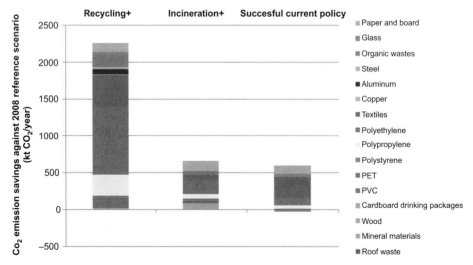

FIGURE 33.1 CO_2 emission reduction potentials of three scenarios for waste management in the Netherlands in 2008. The Recycling+ scenario assumes increased recycling. The Incineration+ scenario assumes similar recycling targets as current policy, while remaining waste is incinerated in high-efficiency waste-to-energy facilities. The successful policy scenario assumes successful implementation of the LAP2 plans. *Corsten et al., 2013.*

depicts the results for the three scenarios. The main contributors to the CO_2 emission reduction potential are found in the optimization of the recycling of plastic (PET, PE/PP), textiles, paper and organic waste.

The study by Corsten et al. shows that, despite many decades of waste management policy, there is still a potential for further improvement of the sustainability of waste management in the Netherlands. Especially in a scenario focusing on product and material reuse and high-quality recycling (*Recycling+*), large potential savings are identified. The key materials that will play an important role in achieving the full savings potential are plastics (PET, PE, PP), textiles, paper and organic waste. A scenario focusing on incineration with higher energy conversion efficiency (*Incineration+*) has the potential of only saving one third of the CO_2 emission savings achieved in the *Recycling+* scenario. The results confirm that, in terms of energy consumption and CO_2 emission reduction, the waste hierarchy as used for the basis of European waste management policy, is still valid in prioritizing waste disposal options, but only when a clear distinction is made between high- and low-quality recycling. The different qualities of recycling should be taken into account when designing (national) waste management policy.

The results of the case study of The Netherlands are not unique. Other analyses and models (see e.g. Morrissey and Browne (2004) for an older review) for other countries show similar results, even though system boundaries (e.g. types of wastes included) may vary in the studies. For example, a study comparing different life-cycle analyses of managing paper waste has shown that recycling results in larger environmental benefits when compared to incineration or landfilling (Villanueva and Wenzel, 2007), and which has been confirmed for emissions of greenhouse gas emissions by a study of Laurijssen et al. (2010) using extended system boundaries, including recycling, incineration and potential uses of the virgin timber when paper is recycled.

A comprehensive and integrated waste management policy is necessary to achieve the full potential energy savings and CO_2 emission reductions identified, which should give priority to high-quality recycling. This is best achieved with a menu of policy instruments, accounting for the specific characteristics of a waste stream and recycled materials markets.

References

Corsten, M., Worrell, E., van Duin, A., Rouw, M., 2013. The potential contribution of sustainable waste management to greenhouse gas emission reduction in the Netherlands. Resources Conservation & Recycling 77, 13–21.

European Commission, 2011. A Resource Efficient Europe—Flagship Initiative under the Europe 2020 Strategy. European Commission, Brussels, Belgium.

Laurijssen, J., Marsidi, M., Westenbroek, A., Worrell, E., Faaij, A., 2010. Paper and biomass for energy? The impact of paper recycling on biomass availability, energy and CO_2 emissions. Resources, Conservation & Recycling 54, 1208–1218.

Morrissey, A.J., Browne, J., 2004. Waste management models and their application to sustainable waste management. Waste Management 24, 297–308.

Villanueva, A., Wenzel, H., 2007. Paper waste – recycling, incineration or landfilling? A review of existing life cycle assessments. Waste Management 27, S29–S46.

VROM, 2010. Landelijk Afvalbeheerplan 2009–2021 – Naar Een Materiaalketenbeleid. 'National Waste Management Plan 2009–2021 – towards a Material Chain Policy'. Ministry of Housing, Spatial Planning, and the Environment, Netherlands.

CHAPTER 34

Voluntary and Negotiated Agreements

Ernst Worrell

Copernicus Institute of Sustainable Development, Utrecht University, Utrecht, The Netherlands

34.1 INTRODUCTION

Voluntary or negotiated agreements (VA/NAs) are the third-generation policy instruments applied by governments after regulation and economic instruments. In the 1990s these have become the instrument of choice in many member countries of the Organisation for Economic Management and Co-operation (OECD) for environmental policy, ranging from public health and greenhouse gas mitigation to waste management. In waste management and recycling, extended producer responsibility (EPR) is a special form of NA that has gained ground in many OECD countries. A VA/NA is an agreement between (selected) industries or sectors and governments to achieve a mutually agreed goal within a set time frame. In turn, the government will enact policies to support the participants to the agreement, or it may refrain from alternative policies (e.g. taxation). Agreements can take various forms:

- Unilateral commitments are agreements with firms that are recognized by the government.
- VAs are programs offered by a government to which a company can commit.
- NAs are the result of negotiation between the government and firms or sectors (or associations representing the firms).

The agreements vary also in the light of the goals set; where the first type generally has targets that are little beyond business-as-usual practices. All three types of agreements are found in environmental policy. The advantages of VA/NAs over more traditional policy may include the following:

- They are relatively quick to enact.
- They provide flexibility and the ability to minimize costs to achieve the target.
- They reduce public costs for administering the policy.
- They build trust between the parties to the agreement (but also needed as basis for the agreement).
- They enable an environment for learning by all parties involved.

VA/NAs also have potential disadvantages, such as the lack of enforcement at nonattainment of the targets, the potential for free riders and the potential to use the agreement to postpone or eliminate future policies. Section 34.2

discusses experiences with selected VA/NAs for recycling (or of which recycling was a specific target); Section 34.3 outlines the main lessons learned from experiences in the design, implementation and evaluation of environmental VA/NAs.

34.2 EXPERIENCES IN RECYCLING POLICY

Various countries since the 1990s have introduced VA/NAs to improve recycling of selected waste streams. Many of these also can be categorized as EPR, as the agreements have formed the basis for current EPR practices, and most of these agreements involved producers to manage the end of life of products. The first VA/NAs were aimed at general waste management and at specific product categories with a high environmental impact (e.g. batteries), electronics (Waste Electrical and Electronic Equipment, WEEE), cars (end-of-life vehicles, ELV) and packaging wastes. Table 34.1 provides an incomplete overview of selected VA/NAs that included goals to promote recycling.

In some countries and regions, VA/NAs have become the instrument of choice to manage waste streams (e.g. Flanders, The Netherlands). These countries generally have followed a model of close collaboration among the different stakeholders, providing a fruitful basis for VA/NAs. For example, the Netherlands has used a series of subsequent NAs to manage packaging

TABLE 34.1 Selected VA/NAs Directed at Recycling, by Country

Country	Agreement	Product or Material	Agreement Period
Australia	National Packaging Covenant	Packaging	1999
Belgium	Bebat	Batteries	1995
	Recytyre	Tires	1998
	Febelauto	Vehicles (End of Live Vehicles, ELV)	1999
	Recupel	Electrical and electronics	2001
France	EcoEmballage	Packaging	
Germany	DSD	Household packaging	
		Vehicles (ELV)	
Netherlands	Packaging Covenant	Packaging	1991
		Batteries	
		Electrical and electronics	
United Kingdom		Plastic films in agriculture	
United States	WasteWise	Waste management	1992
	Plug in to e-cycling	Electronics	2003

ELV, end-of-life vehicles.

waste since the 1990s. In 1988, the Policy Memorandum Prevention and Reuse of Waste Materials set targets for the priority groups and started a series of policies to reduce packaging and packaging waste in the Netherlands. Shortly after the Policy Memorandum, around 300 companies and the government agreed to a VA (Packaging Covenant I), which exempted producers and importers from individual company responsibility and future national waste policies. The European Commission also introduced its first Packaging Directive to harmonize packaging regulations among member states, ensure high environmental protection levels, guarantee the internal market forces and reduce trade barriers. The implementation of the European Directive in Dutch national regulations was done by introducing the Directive on Packaging and Packaging Waste, which became active on August 1, 1997. From this point on, five policy periods can be distinguished in Dutch packaging policy history: three VAs (1991–2005) and two agreements implemented within existing regulation (2006–2012) (see Table 34.2) (Rouw and Worrell, 2011).

- **Packaging Covenant I (PC I)**. To stimulate packaging prevention, the target of the first covenant was to reach 3% less packaging newly introduced on the market in 1997 and 10% less packaging material in 2000, both compared with 1986.
- **Packaging Covenant II (PC II)**. The first prevention and recycling was relatively easily achieved. A new covenant was implemented by the end of 1997. This covenant aimed to realize the targets and agreements made in the Packaging Directive. The amount of packaging newly on the market in 2001 was allowed to grow with a factor equal to 90% of gross domestic product (GDP) using 1986 as the reference year. The Packaging Covenant II included all companies, except those that had fewer than four employees or produced less than 50 t of packaging material.
- **Packaging Covenant III (PC III)**. The third covenant was signed in December 2002. The growth of packaging material use was not allowed to exceed more than two-thirds of GDP growth in 2005, compared with the new reference year 1999. It contained targets to accomplish the reduction of litter.

TABLE 34.2 Overview of Packaging Waste Policy History in the Netherlands

Date	Dutch Policy	Date	European Policy
1988	Memorandum Prevention and Reuse of Waste Materials	1985	European Directive (85/339)
1991–1997	Packaging Covenant I	1994	European Directive (94/62)
1998–2002	Packaging Covenant II		
2003–2005	Packaging Covenant III		
2006–2007	Decree on Packaging and Paper and board Management		
2007–2012	Framework Agreement		
2008–2009	Packaging Tax (>15,000 kg)		
2010–present	Packaging Tax (>50,000 kg)		

Source: Based on Rouw and Worrell (2011).

At the time that the third packaging covenant was introduced, the government was preparing regulatory measures to safeguard the accomplishment of the packaging targets. The effect of the regulatory measures is said to be limited (Schroten et al., 2010).

- **Packaging Decree.** The government doubted the implementation of a subsequent covenant, but the European Commission eventually pressed to realize the goals set out in the Decree on Packaging and Paper and board (i.e. the Packaging Decree). Nedvang, a collective organization of producers and importers organizing collection and processing of packaging waste, was assigned as the executive organization to implement the Decree. An important change was that companies became individually responsible for prevention, collection and recycling of their packaging products brought to market (or so-called Extended Producer Responsibility). Financial responsibility was introduced to realize the "polluter pays principle". The fiscal incentives, however, had little impact on the packaging industry to change production methods considerably, as it did not affect price-sensitive or change-demanding packaging types (e.g. by exempting business-to-business packaging and return-trip packaging, or by taxing only the main component in composites) (Schroten et al., 2010).
- **Framework Agreement.** After the implementation of the Packaging Decree, it became obvious that the packaging industry was not able to implement measures to collect and process domestic packaging waste without the cooperation of municipal authorities. Therefore, the packaging industry, the Association of Dutch Communities (VNG) and the government signed a Framework Agreement in 2007, for packaging waste generated by households. Packaging companies still need to fulfill the requirements of the Packaging Decree.

All the agreements included recycling targets. Total recycling targets (see Table 34.3) set during PC I were both met (see Figure 34.1). Subsequent targets have never been met. The curve depicts an increasing recycling percentage until 1999, when recycling starts to decline and stabilizes at 58%. Despite a lower target of 65% for the Packaging Decree, recycling targets were not met. Total recycling by households during the period 1993–2001 fluctuates around 40%. Except for the first target (40% in 1995), total recycling targets were not achieved for household packaging wastes. On the other hand, some companies demonstrated an increasing recycling rate, so that targets during PC I and II were achieved.

Paper and Board recycling targets have not been met during PC II and III. Only in the years 1995–1997 targets have been met. The Paper Recycling Foundation reported that the 75% recycling target in 2008 has been achieved. While recycling targets decreased by 10% in PC III, subsequent recycling increased during following policy periods. Households did not achieve the targets during PC I (60%) and PC II (85%). Companies did achieve targets from the year 1995.

Glass recycling targets have been very high since the beginning of the implementation of packaging policy measures in the Netherlands. Figure 34.1 shows that these targets were not met during the analyzed policy periods. Glass recycling needs to increase by at least 10%, as 90% recycling is required by 2012. After 1997, household recycling declined while company recycling rapidly increased.

Metal recycling doubled from nearly 40% in 1993 to 80% in 1998. The PC I target (75%) is achieved in the end of the covenant. Targets during the next two covenants are reached almost every year, although recycling targets have increased by 5% every covenant. During the Decree and the Framework, metal recycling targets remain stable at 85%. Companies realized high metal packaging recycling rates. Households realized much lower recycling rates. Recycling numbers from households increased to 66% in 2001.

Plastic recycling targets changed a lot over time (see Table 34.3). Targets were never met

TABLE 34.3 Overview of Packaging Policy Targets in the Netherlands, Defined as Reduction or Increase in Recycling Rate Based on Total Mass of Packaging Material[1]

		PC I (1991–1997)	PC II (1998–2002)	PC III (2003–2005)	Decree (2006–2007)	Framework (2008–2012)
Prevention	kilotons	2063 (1997)	2128 (2001)	2676 (2005)	—	—
		1914 (2000)				
Recycling						
Total	%	40 (1995)	65 (2001)	70 (2005)	65 (2006)	65 (2008)
		60 (2000)				70 (2010)
Paper and board[2]	%	60	85	75	75	75
Glass	%	80	90	90	90	90
Metals	%	75	80	85	85	85
Synthetic	%	50	27	43	95/55/27[4]	38 (2009)
			35[3]			42 (2012)
Wood	%	—	15	25	25	25
Recovery	%	—	—	73 (2005)	70 (2006)	70 (2008)
						75 (2010)
Disposal	kilotons	—	940 (2001)	850 (2005)	—	—

[1] The European Directive requires a minimum packaging recovery of 50% and a maximum of 65%. It was decided that between 25 and 45% of the recovered packaging material needs to be recycled. A minimum recycle target of 15% per specific packaging material is required.
[2] Also beverage cartons are included since Covenant III.
[3] "Obligation of effort" means that this percentage can be reached but is not obligatory.
[4] 95% material recycling of synthetic drinking packaging >0.5 l; 55% material recycling of synthetic drinking packaging ≤0.5 l; 27% material recycling of other synthetic packaging; 45% recovery of other synthetic packaging.

during any of the policy periods. Figure 34.1 shows that recycling is growing slowly, increasing by approximately 1% per year. The recycling percentage of plastic packaging materials by companies is low compared with the recycling of other packaging materials. The total recycling rate of plastics remained less than 20%. This is explained by the minimal recycling by households. The campaign Plastic Heroes and the obligation for local authorities to set up separate collection of plastic packaging from households likely will increase plastic recycling by households.

34.3 LESSONS LEARNED

The example of the subsequent NAs to manage packaging waste and recycling in the Netherlands shows mixed results. In some periods targets for some materials were achieved, but they were not achieved for others. Similarly, evaluations of VA/NAs show that results have been varied, with some programs appearing to just achieve business-as-usual results. The effective agreements are thought to have long-term impacts, including changes of attitudes and awareness, addressing barriers to technology

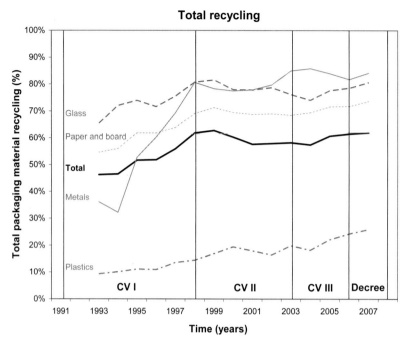

FIGURE 34.1 Achieved recycling rates of packaging materials for the period 1993–2007 over the five agreement periods. The achieved recycling rates can be compared to the targets in Table 34.3 for each material and agreement.

adoption and innovation (although the latter has not been proven in the literature), promoting positive interactions among different stakeholders and facilitating learning. The most effective agreements are those that are legally binding, set realistic targets, include sufficient government support and include a real threat of increased government regulation or taxes if targets are not achieved (e.g. Krarup and Ramesohl, 2000). This demonstrates that VA/NAs should be designed carefully. Various authors have evaluated and compared the design and implementation of environmental VA/NAs in various OECD countries (e.g. Bressers and de Bruijn, 2005; De Clerq and Bracke, 2005; Price, 2005). On the basis of these evaluations, a number of general lessons learned can be distilled. Note, however, that it is the combination of factors that makes a VA/NA a success, not an individual factor (De Clerq and Bracke,

2005). The key success factors supported by evaluations are as follows:

- The **policy climate** should be conducive for a VA/NA, building on a basic level of trust among the stakeholders. The sector should be ready to take on the responsibilities of the agreement. Inviting nongovernmental organizations to participate in the negotiations can help to build a climate of trust between all stakeholders. All major disagreements should be solved within the negotiation process.
- The **negotiation position** of the government should be strong, both before and after the negotiation process. This includes a serious threat of regulation (e.g. tax or otherwise) before the negotiations and also after implementing the agreement in the case of noncompliance. This also means that the

government is well informed on the opportunities to achieve certain targets.
- The sector has the potential to **negotiate as one sector** to reduce the risk of nonparticipation by individual companies and reduce the risk of free riders. It seems that agreements with a smaller number of participants are more successful than others with hundreds of participants.
- The VA/NA should have **clear targets, timetables and responsibilities** assigned to all parties in the agreement. The economic efficiency of the VA/NA depends on the flexibility to achieve the targets. This also will affect the timetable and how targets are defined.
- There is a **potential for market success** for the participants in the agreement. This includes long-lasting government support for the policy goals set out in the VA/NA.
- Monitoring and evaluation receives serious attention in the negotiation and implementation of the VA/NA. This means that clear unambiguous monitoring guidelines are established and that the results are verified independently.

References

Bressers, H., de Bruijn, T., 2005. Environmental voluntary agreements in the Dutch context. In: Croci, E. (Ed.), The Handbook of Environmental Voluntary Agreements. Springer, Dordrecht, The Netherlands, pp. 261–282.

De Clerq, M., Bracke, R., 2005. On the assessment of environmental voluntary agreements in Europe. In: Croci, E. (Ed.), The Handbook of Environmental Voluntary Agreements. Springer, Dordrecht, The Netherlands, pp. 239–260.

Krarup, S., Ramesohl, S., 2000. Voluntary Agreements in Energy Policy—Implementation and Efficiency. AKF Forlaget, Kobenhavn, Denmark.

Price, L., 2005. Voluntary Agreements for Energy Efficiency or GHG Emissions Reduction in Industry: An Assessment of Programs around the World. Lawrence Berkeley National Laboratory, Berkeley, CA (Report LBNL-58138).

Rouw, M., Worrell, E., 2011. Evaluating the impacts of packaging policy in the Netherlands. Resources, Conservation and Recycling 55, 483–492.

Schroten, A., Nelissen, D., Bergsma, G.C., Blom, M.J., 2010. De milieueffecten van de verpakkingenbelasting. CE, Delft, The Netherlands.

CHAPTER 35

Economic Instruments

Maarten Dubois[1], Johan Eyckmans[2]

[1]Policy Research Center for Sustainable Materials, KU Leuven – University of Leuven, Leuven, Belgium;
[2]Center for Economics and Corporate Sustainability (CEDON), KU Leuven – University of Leuven, Campus Brussels, Brussels, Belgium

35.1 INTRODUCTION

Over the past 10 years, the focus in waste management has shifted from solving environmental impacts of waste disposal to managing materials' cycles in a sustainable way. Recycling materials from waste streams plays an important role in the new paradigm of sustainable materials' management and resource efficiency. In particular, it is argued that recycled materials can substitute for the extraction of exhaustible resources, thereby alleviating problems like materials' scarcity and dependency. In addition, they can reduce environmental impacts related to mining, processing and production of virgin materials (Damgaard et al., 2009; European Commission, 2011; Acuff and Kaffine, 2013). This chapter will focus on economic drivers and impediments to recycling. In particular we will discuss how recycling incentives of consumers and producers are affected by policy instruments like recycled content standards, taxes on virgin materials, waste disposal fees, recycling subsidy mechanisms, extended producer responsibility (EPR or deposit-refund in the sequel) and recycling certificates schemes.

35.2 CRITERIA TO COMPARE POLICY INSTRUMENTS

Typically in the environmental economics literature, policy instruments are compared with respect to their achievements on different criteria such as efficiency, effectiveness, equity or political feasibility (see among others Bohm and Russell, 1985; Callan and Thomas, 2000 or Hanley et al., 2007 for a more elaborate discussion and definition of these criteria). This chapter will focus on three fundamental notions of efficiency. First, the idea of efficiency refers to whether a particular policy instrument is capable of achieving the welfare optimal allocation of resources (notion of *allocative efficiency*). This notion of efficiency includes internalization of all relevant environmental external costs and the allocation over time of nonrenewable resources (i.e. the extraction path of the resource through time). Second, does the policy instrument induce a cost-efficient allocation of efforts if actors are characterized by different costs for sorting and recycling (the idea of *cost* or *static efficiency*)? In order to achieve a given target at the lowest possible cost, efforts should be differentiated. In particular, actors with low costs should contribute more than their counterparts. Third, does the instrument foster *dynamic efficiency* (i.e. does it give incentives over time to improve recycling technologies and upstream green product design)? This chapter will review some of the most important contributions to the environmental and resource economics literature, focusing in particular on the interaction of different policy instruments and their effects on recycling behavior.

35.3 BASIC ENVIRONMENTAL POLICY INSTRUMENTS AIMED AT STIMULATING RECYCLING

Dinan (1993) was one of the first to use a stylized theoretical economic model to compare efficiency of three economic instruments that aim to internalize waste disposal externalities: a *tax on virgin materials*, a *deposit-refund* scheme (combination of a consumption tax and reuse subsidy), and a household *disposal fee*. The household disposal fee is a weight-based fee paid by households to collect waste. Externalities resulting from virgin materials extraction or production are not taken into account. In the model with perfect competition and with two goods that have different constant unit disposal costs (private and external cost), the deposit-refund scheme is shown to be the most efficient. The virgin material tax falls short because it does not allow differentiation between products that give rise to different waste management costs. In a similar way, the household collection fee does not allow differentiation between disposal costs for different products or material streams.

The article by Fullerton and Kinnaman (1995) is generally considered the seminal contribution in the environmental economics literature on the comparison of policy instruments for waste management and their effect on *illegal disposal*. They modeled consumption of a single consumption good that, at its end-of-life stage, can be recycled, disposed of in a controlled way or disposed of in an illegal way. Illegal disposal is assumed to generate the most adverse environmental impact. The model takes into account the environmental externalities of virgin material extraction but abstracts from import/export flows (closed economy). Consumers are assumed to be identical utility maximizing subjects faced with a limited budget. Firms produce consumption goods using a constant returns to scale technology (i.e. unit production costs are constant and independent of production volume) under perfect competition. The model shows that due to the presence of illegal disposal, the social welfare maximizing outcome cannot be achieved by household *disposal fees* only. If the disposal fee were set at a level to correctly internalize all environmental externalities of disposal, the so-called Pigovian externality tax in environmental economics (see for instance Porter, 2002; Hanley, et al., 2007 or Kolstad, 2011), households would face strong incentives to dispose of their waste in an illegal and uncontrolled way. If, instead, households were not charged for disposing their waste, they would not face the full social cost of their actions and tend to overconsume wasteful products. In this case, a *consumption tax* is required to compensate for the implicit consumption subsidy for waste-generating goods. Although an *extraction tax* can internalize the externalities of virgin materials extraction, it should not be used to internalize disposal externalities as long as a consumption tax is feasible. Overall, an important message of this article is that recycling behavior can be affected by several waste management policy instruments that can apply at different stages of materials' lifecycle. The net effect is often difficult to predict because of the interacting instruments, externalities and the possibility of illegal dumping.

Palmer et al. (1997) use a partial equilibrium economic model to assess cost efficiency of three different economic instruments aimed at reducing the amount of disposed waste: a *deposit refund*, an *advanced disposal fee* and a *recycling subsidy*. The model uses mass balances for material flows used in consumption, recycling and disposal. Only products with short lifecycles are taken into account. In order to avoid illegal disposal, the household disposal fee is set equal to zero. A numerical simulation shows that the deposit-refund schemes are most desirable because the refund succeeds in stimulating the sorting and recycling in a cost-efficient way, while the consumption tax reduces incentives for consumption of waste-generating products. As an advance disposal fee works in the same way as a consumption tax or deposit, it effectively reduces consumption. However, if sorting and recycling are not stimulated, the incentive to reduce waste disposal is smaller. Finally, the recycling subsidy in combination with low or even zero household fees for waste collection is shown to be least cost efficient. Although sorting and recycling of waste may be organized in an efficient way, consumption of waste generating products is implicitly subsidized. In order to reduce total disposal, the subsidies for recycling should be increased to an excessive level.

Using a partial equilibrium model with identical consumers and perfect competition between identical firms, Palmer and Walls (1997) compared efficiency of *deposit-refund systems* with *recycled content standards* (i.e. standards that impose a fraction of recycled material in new products). Both firms and consumers can dispose waste at constant unit disposal costs (including private and environmental disposal costs). The model uses a mass balance approach for virgin materials going in and disposed materials going out. The model confirms that deposit-refund systems can achieve the social welfare maximizing outcome. In contrast, recycled content standards fail in four aspects. First, standards are typically less efficient than taxes because they fail to internalize externalities of residual emissions. Consequently, if only a recycled content standard is used, overconsumption of wasteful products will prevail. Second, information requirements to impose the efficient level of recycled content are extremely high. It is unrealistic to expect that governments have this detailed information at their disposition. Third, it is well established in the literature (Kolstad, 2011; Tietenberg and Lewis, 2010) that uniform standards are not cost efficient because they do not discriminate between high- and low-cost technologies. Finally, static standards give insufficient incentives for adopting new and more efficient recycling technologies. The intuition for this is that once the standard is reached, there is no incentive for going further.

The review on *virgin material taxes* for aggregates of Söderholm (2011) discusses the theoretical efficiency and effectiveness of implemented taxes in Europe. Although implemented taxes have been effective in reducing the extraction of virgin materials, the author points to important weaknesses. First, virgin material tax rates are typically too low to stimulate recycling from waste products. Other instruments such as disposal taxes or deposit refunds are needed (see Dijkgraaf and Vollebergh (2004) and Dubois (2013) for a discussion of current disposal taxes for landfilling and incineration in Western Europe). Second, the tax can reduce the overall volume extracted but does not give incentives to reduce extraction externalities such as air and water pollution or waste generation. Third, the tax can induce imports or extraction of untaxed resources that also generate environmental effects. The author concludes that a virgin material tax is above all a cost-efficient second-best policy because domestic extraction is typically easy to monitor.

35.4 INCENTIVES FOR UPSTREAM GREEN PRODUCT DESIGN

Fullerton and Wu (1998) assessed efficiency of an extensive list of taxes and subsidies if households are not charged for waste collection. Their article also considers incentives for upstream *green product design*. In their model, three variables should be optimized to achieve a social welfare maximizing outcome: the amount of consumption, the amount of packaging and the recyclability of packaging. Combinations of taxes on consumption, packaging recyclability, amount of packaging and recycled inputs were investigated. Identical consumers and producers, perfect competition and constant returns to scale in the production and waste collection industry were assumed. Firms adapt the attributes of their products (amount of packaging and recyclability) to the demand of consumers (i.e. consumers' willingness to pay for a product with specific packaging attributes has to be taken into account by the firms). Two combinations of instruments were shown to achieve an efficient allocation. First, a tax on consumption combined with a tax on the amount of packaging plus a subsidy for the level of green product design. Second, a tax on consumption—a deposit—combined with a tax on the amount of packaging plus a subsidy for recycled goods—a refund. If the level of green product design cannot be easily observed, the deposit-refund plan is shown to be preferable.

Also Calcott and Walls (2000) investigated which economic instruments could provide incentives for green product design. They compared a household disposal fee with a deposit-refund plan when heterogeneous products were collected by competitive recycling firms. Transaction costs not only make it impossible to diversify the household collection fee per material stream, but also make subsidies directly to households impossible. These authors used recyclability as an attribute of products and used a mass balance approach so that all selectively collected items were recycled and used as production inputs. Perfect competition applies for producers. If pricing of household waste collection cannot be diversified by the type of waste, the household collection could never give appropriate incentives for green product design. In contrast, a scheme with a consumer tax (deposit) and product-specific subsidy to recycling firms (refund) can reach the second-best level of green product design and efficiency. Calcott and Walls (2005) also developed a more complicated model that introduces *transaction costs* into the recycling market for imperfect sorting by consumers. The level of green product design is defined by one variable that is unobservable to policymakers. They concluded that a second-best outcome could be achieved with a deposit-refund system or a combination of the deposit-refund system and household disposal fee.

35.5 MULTIPRODUCT AND MIXED WASTE STREAMS

Aalbers and Vollebergh (2008) focused on the sorting effort required from households to separate waste into homogeneous streams that could easily be recycled. Their article generalizes the work of Fullerton and Kinnaman (1995) and Choe and Fraser (1999) for *multiproduct streams* that can be mixed in the waste stage by introducing the possibility of a refund or sorting subsidy. Green product design is not incorporated in the model. Consumers purchase portfolios of diversified products. Production, landfilling and recycling are modeled in a straightforward way with perfect competition and constant returns to scale. The model confirms that deposit-refund schemes work efficiently if they differentiate between waste streams with different external effects. The need for differentiation grows when differences in environmental effects are larger and when the share of mixed disposal is larger. Indeed, a small fraction of hazardous waste can lead to contamination of large quantities of potentially nonhazardous waste. Therefore, policymakers should primarily focus on adequate separation of highly toxic wastes and easily recyclable wastes using differentiated deposit-refund schemes. Maximal extraction of these potentially contaminating waste streams allows for setting up an undifferentiated deposit-refund scheme that limits administrative costs for the remaining waste streams.

Acuff and Kaffine (2013) used the simulation model of Palmer and Walls (1997) but focused on recycling as a way to reduce externalities in production. For example, since the carbon emissions from recycling metals or plastics are significantly lower than the emissions of virgin materials, appropriate waste management policies can contribute to carbon emissions mitigation. Acuff and Kaffine (2013) showed that welfare gains from recycling materials can be significantly bigger than welfare gains from avoiding disposal. Although instruments can simultaneously give incentives to reduce disposal and stimulate recycling, the use of instruments significantly differs if multimaterial goods are considered. Whereas the externality of disposal is typically modeled as a homogeneous externality, the positive externalities of recycling materials are often heterogeneous across materials. These authors therefore conclude that deposit-refund schemes should be diversified per product or waste stream.

35.6 EPR AND RECYCLING CERTIFICATES

In recent decades, EPR has been increasingly implemented, especially in Europe as an instrument to reduce disposal and stimulate recycling. Although implemented EPR schemes resemble the deposit-refund scheme described in economic literature, Dubois (2012) highlighted the difference. Because implemented EPR schemes typically contain static recycling targets, disposal externalities of nonrecycled waste fractions are not correctly internalized. As a result, incentives for green product design, prevention and technological innovation are suboptimal. The author concluded that an additional excise tax is required to achieve the welfare maximizing outcome.

In order to deal with EPR legislation, producers typically establish and finance a collective producer responsibility organization (PRO) that takes the actions needed to achieve the legal collection and recycling targets. For packaging waste, however, the United Kingdom uses *Tradable Recycling Certificates (TRCs)* to implement EPR in its Packaging Recovery Notes scheme. In a TRC scheme, recyclers can obtain recycling certificates if they meet certain quality standards. Producers of consumption goods, on the other hand, are obliged to hand over recycling certificates to the environmental authorities in proportion to their production volume. The idea is that a certificate market will emerge in which producers buy certificates from recyclers. Ino (2011) used a partial equilibrium model inspired by Palmer and Walls (1997) to show that the current targets for waste collection should increase further to achieve the welfare maximizing level.

Matsueda and Nagase (2012) focused on *TRCs in combination with landfill taxes*. They showed that a landfill tax reduces the effectiveness of EPR targets when used in combination with TRCs. The comparison of TRC with other policy instruments highlights that TRCs are not always the most appropriate instrument (Dubois, et al., 2013). EPR with a PRO, refunded taxes and disposal taxes can induce similar effects while avoiding the risks associated with certificate trading, such as, for instance, price volatility. In contrast to disposal taxes, both TRCs and refunded taxes can strengthen international competitiveness of domestic waste treatment companies, which is why the sector often lobbies in favor of these instruments. In order to determine whether EPR is most efficient with a PRO or with TRC, more research is needed.

35.7 DURABLE GOODS

The works discussed so far do not consider explicitly the complications that arise with durable goods (i.e. goods that are long lived, such as cars, refrigerators or washing machines). Runkel (2003) assessed the efficiency of introduction of EPR for *durable goods* under both perfect and imperfect competition. The model makes allowances for different vintages (i.e. generations) of products. Under perfect competition, the model confirms that implementation of EPR is desirable if illegal disposal and bounded consumer rationality are an issue. By internalizing the future cost of waste management, EPR not only reduces consumption of waste-generating products but also improves durability (defined as average lifetime of durable goods). However, under imperfect competition with high monopoly rents, the effect is ambiguous because EPR improves incentives for durability but may worsen welfare losses due to noncompetitive behavior by producers.

Runkel's model was extended in Eichner and Runkel (2005) to assess the effect of *recyclability* rather than durability of durable consumer goods. The model indicates that the efficiency of EPR may differ in the short and long run. EPR gives incentives for increased recyclability, but at the same time this may lead to more heterogeneity of discarded goods in the short run. The gain in recyclability may therefore be reduced by the increase in heterogeneity. In the long run, however, vintages and recyclability evolve to a steady state, and EPR is once again shown to improve social welfare.

Shinkuma (2008) also focused on the role of EPR for durable goods and on incentives for *reuse* in particular. In a world without transaction costs and illegal disposal, an advance disposal fee is the least efficient instrument because it gives insufficient incentives to *repair* durable goods and the lifecycle of products would be shorter than optimal. In a world with illegal disposal, disposal fees become difficult to implement, but the advance disposal fee with subsidies is shown to be a second-best solution.

35.8 IMPERFECT COMPETITION IN PRODUCT AND RECYCLING MARKETS

Eichner and Pethig (2001) used a general equilibrium model including production, consumption and waste management and focused on imperfect markets for industrial recycling and treatment of recycling residues. In contrast with other contributions, they assumed that illegal disposal of households was not an issue. They assessed how a variety of tax—subsidy combinations affect incentives for green design under a wide range of conditions. Fleckinger and Glachant (2010) assessed the interaction of EPR and market power in the production and waste management market. They compared an EPR scheme characterized by individual take-back obligation for producers with collective EPR schemes that allowed for collusion. Under imperfect competition in production and perfect competition for waste management, both individual and collective EPR schemes failed to achieve the welfare maximizing outcome. Under imperfect competition in the waste management sector, a collective EPR scheme leads to a superior outcome because *collusion of producers* increases the negotiation power of producers. This drives down the monopoly rents from the waste management sector and brings consumer welfare closer to social optimum. In general, models with imperfectly competitive product or waste and material markets give rise to more ambiguous results as they are very sensitive to modeling assumptions. An often recurring finding, however, is that noncompetitive behavior can alleviate environmental externalities because a *monopolist* tends to restrain output to drive up prices (see also Kolstad, 2011; Requate, 2005). The final welfare impact depends on the relative importance of environmental externalities versus the loss of consumer welfare due to excessive pricing.

35.9 POLICY INSTRUMENTS IN AN INTERNATIONAL MARKET FOR WASTE AND MATERIALS

An important evolution in the waste market is the growth of European and *international trade* in waste (EEA, 2012; Kellenberg, 2012). National policymakers, however, still have the authority to determine local waste taxes. The variety of implemented waste policies in Europe is significant as highlighted by the surveys of national waste policies such as Bio IS (2012) and ACR+ (2013). In such a diversified environment, regional policymakers may be tempted to use environmental policy as a strategic instrument to promote domestic interests like protecting local jobs or tax revenues (Kennedy, 1994; Cassing and Kuhn, 2003). If revenues and employment from the recycling sector are important, countries may end up in a "race to the bottom" where countries gradually reduce disposal taxes. Anecdotal evidence suggests that the risk for a downward tax spiral is present in Europe. For example, Sweden and Norway first installed an incineration tax in an open border system but abolished it in 2010 due to international waste shipments (Bjorklund and Finnveden, 2007; Dubois, 2013).

35.10 RECYCLING AND NONRENEWABLE RESOURCES IN A MACRO ECONOMIC PERSPECTIVE

The literature we reviewed so far did not consider explicitly the fact that recycling can alleviate the problem of *exhausting nonrenewable resources*. Smith (1972) was one of the first authors to investigate the impact of recycling on the extraction rate of nonrenewable resources, followed by, among others, Hoel (1978) and more recently André and Cerdá (2006). There is also a growing literature that studies the impact of recycling on the growth rate of an economy (see for example Di Vita, 2007; Pittel et al., 2010) in which the substitutability between recycled and virgin material plays a crucial role. Finally, Seyhan et al. (2012) presented a numerical application of the theoretical models for the case of phosphorous extraction.

Although looming *material scarcity* is an often-used argument for promoting public policy intervention fostering recycling and, more in general, material efficiency (see among other Allwood et al., 2011), many resource economists question this argument. According to the economics perspective, policy interventions should primarily aim at fixing market failures, such as environmental externalities, of which the social costs are reflected inadequately in unregulated market prices. According to Söderholm and Tilton (2012), scarcity is probably less of a concern, even in the long run, than environmental and technological externalities and information asymmetries. They also argued that remediation of these externalities should be done rather with implementation of cleverly designed policy mixes that incentivize firms, consumers and recyclers rather than picking and supporting particular technology winners.

35.11 CONCLUSION

Recycling can help to reduce the impact of negative environmental externalities related to waste disposal and virgin material extraction and production. Given that environmental externalities are often underpriced in market economies, there remains ample scope for further stimulating recycling. Although the case for public intervention is clear, the remedy is often complicated and involves a clever mix of policy instruments to incentivize the different actors (producers, recyclers, consumers, etc.). A particular concern is that charging consumers for disposal of their waste may lead to illegal waste dumping and even greater negative environmental impacts. Combinations of taxes on consumption of waste-generating products and subsidies for recycling are often recommended. EPR schemes are particular examples of such approaches. Regarding the capability of recycling to overcome material's scarcity, the environmental and resource economics literature is less univocal. Much seems to depend on the substitutability of recycled versus virgin inputs in production. If this substitution is easy, recycling can significantly contribute to a transition toward a more sustainable materials' management. Resource economists recommend stimulating recycling to reduce environmental externalities related to mining and production of virgin materials. They are, however, more skeptical about the argument that recycling should be stimulated for reasons of alleviating material scarcity.

References

Aalbers, R., Vollebergh, H., 2008. An economic analysis of mixing wastes. Environmental and Resource Economics 39, 311–330.

Acuff, K., Kaffine, D.T., 2013,. Greenhouse gas emissions, waste and recycling policy. Journal of Environmental Economics and Management 65, 74–86.

ACR+, 2013. Plastic bags: an inventory of political instruments, Brussels. http://www.acrplus.org/index.php/fr/2013-06-11-10-23-55/actualites-de-l-acr/316-plastic-bags-report (accessed 25.01.14.).

Allwood, J.M., Ashby, M.F., Gutowski, T.G.,, Worrell, E., 2011. Material efficiency: a white paper. Resources, Conservation and Recycling 55 (3), 362–381.

André, J.F., Cerdá, E., 2006. On the dynamics of recycling and natural resources. Environmental and Resource Economics 33, 199–221.

Bio IS, 2012. Use of Economic Instruments and Waste Management Performances, Report for the European Commission DG ENV. Bio Intelligence Service, Paris.

Bjorklund, A., Finnveden, G., 2007. Life cycle assessment of a national policy proposal—the case of a Swedish waste incineration tax. Waste Management 27, 1046–1058.

Bohm, P., Russell, C.S., 1985. Comparative analysis of alternative policy instruments. In: Kneese, A.V., Sweeney, J.L. (Eds.), Handbook of Natural Resource and Energy Economics, vol. 1. Elsevier, Amsterdam, pp. 395–460.

Calcott, P., Walls, M., 2000. Can downstream waste disposal policies encourage upstream 'design for environment'? American Economic Review 90, 233–237.

Calcott, P., Walls, M., 2005. Waste, recycling and 'design for environment': roles for markets and policy instruments. Resource and Energy Economics 27, 287–305.

Callan, S.J., Thomas, J.M., 2000. Environmental Economics and Management: Theory, Policy and Applications. The Dryden Press, Harcourt College Publishers, pp. 708.

Cassing, J., Kuhn, T., 2003. Strategic environmental policies when waste products are tradable. Review of International Economics 11 (3), 495–511.

Choe, C., Fraser, I., 1999. An economic analysis of household waste management. Journal of Environmental Economics and Management 38, 234–246.

Damgaard, A., Larsen, A.W., Christensen, T.H., 2009. Recycling of metals: accounting of greenhouse gases and global warming conditions. Waste Management and Research 27, 773–780.

Di Vita, G., 2007. Exhaustible resources and secondary materials: a macroeconomic analysis. Ecological Economics 63, 138–148.

Dijkgraaf, E., Vollebergh, H., 2004. Burn or bury? A social cost comparison of final waste disposal methods. Ecological Economics 50, 233–247.

Dinan, T.M., 1993. Economic efficiency effects of alternative policies for reducing waste disposal. Journal of Environmental Economics and Management 25, 242–256.

Dubois, M., 2012. Extended producer responsibility for consumer waste: the gap between economic theory and implementation. Waste Management and Research 30 (9 Suppl.), 36–42.

Dubois, M., 2013. Towards a coherent European approach for taxation of combustible waste. Waste Management 33, 1776–1783.

Dubois, M., Hoogmartens, R., Van Passel, S., Vanderreydt, I., 2013. Opportunities and drawbacks of tradable recycling certificates. In: Proceedings of the 14 International Waste Management and Landfill Symposium, 1–4 October, Sardinia.

EEA, 2012. Movements of Waste across the EU's Internal and External Borders. European Environment Agency, 2012/7, Copenhagen.

Eichner, T., Pethig, R., 2001. Product design and efficient management of recycling and waste treatment. Journal of Environmental Economics and Management 41, 109–134.

Eichner, T., Runkel, M., 2005. Efficient policies for green design in a vintage durable good model. Environmental and Resource Economics 30, 259–278.

European Commission, 2011. Roadmap to a Resource Efficient Europe. COM, 571.
Fleckinger, P., Glachant, M., 2010. The organization of extended producer responsibility in waste policy with product differentiation. Journal of Environmental Economics and Management 59, 57–66.
Fullerton, D., Kinnaman, T.C., 1995. Garbage, recycling and illicit burning or dumping. Journal of Environmental Economics and Management 29, 78–91.
Fullerton, D., Wu, W., 1998. Policies for green design. Journal of Environmental Economics and Management 36, 131–148.
Hanley, N., Shogren, J., White, B., 2007. Environmental Economics in Theory and Practice, second ed. Palgrave Macmillan, Basingstoke, Hampshire, UK.
Hoel, M., 1978. Resource extraction and recycling with environmental costs. Journal of Environmental Economics and Management 5, 220–235.
Ino, 2011. Optimal environmental policy for waste disposal and recycling when firms are not compliant. Journal of Environmental Economics and Management 62, 290–308.
Kellenberg, D., 2012. Trading wastes. Journal of Environmental Economics and Management 64, 68–87.
Kennedy, 1994. Equilibrium pollution taxes in open economies with imperfect competition. Journal of Environmental Economics and Management 27, 49–63.
Kolstad, C.D., 2011. Intermediate Environmental Economics. Oxford University Press, Oxford.
Matsueda, N., Nagase, Y., 2012. An economic analysis of the packaging waste recovery system in the UK. Resource and Energy Economics 34, 669–679.
Palmer, K., Walls, M., 1997. Optimal policies for solid waste disposal: taxes, subsidies and standards. Journal of Public Economics 65, 193–205.
Palmer, K., Sigman, H., Walls, M., 1997. The cost of reducing municipal solid waste. Journal of Environmental Economics and Management 33, 128–150.
Pittel, K., Amigues, J.-P., Kuhn, T., 2010. Recycling under a material balance constraint. Resource and Energy Economics 32, 379–394.
Porter, R., 2002. The Economics of Waste. Resources for the Future. WashingtonPRO Europe (2010) Producer Responsibility in action, pp. 83.
Requate, T., 2005. Environmental Policy under Imperfect Competition—A Survey. Economics working paper, Christian-Albrechts-Universität Kiel, Department of Economics, No. 2005-12. http://econstor.eu/bitstream/10419/22000/1/EWP-2005-12.pdf.
Runkel, M., 2003. Product durability and extended producer responsibility in solid waste management. Environmental and Resource Economics 24, 161–182.
Seyhan, D., Weikard, H.-P., van Ierland, E., 2012. An economic model of long-term phosphorus extraction and recycling. Resources, Conservation and Recycling 61, 103–108.
Shinkuma, T., 2008. Reconsideration of an advance disposal fee policy for end-of-life durable goods. Journal of Environmental Economics and Management 53, 110–121.
Smith, V.L., 1972. Dynamics of waste accumulation: disposal versus recycling. Quarterly Journal of Economics LXXXVI, 600–616.
Söderholm, P., 2011. Taxing virgin natural resources: lessons from aggregates taxation in Europe. Resources, Conservation and Recycling 55, 911–922.
Söderholm, P., Tilton, J.E., 2012. Material efficiency: an economic perspective. Resources, Conservation and Recycling 61, 75–82.
Tietenberg, T., Lewis, L., 2010. Environmental Economics and Policy, sixth ed. Pearson Addison-Wesley, Boston.

CHAPTER 36

Information Instruments

Ernst Worrell

Copernicus Institute of Sustainable Development, Utrecht University, Utrecht, The Netherlands

36.1 INTRODUCTION

Policy instruments are implemented to achieve a set target (e.g. increasing the share of recycling in waste management) by reducing the barriers that inhibit stakeholders to achieve the results, or stimulating stakeholders to contribute to achieving the target. In practice, many barriers limit the uptake of actions by households, companies, and waste processors to increase recovery of recyclable materials and/or products. These barriers may be of technical origin (e.g. technologies to recover materials from a waste stream), lack of knowledge (e.g. recognizing recyclable materials, or knowing how or where recyclable materials are collected), economic (e.g. the costs of separately collecting recyclables, or the low market price of recycled materials), organizational (e.g. the infrastructure within an organization or society to collect recyclables), a principal agent problem (also known as the renter–landlord problem, e.g. misallocation of investments and benefits), or lack of interest (e.g. in environmental consciousness, or lack of time or attention owing to other needs). Many of these barriers are interrelated, and often multiple barriers may be identified simultaneously. Whereas policies may be able to reduce the impact or magnitude of the barrier, other barriers may still inhibit the expected outcome. You could see the whole supply chain of recyclables as a real chain, with the individual links representing stakeholders or actions. A chain is as strong as its weakest link. Hence, to achieve the desired goal, a good understanding of the barriers throughout the supply chain is needed, as well as an understanding of the drivers of the individual stakeholders. This also implies there is no silver bullet or single policy that will be effective under all circumstances. Hence, generally a portfolio of policy instruments will be needed to achieve a defined (recycling) target.

Policy instruments can generally be divided into various categories: i.e. communication (this chapter), economic (Chapter 35), regulatory (Chapter 37), or negotiated agreements (Chapter 34); the latter could include a variety of instruments discussed in the other chapters. Communication or information instruments assume that stakeholders will change the behavior when they are better informed about the benefits and needs of a given activity, e.g. separate collection or recycling. This assumes that there is a knowledge barrier, e.g. that stakeholders are not aware of the benefits to society or to themselves of sorting wastes to allow improved recycling. Virtually any policy instrument that is implemented needs some sort of communication to stakeholders, because they need to be informed about the background and benefits of these policies to generate buy-in, and to make implementation effective and efficient in achieving the policy goals. Therefore, education, publicity, and promotion are essential for the success of any recycling scheme (Evison and Read, 2001). As such, information or communication instruments are rarely implemented in isolation, but are often part of a larger policy portfolio (or mix). This also makes it difficult to assess the effectiveness and efficiency of a single policy instrument, including the role of information instruments. In fact, it can be argued that, given the importance to affect the full chain of stakeholders and actions, evaluation of an instrument in isolation is not needed. Rather, it is more important to understand the role of the various instruments within the portfolio to improve the overall effectiveness of the portfolio.

Quality promotion and publicity on a regular basis will produce better recycling performance figures, whereas poor-quality promotion, or none at all, will result in low recycling rates (Evison and Read, 2001). Hence, when planning or changing a recycling service, it should include full education and publicity elements. Many studies have identified the factors that contribute to improved recycling behavior, and have found those factors to include knowledge (on environment and recycling), moral standards (Halvorsen, 2012), socioeconomic situation, economics (of waste disposal), convenience of recycling (e.g. type, distance, frequency), and behavior of others (e.g. Do the neighbors also recycle?). This stresses the role and importance of information to increase the effectiveness, while also supplying the infrastructure and other (e.g. economic) incentives to recycle. In designing an information campaign to support recycling, it is important to understand the target group, the different means of dissemination and communication, and the messaging. In this chapter, based on experiences in various countries and settings, these three elements are discussed.

36.2 TARGET GROUPS/AUDIENCE

In communicating recycling policy, as in marketing and advertising, a target audience is a specific group of people within the target market at whom the message is aimed. A target audience can be formed of people, e.g. a certain age group, gender, marital status, area (neighborhood), or groups demonstrating common behavior. A combination of factors is a common target audience. Other groups, although not the main focus, may also be interested. Determining the appropriate target audience is an important activity in designing an information campaign. As shown earlier, a number of factors affect recycling behavior. These factors can be used to determine and understand the barriers and subsequent information needs of the target group, and design the message of the campaign and the means of best reaching the target audience. There is often a need to develop specific targeted information campaigns, especially for the lower-recycling groups in society. These groups may be found in certain neighborhoods and social or age groups. For example, it has been shown that improving adolescents' knowledge about what is recyclable is of considerable importance to campaign designers. However, even high school students with strong pro-environmental attitudes were uncertain about what is recyclable (Prestin and Pearce, 2010). In this case, an effective campaign could begin with messages that increase knowledge on recyclables and ways to collect, as was also demonstrated in the following example from Poland. In 1995, an integrated program was launched by the aluminum industry to increase recycling of aluminum cans (RECAL). The targeted School Can Program provides technical and educational assistance to schools municipalities, including educational materials on waste management, containers and bags for aluminum cans, selection magnets, and lists of local collecting depots. The RECAL Program has succeeded in raising public awareness and increasing recovery of aluminum cans in various regions of Poland (Grodzinska-Jurczak and Bartosiewicz, 2001).

Staff within a company is another special target group to increase separation, recovery, and recycling of internally generated production and office waste. Within the company, various audiences may exist (e.g. production sites, offices, production floor, staff, management) and the communication plan should be part of an integrated waste management strategy. Next to providing a recovery and recycling infrastructure, communication is key to success. Experience shows that it is important for a clear message and target to be communicated from the highest levels of the company, for a common language to be developed to ease communication, and for recycling to become a common aspect of training of (new) staff, repeated at least annually. Feedback on the results and success of the recycling program to staff is an important element in corporate programs (see also subsequent sections).

36.3 COMMUNICATION TOOLS

An effective communications program can choose from and combine some or all of the following tools to communicate the intended recycling message to the target audience (see earlier discussion): the Internet, specific publications and print materials, public service broadcasts, media relations, spokespersons, and events (e.g. special days, awards, exhibitions, seminars). The combination of tools used will depend on the target audience and strategic goals, the advantages and disadvantages of each tool, and the available budget. Quality promotion and publicity on a regular basis will produce better recycling performance figures, whereas poor-quality promotion, or none at all, will result in low recycling rates; thus, when planning the provision of a recycling service, it should include full education and publicity elements (Evison and Read, 2001). When considering which media to use for promotions, local newspapers are not always the best means of informing the public, but a special (regular) environmental newsletter can effectively put forward policies and strategies. Regular leaflets help to maintain public awareness, and knowledge will decline if frequent reminders are not used (Evison and Read, 2001). A survey of recycling programs in the United Kingdom indicated that the publicly preferred option for communication was leaflets (79%), followed by newspapers (34%) and personalized letters (33%) (Mee et al., 2004). For the period of study, a relatively high proportion of residents were found to have access to the Internet at home (66%), but only 15% were prepared to use this as a communications channel. It was claimed that marketing and communications activities had influenced some 75% of them to recycle more, and newsletters (70%) were the most effective communications method. Recycling rates increased from a mere 10% to nearly 50% in 2 years (Mee et al., 2004). A study aimed at understanding the factors to increase compact fluorescent lamp (CFL) recycling in the United States has shown that a lack of knowledge adversely affects the CFL recycling rate. However, it was excessively difficult to find knowledge on recycling (e.g. difficult to locate on web sites) and the collection system was not sufficiently convenient. The analysis showed that information should be provided in a convenient and simple manner when designing or improving collection and recycling from households (Wagner, 2011).

36.4 MESSAGING: INFORMATION AND COMMUNICATION

A recycling program needs a well-designed marketing strategy and campaign to succeed, targeted at the selected audiences. It is important to maintain consistency in the appearance, style and tone, and message in all communications tools of the campaign. Without this consistency, the message will not be as effective, even if the target audience is exposed to several of the communications products. Worse still, a lack of consistency may confuse or distract the target audience. The message consists of the type of information included, as well as the way it is communicated. As discussed earlier, information regarding what should be recycled, as well as how and where it can be recovered/collected, are key elements. This may also include labeling materials and products for recycling (e.g. the recycling logo, the labeling of plastic types), to allow stakeholders to effectively separate products for recycling. Feedback is another important element of a successful communication strategy. Perrin and Barton (2001) showed that providing informative feedback to households increased recycling volumes, participation, and weekly set-out levels. Recoveries of all fractions increased, particularly the packaging fraction following feedback (Perrin and Barton, 2001).

An analysis of slogans for battery collection and recycling in Switzerland provides an interesting example of the importance of the message itself. Hansmann et al. (2009) surveyed overall impressions of 10 different slogans for the promotion of batteries, and subsequently tested two selected slogans in a field experiment at different supermarkets. Slogan effectiveness was positively related with ratings for good and ecological argumentation, creativity, humor, and easy comprehensibility, and it was negatively related to authoritative wording. The behavioral effectiveness of a humorous slogan urging people to return used batteries, and a factual slogan informing people that batteries were separately collected, were investigated in a 9-week field experiment in 21 supermarkets. The informative and easily comprehensible factual slogan achieved an increase of 36% in the weight of returned batteries. In contrast to the initial survey, the prescriptive humorous slogan did not show a positive effect compared with the baseline. The results confirm that factual communication avoids negative reactions, increasing effectiveness (Hansmann et al., 2009).

Negative or unclear information as well as program changes for specific materials can also negatively affect the behavior of households in separating all types of wastes. A study (Clarke and Maantay, 2006) of the impact of publicity around program changes in collection of recyclables in New York City, showed that in the worst recycling districts of the city, the rate dropped from a high of 35% to a high of 21% within a single year. The 12 best recycling districts also experienced a drop in capture rates, but not to the degree in the low-diversion districts. The strong drop in participation probably resulted from the decision to stop collecting plastic, glass, and wax paper containers in July 2002, and to considerable negative publicity about the recycling program before the reduction in service. This illustrates that recycling behavior depends to a great degree on attitudes toward the program, and that program changes and negative publicity can be disruptive, perhaps for a long time. This example illustrates that communication of program changes should be done carefully to avoid negative impacts on overall program performance.

36.5 CONCLUSION

Information is crucial to the success of any recycling and recovery program, be it within a country, municipality, or company. Information should be tailored to the barriers and needs of the specific audience to be effective. Hence, an information campaign needs to start with a careful analysis of the stakeholders targeted, and may need to differentiate groups within the total population to increase the effectiveness of communication. This also affects the means to communicate the message and the formulation of the message itself. Research in various countries has demonstrated that information should be easily accessible, and preferably in the form of dedicated newsletters with a recognizable appearance, style, and message. It is also essential to include and repeat simple and basic information about what and how/where to collect the recyclables. Regular information and feedback on program performance and success are essential to maintain the effectiveness of the program over time. Information on changes in the program needs to be carefully managed to maintain the overall effectiveness of the program. Although messaging strongly depends on the culture within a country, region, or country, factual information generally provides the best results.

References

Clarke, M.J., Maantay, J.A., 2006. Optimizing recycling in all of New York City's neighborhoods: Using GIS to develop the REAP index for improved recycling education, awareness, and participation. Resources, Conservation and Recycling 46, 128–148.

Evison, T., Read, A.D., 2001. Local authority recycling and waste—awareness publicity/promotion. Resources, Conservation and Recycling 32, 275–291.

Grodzinska-Jurczak, M., Bartosiewicz, A., 2001. The RECAL Foundation Programme; an example of ecological education in Poland. Resources, Conservation and Recycling 34, 19–31.

Halvorsen, B., 2012. Effects of norms and policy incentives on household recycling: an international comparison. Resources, Conservation and Recycling 67, 18–26.

Hansmann, R., Loukopoulos, P., Scholz, R.W., 2009. Characteristics of effective battery recycling slogans: a Swiss field study. Resources, Conservation and Recycling 53, 218–230.

Mee, N., Clewes, D., Phillips, P.S., Read, A.D., 2004. Effective implementation of a marketing communications strategy for kerbside recycling: a case study from Rushcliffe, UK. Resources, Conservation and Recycling 42, 1–26.

Perrin, D., Barton, J., 2001. Issues associated with transforming household attitudes and opinions into materials recovery: a review of two kerbside recycling schemes. Resources, Conservation and Recycling 33, 61–74.

Prestin, A., Pearce, K.E., 2010. We care a lot: formative research for a social marketing campaign to promote school-based recycling. Resources, Conservation and Recycling 54, 1017–1026.

Wagner, T.P., 2011. Compact fluorescent lights and the impact of convenience and knowledge on household recycling rates. Waste Management 31, 1300–1306.

CHAPTER

37

Regulatory Instruments: Sustainable Materials Management, Recycling, and the Law

Geert van Calster[1,2]

[1]Department of International and European Law, University of Leuven, Leuven, Belgium;
[2]Brussels Bar, Brussels, Belgium

37.1 INTRODUCTION

Recycling is one of the more established parts of sustainable materials management (SMM). Jurisdictions worldwide are fast moving away from a wasteful society to one of resources. SMM has caught the fancy of many a regulator worldwide and has been defined as

> an approach to promote sustainable materials use, integrating actions targeted at reducing negative environmental impacts and preserving natural capital throughout the lifecycle of materials, taking into account economic efficiency and social equity.[1]

The Environment Council of the European Union (EU) adopted SMM at its meeting in December 2010,[2] linking it to sustainable production and consumption. SMM has many aspects that altogether represent almost the entire plethora of legal challenges associated with a successful environmental policy. This contribution highlights the legal context for recycling, taking EU waste law as a case study: selection is of the essence. International environmental law does influence national waste management policies (as does international trade law[3]). However, the most important directions

---

[1] OECD, Working Group on Waste Prevention and Recycling, Outcome of the first workshop on SMM, Seoul, November 28–30, 2005.

[2] http://www.consilium.europa.eu/uedocs/cms_data/docs/pressdata/en/envir/118642.pdf, last consulted February 20, 2013.

[3] See for instance the author's "Faites vos jeux: regulatory autonomy and the World Trade Organisation after Brazil tyres", *Journal of Environmental Law* 20 (1), 121–136; and "China, minerals export, raw and rare earth materials: a perfect storm for WTO dispute settlement", *Review of European, Comparative & International Environmental Law* 2013 (1), 117–122.

in waste law and policy continue to be decided by national priorities (in the case of European Union countries, by European priorities).

The analysis below is not meant to be conclusive.[4]

37.2 RESOURCE EFFICIENCY AND WASTE STRATEGY—THE BLURB

At the end of June 2013, the EU Institutions adopted the 7th Environmental Action Plan (EAP). The plan (officially known as the General Union Environment Action Program to 2020, "Living Well, Within the Limits of our Planet"[5]) and other documents, such as the Europe 2020 Strategy,[6] the Commission Communication on a Roadmap for Resource Efficiency,[7] and the Commission Action Plan for 2013,[8] highlight the targets and goals of the EU in terms of resource efficiency and waste strategy.

A common thread throughout these documents is recognizing waste as a commodity and as an important source of energy. In line with the most recent EU waste framework Directive,[9] they aim at creating a more competitive EU market for recycling and recovery, using waste as a resource to relieve the pressure on finite raw materials.

The main EU objectives in the field of waste regulation are defined by the EAP as follows:

> There is also considerable potential for improving waste management in the EU to make better use of resources, open up new markets, create new jobs and reduce dependence on imports of raw materials, while having lower impacts on the environment. Each year in the EU, 2.7 billion tonnes of waste are produced, of which 98 Mt is hazardous. On average, only 40% of solid waste is reused or recycled. The rest goes to landfill or incineration. In some Member States, more than 70% of waste is recycled, showing how waste could be used as one of the EU's key resources. At the same time, many Member States landfill over 75% of their municipal waste. Turning waste into a resource [...] requires the *full implementation of EU waste legislation* across the EU, based on strict application of the waste hierarchy and covering different types of waste

It is suggested by these documents that additional efforts are needed to reduce per capita waste generation in absolute terms, limit energy recovery to nonrecyclable materials, phase out landfilling, ensure high quality recycling, and develop markets for secondary raw materials. To achieve this, market-based instruments that privilege prevention, recycling, and reuse are said to need encouragement across the EU. Barriers facing recycling activities in the EU internal market should be removed and existing prevention, reuse, recycling, recovery, and landfill diversion targets should be reviewed so as to move toward a circular economy, with a cascading use of resources and residual waste close to zero.

A Commission study carried out in all Member States shows the importance of using smart economic instruments for creating a competitive market in waste management and recycled products, while reducing the regulatory burdens on the industry.[10] The study analyzed the regime of charges for waste disposal and treatment (landfill

---

[4] More detail in van Calster, G., *Handbook of EU Waste Law*, forthcoming with Oxford University Press in 2014.

[5] Commission proposal for a General Union Environment Action Program to 2020: "Living well, within the limits of our planet", 2012/0337 (COD), adopted in November 2011.

[6] Communication from the Commission "Europe 2020: a strategy for smart, sustainable and inclusive growth", COM (2010) 2020.

[7] Communication from the Commission on a "Roadmap to a Resource Efficient Europe" COM (2011) 571.

[8] Communication from the Commission "Commission Work Program 2013", COM (2012) 629.

[9] Directive 2008/98, (2008) OJ L312/3.

[10] For example, by introducing a catch-all permit for the operating plants that are likely to produce waste.

taxes and restrictions), incineration taxes and restrictions, pay-as-you-throw (PAYT) schemes, and harmonized producer responsibility for specific waste streams.

Some suggested economic instruments were recycling partnerships with the countries that currently absorb most of the resources; harmonization of the methods for calculating a minimum tax level for landfills and incineration (taxes could be more strongly encouraged in the worst performing Member States); introduction of responsibility schemes based on successful models already used in the EU; and recognizing the importance of recycling under the EU Emission Trading Scheme for the reduction of air emissions and the saving of primary resources.[11]

The blurb is definitely strong. However, its realization is often blocked by a lack of clarity in the actual legal framework.[12]

37.3 THE EU FRAMEWORK DIRECTIVE ON WASTE, AND ITS VIEW ON RECOVERY AND RECYCLING

37.3.1 The Core Definition of Waste

In the new Directive, waste is defined in Article 3(a) as

> any substance or object which the holder discards or intends or is required to discard.

Following the court's guidelines in several cases, the Directive notably also includes a procedure that eventually should lead to a clarification of the concept of so-called full recovery. Article 6 deals with the end of waste status and establishes criteria for determining the moment when a waste has been fully recovered and hence ceases to be waste. The end-of-waste concept in the Directive sets out the conditions based on which substances or products which had fallen under the waste definition within the meaning of the Directive no longer do so. The specific conditions set forth in the Directive (following the criteria developed in case law) that a substance or object needs to fulfill after undergoing a recovery or recycling process, are as follows:

1. The substance or object is commonly used for specific purposes
2. A market or demand exists for such a substance or object
3. The substance or object fulfills the technical requirements for the specific purposes and meets the existing legislation and standards applicable to products (i.e. virgin materials)
4. The use of the substance or object will not lead to overall adverse environmental or human health impacts (i.e. in this case, a good indication may be comparing the use of the material under the relevant product legislation with the use of the same material under the waste legislation)

The European Commission is instructed to supplement the abstract definition with specific technical criteria to be adopted in co-operation with the Member States. For example, in March 2011, a regulation[13] was adopted for setting the criteria whereby iron, steel and aluminum scrap cease to be waste. The document also sets forth a system of conformity certification and a quality management system. A similar document was adopted in December 2012 for glass cullet.[14] In

---

[11] Use of Economic Instruments and Waste Management Performances, European Commission (DG ENV) April 10, 2012, available at http://ec.europa.eu/environment/waste/pdf/final_report_10042012.pdf.

[12] See more detail in the author's "Opportunities and pitfalls for sustainable materials management in EU waste law", *Revue des Sciences humaines, éthique et société*, 2013, forthcoming.

[13] Regulation 333/2011, (2012) OJ L94.

[14] Regulation 1179/2012, (2012) OJ L337.

2013, the Commission lodged a proposal of a regulation establishing criteria for determining when copper scrap ceases to be waste.[15] According to the Thematic Strategy on the Prevention and Recycling of Waste of January 2011, the Commission will establish such criteria for paper and compost as well.[16]

Another important aspect underlined by the Waste Directive Guidelines, which the Commission issued in assisting with the implementation of the Directive, is that the person who places the material on the market for the first time after it had achieved the end-of-waste status must ensure that the material meets the relevant requirements under the chemical legislation. Recovered substances are exempted from registration, but only if certain conditions are met.[17] Following a recent reference for a preliminary ruling by Finnish High Court on the question as to what extent hazardous waste also ceases to be waste when it fulfills the end-of-waste requirements laid down in Article 6(1),[18] the European Court of Justice confirmed in its judgment of March 7, 2013 that hazardous waste may be returned as secondary raw materials, and that REACH may play an important role in this respect.

37.3.2 Recovery and Disposal within the New Waste Hierarchy

Article 4 of the Waste Framework Directive provides for a waste hierarchy, with prevention as the first policy tier, followed by preparing for reuse, recycling and other recovery (e.g. energy recovery), and disposal.

However, life-cycle thinking[19] on the overall impacts of the generation and management of waste can determine specific waste streams to depart from this hierarchy for the higher purpose of achieving the "best overall environmental outcome". This is reflected in the current wording of paragraph 2 of Article 4, with Member States having the obligation to consider the downstream and upstream benefits when establishing the legal and policy approach toward certain waste streams.

The Commission has put together an online platform entirely dedicated to life-cycle thinking, reuse, recycling, energy recovery and, ultimately, disposal, which provides a collection of studies and online tools aimed to help the industry and regulators to work together for preventing burden shifting in resource, product, and waste generation and management.[20]

37.3.2.1 *Recovery*

Recovery operations are defined as "any operation the principal result of which is waste serving a useful purpose by replacing other materials which would otherwise have been used to fulfill a particular function, or waste being prepared to fulfill that function, in the plant or in the wider economy".

This definition was developed based on the court's case law. According to the Waste Directive Guidelines, the new recovery definition was also

---

[15] Proposal for a Council Regulation establishing criteria determining when copper scrap ceases to be waste under Directive 2008/98/EC of the European Parliament and of the Council, (2012) OJ L337.

[16] Report from the Commission on the Thematic Strategy on the Prevention and Recycling of Waste COM (2011) 13.

[17] See for instance, Article 2(7)(d) of the REACH Regulation setting forth the conditions for the exemption of recovered substances or mixtures.

[18] Case C-358/11. See outcome on the author's blog, www.gavclaw.com.

[19] This concept should be distinguished from other concepts such as life-cycle assessment, cost-benefit analysis, life-cycle costing, which are only nonbinding available support methods for life-cycle thinking (Waste Directive Guidelines, page 51).

[20] The online platform is available at http://lct.jrc.ec.europa.eu/.

aimed to facilitate the classification of waste incinerators with efficient energy generation as recovery operation, since the substitution can take place not just in the plant where waste is being treated, but also in the "wider economy". Within the same line of thought, the definition should be read as also applying to processes involving the transformation of the waste material into a product that can be used as raw material in other processes.

Any recovery operation is completed when the resulting product does not bear the waste-related risks and complies with Article 13 (protection of human health and the environment), which is largely the same as the former Article 4(1) of the old Directive. The Commission specifically states that preparing for reuse, recycling, and other recovery are subcategories of the overarching concept of recovery, and the Waste Directive Guidelines provide closer scrutiny of these concepts.

37.3.2.2 Preparing for Reuse

The main difference between preparing for reuse (which is defined in Article 3(16) of the Directive as "checking, cleaning or repairing recovery operations, by which products or components of products that have become waste are prepared so that they can be re-used without any other pre-processing") and reuse is that the case of reuse the material has not become waste. An example of preparing for reuse would be mending a broken object.

In practice, problems can also arise due to the correlation between waste legislation and other special regulation, such as REACH.[21]

37.3.2.3 Recycling

Recycling allows waste material to be processed in order to alter its physico-chemical properties, allowing it to be used again for the same or other applications. According to Article 3(17) of the Directive, this includes, for example, the reprocessing of organic material but does not include energy recovery (specifically qualified as "other type of recovery") and the reprocessing into materials that are used as fuels or for backfilling operations.[22]

Therefore, only the reprocessing of waste into materials, products, or substances can be accepted as recycling. As an example, the Waste Directive Guidelines highlight that the biological reprocessing of waste before backfilling is to be classified as a pretreatment operation prior to other recovery (as backfilling is considered to fall under such category) and not as a recycling operation.

The Directive also requires that separate collection of at least paper, metal, plastic, and glass be established by 2015 by the Member States (with the purpose of achieving high-quality recycling), if technically, environmentally and economically practicable (the proportionality test) for the specific Member State. A Member State may therefore justify its failure to reach the imposed targets by demonstrating that they were not technically, environmentally and economically practicable for the Member State concerned.[23]

In the Commission's view, "technically practicable" means that the separate collection may be implemented through a system which has been developed and proven to function in practice; "environmentally practicable" means that the added value of ecological benefits justify

---

[21] Case C-358/11.

[22] Backfilling operations are not defined in the Directive itself but in the Commission Decision of November 18, 2011 establishing rules and calculation methods for verifying compliance with the targets set in Article 11(2) of the Directive, as being "a recovery operation where suitable waste is used for reclamation purposes in excavated areas or for engineering purposes in landscaping and where the waste is a substitute for nonwaste materials".

[23] See also the High Court (Wales) in Campaign for Real Recycling, March 6, 2013, (2013) EWHC 425 (Admin).

possible negative environmental effects of the separate collection; and "economically practicable" separation should not cause excessive costs in comparison with the treatment of a non-separated waste stream, while considering the added value of recovery and recycling and the principle of proportionality.

Therefore, the Waste Directive Guidelines do not show how it will be decided, in practice, what technically, environmentally, and economically practicable actually is, since across the Member States there are significant differences when it comes to waste management and policy approaches toward recycling and separate collection.[24] For example, the most advanced Member States (Belgium, Denmark, Germany, Austria, Sweden, and The Netherlands) landfill less than 3% of their municipal waste. At the other extreme, nine Member States landfill more than 75% of their municipal waste.[25]

Other special Directives supplement the requirements of separate collection in the new Directive (e.g. Article 5 of the WEEE Directive 2002/96/EC; Article seven of the Batteries Directive 2006/66/EC; Article 6(1) and (3) of Annex I in the End-of-Life Vehicles Directive 2000/53/EC; and Article 6(3) of Directive 96/59/EC on PCB/PCT).

Comingled collection of more than one separate waste streams is also acceptable in the Commission's view, but only insofar as the benchmark of high-quality recycling is observed (meaning that there should be no significant differences between the recycling outcome of comingled waste and that of separately collected streams).

37.3.2.4 Recycling Targets

Dr Caroline Jackson, MEP, the rapporteur for the European Parliament on the review of the Directive, was adamant that the Directive should include recycling targets for household and other wastes. Eventually, the council accepted the demands for a recycling goal for 2020 of 50% of combined glass, paper, metal, and plastic waste from households or other origins. It also agreed to a recycling target of 70% for construction and demolition waste. Waste prevention targets will not be set until 2014.

Other targets in connection with recycling and reuse were established by Article 11(2) of the Directive until 2020. On November 18, 2011, the Commission adopted a decision establishing rules and calculation methods for verifying compliance with these targets.[26]

Other specific EU legislation on waste also sets a number of targets for reuse, recycling, and recovery for different waste streams including end-of-life vehicles, packaging, and waste electrical and electronic equipment. Each year (or second year, in the case of electrical and electronic equipment), the Commission receives data on the national reuse, recycling, and recovery figures for these waste streams.

37.3.2.5 Other Recovery

Other recovery is mentioned as one of the tiers of the waste hierarchy in Article 4, after recycling and before disposal. One noteworthy inclusion in the list of other recovery operations are highly energy efficient municipal waste incineration facilities dedicated to the processing of such waste only, as opposed to facilities that were built with a different purpose and that use inter alia municipal waste as a source of fuel; these facilities typically, even under the old Directive, qualify as recovery facilities.

---

[24] Study on coherence of waste legislation, European Commission (DG ENV) August 11, 2011 available at http://ec.europa.eu/environment/waste/studies/pdf/Coherence_waste_legislation.pdf.

[25] Available at http://europa.eu/rapid/press-release_STAT-12-48_en.htm?locale=en.

[26] Commission Decision of November 18, 2011.

When adopting the new Directive, members of Parliament were divided whether incineration should be classified as recovery or disposal and under what conditions. An annex now specifies the targets that need to be reached for the facilities concerned to be qualified as carrying out recovery operations. This proviso is the result of a series of landmark Court of Justice judgments,[27] which ruled out incineration of municipal waste as being a recovery operation, if the installations concerned were built with the primary purpose of waste incineration.

37.3.3 Goods with a Positive Economic Value

Goods with a positive economic value are not excluded from the waste definition. The Court of Justice has consistently held national legislation to be incompatible with the waste Directives, if it excludes from the concept of waste certain substances that formed part of an economic chain—in other words, substances that had a positive economic value for someone. A substance of which its holder disposes may constitute waste even when it is capable of economic reutilization.[28] National legislation defining waste as excluding substances and objects that are capable of economic reutilization is therefore not compatible with the waste framework Directive.[29] The EU definition of waste does not exclude substances and objects that are capable of economic reutilization, even if the materials in question may be the subject of a transaction or quoted on public or private commercial lists.[30,31]

37.3.4 Recovery and Disposal—The Role of the Market in Encouraging Recycling

The essential characteristic of a waste recovery operation is that its principal objective is that the waste serves a useful purpose in replacing other materials that would have had to be used for that purpose, thereby conserving natural resources. Whether this is in practice the case is up to national courts and waste agencies to decide. The 2005 thematic strategy for prevention and recycling of waste already flagged the need to use market forces to recycle more in the EU.

It was said that EU recycling policy could be improved in several ways, including the following:

- *Setting recycling targets for materials.* Currently, EU law requires the recycling of materials from certain wastes (e.g. packaging, cars, electronics), but does not require the recycling of these same materials when they are used in other products. For example, packaging cardboard has to be recycled, but office paper or newsprint does not; the same goes for aluminum, plastics, and other materials. A more coherent approach to recycling could result in greater environmental benefits.
- *Getting the prices of the different waste treatment options right.* Despite strict EU legislation, disposing of waste in landfills and incinerators is often still cheaper than recycling. This could be corrected through

---

[27] See the author's "Waste incineration cases spark heated debate on waste management priorities" (2003) Review of European Community and International Environmental Law (RECIEL) 340–344.

[28] Joined Cases C-206/88 and C-207/88 Criminal proceedings against Vessoso and Zanetti (1990) ECR 1461, para 8.

[29] Case C-422/92 Commission versus Germany (1995) ECR I-1097, para 22 and Case C-359/88 Criminal proceedings against E. Zanetti and others (1990) ECR 1509, para 13.

[30] Joined Cases C-304/94, C-330/94, C-342/94, and C-224/95 Criminal Proceedings against Euro Tombesi and others (1997) ECR I-3561, paras 47–52.

[31] Case C-457/02 Criminal proceedings against Antonio Niselli.

tradable certificates, the coordination of national landfill taxes, promoting PAYT schemes, and making producers responsible for recycling.

It is not always easy to see the forest for the trees in European Commission documents on recycling. A number of issues, however, were quite striking in the 2005 strategy. First, the emphasis was often placed by the Commission on materials, rather than on specific types of waste. In general, comments have expressed some sympathy for this approach, which is not that surprising in the light of the figures quoted. For instance, the Commission argues that the packaging and packaging waste Directive applies to about 5% of all waste in the EU, and the end-of-life vehicles Directive just 1%. Focusing on the materials would therefore seem to be of some value. Conversely, however, the Commission itself indicates that a shift of focus toward materials rather than types of waste would endanger the concept of producer responsibility. Indeed, comments received so far indicate that producer responsibility acts as a strong signal in the marketplace, by focusing on an identifiable group within the waste stream. This may be in danger should one leave this focus and opt for a more diffuse approach.

A further point to note is the Commission's taste for identifying recycling targets per material, to be set not at the national level but at the EU level. This should enable the recycling industry to operate flexibly in a pan-European recycling area. This is a proposal that is very market focused and has led to a number of negative reactions. In particular, commentators fear that should the approach entail the EU as a whole to reach certain targets, rather than each Member State individually, this may tempt the environmentally weaker Member States to withdraw from the stricter obligations without fear of backlash. Moreover, it is argued by some that the approach would also lead to increased trade in waste, in particular to Member States with weaker standards for the recycling of waste.

The Commission also expressed a strong preference for economic instruments to reach the targets put forward. This may, for instance, lead to tradable waste rights. Comments received with respect to this route (focus on economic instruments) are overwhelmingly positive, amongst others, because the view would seem to be that recycling waste is expensive and suffers from a comparative disadvantage compared to waste processing techniques, which do not entail the internalization of environmental costs. The option of tradable waste rights, however, is generally received as being too radical. It may also be noted that the option of economic instruments can now hardly be labeled as new, given that the Commission has been putting it forward for some years now; the crucial question is how to translate it into practice.

The Commission remains extremely pessimistic with respect to any substantial EU role in the prevention of waste.

The progress toward the objectives set out in the strategy was reviewed in 2011 when the Commission adopted a report on the thematic strategy and a staff working document. It includes a summary of the main actions taken by the Commission, the main available statistics on waste generation and management, a summary of the main forthcoming challenges, and recommendations for future actions. The Commission concluded that main objectives of the strategy still remained valid. It underlined that it has played an important role in guiding policy development and that significant progress has been achieved in the improvement and simplification of legislation, the establishment and diffusion of key concepts, such as the waste hierarchy and life-cycle thinking, on setting focus on waste prevention, on coordination of efforts to improve knowledge, and on setting new European collection and recycling targets.

The European waste recycling targets are currently undergoing a process of review and reassessment. In June 2013, the Commission started a large consultation process aiming to identify the issues and solutions to the targets in the waste framework Directive, the landfill Directive and the packaging and packaging waste Directive. The consultation procedure lasted until September 2013. According to the Commission,[32] the review of the targets has two purposes: to respond to the review clauses set out in the Directives and to bring these targets in line with the Commission's ambitions of promoting resource efficiency as detailed in the Roadmap on Resource Efficiency,[33] which have also been included as part of the Commission's proposal for a 7th Environmental Action Program.

---

[32] More information at http://www.wastetargetsreview.eu/section.php/4/1/consultation.
[33] Communication "Roadmap to a Resource Efficient Europe" COM (2011) 571.

APPENDIX 1

Physical Separation 101

Kari Heiskanen
Aalto University, Espoo, Finland

A1.1 BREAKAGE

Breakage at recycling is a composite outcome of brittle, ductile or viscous behavior.

The brittle particles can be broken by compression or impact. Most slags, minerals and rocks are brittle, as often are many types of coating. Fine particles (smaller than 50 μm) are easy to produce. Particle size distributions tend to follow roughly simple distributions (log-normal or Weibull). Ductile materials, like metals, need to have a shear force for breaking. It can be created also by impacting crushers. Particles tend to become bent and folded depending on their dimensions and plasticity. Compressive stress results in major deformation before breakage. Very low temperature will cause the particles to behave in a more brittle way. Production of fine particles is difficult. Viscous behavior is time and temperature dependent (glass and plastics) and will at lower temperatures and at high impact velocities behave in a way that resembles brittle behavior. However, no fines by a high compression zone close to the impact point do form. Particles break at the time of pressure release. At cryogenic temperatures, the behavior is almost brittle and energy consumption in breakage (not cooling) is low. At high temperatures, the behavior is different. An impact causes substantial deformation but may not produce breakage. If breakage occurs, long fiber like particles often appear. Fine particle production needs cooling.

A problem arises when recycled products are made of materials showing various behaviors. Especially in ductile materials the type of joints will dictate very much of the outcome. Random breakage models, which are useful in brittle breakage, are not valid for ductile breakage.

A1.2 SIZE CLASSIFICATION

Size classification is a prerequisite for many mechanical separations for recycled materials. In the coarser end, it is conducted by screens, and in the finer end, by dynamic size separators.

A1.3 SCREENS

Screens are suitable separators for sizes often encountered in recycling. They are applicable from several hundreds of millimeters to about 100 μm. Most common ones are vibrating deck screens, roller, gyratory and trommel screens. Deck materials can be made of steel wire, rubber or polyurethane. Vibration is most commonly created by one or two (seldom three) shafts with unbalanced weights. Vibration can also be induced by electromagnetic exciters. Vibration pattern is most often elliptical in a direction to move material on the screen deck.

When material enters the screen surface, it stratifies and the finest material passes through rapidly as a zero order process $dC/dt = k$. After the finest particles, finer than about $0.5\,D_a$, have passed, the process becomes a probability process and of first order $dC/dt = -gC$. Each particle has a probability to pass the screen at each vibrational cycle. The probability goes to zero when two of the main dimensions are larger than $\sqrt{2} \times D_a$ for square openings.

A1.4 DYNAMIC SEPARATORS

The most commonly used device is the hydrocyclone, where the flow is fed tangentially to the outer perimeter wall of a cylindrical body. The cylindrical body is connected to a conical bottom. Flows enter from the cyclone on the centerline. The vortex finder is placed on the top of the cylindrical body, extending into the cyclone deeper than the feed point at the wall. The apex is at the bottom of the conical part (Figure A1.1).

The tangential flow induced enters at the outer wall and exits at the centerline causing a three-dimensional flow pattern. The flow needs to go from perimeter to the centerline causing a radial component of flow directed to the centerline. The tangential movement of flow causes a centrifugal force. As the centrifugal force is related to the mass of a particle and flow drag force to the area of the particle, large particles are more affected by the centrifugal force and the fine particles by the drag force. As a result, the coarse particles will remain at the perimeter and fine particles move toward the centerline. The vertical movement will transport the particles either up to the vortex finder or down the cone and the apex. As a result we have downward spiraling flow at the perimeter and an upward flow at the center of the cyclone. Fines will come from the vortex finder and the coarse out from the apex. By changing the orifice sizes, feed volume and cyclone curvature (diameter), we can obtain a required performance (Nageswararao et al., 2004) (Figure A1.2).

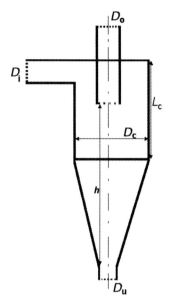

FIGURE A1.1 Hydrocyclone.

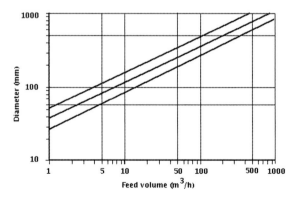

FIGURE A1.2 Hydrocyclone capacities.

$$P_i = K(D_c)^a \left(\frac{D_o}{D_c}\right)^b \left(\frac{D_u}{D_c}\right)^c \left(\frac{D_i}{D_c}\right)^d$$
$$\times \left(\frac{L_u c}{D_c}\right)^e \theta^f \left(\frac{P}{\rho_p g D_c}\right)^g \lambda^h \quad (A1.1)$$

For cut point x_{50} this becomes

$$\frac{d_{50}}{D_c} = K(D_c)^{-0.65} \left(\frac{D_o}{D_c}\right)^{0.52} \left(\frac{D_u}{D_c}\right)^{-0.5} \left(\frac{D_i}{D_c}\right)^{0.2}$$
$$\times \left(\frac{L_u c}{D_c}\right)^{0.2} \theta^{0.15} \left(\frac{P}{\rho_p g D_c}\right)^{-0.22} \lambda^{0.93}$$
$$(A1.2)$$

A1.5 GRAVITY SEPARATIONS

Gravity separations are feasible if the concentration criterion between particles to be separated is

$$Cr = \left|\frac{\rho_{heavy} - \rho_{fluid}}{\rho_{light} - \rho_{fluid}}\right| > 2.5 \quad (A1.3)$$

Gravity separation needs controlled particle size distributions for efficient operations.

A1.6 WATER MEDIA SEPARATIONS

Jigs can be used to separate closely sized coarse material. The smaller the concentration criterion, the narrower size fraction that must be utilized. Largest sizes usable depend on density but are typically less than 10 cm. The jigging action sinusoidally pulsates water through a particle bed lying on a screen. At the upward pulsing, the bed loosens and an initial acceleration takes place. This is almost relative to density as fluid drag forces are small due to small velocity. As the fluid movement increases, the particles settle down in a hindered way in this upward stream. Hindered settling velocity is a function of both size and density. This causes fine light particles to come to the top of the loose bed. As the flow velocity starts to slow, large coarse grains settle faster toward the screen and bottom of the bed. At a certain point the bed becomes consolidated and trickling of fine particles takes place. As this pulsation goes on the bed stratifies so that light particles are at the top and heavy particles at the bottom. To operate a jig properly, a ragging (coarse heavy particles) needs to be introduced on the screen before the feed stream is introduced (Figure A1.3).

Another common gravity separator in recycling is the shaking table, where a thin film of suspension at about 25% solids by weight is introduced with dressing water to an inclined surface. Due to the friction, the flow on the surface becomes a Couette flow with a velocity profile increasing from zero at the surface to a larger value at the top (liquid—air surface). This causes stratification by density and size. The heavy and

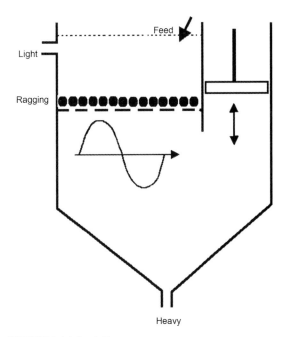

FIGURE A1.3 A jig.

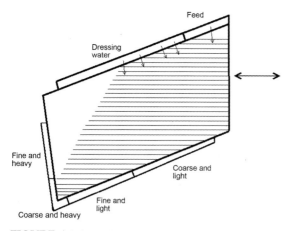

FIGURE A1.4 A shaking table.

fine particles are found at the bottom and large and light on the top. The stratified bed is moved along by longitudinal strokes with a slow forward and a rapid return movement. The table is equipped with riffles, which become progressively lower in the direction of flow. As the stratified particle bed moves slowly forward, particle layers are exposed to the dressing water, which sweeps first coarse light particles away from the riffles. Only fine heavy particles remain at the point where riffles taper off (Figure A1.4).

Shaking table feed sizes can vary from 100 μm to 15 mm. Capacity varies for a standard 72 in table from $0.5\,t/h$ to $11\,t/h$ depending on particle size. Tables can also be operated by air. Then the bottom of the table is aerated to provide fluidization.

Other gravity separators include spirals and Mozley multigravity separators and centrifugal gravity separators like the Knelson and the Falcon separators.

A1.7 DENSE MEDIA SEPARATIONS

Particles are introduced to media that has an (apparent) density between the densities of the material to be recovered and the non-recovered stream. Typical separations are plastics, rubber and textiles from metals and aluminum and magnesium from other metals.

Industrial dense media separations (sink-float) are normally performed in ferrosilicon $\rho = 6.7-6.9$) slurries, magnetite or with the heavy material itself or in the case of plastics and textiles in salt solutions. Atomized ferrosilicon (16–17% Si) can be used at high densities up to $3700-3800\,kg/m^3$.

After separation, both the sink and the float streams must be washed to minimize medium losses. The separation is quite sensitive to fines and needs to be washed free from them, as such material increases the medium apparent viscosity too much. Viscosity must be controlled by cleaning a part of the heavy media flow with magnetic separators and by thickening the pulp to counteract the dilution caused by washing.

A1.8 FLOTATION

Flotation in recycling has a limited use in separating shredded materials, but has more use in separating valuable materials from granular recycling feeds like slags. Typical particle size for flotation is for light solids from a few millimeters and for heavy solids ($\rho > 2700\,kg/m^3$) from 300 μm to about 10 μm.

Flotation is physicochemical process where gas bubbles are attached to the particles in a fluid mixture and lifted by bubble buoyancy to the top of the reactor (flotation cell), forming a froth, which is then collected.

Flotation is based on the idea of manipulating the apparent free surface energy with reagents and choosing the operational conditions in a way that the difference between the particles to be separated and the rejected particles becomes as large as possible. The difference must be adjusted relative to the surface tension of the liquid used in flotation (most typically water, but not necessarily). If cohesion in the fluid is larger than the adhesion between the surface and the fluid, the fluid tends not to wet the surface and gas bubbles will adhere to the surface. Such surfaces are hydrophobic. Plastics have free surface energy in the range of 30–45 mJ/m^2, which renders them all hydrophobic and makes the separation of them by flotation very difficult. Most minerals tend to have free surface energies in the range of 300–500 mJ/m^2, causing them to become easily wetted, or hydrophilic.

The contact angle[1] for flotation must typically be greater than 60° for flotation.

$$\gamma_{SG} - \gamma_{SL} = \gamma_{LG} \cos(\theta) \quad (A1.4)$$

Reagents used for the manipulation are called collectors. They contain a polar part that contacts to the mineral surface and a non-polar end that causes the change in the free surface energy. The longer the hydrocarbon chain length, the stronger is the collector. Most common ones are thiols, fatty acids and amines. The two first groups are anionic and the third cationic. Most common of the thiol group are xanthates and dithiophosphates. They are used for sulfides and metals, where electrochemical reactions can take place. Mixed potentials vary from +100 to −300 mV for sulfides. Also physical adsorption can take place. Fatty acids are used for oxides, carbonates, phosphates etc. Their dissociation is controlled by pH. Cationic amines are used for silicate materials. Due to the similarity of surfaces with silica tetrahedrons, selectivity is difficult to obtain.

For systems with electrochemical collector reactions, pH is a major controlling variable. Selectivity can be obtained by floating minerals at given pH values. This can be done by different acids (e.g. sulfuric acid) and different bases (e.g. milk of lime, soda). Activators can enhance collector adsorption and depressants inhibit collector adsorption. Most common ones are potential controlling reagents (e.g. oxygen, CN^- and HS^- ions). Metal ion concentrations can also be controlled by addition of metal salts or by complexing then out from the suspension.

Flotation machines can be mechanical, pneumatical or dissolved air devices. In mechanical flotation cells, air is introduced via a mechanical stirrer (rotor), which both mixes the suspension and disperses air into fine bubbles. The turbulent mixing causes particles and bubbles to collide and attach to each other if the contact angle is adequate. Mechanical cells come in sizes from 1 l to 500 m^3/axle. Pneumatical cells have separate mixing chambers where bubbles dispersed by either a sparger or a porous plate are contacted with the suspension. Dissolved air flotation dissolves air into water at higher pressure and mixes it into a suspension with particles and then releases the pressure to create a large number of fine bubbles at material surfaces.

A1.9 MAGNETIC SEPARATIONS

Magnetic separation is based on the difference on magnetic susceptibility (S). For recycling purposes we can distinguish three regimes depending on how they react to a magnetic field.

$$S = \frac{M}{H}, \quad (A1.5)$$

where M is intensity of magnetization and H is field intensity. Diamagnetic materials have a small negative susceptibility, meaning that the field inside a particle is less than outside and thus they are not amenable for magnetic separation (quartz $S \approx 0.001$). Such materials include rubber, plastics, glass, most metals, slags and dusts. Paramagnetic materials have a small positive susceptibility and can be separated from diamagnetic matrix by a high field intensity. Field intensities can be up to 2 T (tesla). Some metals like Ti, V, Cr, Mn and platinum group metals are paramagnetic. Also many Fe-containing compounds exhibit a paramagnetic property. The third group is ferromagnetism, where the susceptibility is high. Of pure metals,

---

[1] Convention is that contact angle is always measured via the liquid phase.

Fe, Co and Ni are ferromagnetic. Most steels (not austenitic stainless steels) are ferromagnetic. The magnetic field intensity is much higher inside a particle and thus give rise to high magnetic force. Used separator field intensities are typically 0.1–0.2 T. If ferromagnetic particles are separated in a high intensity field, the particles will magnetize and will remain stuck in the separator.

Magnetic force F_M is related to particle volume, its susceptibility, magnetic field intensity H and its local spatial gradient. This is important for separating paramagnetic materials, where the force is weak without the gradient effect. For cobbing, particle size can be several hundred millimeters. For other magnetic separations, particle sizes down to a few micrometers can be used.

A1.9.1 Low Intensity Separation

Low intensity separators are suitable for a large particle size range. Dry cobbers (belt separators) can separate large particles up to a few hundred millimeters. Permanent magnets are installed inside a belt pulley. Wet separators are usually meant for fine ferromagnetic materials. They consist of a rotating drum with stationary permanent magnets inside. In concurrent separators the flow of slurry is concurrent with the separator. It is a typical pre-separation device. A counter-rotation device has the non-magnetic flow direction counter-current, but the flow of magnetic material is concurrent for long pick-up zone and thus a high recovery. The counter-current separator is almost similar but with a feed point closer to the final magnetic material discharge. The concurrent devices can tolerate up to 8 mm particles. A typical diameter for these separators is 1200 mm with varying lengths up to 3.6 m.

A1.9.2 High Intensity Separation

High intensity separators have field strength of more than 1 T.

A1.10 EDDY CURRENT SEPARATION

An alternating magnetic field (change of magnetic flux (dH/dt)) induces a circulating (eddy) current to conducting particles. This in turn creates a force that is perpendicular to the flux and the direction of the eddy current. The force is

$$F = \frac{1}{2}m\frac{\sigma}{\rho}S\left(\frac{B}{h}\right)^2 Qv, \qquad (A1.6)$$

where m is particle mass, σ = conductivity, ρ = density, S = particle shape factor, B = magnetic field intensity, h = distance to magnets, v = relative velocity (particle-magnet) and Q = constant. The ratio between conductivity and density is proportional to the force. Aluminum and magnesium have high ratios and can be separated easily, copper and silver are medium high and can be separated with high relative magnet speeds, zinc and gold are quite low and lead and stainless very low, causing their separation to be almost impossible; glass, rubber and plastics have zero and are not affected by the eddy current.

A1.11 ELECTROSTATIC SEPARATIONS

Electrostatic separation uses electric conductivity for separation. Particles can be charged by high voltage corona charging with a high electron flow. When the feed passes to an earthed roll, conductors will lose their charge and fly off the roll by centripetal force, while non-conduction particles are attached to the roll and are brushed off to a separate bin. Particles can also be charged by letting them pass by charged plates (anode). This will change the trajectories of conductors to those of the non-conductor particles.

Feed needs to be dry and dust-free. As a single pass is not very efficient, separators are often arranged in cascades. They are suitable for millimeter size and finer feeds.

A1.12 SORTING

Sorting consists of identifying the particle and removing it from the stream if it fulfills preset criteria. The simplest way of doing this is manual sorting, not discussed here. Mechanical sorting has advanced on several fronts. The whole electromagnetic spectrum from X-rays to infrared can be used for detection. Also spectral analysis is commonly used to identify materials after they have been excited by e.g. a laser pulse. Analysis time and particle size are crucial. A 200 ms identification time allows treating 18,000 particles per hour/sensor. If every tenth particle needs to removed and if they are, say square particles of 100 × 100 mm with a thickness of 5 mm having a density of 2.7, the capacity is about 240 kg/h for the removed stream. A 80 × 80 mm size reduces the capacity to 150 kg/h.

If a particle needs to be removed from the stream, it can be blown by a directed air blast/blasts from the stream. It can also be picked from the stream by a robotic arm. Sorting technologies are limited at the upper limit by the weight of particles and at the lower limit by particle size. If particles are large, the air blasts do not have enough energy to remove unwanted particles from the streams. If gates or moving flaps are used, their slow movement limits capacity. At the finer end they are limited by CAPEX (Capital expenditure) costs becoming high compared to the capacity.

Reference

Nageswararao, K., Wiseman, D.M., Napier-Munn, T.J., 2004. Two empirical hydrocyclone models revisited. Minerals Engineering 17, 671–687.

King, R.P., 2001. "Modeling and Simulation of Mineral Processing Systems" Butterworth Heinemann Publications, Oxford, Great Britain. ISBN 0750640040

Reuter, M.A., Heiskanen, K., Boin, U., van Schaik, A., Verhoef, E., Yang, Y., 2005. The Metrics of Material and Metal Ecology, Harmonizing the Resource, Technology and Environmental Cycles. Elsevier BV, Amsterdam, 706 pp. (ISBN: 13 978-0-444-51137-9, ISBN: 10: 0-444-51137-7, ISSN: 0167-4528).

APPENDIX 2

Thermodynamics 101

Patrick Wollants

KU Leuven - Faculty of Engineering, Department Metallurgy and Materials Engineering (MTM), Kasteelpark Arenberg, Leuven, Belgium

A2.1 ON THE CONSUMPTION AND AVAILABILITY OF METALS

It is hard to imagine today's life going on without the extensive use of a wide diversity of metals in all kinds of applications, very hard indeed. You could give it a try and tell me how. I do not expect an easy answer. The reason is not that living without using (much) metals would not be possible. History teaches us that our ancestors did. But also that knowledge of producing metals and using them in a variety of applications was fatally and undeniably linked to power and to some extent to welfare as well. How willing are you to give up part of your welfare and power…?

So our present society needs metals. Facing scarcity of energy and metals, we must produce metals at as low an energy cost as possible (optimizing production processes from an energetic point of view is the message) and we must recover as much as possible metals from their end-of-life applications to use them again and again and again… (recycling is the message, and also avoiding use of metals in dissipative applications).

Most probably, the concept of closing the metals cycle as part of the hope to be able to create a sustainable world is a beautiful idea of the good mind that can only partially be realized. The real mission is: use less. Use less non-renewable energy, use less materials, metals in particular. Forget the nightmare of growth scenarios and reflect on the essentials. This is the real challenge.

Figure A2.1 gives an overview of the abundance of the chemical elements in Earth's upper continental crust. Metals seldom occur in their native state in nature. Some of the precious metals, Au, Ag, Pt… do, but only in limited amounts. Most of the common metals are found in a wide variety of stable minerals: oxides, sulfides, fluorides, halides… Part of them is present in ores, containing concentrations of minerals near the surface of the Earth that can be mined economically. The relative abundance of metals in Earth's crust ranges from extremely low (0.001–0.01 ppm for Au, Os, Pt, …) to very high (Si 27.7 wt%, Al 8.1 wt.%, Fe 5.0 wt%, …).

Despite the vast reserves of a number of industrially important metals, it is clear that the growing world population—welcoming very recently number 7 000 000 000—cannot hope to be able to consume metals and energy at the rate that became standard for western industrialized society. Actually, with the growth of both population and prosperity, especially in

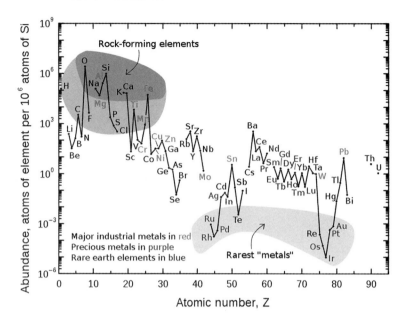

FIGURE A2.1 The abundance (atom fraction) of the chemical elements in Earth's upper continental crust is given here as a function of their atomic number. The rarest elements in the crust (shown in yellow) are the densest. The major industrial metals are shown in red. *Figure taken from Wikipedia, the free encyclopedia.* (For interpretation of the references to color in this figure legend, the reader is referred to the online version of this book.)

developing countries, the prospect of much higher resource consumption levels is "far beyond what is likely sustainable" if realized at all given finite world resources, as stated in a recent report by UNEP's International Resource Panel.

Actually, some numbers are so hard to imagine that it becomes difficult to see the truth behind. Take as an example the production of crude steel. The total crude steel production for the first eight months of 2011 was just beyond 1 milliard tons. A reasonable estimate of the total production in 2011 might be 1.570 milliard tons. With a specific density of 7.85 kg/dm^3, these 1.570 milliard tons would have a volume of $200 * 10^6 \text{ m}^3$, a mountain with a surface of 1 km^2 and a height of 200 m. Another way of looking at these figures is to assume that all this steel would be converted into plate of 1 mm thick. The resulting surface would be $2 * 10^5 \text{ km}^2$, which corresponds to about a 5 km wide strip along the Earth's equator. Also, the global warming potential is estimated at 2.3 kg CO_2-equivalent/kg or in total $3.611 * 10^{12}$ kg of CO_2. The molar mass of CO_2 being 44 kg/kmol, this corresponds at room temperature and normal pressure to some $1.8 * 10^{12} \text{ m}^3$ of CO_2 or a cube with an edge of approximately 12.2 km.

In our state of transition we must realize that for many reasons a dramatic change in living conditions and expectations is not likely to occur all of a sudden. In the meantime we must put all our efforts together to safeguard natural resources as much as possible, to avoid crossing the point of no return beyond which hope for future generations might be lost.

A2.2 RECYCLING AND EXTRACTIVE METALLURGY: AN ENERGY ISSUE

The amount of available resources is limited. This is a very strong argument in favor of recycling, as much as possible, end-of-life products. However, recycling does not come without a cost. Dependent on the chemical properties and the degree of mixing, more or less energy is required to recover metals in their pure or alloyed state from end-of-life applications. It should be clearly understood that recycling is not creating new resources. Recycling simply postpones the moment at which some metals will be no more available in the quantities that humankind requests.

The economic production of metals is not only a question of abundance in the Earth's crust. The continued increase in the use of metals over the twentieth century has led to a substantial shift from geological resource based production, to recovery of metals stored in waste deposits and end-of-life products. It has increasingly become an energy issue and, hence, largely depends on the thermodynamic stability of the primary minerals and on the way they are distributed in the ores, as increasingly on the recyclability of end-of-life used metals.

So we must minimize energy consumption to produce the metals we need and we must minimize the consumption of these metals, certainly in dissipative applications.

A2.3 THE SECOND LAW OF THERMODYNAMICS DEVIL: AN ENTROPY ISSUE

According to the first law of thermodynamics, energy cannot be created or destroyed. Nice. However, the second law of thermodynamics, the "entropy law" makes clear that the quality of this energy in every process decreases: always useful energy is dissipated as heat in the environment, or, in every real process entropy is created. In the language of thermodynamics, the entropy of an isolated system (no exchange of heat, work or matter at its boundaries) can only increase. This introduces the concept of time: an isolated system can never return to a state it was in before. This principle is known as the arrow of time.

From the recycling point of view, this means that a perfect "closing the loop" scenario cannot be written. Recycling requires energy, and in any energetic process energy is dissipated. One could argue that the planet we live on is an open system from the point of view of energy exchange: the sun provides us with "renewable energy". This is a falsification. If we want to convert this solar energy into, e.g. electric power, we need suitable infrastructure to realize this. For the construction of this infrastructure, energy and materials are required.

When mixing different metals or compounds (as in waste streams, end-of-life products), entropy always increases. Therefore, to un-mix a mixture, one must decrease the entropy of that system. This can only be realized by using energy from the environment. Thus, recycling is not cost-free: we need energy. We need more energy the more intimate materials are mixed, and the stronger the chemical, and physical interactions between these materials are.

So, actually, sustainability does not refer to safeguarding natural resources into eternity, but to extending availability for future generations as long as possible. We play a game we cannot win, but hope is there that it might take a (very) long time to finish the game. From this point of view, the whole concept of recycling is extremely important. Recycling can lead to less energy-intensive processes to produce the materials our society seems to need, and extends the horizon of available time left. But the second law cannot be beaten: energy is dissipated all the time, and increased mining and recycling activities, to satisfy growing market demands, only accelerate this irreversible process. So, the real message of importance is: use less of everything, use less energy, use less materials, be it recycled or not. This mind-change will most probably show to be much more important—in the long term—than optimizing recycling processes. But for the moment being, we must. And we must invest in the education of our future generations. They will decide how long we can abuse our planet, and how the gift of life on this beautiful planet can be extended.

A2.4 CHEMICAL THERMODYNAMICS AND REACTION EQUILIBRIUM

Concerning the primary production of metals and their recovery from end-of-life applications and all kinds of waste streams, some principles of chemical thermodynamics should be understood.

The simple fact that most metals in nature are not pure but are bound in minerals to oxygen, sulfur, chlorine, etc. reflects that the Gibbs free energy of these minerals is lower than the corresponding Gibbs free energies of the unreacted elements.

Metal ores include mainly oxides, sulfides, carbonates and halides. As sulfides and carbonates can be easily converted into oxides by roasting or calcination, for example:

$$ZnS + \tfrac{3}{2}O_2 = ZnO + SO_2,$$

extraction from metals is mainly from oxides by chemical or electrolytic reduction.

According to the principles of equilibrium thermodynamics, every system tries to minimize its Gibbs free energy, G, on its way to equilibrium. Consider, for example, the reaction

$$Si + O_2 = SiO_2.$$

The standard free energy of formation of SiO_2, $\Delta G^0_{SiO_2}$, is the difference between the standard Gibbs free energy of the reaction product (1 mol of pure SiO_2) and the standard Gibbs free energy of the reactants (1 mol of pure Si and 1 mol of pure O_2 at a pressure of 1 atm). The Gibbs energy of an element or compound is expressed with respect to a standard state, often the state of the pure element in its stable crystal structure at the given temperature (liquids and solids) and for gases the perfect gas at the given temperature. For any chemical reaction, hence also for the formation of oxides, sulfides..., the actual reaction free energy can be expressed, referring to the standard states, as:

$$\Delta G_T = \Delta G^0_T + RT \ln Q$$

where the reaction quotient Q (take the example of the formation of SiO_2) refers to the actual activities of the reactants and products:

$$Q = \frac{a_{SiO_2}}{a_{Si} \cdot a_{O_2}}.$$

The activities of pure solids and liquids are the activities in the corresponding standard states, and are equal to 1. The activity of the gas at moderate pressures is equal to the actual pressure of the gas:

$$Q = \frac{1}{p_{O_2}}.$$

A reacting system comes to equilibrium when its Gibbs free energy reaches its minimal value, thus when the driving force for the reaction becomes zero:

$$\left(\frac{\partial G}{\partial \lambda}\right)_{T,P} = \Delta G_{T,P} = 0.$$

λ is the so-called progress variable. We conclude that at equilibrium

$$\Delta G_T = \Delta G^0_T + RT \ln Q_{eq} = 0,$$

and hence,

$$\Delta G^0_T = -RT \ln Q_{eq} = -RT \ln \left(\frac{1}{p_{O_2}}\right)_{eq}$$

$$= RT \ln p_{O_2,eq}.$$

In the case of the formation of SiO_2 the standard reaction free energy predicts the equilibrium pressure of oxygen as a function of temperature. The more negative ΔG^0_T, or the more stable the oxide, the lower will be this equilibrium pressure of oxygen: the oxide hardly decomposes. The negative ΔG^0_T for the formation of a compound indicates directly how much energy is at least required to decompose the compound in its elements, hence to produce the pure metal out of its mineral.

A2.5 ON THE STABILITY OF OXIDES AND OTHER METAL-CONTAINING MINERALS

The Gibbs free energy is composed of two contributions: enthalpy, H, and entropy, S:

$$G = H - TS.$$

Thus, for the formation of the above-mentioned compounds:

$$\Delta G_T^0 = \Delta H_T^0 - T\Delta S_T^0.$$

Ellingham was the first to plot ΔG_T^0 as a function of temperature for oxidation and sulfidation reactions for a series of metals, relevant to the extraction of metals from their ores. Figure A2.2 shows an Ellingham diagram for oxides. Between temperatures at which phase transitions of reactants or products occur, the relation between ΔG_T^0 and T is approximately linear, the slope being equal to the average value of $-\Delta S_T^0$.

From this figure it is clear that the slopes are almost the same for all oxides. This is because the entropy changes in all these cases are similar, being almost entirely due to the condensation of 1 mol of oxygen:

$$-\Delta S_{298}^0 = \left(\frac{2x}{y}\right) S_{298}^0[M] + S_{298}^0(O_2) - \left(\frac{2}{y}\right) S_{298}^0[M_xO_y]$$

This Ellingham diagram shows the relative stability of the different oxides and the minimal amount of energy required to produce the metal. The more negative the ΔG_T^0 at a given temperature, the more stable the corresponding oxide and hence the more energy will be required to extract the metal from it. A metal oxide can be reduced by any metal or element that itself forms a more stable oxide.

In a similar way diagrams can be constructed for the formation of sulfides, halides, nitrides, carbonates,... All these diagrams give a good view on the relative stability of the compounds involved.

A2.6 THE CARBON TRAGEDY

The unique position of carbon in the Ellingham diagram for oxides appears when we consider the respective reactions

$$2C + O_2 = 2CO$$
$$C + O_2 = CO_2$$
$$2CO + O_2 = 2CO_2$$

We see that in reaction 3 there is a net loss of 1 mol of gas (slope of ΔG_T^0 versus T curve similar as for most metal oxides), in reaction 2 the number of moles of gas remains the same (entropy change close to zero and hence horizontal line in the Ellingham diagram) and in reaction 1 there is a net production of 1 mol of gas (slope of ΔG_T^0 versus T curve similar in magnitude as for most metal oxides, but opposite in sign). Actually this means that CO becomes more stable than a large number of important metal oxides at sufficiently high temperatures. Hence the key role that carbon plays in the production of metals. CO has a high calorific value: it is further oxidized to CO_2, the heat of combustion being used, for example, to produce electric power, but this at the cost of the unavoidable production of CO_2, which makes the C-based production of steel (and many other metals) in the midterm ecologically unsustainable.

A2.7 H_2 IS AN ALTERNATIVE REDUCTOR

From the Ellingham diagram we also see that H_2 is great from a reducing point of view, leading to the formation of harmless H_2O as reaction product. From an ecological point of view it

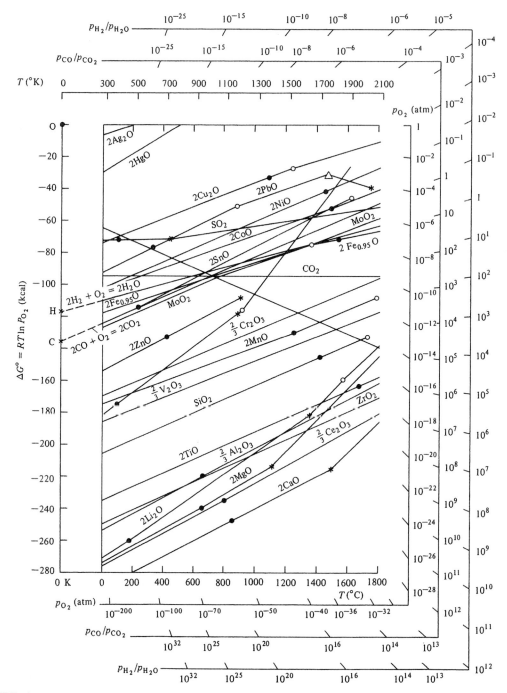

FIGURE A2.2 Standard free energy of formation of some selected oxides, corresponding to the general reaction $(2x/y)M + O_2 = (2/y)M_xO_y$. The symbols • and * refer to the melting and boiling points of the metals; the symbols ○ and △ refer to the melting and boiling points of the oxides (Lupis, 1983).

would be great to have access to cheap and vast amounts of hydrogen. There are not that many solutions. Using renewable solar energy to produce H_2 by the electrolytic dissociation of H_2O is worth further being investigated and evaluated.

A2.8 VERY STABLE OXIDES

Very low in the diagram we find the lines of very stable oxides. LiO_2, MgO and Al_2O_3 are amongst them. They are so stable that C, at acceptable temperatures, cannot be used to reduce them to produce the corresponding metals. For the production of Al, for example, the Hall-Héroult process is commonly used. It is the major industrial process for the production of aluminum. It involves dissolving alumina in molten cryolite, and electrolyzing the molten salt bath to obtain pure aluminum metal.

The production of aluminum from primary ores requires about 10 times as much energy per kg as the production of iron from its primary ores. On the other hand, melting well-recycled scrap is much cheaper in terms of energy consumption. Melting aluminum requires only 5% of the energy required for the production of aluminum starting from primary bauxite. But, the issue of purity and composition of the recycled metals and alloy, remains important. Iron as an impurity in recycled aluminum, and copper as an impurity in recycled steel, can hardly be removed. For many applications they badly influence the properties of the recycled metals. It is a major problem. The second law looks around the corner.

A2.9 ABOUT SOLUTIONS AND DESIRED PURITY LEVELS

A2.9.1 Free Energy of Mixing and Entropy of Mixing

It is of utmost importance to avoid mixing of materials (components, elements). Separation always is energy-intensive. Dissipative applications of materials should be avoided as well. Dissipated materials cannot be recovered, unless by extremely energy-intensive processes.

The partial molar free energy or chemical potential of an element i, in a liquid or solid solution, is expressed as

$$\mu_i = \mu_i^0 + RT \ln a_i$$

wherein

$$a_i = \frac{p_i}{p_i^0},$$

with p_i the vapor pressure of component i at the concentration of the given solution at fixed temperature and pressure, and p_i^0 the vapor pressure of component i in its standard state, which in case of the Raoultian standard state is pure i in its stable crystal structure at the given temperature and pressure. Because partial molar properties are additive, the integral molar free energy of a solution is given by the equation

$$G_m = \sum_i x_i \mu_i^0 + RT \sum_i x_i \ln a_i.$$

Hence, the integral molar free energy of mixing is

$$G_m^M = RT \sum_i x_i \ln a_i.$$

As Raoultian activities range from 0 at zero concentration to 1 for the pure component, the integral molar free energy of mixing is always negative: mixing is a spontaneous process, as also illustrated by the fact that the integral molar entropy of mixing is always positive:

$$S_m^M = -\left(\frac{\partial G_m^M}{\partial T}\right)_P = -R \sum_i x_i \ln a_i.$$

where we assumed, for the sake of simplicity, that a_i is temperature independent in the temperature range under consideration.

Note that the value of the chemical potential becomes $-\infty$ at zero concentration of i. This means that the removal of impurities i out of any solution becomes more energy-intensive the lower the residual concentrations are.

A2.9.2 Ideal and Real Raoultian Activities and Raoultian Solutions

The activity a_i is proportional to the mole fraction x_i: the proportionality factor γ_i being the activity coefficient:

$$a_i = \gamma_i x_i.$$

For solutions of metals, the activity coefficient γ_i is a measure for the interactions of metal i with the other metals present in the solution. When γ_i equals 1, there is no interaction. Then we have a so-called perfect Raoultian solution. When γ_i is much smaller than 1, there are strong attractive interactions between the metal i and the other metals present in the solution, hence separation becomes increasingly difficult (Figure A2.3).

In order to be able to describe real solution behavior in a quantitative way, one should know how the molar Gibbs free energies of mixing of the phases involved change with composition, or, *how the chemical potentials of the different components depend on the composition*. Thus, one should know how the activity coefficients of the different components in the solution depend on temperature, pressure and composition. This information can be obtained by experiments or by using theoretical models.

Theoretical models for a description of the behavior of different components in say a liquid solution, e.g. a liquid metal phase, a liquid metallurgical slag containing a mixture of oxide phases or a mixture of sulfides requires insight in the behavior of matter at the atomic level. Therefore we need strong simplifying assumptions concerning this behavior. For example, in a so-called ideal Raoultian solution, γ_i is equal to 1 over the entire composition range. All components of the solution are then supposed to have identical properties, which actually is never the case (the components involved would not be different species then). The molar free energy of mixing of an ideal solution is then

$$\Delta G_m^M = RT \sum_i x_i \ln x_i.$$

Then the properties of the solution can be calculated in a straightforward manner, not requiring much experimental input.

Unfortunately, *ideal behavior* never exists, but as a limiting case is approached in certain composition ranges. Yet the concept of ideal behavior is important, as it provides a *reference* for the description of the behavior of mixtures of real gases and of real liquid and solid solutions as well. The difference between real and ideal behavior is the key issue in the thermodynamic description of the composition dependency of the chemical potentials of components in mixtures of real gases and of real liquid and solid solutions.

Many theoretical models have been suggested to describe the composition dependency of the activities and activity coefficients of components in non-ideal solutions These models

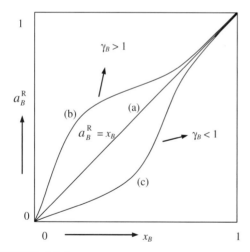

FIGURE A2.3 Schematic representation of the relation between a_B^R and x_B in a binary solution $A - B$: (a) ideal Raoultian behavior, (b) positive deviation from ideal Raoultian behavior, and (c) negative deviation from ideal Raoultian behavior.

range from very simple, e.g. the so-called regular solution model, according to which the molar free energy of mixing of a binary solution is given by the equation

$$\Delta G_m^M = RT(x_A \ln x_A + x_B \ln x_B) + \Omega x_A x_B,$$

wherein Ω is the regular solution parameter, which, in the basic regular solution model, is a temperature-independent constant. For some solutions, containing components showing very similar properties, the agreement with experimental observations is quite reasonable and good enough to describe the properties of the solution in question with an accuracy that falls within the range of experimental uncertainty. However, there is not a single model, by which the properties of all solutions can be described. One can extend the applicability of theoretical models by introducing temperature and concentration dependent parameters, and by introducing always more parameters. But finally, anyhow, experimental data are needed to get a "perfect" quantitative description of solution behavior.

When a reliable description of the molar Gibbs free energies of mixing of all the possible phases in which the different components of the system, say, CaO, SiO_2 and Al_2O_3, can occur, is available, the relative amounts and the composition of the different phases coexisting at equilibrium can be calculated as a function of temperature and pressure. Phase diagrams are the graphical representation of these phase equilibria. They show in a relatively simple way, for example as a function of temperature and concentration, which phases coexist at equilibrium and what is their composition.

Theoretical models are helpful to describe at least in a qualitative way the behavior of solutions, especially when different solutions or phases are contacting each other. The distribution of elements or components between these phases is bound to thermodynamic equilibrium conditions. These conditions learn that elements (metals to be recovered) distribute themselves between different phases. In stainless steel making, e.g. Cr, to a certain extent, is picked up by the steelmaking slag. So, Cr is "dissipated". Eventually the slag can be recycled, say as component for cement production. From the Cr point of view, the remaining Cr in the slag is lost.

Always, the chemical potential of any component i decreases strongly at low concentrations of i. As the chemical potential is a measure for the amount of energy needed to remove remaining i out of the given solution, it becomes clear that the refining of metals can be very expensive in terms of energy and that it is mandatory to avoid unwanted mixing of elements during recycling, and, far better, early in the production process.

A2.10 SOME CONCLUSIONS

- The production of metals from primary ores is energy intensive.
- Sometimes the production of metals from recycling is less energy intensive than the production of metals from primary ores.
- Recycling of metals, trying to close the materials cycle is of the highest priority, both from an energy use and a materials supply perspective.
- Thermodynamics learns that, anyhow, mixing is a spontaneous process by which the entropy increases.
- Based on the principles of equilibrium thermodynamics the distribution of components between different coexisting phases can be calculated.
- Any activity, also recycling, leads to the dissipation of energy.
- Recycling of metals from very dilute mixtures and solutions is very expensive and by all means should be avoided.

- Careful and selective collection of waste metals is mandatory in order to avoid unwanted contamination and additional expensive (eventually impossible) refining.
- Dissipative applications should be avoided. This is a top priority.
- C-based reduction processes in the midterm become ecologically unsustainable, due to the massive production of CO_2 and the depletion of fossil fuels.
- H_2-based reduction processes based on the use of "renewable" solar energy in electrolysis should be explored in detail.
- The key to a more sustainable use of metals is to use less of them. Much less. Less is beautiful.

Reference

Lupis, C.H.P., 1983. Chemical Thermodynamics of Materials. North Holland, New York, p. 134.

APPENDIX 3

Life-Cycle Assessment

Johannes Gediga

PE International AG, Leinfelden-Echterdingen, Germany

A3.1 LIFE-CYCLE ASSESSMENT

Life-cycle assessment (LCA) addresses the environmental aspects and potential environmental impacts (e.g. resource use and environmental consequences of emission and waste releases) throughout a product's life cycle from raw material extraction through production, use, end-of-life treatment and disposal (i.e. cradle to grave).

LCA can assist in identifying opportunities to improve the environmental performance of products at various points in their life cycle. The results of LCA's can inform decision-makers in industry for, e.g. strategic planning, product or process design or redesign. Communication to different stakeholder groups like governmental or non-governmental organizations is one important point of LCA to understand the environmental impact over the life cycle of the product under consideration. ISO 14040:2006 Environmental management—Life cycle assessment—Principles and framework describes the approach how to carry out an LCA under the guidance of the ISO 14040 series (ISO 14040, 2006). It is a methodological guideline that can be applied to any product system in our life. The framework is described in Figure A3.1.

LCA addresses environmental impacts of the system in the areas of ecological health, human health and resource depletion. In Figure A3.1 the main stages (including their interconnections) of an LCA are shown:

- Goal and scope definition (functional unit, system boundaries, in- and exclusions, applied method, intended audience, etc.)
- Life-cycle inventory (data collection, data quality check, LCA modeling and inventory generation)
- Life-cycle impact assessment (calculation of impacts based on inventory like global warming potential (GWP in kg CO_2 equivalent), acidification potential (AP in kg SO_2 equivalent, UseTox, etc.) (Guinée et al., 2002; Goedkoop et al., 2009))
- Interpretation of the results

It does not address in the first place economics or social effects. There are methodologies available to take the life-cycle cost (Rebitzer and Seuring, 2003) into account, but that is not under consideration. There are methodological approaches to integrate social aspects like working environment into LCA, which are not described in this appendix (Poulsen and Jensen, 2005).

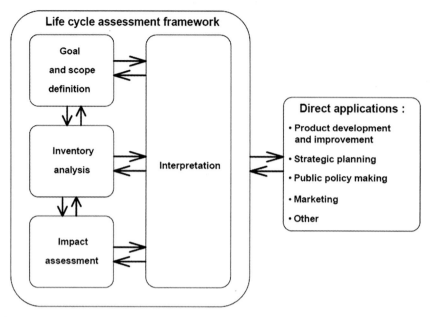

FIGURE A3.1 Framework of life-cycle assessment following ISO 14040/14044 (ISO 14040, 2006).

LCA quantifies the environmental impact of products over the life cycle to understand the main impacting stages of the life cycle and to be able to improve the material sourcing, production, the supply chain, the use phase and the end of life.

A3.2 LIFE-CYCLE ASSESSMENT IN THE MINING AND METALLURGY

The main focus of this book is on recycling, which covers the end of life of products, and therefore a multi-metal approach is sometimes given because of the mixture of different metals at that stage of the life cycle. The International Council on Mining and Metals (ICMM) is a CEO-led organization dedicated solely to sustainable development. An understanding of LCA is central to identifying points of focus for stewardship within the life cycle in terms of environmental impact. ICMM highlights the importance of taking "a holistic systems approach" whereby upstream and downstream users in the value chain are considered in materials-related decisions. This contributes to a better shared understanding of the roles and responsibilities of the various actors along the value chain leading to enhanced sustainability. LCA is increasingly seen as tools for materials choice decision-making in the design and marketing of sustainable products (Atherton and Davies, 2006) besides the other components of economic and social aspects Figure A3.2.

Considering the ICMM a CEO-led organization for sustainability, almost all metal associations performing LCA's for their products show environmental improvement over time by using the transparent methodology of LCA. It is obviously used in the communication to stakeholders like governments and downstream users, which basically follows the material stewardship thinking of the ICMM. Metal associations are asked for environmental profiles of their products because downstream users nowadays in different sectors include the

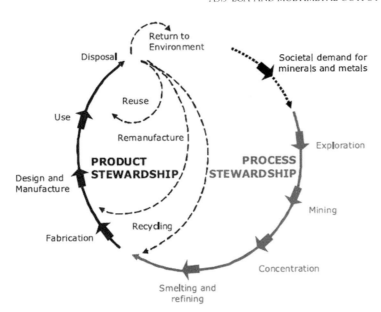

FIGURE A3.2 The scope of materials stewardship encompasses process and product stewardship (Atherton and Davies, 2006).

LCA approach into the design cycle of their products. One major sector applying the LCA methodology for design for environmental risk is the automotive industry (Finkbeiner and Hoffmann, 2006). Here the whole life cycle of cars will be considered in applying the LCA approach to identify potential environmental risks as well as improvement options. Other sectors that are, for instance, in the focus of legislators because of health-harming emissions performing LCA's to proof the benefits over the complete life cycle including the recycling of the products at the end of life. In this context the focus is on recycling how the impact is reduced compared to the primary production, but the primary production cannot be neglected either since lead, for instance, is a byproduct of zinc production as well as vice versa. This means that there are mainly multielement resources from which different materials are produced. This is an important point in terms of LCA since to allocate environmental impacts to the different materials, an allocation method needs to be applied. Different allocation methods are recommended in the ISO 14040 series and have a huge impact on the results of an LCA, which is described in specific for gold as a byproduct of different multielement resources. At the moment a consortium of representatives of different commodities is meeting on a regular basis to harmonize the methodologies for LCA, with a special focus on allocation of byproducts as well as system boundary setting in general. At the end of life there are products with an agglomeration of multimetals and therefore the applied method needs to be chosen in the right way. In the following section the difficulties of such an approach will be shown and a solution discussed for primary metal production as well as end of life.

A3.3 LCA AND MULTIMETAL OUTPUT

On the primary extraction side, there are interconnected carrier metals that are producing similar byproducts and products. As an example, nickel and copper are mentioned. Nickel as carrier metal produces copper, cobalt,

FIGURE A3.3 Example for different allocation methods (GaBi 6, 2012).

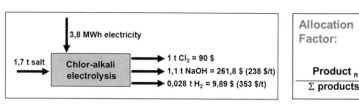

precious group metals (PGM), and copper as carrier material produces molybdenum, PGMs and zinc. Based on that natural pre-set, it is important for an LCA result that the impacts for the same metal from different carrier metals have no orders of magnitude difference (of course it depends on the type of production and type of mine underground/open pit/open cast). What is meant here is that the choice of the allocation of the environmental impact to the metals needs to be harmonized between the different multimetal carriers. An example of different allocation methods is shown in Figure A3.3 on the chlorine-alkali-electrolysis.

This figure makes obvious how different the results are considering different allocation method. In Table A3.1 results of allocation to copper as carrier metal as well as to nickel as carrier metal is shown. The basis material production data are from the public domain (MIM Holdings Limited, 2002; Goldfields Limited, 2006) and the LCA model is built in the standard LCA software GaBi 6 (GaBi 6, 2012). It is obvious that the kg CO_2 equivalent per kg for copper from nickel and copper carrier materials differ by a factor of 2 in applying mass allocation and for market value allocation it is only a factor of 1.3. This is one aspect, but if the choice is on one allocation method, it should also be applied to the byproduct. Table A3.1 also shows the results of kg CO_2—equivalent per kg of gold. Here the difference between gold from nickel and copper carrier metal is obviously a factor of 18 for mass allocation and a factor of 3 for

TABLE A3.1 Example on Different Allocation Applied to different Carrier Materials

kg CO_2-equivalent (GHG Emission Equivalent) per kg Product	Mass Allocation	Market Value Allocation	No Allocation
Copper from copper ore	4.2	4.1	
Copper from nickel route	7.2	3.1	
Gold from copper per kg Au	12,791	29,773	
Gold from nickel PGM ore	689	10,325	
Gold from mine			63,333

market value allocation. Comparing such results with gold production from gold as carrier metal, the difference is even higher compared to mass allocation.

As mentioned previously, a group of metal commodity representatives is meeting on a regular basis to develop a guideline for harmonization of allocation and other LCA-related methodology in the metal context. This is very important to avoid competitive advantage in the environmental meaning for same metals from different carrier materials by using different allocation methods.

A3.4 END-OF-LIFE TREATMENT IN THE LCA CONTEXT

There are different concepts for end of life discussed in the metal sector. In the following the extremes of the treatment of end of life recycling are described shortly.

There is extensive literature on the issue of the treatment of recycling processes in general. In a nutshell, the LCA calculation rules for recycling need to address whether and how the environmental burden of the primary production of a material is shared between the first user and the subsequent users of that material. The proposed methodologies can roughly be classified into two main approaches that are used in daily LCA practice:

1. the recycled-content approach (known as the cut-off approach) based on the principle of the first responsibility, and
2. the end-of-life recycling approach (known as the avoided-burden approach) based on the principle of the last responsibility.

These approaches represent the two extremes of how recycling can be treated in LCA. The recycled-content approach means that a product has to carry the full environmental burden of the production of its primary material, even if it is subsequently recycled. This approach provides incentives to use recycled materials due to the lower burden, but provides no incentive to develop recyclable products because there will be no benefit in terms of environmental impacts for the primary product designed for recycling.

The end-of-life recycling approach gives a benefit to the product if a recyclable material is produced from the end-of-life product; i.e. it gets a credit. This approach means that the environmental burden of the primary material production does not remain with the original user of that particular material because the burden is transferred to the subsequent users of the recycled material. This type of modeling provides incentives for recyclable products, but no incentives to use secondary materials. It may come with the risk that primary materials are used excessively, if the recycling potential is overestimated. Several approaches have been proposed that sit between these two extremes, i.e. partitioning the burden of primary material production between different life cycles. Examples are the approach that the primary material burden is to be shared equally among all life cycles, cascade recycling approaches, economic partitioning approaches or approaches that try to use the material quality as partitioning criterion. While there is currently no consensus on which of these methods is generally most suitable, the different options are widely acknowledged. This is an important point that is under discussion between the different commodities producers on the level of the group described earlier to find a consensus to deal with end of life.

A3.5 CASE STUDIES ON LCA RESULTS FOR MULTIMETAL OUTPUTS

LCA shows you the environmental impact of the considered metal production. In Table A3.2 the impact of copper production based on market allocation is shown. The considered country mix is production in Australia (13%), Indonesia

TABLE A3.2 Selected Impact Categories for the Primary Copper Production Mix from Australia, Indonesia and Chile Based on Publicly Available Literature (MIM Holdings Limited, 2002; Goldfields Limited, 2006)

CML2001-Nov. 2010	Copper Mix (From Electrolysis)
Abiotic depletion (ADP elements) (kg Sb-equiv.)	0.003
Abiotic depletion (ADP fossil) (MJ)	42
Acidification potential (AP) (kg SO_2-equiv.)	0.07
Eutrophication potential (EP) (kg phosphate-equiv.)	0.001
Global warming potential (GWP 100 years) (kg CO_2-equiv.)	4.2
Photochem. ozone creation potential (POCP) (kg ethene-equiv.)	0.003

do LCAs. For the mining industry, these numbers are an indication.

Based on the granularity of data collection, the whole production process can be broken down into single process steps, and therefore it is very useful to the metal producing companies as it is shown for the different copper smelting technologies used in different countries. This assessment can support, for instance, the design decisions to use certain technologies in different countries. It can be seen that the power provision has different impacts to the GWP for different technologies in different countries Figure A3.4.

Taking this granular information not from the existing plants but using LCA information already during process and plant design (from ore or end of life), the advantage for a sustainable material production business is obvious. There are already interfaces between plant/process design software systems like HSC and LCA software systems, as shown in Figure A3.5.

The application of such process design tools can validate existing data of metallurgical

(17%) and Chile (70%). This could be a result that could be used by downstream users in different sectors (e.g. automotive, electronic) to

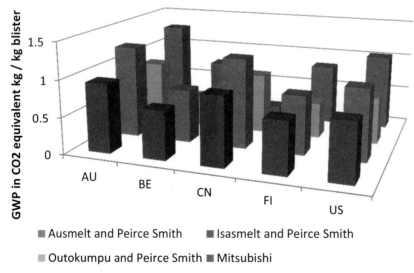

FIGURE A3.4 Global warming potential (GWP in kg CO_2—equivalent per kg matte output) (Matusewicz and Reuter, 2008).

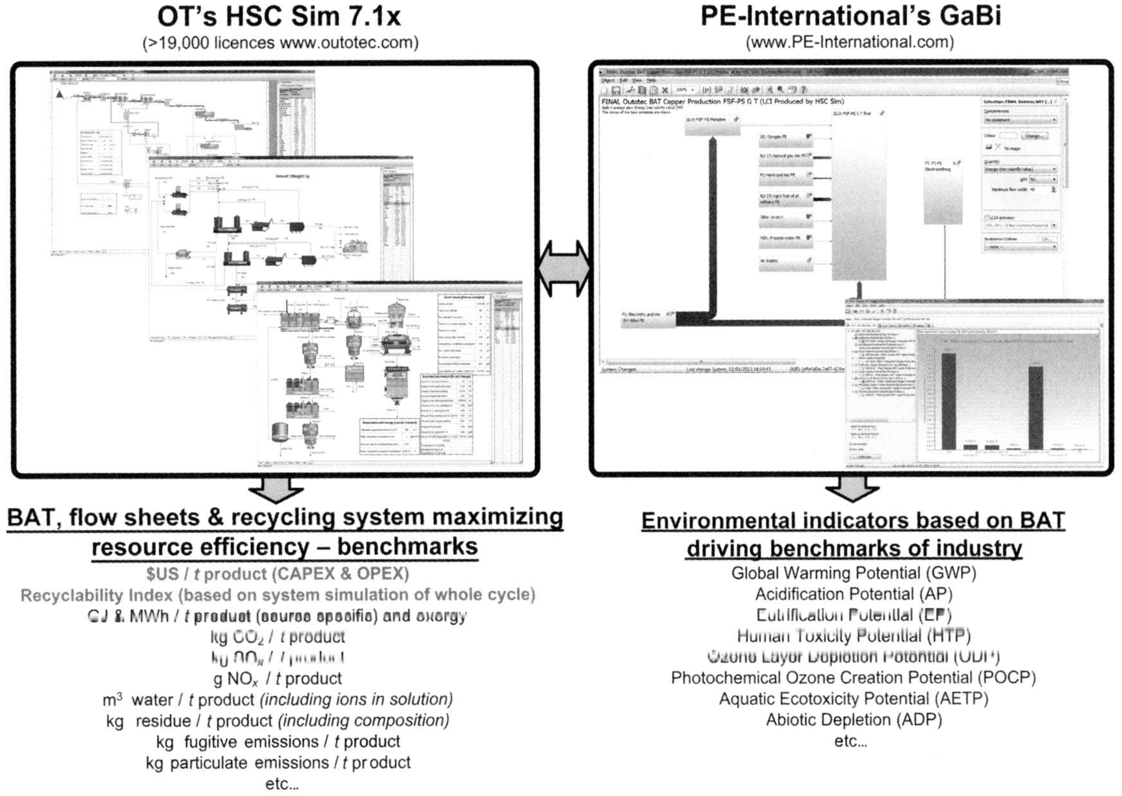

OT's HSC Sim 7.1x
(>19,000 licences www.outotec.com)

PE-International's GaBi
(www.PE-International.com)

BAT, flow sheets & recycling system maximizing resource efficiency – benchmarks
$US / t product (CAPEX & OPEX)
Recyclability Index (based on system simulation of whole cycle)
GJ & MWh / t product (source specific) and exergy
kg CO_2 / t product
kg SO_x / t product
g NO_x / t product
m³ water / t product *(including ions in solution)*
kg residue / t product *(including composition)*
kg fugitive emissions / t product
kg particulate emissions / t product
etc...

Environmental indicators based on BAT driving benchmarks of industry
Global Warming Potential (GWP)
Acidification Potential (AP)
Eutrification Potential (EP)
Human Toxicity Potential (HTP)
Ozone Layer Depletion Potential (ODP)
Photochemical Ozone Creation Potential (POCP)
Aquatic Ecotoxicity Potential (AETP)
Abiotic Depletion (ADP)
etc...

FIGURE A3.5 Interface between a process/plant simulation software and existing LCA software is shown (HSC Sim 7.1 (c) Outotec, Research Centre, 1974–2012; www.outotec.com).

processes in a scientific and rigorous way by considering slag chemistry (slag is in a liquid status) as well as thermodynamic and mass balances for all elements entering and leaving the processes/production under consideration.

Additionally, this type of granularity will become more important in LCAs of metallurgical processes since the allocation of input materials, energies, etc. can be more accurate if single process steps are better understood and intermediate process steps can be identified where byproducts are produced. As an example, precious metals are produced as a byproduct of base metal production routes (e.g. nickel, copper). The precious metals go in small streams through all different processes. With the application of a simulation tool like HSC (Pertti Lamberg et al., 1974–2012), the allocation of the environmental impact of upstream materials and energy can be performed in a related way since processes can be described as to how they are responsible for the production of the product itself (in terms of metallurgy). To explain it in a more pictorial way, you may take the process that is producing a base metal (main purpose of the process step). The precious metals are not even considered metallurgical in that process, but the precious metals are released into a separated "waste stream". This "waste stream" containing the precious metals will be further treated to separate the precious metals from each other as well as other materials like selenium oxide or others. That means in the future a black box allocation might be replaced with a real process-specific allocation based on metallurgical expertise.

Such a simulation tool supports as well the LCA results at the end of life of a product in a positive way where different multimetals can be combined that have no natural relation in the ore at the beginning of the life cycle of the considered product. To identify best practices in separation of such metals, the simulation tool in combination with LCA is one of the major decision-making tools to identify the processes with the lowest environmental impact.

A3.6 SUMMARY AND OUTLOOK

LCA addresses the environmental aspects and potential environmental impacts throughout a product's life cycle from raw material extraction through production, use, end-of-life treatment and disposal. It supports a designer to make the right choice in terms minimizing the environmental impact of a product over the whole life cycle. LCA is able to analyze life-cycle phases separately and gives indications where environmental impact reductions are possible. This in combination with the HSC process and plant simulation tool offers huge potential in scenario calculation of reduction measures.

As described in this appendix, LCA can support the plant designer to optimize in the design/development of the production process of metals to be best in class from the design phase on. This combination of the two methods incorporated in software tools like GaBi and HSC opens new possibilities in the future to develop best-practice benchmark libraries for different metals referring to different production routes to show the reduction potential between existing and best practices.

References

Atherton, J., Davies, B., 2006. Materials stewardship—towards the sustainable use of minerals and metals. World of Metallurgy—ERZMETALL 59 (1).

Finkbeiner, M., Hoffmann, R., July 2006. Application of Life Cycle Assessment for the Environmental Certificate of the Mercedes-Benz S-Class (7 pp.). International Journal of Life Cycle Assessment 11 (4), 240—246.

GaBi 6, 2012. Software und Datenbank zur Ganzheitlichen Bilanzierung. PE International AG, Leinfelden-Echterdingen.

Goedkoop, M.J., et al., January 2009. ReCiPe 2008; A Life Cycle Impact Assessment Method which Comprises Harmonised Category Indicators at the Midpoint and the Endpoint Level (Report I: Characterisation; 6), first ed.

Goldfields Limited, 2006. Sustainable Development Annual Report.

Guinée, J.B., et al., 2002. Handbook on Life Cycle Assessment. Operational Guide to the ISO Standards. Kluwer Academic Publishers, Dordrecht. ISBN: 1-4020-0228-9.

ICMM (International Council on Mining & Metals), 2013: http://www.icmm.com/.

ISO 14040, 2006. Environmental Management—Life Cycle Assessment—Principles and Framework.

Matusewicz, R., Reuter, M.A., 2008. The role of Ausmelt's TSL in recycling and closing the material loop. In: Mishra, B., Ludwig, C., Das, S. (Eds.), Proceedings REWAS2008, Cancun, Mexico. The Minerals, Metals & Materials Society (TMS), Warrendale, Pennsylvania, pp. 951—962. ISBN: 978-0-87339-726-1.

MIM Holdings Limited, 2002. Environment and Community Report.

Pertti Lamberg, A.R., Mansikka-aho, J., Björklund, P., Pentala, J.-P., Talonen, T., 1974—2012. HSC 7.1, Outotec. Research Centre.

Poulsen, P.B., Jensen, A.A., January 31, 2005. Working Environment in Life-Cycle Assessment. SETaC Foundation.

Rebitzer, G., Seuring, S., March 2003. Methodology and application of Life Cycle costing. International Journal of Life Cycle Assessment 8 (2), 110—111.

Index

Note: Page numbers with "f" and "t" denote figures and tables, respectively.

A

Abbey Steel, 463
Abrasive media, glass as, 201
Acid drainage, batteries, 97–98, 98f
Acid leaching, 76–77, *see also* Leaching
Acid production, 260
Adsorption deinking, 177
Aeronautics industry, rCFRPs in, 277
Aerospace industry, 141
Advanced disposal fee, 512–513
Agglomeration, 185
Aggregate
  glass
    characteristics, 202t
    light-weight aggregate, 203–204
    unbound, 201
    use in concrete, 202–203, 203f
  use of by-products as, 241
Agreements, voluntary and negotiated, 503
Agriculture by-products, 245
Air/oxygen injection, 108
Air-sitting in construction and demolition, 392, 395
Aligned preforms impregnation, 276
Alkali waste neutralization, 97–98
Alkali-activated cements, 225–227
Alkali–silica reaction (ASR), 202–204
Allocative efficiency, 511–512
Alloys
  battery, 132
  indium-based, 30–31
  -specific recycling, 18–19
Alternative cements, 225–227
  alkali-activated cements, 225–227
  calcium sulfoaluminate cements, 225
Alternative fuels, 220–221
Alternative raw materials, 221–222, 492

Aluminothermic reduction (ATR), 144–145
Aluminum, 18–19, 36, 56–57, 68–70, 78, 85, 92–93, 96–97, 115–116, 129, 300–302, 307–310, 464, 471–473, 472f
  production, 262
Aluminum gallium arsenide (AsGaAl), 32
Aluminum oxide, 222t
Aluminum sulfide–iron sulfide, 78
Amber glass, 191, 194
Ammonia, 121
Ammonia chloride, 107
Ammonium carbonate, 99
Ammonium chloride, 267
Ammonium sulfate, 99–100
Anaerobic digestion, 465
Anthropogenic/in-use stocks, 494–495
Antimony, 23, 85, 92, 96, 105, 107, 109
  circulation in Europe, 147f
  recycling, 147–148
  removal of, 108
Apparent density, 44
Apparent viscosity, 44
  parameters, 44t
Apple Inc., 463
Arch foundation, 289
Arrow of time, 547
Arsenic, 23, 74, 76, 92, 96, 107, 109, 156
  removal of, 108
Ash
  bottom ash, 233t–238t
  fly ash, 223, 225, 227, 241, 246, 286
  lumber, 156–157, 245
Asia, paper consumption in, 166, 167f
Asphalt, 293t
Aurubis, 88–90, 129
Ausmelt furnace, 88, 89f

Ausmelt Top submerged lance smelting, 103
Austenitic stainless steel, 73–75
Austerity, 450–453
Auto shredder residue (ASR), 77
Automated electronic eye sorting, 194
Automobile catalytic converters, 35
Automotive catalysts, 132
Automotive recycling, 344–347
  application and results of process simulation/recycling, 345–347, *see also* End-of-life vehicles (ELVs)
Avoidance, defined, 10

B

Bakelite, 179
Balancing, material, 51–54
  linear data reconciliation, 51–54
  nonlinear data reconciliation, 54
Ballotini, 198–199
Bamboo mat corrugated roofing sheet, 287
Barium, 23, 192
Barium magnesium aluminates, 433
Barriers
  policy instruments, reducing, 521
  to recycled materials use, 426f
Base metals operations (BMO), 129–130
Basel Convention on the Transboundary Movements of Hazardous Wastes and Their Disposal, 481
Basic oxygen furnace (BOF), 65–66, 72
  behavior of impurity and tramp elements during steel decarburization, 75t
Bastnäsite, 31
Bath smelting, 88

563

Bath smelting (*Continued*)
  plus TBRC, 88–90, 101–102
Batteries, 321–328
  alloys, 132
  in electric vehicles, 327–328
  lead–acid, 96–97, 327
    antimony recycling from, 148
    composition, 97t
    processing, 97–100, 98f
    schematic, 97f
  lithium ion batteries, 145t, 325–327
  material compositions by weight, 326t
  rechargeable batteries, 325–327
  recycling technologies, 359–360
  SLI-type battery, 96
  zinc/alkaline batteries, 325
Bauxite, 32
Belt filters, 175
Benzyl-alcohol, 273
Beryllium, 24
Berzelius Stolberg GmbH, 102
Best available technology (BAT), 344, 356, 360–361, 363–367, 370–373
Best-of-2-Worlds philosophy, 443
Betts process, 109
Binary particles, 55
Bio-based plastics, 182
Bioenergy, wood as, 152–154
Biomass, woody, 152, 155–158
Bismuth, 92, 96, 107, 109, 115
Bismuth telluride, 139
Bituminous highway pavements, use of glass in, 201–202
Black copper, 90
Blast furnace (BF), 72, 100, 129–130, 183–184
Blast furnace slag (BFS), 222–224, 227, 241, 245
  as industrial by-product, 233t–238t
Bleaching, in paper recycling, 175–176
Blended cements, residues in, 222–225, *see also* Cementitious binders
Blow molding, 186
Boards, 165, 293t
  recycling, 506
Boliden, 129
Boliden Rönnskär smelter, 90
Boron-based solutions, 156

Borosilicate glass, 192, 204
Bottle banks, 193
Bottle-to-bottle recycling, 188–189
Bottom ash, as industrial by-product, 233t–238t
Boudouard reaction, 122
Boulding, Kenneth, 447–448
Brass, 85–86
Breakage, 537
  at recycling, 537
Breaking of batteries, 98–99
Bricks, 293t
  as industrial by-product, 233t–238t
Bronze, 85–86
Brown glass, *see* Amber glass
Budget airlines, 453
Building foundation, 289
Building industry by-products, 239t–240t
Buildings
  carbon balance in, 153
  design for disassembly of, 154
  materials production for, 160
    alternatives, 161t
    primary energy and GHG implications, 159t
Bulbs, light, 192
Bulk molding compound (BMC) compression, 275
Buoyancy, 42
Bushing, 192
By-product, *see* Industrial by-products

C

Cadmium, 74, 115
Cadmium–tellurium-based solar cells, 139
  recycling methods for, 139t
Calcination, 133–134
Calcium, 96, 156–157
Calcium carbonate, 176, 222t
Calcium silicate hydrate (CSH) gel, 204
Calcium sulfoaluminate cements, 225
Calcium–metal halide fluxes, 78
California bearing ratio (CBR), 201
Cameras, disposable, 462
Canada, paper recovery in, 166, 166f
Capacitors, 141
Carbon, 72–73, 78–79
  balance in wood- and concrete-frame buildings, 153

Carbon dioxide emissions, 483–484, 500–501, 500f
Carbon fiber reinforced polymer (CFRP) *see* Carbon fibers recycling
Carbon fibers recycling, 269–284
  mechanical properties of rCFs, 273f
  applications for rCFRPs, 277–278
    nonstructural applications, 277–278
    structural applications, 277
  challenges
    mechanical response, understanding of, 279–280
    scalability and optimization of recycling operations, 280–281
    successful carbon fiber reinforced polymer recycling chains, 278–279
  composites remanufacturing, 274–277
    advantages and drawbacks, 274t
    aligned preforms impregnation, 276
    direct molding, 275
    nonwoven preforms impregnation, 275–276
    woven fabrics impregnation, 276–277
  life-cycle analysis of CFRPs, 278
  processes, 270–274
    advantages and drawbacks, 272t
    mechanical recycling, 270
    thermochemical fiber reclamation, 270–274
Carbon tragedy, 550
Carbothermal reduction, 119
Carbothermic copper slag reduction, 90
Cardboard recycling, 293t, 533, *see also* Boards, *see also* paper
Carpet, conversion of, 215
Carrolite, 27–28
Cars
  car-breakers, 462
  bodies, designing, 368–370
  fuel consumption in, 467–468
  SuperLight Car project, 368–370
  technology of, 462–463
Cascade circuits, 172
Cascading, wood, 155–156, 160
Ca-Si-Al-oxides, 222t
Ca-Si-Fe-(Mg)-oxides, 222t

INDEX

Catalysts, 126, 132, 141
  compositions in product-centric recycling, 328–329
  recycling technologies, 359–360
  spent oil refining catalysts, treatment and recycling activities, 263–265
  wash-coating catalysts, 35
Catalytic converters, 35
Cathode fluorescent lamps (CCFLs), 31–32
Cathode ray tubes (CRTs), 192, 203
CE mark, 247
Cellular concrete technology, 288
Cement, 246
  concrete, 289
    glass as component in, 204
    reinforced, slab and beam, 289
  ferro cement roofing components, 287
  recycling, 468–469
Cementation, 120–121
Cementitious binders, 219–230
  alternative cements, 225–227
    alkali-activated cements, 225–227
    calcium sulfoaluminate cements, 225
  clinker
    and cement, 222–225
    production, 220–222
  EN 197-1 cement type specifications, 223f
Centrifugal cleaning, in paper recycling, 172–173
Ceramics, 141
  use of glass in, 199–201, 200f, 200t
Cerium, 132–136
Cerium oxide, 132
Certification, of by-products, 247
Chalcopyrite, 91–92
Charges for waste collection and processing, 486
Charity shops, 473
Chemical fiber pulps, 165
Chemical industry by-products, 233t–238t
Chemical preservatives for wood, 156
Chemical process industry, 141–145
Chemical recycling
  of carbon fibers, 273–274
  of plastics, 184, 188–189
Chemical vapor transportation process, 266–267
China clay (kaolin) waste, 176, 287

  as industrial by-product, 233t–238t
Chlorine, 77
Chlorine-alkali-electrolysis, 557–558
Chromated copper arsenate (CCA), 156
Chromium, 67–68, 72–77, 79, 79t, 140
Chromium oxide, 73
Circular economy, 4f, 10, 445–478
  aluminum, 471–473
  cement, 468–469
  clothing and textiles, 473
  discussion, 474–476
  energy requirement, 446
  glass, 473–474
  iron and steel, 466–468
  material conservation, hierarchies of, 449–465
    decomposition, 465
    disposal, 465
    downcycling, 465
    recycling, 464
    reducing demand, 450–459
    reuse of materials, 459–464
  paper, 470–471
  plastics, 469–470
  possibility/desirability, 445–449
Clay-based fired ceramics, 199
Cleaning, in paper recycling, 172–173, 174f
Clear glass, see Flint glass
Climate change, 483–484
  impact on paper recycling, 177–178
Clinker, 219
  and cement, 222–225
  production, 220–222
    alternative fuels, 220–221
    alternative raw materials, 221–222
Closed economy, 445
Closed loop industrial ecosystem, 481
Closed-loop recycling, 188–189
  of glass, 191, 194–195
    environmental benefits, 195–196, 196f
Clothing recycling, 473, see also Textile recycling
Coal combustion residues, 246
Cobalt, 27–28, 36, 107–108, 134, 146, 557–558
Co-evaporation, 33
Cogeneration plants, 261–262
Coins, 127
Cold forming/drawing, 76

Cold hearth melting, 144
Collection
  of glass containers, 193
  in informal sector, 442
  of recovered paper, 165–166
  systems, 167–168
  recycling, 40
Collectors, 541
Collusion of producers, 516
Color sorting of glass, 194–195
Co-mingled/mixed collection for recycling, 303t, 304
Commingled collection, 167
Communication
  instruments, 521–522
  tools, 523
Compact fluorescent lamp (CFL) recycling, 523
Complex multimaterial consumer goods recycling, 329–344
  complex product and material mixtures from EoL sources, 330
  first principles, 330–340
    metallurgical processing infrastructures and process thermodynamics, 336–338
    nonlinear interconnected nature of design and recycling, 333–336
    process and recycling system optimization, 342–344
    product design, liberation, recyclate quality and losses in recycling, 332–333
    recycling technology developments in, 340–342
    resource efficiency maximization, thermodynamic detail for, 338–340
Component reuse, 462–463
Composites remanufacturing, carbon fibers recycling, 274–277
  advantages and drawbacks, 274t
  aligned preforms impregnation, 276
  direct molding, 275
  nonwoven preforms impregnation, 275–276
  woven fabrics impregnation, 276–277
Compressive stress, 537
Computer processing units (CPUs), 455–456
Concentration limit, 433–434

Concrete, 293t
  glass as cement component in, 204
  as industrial by-product, 233t–238t
  use of glass aggregate in, 202–203, 203f
Concrete recycling methods, 386–387
  cost and benefit analysis, 393–395, 389t–392t
  gravity concentration method, 387
  heating and grinding method, 386
  mechanical grinding method, 386–387
  screw grinding method, 386
Constraints, recycling rate, 22–23
Construction (civil engineering) by-products, 239t–240t, 241–245
Construction of Earth
  technical and environmental features and their measurement, 248t–250t
  technical requirements for by-products used in, 242t–244t
Construction and demolition, 285–296, 385–396
  by-products, 233t–238t
  case study
    low-cost construction technologies, 289–291
    traditional construction methods, 289
  concrete recycling methods, 386–387
    gravity concentration method, 387
    heating and grinding method, 386
    mechanical grinding method, 386–387
    screw grinding method, 386
  cost and benefit analysis, 388–395
    concrete recycling method, 393–395, 391t–392t
    current method, 389–393, 389t–390t
    current method vs. concrete recycling method, 395
  earth/mud building, 287
  lightweight foamed or cellular concrete technology, 288
  low-cost housing technologies
    cost-effectiveness of using, 291–294
    existing, 286–287
  prefabrication method, 287–288
  recycled aggregate concrete applications, 385–386

recycling technologies and practice, 294
reuse of demolished concrete, 386t
stabilized earth brick technology, 288
Consumer constraints, 431
Consumption
  of energy for paper types, 169t
  of materials, 5, 9
  of paper per capita, 166, 167f
  tax, 512
  of water in paper recycling, 177
Container glass, 191–193
  -deposit legislation, 193
  reuse of, 194
Contimelt process, 88, 88f
Continuous-filament glass fiber, 192
Conventional pulpers, 171–172
Cookware glass, 192
Cooperite (PtS), 33–34
Copper, 24, 27, 33–34, 36, 40, 45, 68–70, 74–76, 78–79, 79t, 81, 95–96, 107, 109, 121, 129–130, 132, 557–558
  -containing residues, 116–119
  decoppering, 108
  industry dusts
    flue dusts, 120t
    zinc recycling from, 119–121
  recycling, 85, 311
    challenges in, 91–93
    processes for, 87–91
    raw material for, 86–87
    routes for different types of materials, 93f
  secondary, raw materials, 120t
Copper indium gallium diselenide (CIGS) solar cells, 138–139
Copper-based solutions, 156
Copper-indium-gallium-diselenide (CIGS), 31–33
Cost and benefit analysis, in construction and demolition, 388–395
  concrete recycling method, 393–395, 391t–392t
  construction waste, 393
  crushing, 394
  final product stage, 395
  manual removal process, 394
  sorting, 393–394
  stockpiling, 393

washing, screening or air-sitting, 395
current method, 389–393, 389t–390t
  blasting, 392
  construction waste, 389–392
  crushing, 392
  final product stage, 393
  sorting, 392
  stockpiling, 392
  stripping, 392
  washing, screening or air-sitting, 392
current method vs. concrete recycling method, 394t, 395
Cost performance, waste triangle, 414
Cost/static efficiency, 511–512
Cost-to-performance ratio, 422
Council for Textile Recycling, 211–212
Courtyards, technical requirements for by-products used in, 242t–244t
Cracks, steel, 75–76
Creosote, 156
Crumbs, 185
Cullet, defined, 194
Cumulative distribution, 41
Curbside collection systems, 412
  glass collection, 193
  in packaging recycling, 304
Cutting bits out, 463
Cylindrical vessels, pulping in, 172

D
Data reconciliation
  linear, 51–54
  nonlinear, 54
Decomposition, 465
Decoppering, 108
Decoupling, 298–299
Decree, 506
Deinking, 168, 170–171, 173, 177
  adsorption, 177
  ultrasound, 177
De-inking sludge, as industrial by-product, 233t–238t
Demand, reduction of, 450–459
  improving yield, 457–459
  intensifying use, 453–454
  life extension, 454–456
  lightweight design, 456–457, 469f
  simplicity, austerity, poverty, 450–453

Dematerialization, 298–299
Demolition, *see* Construction and demolition
Demolition wood, 154
Dense media separations, 540
Deposit-refund schemes (DRS), 487, 512–515
Design for disassembly, of buildings, 154
Design for recycling and resource efficiency, 14f, 360–376, 421–422
  eco-design, 421, 422f
  limits, 423–425
  electrical and electronic equipment, 370–373
  lighting, 373–374
    alternatives for use of RE(O)s in fluorescent powder in lighting (and TVs), 374
    multimaterial lightweight automobiles, 367–370
    designing car bodies (example), 368–370
  product composition data requirements, 374–376
  recyclability index and eco-labeling of products, 362–367
    DfR rules and guidelines, 364–367
    requirements for the application of tools and rules, 374
  steel, 80
Desulfurization of battery paste, 99–100
Detection limit (DL), 433–435
Detection system, 434f
Detinning of steel scrap, 78
Developing countries, *see* Informal e-waste recycling
Dewatering, in paper recycling, 174–175, 176f
Dezincing of galvanized steel scrap, 76–78
Diamonds, 216–217
Differential and variable rate (DVR)-based schemes, 486
Digital printing inks, 177
Dimethyl terephthalate (DMT), 188
Direct molding, 275
Direct reduced iron (DRI), 70–72, 81
Direct scrap, *see* Home scrap
Direct smelting processes, lead, 100–101, 102f

Directive on Packaging and Packaging Waste, 504–505
Disamenity effects of waste-related processes, 484
Disassembly, costs of, 462
Discussion recycling, 474–476
Diseconomies of scope, 480–481
Disk filters, 175
Dispersing, in paper recycling, 175–176
Disposal fee, 512
Dissipated materials, 551
Distillation, 132
Door and window frames, 291
Door-to-door collection system, 441–442
Downcycling, 19, 465, 499
Drag coefficient (C_d), 42–44
Drag diameter, 40–41
Drag force, 42
Drop-off system, 412
  in packaging recycling, 304
Drosses, zinc oxide production from, 114–116
Drum pulpers, 169–172
  high-consistency, 173f
Drum separator/screen, 184–185
Durable goods, 515–516
  extended producer responsibility (EPR) for, 515–516
Dusts, 262
  copper recovery from, 87, 90–91
  flue dusts, 120t
Dynamic efficiency, 511–512
Dynamic separators, 538–539
Dynamics, of recycling process, 50–51
Dysprosium, 132–136

E

EAFS, as industrial by-product, 233t–238t
Earth construction
  technical and environmental features and their measurement, 248t–250t
  technical requirements for by-products used in, 242t–244t
Earth/mud building, 287
Eco-Cement, 222
Eco-design, 421
  of materials, 422f
  recycling strategies, 422

  limits, 423–425
  wheel of, 422f
Ecoinvent data, 424–425
Economic instruments, 511
  durable goods, 515–516
  environmental policy instruments, 512–513
  EPR and recycling certificates, 515
  international market for waste and materials, policy instruments in, 517
  multiproduct and mixed waste streams, 514–515
  nonrenewable resources, 517
  policy instruments, comparing, 511–512
  product and recycling markets, imperfect competition in, 516
  upstream green product design, incentives for, 513–514
Economic recycling, 5
Economics of recycling, 479–490
  economic instruments, 485–487
    deposit-refund schemes (DRS), 487
    instrument mix design, importance of, 487
    subsidies and public facilities, 486–487
    taxes and charges, 486
  economic trends and drivers, 479–481
    generation of waste, 480
    international trade, 481
    recycling, 480–481
  environmental and social costs and benefits, 481–485
    economic valuation, 482–483
    external costs and benefits, 483–484
    private and external costs, 484–485
Economies of scale, 480–481
Eco-profiles, 424–425
Eddy current separator, 184, 542
Efficiency, *see* Recycling efficiency, *see* Resource efficiency
E-glass, 192
Electric arc furnace (EAF), 65–66, 70–73, 71f, 77, 79
  dust in zinc recycling, 116–119, 117f

Electric equipment, design for recycling and resource efficiency, 370–373, *see also* Waste electric and electronic equipment (WEEE)
Electric motors, 455–456
Electric smelting furnace, 87
Electric vehicles, batteries in, 327–328
  recycling routes for, 146f
Electrical sector
  glass recycling in, 193
  use of PGMs in, 35, 126–127
Electrochemical deposition, 33
Electrolysis, 88–90, 92, 121
Electrolytic detinning of steel scrap, 78
Electrolytic refining, lead, 109
Electron-beam (EB), 33
  evaporation, 30
Electronic equipment, design for recycling and resource efficiency, 370–373
Electronic eye sorting, 194
Electronic metals, 137–140
Electronic sector
  glass recycling in, 192–193
  use of PGMs in, 126–127
Electrorefining, 92
Electrostatic separations, 542–543
Electrowinning, 92, 120, 129
Ellingham diagram, 549–550
End processing, in informal sector, 442–443
End-of-life (EOL) products, 125
  gallium recycling and, 33
  metal recycling, 17–19, 18f, 547
    recycling rate (EOL-RR), 19–20, 20f, 22, 23f
  vehicles, *see* End-of-life vehicles (ELVs)
  wood, 152–154
End-of-life impacts, 424, 424f
End-of-life recycling
  approach, 559
  rate, 256, 258f
End-of-life vehicles (ELVs), 316–318, 317t
  automotive recycling/recycling, 344–347
    application and results of process simulation/recycling, 345–347
  flowsheet and process simulation model, 344f

End-of-Life Vehicles Directive (1997), 193
End-of-pipe policy, 492–493
Energy
  costs, 498
  demand
    in glass recycling, 195–196
    in paper recycling, 177
  importance of, 547
  recovery, 465
  defined, 11
Energy production industry by-products, 233t–238t
Engitec, 99
Enhanced Landfill Mining concept, 224
Entropy
  law, 36, 547
  of mixing, 551
Environment
  benefits of closed-loop glass recycling, 195–196, 196f
  Laxer standards, 481
  recycling and, 422–423
  risks of by-products, 250–251
  policy instruments, 512–513
  technical-economic environment, 430
  United Nations Environment Programme, 20
Environmental consciousness, 484
Environmental Kuznets curve (EKC), 298–299, 480
Environmental performance, waste triangle, 414
Equilibrium thermodynamics, 548
Erbium, 133–136
EReaders, 470–471
Estimability, 52
Ethylene glycol (EG), 188
Eurecat Inc., 264–265
European List of Standard Grades of Recovered Paper and Board (EN 643), 167–168
European Union (EU), 152–153, 180–183
  Biocidal Products Directive (98/8/EC), 156
  EU Batteries Directive, 134–135
  EU Packaging and Packaging Waste Directive (1994), 196–197
  plastics

  distribution of applications in consumption and waste in, 183f
  recovery and recycling rates, 183
  recycling, 184
  usage in, 180f
  Raw Materials Initiative, 255
  waste framework Directive, 528–535
    core definition of waste, 529–530
    recovery and disposal, 530–535
    goods with positive economic value, 533
  WEEE Directive, 398–400, 400f
Europium, 31–32, 133–136
Evaporation techniques, 78
E-waste, *see* Waste electric and electronic equipment (WEEE)
Excess soil, as industrial by-product, 233t–238t
Expanded polystyrene (EPS), 181t–182t
Experiences, in recycling policy, 504–507
Extended producer responsibility (EPR), 303, 397, 503, 506, 511, 516
  and recycling certificates, 515
External costs and benefits, 483–484
Extraction tax, 512
Extractive metallurgy, 546–547
Extrusion, 185, 188
EZINEX process, 119, 121

F

Factual communication, 524
Fast spectrometric sorting, 424t
Fat hardening, 261
Feedback, 523–524
Feedstock recycling, 183–184
Feret diameter, 40–41
Ferritic stainless steel, 73–75
Ferro cement roofing components, 287
Ferroalloys, 67–68, 140
  production from waste, 257–265
    secondary raw materials, 259–262
    spent oil refining catalysts, treatment and recycling activities, 263–265
Ferrochromium (FeCr) slag, as industrial by-product, 233t–238t
Ferroplatinum (Fe–Pt), 33–34
Ferrous metal, 293t
Ferrous slag, 246

Fertilizer industry, 245, 250–251, 261
FGD (flue gas desulphurization) slag, as industrial by-product, 233t–238t
Fiber, glass, 192
Fiber consumption, 211
Fiber extrusion, 186
Fiber sludge, 246
Film blowing, 186
Filtration media, glass as, 204–205
Final tailings, as industrial by-product, 233t–238t
Fine crushing, 186–188
Flake-to-fiber recycling, 188
Flat glass, 191–192
  recovery, 193–194
Flat-panel displays (FPDs), 30–31
Flexo ink, 177
Flint glass, 191, 194
Float glass method, 192
Floor space, 462–463
Flooring, 289, 291
Flotation, 90, 540–541
  in paper recycling, 170, 173–174, 175f
Flowsheet, recycling, 51f, 53f
  as signal flow diagram, 51f
Flue dust
  from copper industry, 120t
  as industrial by-product, 233t–238t
Fluidized bed process (FBP), 272–273
Fluorescence duration, 432t
Fluorescence intensity, 432t
Fluorescence signal, 433
Fluorescent lamps, 319–321
Fluorescent lighting tubes, 192
Fluoro-max Jobin Yvon laboratory spectrometer, 433
Fluorophores, 433
FLUREC process, 121
Fly ash, 223, 225, 227, 241, 246, 286
  as industrial by-product, 233t–238t
Foamed glass, 199
Forest industry by-products, 233t–238t
Forest management, 151
Formalization, 441–442
Fossil fuels, 153, 157–158
  resources, 446–447
Foundry sands, as industrial by-product, 233t–238t
Framework Agreement, 506
Free energy of mixing, 551

Frequency distribution, 41, 47
  of single property, 55f, 56
Fuel consumption in cars, 467–468
Fuel gas treatment, by-products in, 246
Fuming
  of copper slags, 90
  of zinc slags, 121–122, 122t
    reactions between gas, char and slag, 122f
Fuming furnace, 73–74
Functional needs analysis, 429–431
  characterization of various elements of system's environment, 430
  functional analysis, 429–430
  general conclusions, 431
  main constraint functions, 431
Functional recycling, 18
Fundamental variance, 47–48
Future sustainable material economy, 448–449

G

GaDi softwares, 242, 246–247, 260, 550
Gadolinium, 132
  recycling, 133–136
Galena, 95, 100
Gallium, 27–28, 32, 137–138, 138f
Gallium arsenide (GaAs), 32–33, 137
Gallium nitride (GaN), 32, 137
Gallium phosphides, 32
Galvanized steel scrap, dezincing of, 76–78
Galvanizing, 36
Garnets, yttrium–iron, 136–137
Gas cleaning, 72–74
Gaseous alternative fuels, 221t
Gates–Gaudin–Schuhmann (GGS) equation, 42
General Union Environment Action Program to 2020, 528
Generic End of Waste (EoW) criteria, 231–232
GeoHay, 213–214, 214t
Geopolitics of resources and recycling, 491–496
  geopolitical context, recycling in, 494–495
  resources, scarcity and geopolitics, 493
Germanium, 23–24, 27, 33
Gibbs energy, 548
Glass, 293t
  collection, 193
  as industrial by-product, 233t–238t
  industry, 132
    use of precious metals in, 127
  manufacture, 192
  optical glass, rare earths recovery from, 358
  and packaging waste, 301–302, 302f, 304
  recovery for reuse and recycling, 192–194
  recycling, 473–474, 506
    closed-loop recycling, 194–196
    growth of, 196–198
    open-loop recycling, 198–205
  reuse of, 194
  types of, 191–192
Glass wool, 192
  insulation, 199
Glycolysis, 188
Gold, 30–31, 33, 37, 78–79, 126–129
Goods with positive economic value, 533
Grade-recovery curves, 57–60
  after breakage, 59f
  liberation based-, 57f
  mechanical separations, 58–60, 59f
Granulated aggregate, 204
Graphic paper, 177
  stock preparation, 170–171, 171f
Gravity
  concentration, concrete recycling method, 387
  separations, 539
Green glass, 191
Green liquid precipitate, 246
  as industrial by-product, 233t–238t
Green product design, incentives for, 513–514
Greenhouse gas (GHG) emissions, 151, 153
  balances, case study, 157–158
  implications
    building materials production, 160, 161t, 177, 188–189, 446, 465
    particleboards production, 158–160, 159t
  reduction of, 500
Greenloop, 214t
Guidelines for Paper Stock: PS 2008, 167–168
Gypsum, as industrial by-product, 233t–238t

H

H1N1 virus, 441–442
Hafnium, 140–141
Hall-Héroult process, 550
Hamburg, 129
Harris process, 108, 129–130, 139
Heating and grinding, concrete recycling method, 386
Heavy rare earth elements (HREE), 132
Hewlett–Packard printer cartridges, 463
Hierarchy of waste management, 449–450
High stiffness fibers, 192
High-alloyed steel, 67–68
High-consistency (HC) pulping, 171–172
  drum pulper, 173f
High-density polyethylene (HDPE), 180, 189, 300–301
High-frequency vacuum melting, 133–134
High intensity separation, 542
Highway pavements, use of glass in, 201–202
Hindered settling, 44–45, 45t
  velocity, 539
Hitchhiker metals, 27
  potential and actual production of, 28t
Hoboken, 129
Holmium, 133–136
Home scrap, 11, 18, 68, 86
Horticulture by-products, 245
Hot briquetted iron (HBI), 70–72, 81
Housing
  flowsheet and process simulation model
    LHHAs, 348f
    SHHAs, 349f
  low-cost housing technologies, 286–287
    cost-effectiveness of using, 291–294
  soil structures in house building, technical requirements for by-products used in, 242t–244t
Household disposal fee, 512
Houseware glass, 192
HSC Sim, 342–344, 344f, 346–347, 348f, 349f, 350f
Hutchinson Correctional Facility (HCF), 215

Hydrochloric acid, 121
Hydrocyclone, 538, 538f
Hydrogen, 550
Hydrogen peroxide, 107
  bleaching, 176
Hydrogen processing, 265–266
  hydrogen decrepitation process, 265f
Hydrogenation catalysts, 130–132
Hydrometallurgy, 92, 107–108, 115, 119–120, 132, 134, 263–264, 311, 338–340, 355–356
Hydrophobic printing inks, 173–174, 177
Hydroprocessing catalysts, 141
  recycling processes for, 142t
Hydrothermal hot pressing, 205
  products from glass cullet using, 206t

I

Idea of efficiency, 511–512
Ideal Raoultian solution, 552
Illegal waste disposal, 512
Impact pathway approach, 484–485
Impurities in steel scrap, 79, 79t
  living through new alloys with, 81
Incineration, textile recycling, 216
Indium, 23–24, 27–31, 33, 36, 93, 125
  losses during life cycle, 29f
  reclamation and recycling routes for, 126f
  recycling, 138–139
Indium antimonide (InSb), 31
Indium Corporation, 29–30
Indium gallium arsenide (InGaAs), 31
Indium gallium nitride (InGaN), 31
Indium gallium phosphide (GaInP2), 32
Indium phosphide, 31
Indium tin oxide (ITO), 138–139
Indiumnitride (InN), 31
Indium-tinoxide (ITO), 30–31
Induction sorting, 184
Industrial by-products, 231–254
  application of, 232–246
    in agriculture, horticulture and landscaping, 245
    in civil engineering, 241–245
    in manufacture of new products, 246
    in metallurgical processes, 245
    in on-site recycling, 245–246
    waste management and wastewater or flue gas treatment, 246

by-product, defined, 231–232
  major by-products and their generic properties, 232
  technical and environmental requirements, 247–251
    environmental and health risks, 250–251
    quality control, 247
Industrial residue types, 261t
Industrial Revolution, 10
Industrial society, 3–5
Industrial sorting system, 434f
Informal e-waste recycling, 442–443
  collection, 442
  connecting informal sector to formal recycling industry, 443
  end processing, 442–443
  preprocessing, 442, *see also* Waste electric and electronics equipment
Informal sector, 439
  defined, 440
Informal solid waste management, 441–442
  example of Zabbaleen in Egypt, 441
  negative effects by formalization approaches, 441–442
  waste as an opportunity for development, 441, *see also* Solid waste management
Information instruments, 521–522
  communication tools, 523
  information and communication, 524
  target groups/audience, 522–523
Injection molding (IM), 185–186, 275
In-line strip production (ISP), 80–81
Inorganic industrial by-products, 232
Instrument mix design, importance of, 487
Insulation, glass wool, 199
Integrated circuits (ICs), 32, 35
Integrated life-cycle flows, post-use wood in, 154, 155f
Integrated waste management, 499–501
International Council on Mining and Metals (ICMM), 556
International Maritime Organization, 462
International market for waste and materials, policy instruments in, 517
International trade, 481

Internet, 461–462
iPads, 37
iPhones, 37
Iridium, 33–35, 126–127
  recycling, 129
Iridium platinum (Ir–Pt), 33–34
Iron, 36, 40, 74, 78, 92–93, 105–107, 109, 115–116, 129, 132, 134, 466–468
Iron calcinate/hematite filler, as industrial by-product, 233t–238t
Iron ore-based steel production, 71f, 72
Iron oxide, 222t
Isasmelt furnace, 88, 89f, 129
ISO standard NF X50-100, 429

J
Jewelry, 127
Jigs, 539
Jimtex for ECO2Cotton, 214t

K
Kaldo furnace, see Top blown rotary converter (TBRC)
Kaldo technology, 312f
Kaolin, see China clay
Kayser, 88–90
Kiln dust, 245–246
  as industrial by-product, 233t–238t
Kivcet, 101, 103
Kosaka Smelter and Refinery, 90–91
Kroll–Betterton process, 109

L
Labeling technique, 429–438, 524
  functional needs analysis, 429–431
    characterization of various elements of system's environment, 430
    functional analysis, 429–430
    main constraint functions, 431
  polymer labeling processes, bibliographical research on, 431–433
  polypropylene samples, first results of detection tests with, 433–436
    concentration limits, 433–436
    experimental conditions, 433
Lagrange multipliers, 53–54, 56
Lamps
  flowsheet and process simulation model, 355f
fluorescent lamps, 319–321
  CCFLs, 31–32
  material composition of, 323t
  quantities and application of critical materials in lighting, 321, see also Lighting recycling
Landelijk Afvalbeheerplan (LAP2), 499–500
Landfill, 13, 415–417, 482
  defined, 11–12
  technical requirements for by-products used in, 242t–244t
  textile recycling, 216
  wood deposition in, 152
Landfill gas (LFG), 152–153, 157–158
Landfill mining, 494
Landscaping, 245
Lansink's ladder, 449–450
Lanthanides, 132–136, 433
Lanthanum, 132
  recycling, 133–136
Laplace transform, 50
Large household appliances (LHHAs)
  flowsheet and process simulation model, 348f
Large municipal solid waste (LMSW), 379–384
  circular process for, 379–380
  collection system of, 380
  future, 383
  recycling efficiency, 382–383
  sorting of, 381
    installation, 381–382
    preconditions for, 380
    process, 382
  See also Municipal solid waste
Larvik furnace, 115–116
Laser crystals, YAl–garnet, 136–137
Laser diffraction, 40–41
Laser induced breakdown spectroscopy (LIBS), 70
Laurite (RuS2), 33–34
Laxer environmental standards, 481
Leaching, 49–50, 76–78, 92–93, 107, 120–121, 129
  behavior of by-products, 250–251
Lead, 29–30, 33, 36, 78–79, 87, 115–116, 120–121, 129–130, 139, 192
  batteries, 327
  -containing residues, 116–119, 117t
metallurgy, zinc slag fuming from, 121–122, 122t
  reactions between gas, char and slag, 122f
  recycling, 95
    alternative approaches, 107
    lead–acid battery, 96–100
    refining, 107–109
    smelting, 100–106
  refined, production and usage of, 95t
Least-square minimization, 53
Legislation, container-deposit, 193
Lego toy series, 462–463
Liberation curve, 39–40, 55–57
  -based grade-recovery curve, 57, 57f
Life-cycle, 530
  metal and product, 18f
Life extension, 454–456
Life-cycle analysis (LCA), 250–251, 342–343, 346–347, 363, 368–369, 370f, 555
  case studies on, 559
  of carbon-fiber reinforced polymers, 278–279
  end-of-life treatment in LCA context, 559
  in mining and metallurgy, 556
  and multimetal output, 557–558
  in plastic recycling, 186, 188
Life-cycle flows of post-use wood, 154, 155f
Light bulbs, 192
Light rare earth elements (LREE), 132
Light-emitting diodes (LEDs), 31–32, 321
  recovery, 356
  indium and gallium recovery from, 357
Lighting recycling, 353–359
  design for recycling and resource efficiency, 373–374
    alternatives for the use of RE(O)s in fluorescent powder in lighting (and TVs), 374
  economics of recovering rare earths, 358–359
  LEDs, see Light-emitting diodes (LEDs)
  material composition of lamps, 323t
  metallurgical recovery of REO-containing fluorescent powders, 357
  rare earth oxides recovery, 357

Lighting recycling (*Continued*)
  physical recycling of fluorescent powders, 355–356
    phosphors recovery, 355–356
    rare earths recovery from optical glass, 358
Light-weight aggregate, 203–204
Lightweight design, 456–457, 469f
Lightweight foamed concrete technology, 288
Lime (calcium oxide), as industrial by-product, 233t–238t
Limestone, 220, 224–225
Linear data reconciliation, 51–54
Linear economy, 3, 4f, 10
Linear low-density polyethylene (LLDPE), 180, 186–188
Liquid alternative fuels, 221t
Liquid crystal displays (LCDs), 29, 31–32, 192, 203
Liquid magnesium chloride, 266
Lithium, 145–147
Lithium ion batteries, 325–327
  metal contents in, 145t
  recycling processes for, 145t
Log-normal distribution, 42
Low-alloyed steel, 67–68, 74–75
Low-consistency (LC) pulping, 171–172
Low-consistency (LC) screening, pressure screen for, 174f
Low-cost housing
  case study
    low-cost construction technologies, 289–291
    traditional construction methods, 289
  cost-effectiveness, 291–294
  earth/mud building, 287
  existing technologies, 286–287
  lightweight foamed or cellular concrete technology, 288
  prefabrication method, 287–288
  stabilized earth brick technology, 288
Low-density polyethylene (LDPE), 180
Low-energy bulbs, 192
Low intensity separation, 542
Lumber recycling, 151
  background, 151–154
  case studies, 157–160
  post-use management of wood, 154–157
Lutetium, 132–136

M
Magnesium, 115, 156–157
Magnesium oxide (MgO), 191–192
Magnetic detection, 432t
Magnetic separation, 45, 541–542
Magnetic susceptibility, 541–542
Magnets, 132
Manganese, 67–68, 81, 115, 140
Manufacture of new products industry by-products, 246
Market
  demand, 425–426
  recycling, 5–6
  price, 6
Markov recycling of metals, 23f
Martin diameter, 40–41
Masonry, 293t
Mass balances, 49–50, 49f
Mass distribution, particle, 47
Material conservation hierarchies, 449–465
  decomposition, 465
  disposal, 465
  downcycling, 465
  recycling, 464
  reducing demand, 450–459
    improving yield, 457–459
    intensifying use, 453–454
    life extension, 454–456
    lightweight design, 456–457, 469f
    simplicity, austerity, poverty, 450–453
  reuse of materials, 459–464
    diverting scrap, 463–464
    reusing components, 462–463
    reusing material, 463
    reusing products, 460–462
Material design principle for recycling, 421–422
Material efficiency, 10–11, 467
Material flow analysis (MFA), 27, 81, 335, 342–343, *see also* Life-cycle analysis (LCA)
Material resources, usage of, 484
Material-centric recycling, 307–311
  aluminum recycling, 307–310
  copper recycling, 311
Materials, 3, 12–13
  balancing, 51–54
  -centric approach, 12, 14, 40
  cycles, managing, 511
  labeling, 524
  global production of, 4f
  reuse and recycling of, 4f
  scarcity of, 491–492, 517
  sorting process, 422, 426
Mattresses, conversion of, 214–215
Mechanical abrasion, 47
Mechanical biological treatment (MBT), 413–414
Mechanical drawing, 192
Mechanical fiber pulps, 165
Mechanical grinding, concrete recycling method, 386–387
Mechanical recycling
  of carbon fibers, 270
  of plastics, 184–186, 188
Mechanical separations, 58–60, 59f, 413–414
Medium-consistency (MC) pulping, 171–172
Medium-term shortages, 423–424
Meretec process, 76–77
Messaging, 524
Metakaolin, 224, 226
Metal industry by-products, 245
Metal life cycle and flow annotation, 256, 257f
Metal wheel, 337f, 370–373
Metallurgical industry by-products, 239t–240t
Metallurgy, 13–15, 14f, 132, 359
  of copper recycling, 90, 92–93
  of lead recycling, 108
  processing infrastructures, 336–338, 337f, 340–341
  recovery of REO-containing fluorescent powders, 357
Metals
  -containing minerals, stability of, 549–550
  consumption and availability of, 545
  recycling, 11, 13–15, 17, 506
    considerations and technologies, 17–19
    metal and product life cycle, 18f
    rare, *see* Rare metals, *see* recycling
    recovery, sophistication in, 6f
    recycling statistics, 19–22
    resource efficiency design, 14f
  stocks and flows of, 494–495
Metal-specific recycling, 18
  old scrap, 24
Metal-unspecific recycling, 24
Metal-unspecific reuse, 19

Methane emissions, 483–484
Methanolysis, 188
Mineral deposit, 493
Mineral reserve, 493
Mineral wool, as industrial by-product, 233t–238t
Mineral-centric approach, 12
Minerals processing, 13–15, 14f
Miniaturization paradox, effects on e-waste streams, 401–402
Mining and enrichment of ores industry by-products, 233t–238t
Mining and metallurgy, life-cycle assessment (LCA) in, 556
Mining industry by-products, 239t–240t
Mirrors, use of precious metals in, 127
Mixed collection, in packaging recycling, 304
Mixed waste
  dry recyclables, source separation of, 413
  mechanical separation, 413–414
  streams, 514–515
Mobile phones, 32
Molten magnesium, 266
Molten-salt electrolysis, 133–134
Molybdenum (Mo), 27–28, 36, 74–75, 79, 79t, 140–142, 258–259, 260t
Mono-modal particle distributions, 42
Monopolist, 516
Mud brick, 287
Multimaterial lightweight automobiles, design for recycling and resource efficiency, 367–370
  designing car bodies (example), 368–370
Multimetal output, life-cycle assessment (LCA) and, 557–558
Multiproduct streams, 514–515
Municipal solid waste (MSW), 183, 186–188, 405–418
  definition, 405–406
  management, 408–417
    assessment framework, 414–417
    mechanical separation of mixed waste, 413–414
    monitoring, 411
    separate collection and waste sorting, 412–414
    source separation of mixed dry recyclables, 413
  treatment, 414–417, 415t
  packaging waste in, 302–303
  quality, 408
  quantities, 406–408
    European MSW generation, 407–408
    global MSW generation, 406–407
  waste generation
    per country income level and projection, 408t
    per region and projection, 406t
  waste types and their sources, 410t, see also Large municipal solid waste (LMSW)

N

National Association of Resale and Thrift Shops, 217
National Board of Forestry, 156–157
National Waste Management Plan, 499–500
Near infrared (NIR) sensor, 185
Near infrared (NIR) spectral domain, 431
Negotiated agreement (NA), 503
Neodymium, 132–136
Neonatal mortality rate, 441
Net-exporter of recyclable materials, 481
The Netherlands
  packaging policy targets in, 507t
  waste and resources management in, 499–500
Neutralization, alkali waste, 97–98
New scrap, 11, 23–24, 86
Newsprint, 168–169
Newton's equation, 43
NF X50-100, 429
Nickel, 27–28, 33–34, 36, 67–68, 72, 74–77, 79, 79t, 81, 85, 107–108, 129–130, 132, 134, 141, 258–259, 260t
NiMH batteries, 327, 327t, 328t
Niobium, 67–68, 140
Nitrogen, 141
Noise barriers and butts structures, technical requirements for by-products used in, 242t–244t
Nonenergy raw materials, 255
Nonferrous metal, 293t
Nonferrous slag, as industrial by-product, 233t–238t
Nonfunctional recycling, 19
Nonlinear data reconciliation, 54
Nonredundant variable, 52
Nonrenewable materials, 9
Nonrenewable resources
  in macro economic perspective, 517
  usage of, 424–425
Non-spherical particles, 44
Nonwoven preforms impregnation, 275–276
Nuclear energy industry, 141–145
Number distribution, particle, 41–42, 47
Nutrient cycling, wood, 156–157

O

Observable variable, 52
Obsolete scrap, see Old scrap
OEMs, 347, 362–364, 369–370, 374–375
Oil-borne preservatives, 156
Old scrap, 11, 68, 86
  recycling
  metal-specific, 24
  metal-unspecific, 24
Old scrap ratio (OSR), 19
  for metals, 21f, 22
  for various metals, 260f
Onion skin models, 455f
On-site recycling, 245–246
Open economy, 445
Open-loop recycling, 188
  of glass, 191, 198–205
    abrasive media, 201
    alternative products, 198–199
    bituminous highway pavements, 201–202
    cement component in concrete, 204
    ceramics, 199–201
    concrete, 202–203
    filtration media, 204–205
    light-weight aggregate, 203–204
    unbound aggregate, 201
    ziolites and silicate products, 205
Optimized recycling scenario, 436
Ordinary Portland Cement (OPC), 219
Organic waste, 408–409, 413, 441–442
Organic-based homogenous spent catalysts, 132

Organization for Economic Cooperation and Development (OECD), 480
Osmiridium (Os—Ir), 33—34
Osmium, 33—34, 126—127
Oxidation in fluidized bed, 272—273
Oxide scrap material, smelting of, 87
Oxides and metalcontaining minerals, stability of, 549—550
Oxygen, 107, 141

P

Packaging, 297—306
  composition, 300—301
  paper, 166, 168—169
    stock preparation, 169—170, 170f
  plastic in, 180—183, 300—301, 304, 413, 470, 506—507
  recovery and collection schemes, 302—305
  recovery and recycling, 300—301
  steel, 73
  waste, 297—300
Packaging and Packaging Waste Directive (1994), 193
Packaging Covenant, 505
Packaging Decree, 506
Packaging Directive, 504—505
Packaging policy targets, in the Netherlands, 507t
Painting, 289
Palladium, 30—31, 33—36, 78—79, 126—127, 129
Paper mill sludge, as industrial by-product, 233t—238t
Paper recycling, 9, 293t, 470—471, 506
  and packaging waste, 301—302, 302f, 304
  collection and sorting systems, 167—168
  consumption per capita, 166, 167f
  energy consumption for paper types, 169t
  raw material efficiency, 168—169
  recovered paper
    collection of, 165—166
    utilization of, 165—166, 168, 169f
  stock preparation, 169—178
Parkes process, 109
Parking lots, technical requirements for by-products used in, 242t—244t
Particle size, 40—44
  distributions, 537
  translational velocity, 42—44
Particle surface, 40—41
Particle volume, 40—41
Particleboards, 154
  production, 158—160
    alternatives, 158f
    primary energy and GHG implications, 159t
Peirce-Smith converter, 87—88, 90
Pelletized aggregate, 204
Pentachlorophenol, 156
Permanent magnets, 132—133
  scraps
    in-plant recycling of, 133f
    recycling methods for, 133—134, 134t
Petroleum catalysts, 141
Petroleum coke, 262
Petroleum industry, 262
Phosphors, 135—137
  indium recovery and, 32
Phosphorus, 81, 156—157
Photography, use of precious metals in, 127
Photovoltaic cells, indium and, 31
Photovoltaic scrap, recycling methods for, 139t, 140
Physical separations
  breakage, 537
  dense media separations, 540
  dynamic separators, 538—539
  eddy current separation, 542
  electrostatic separations, 542—543
  flotation, 540—541
  gravity separations, 539
  high intensity separation, 542
  low intensity separation, 542
  magnetic separations, 541—542
  screens, 537—538
  size classification, 537
  sorting, 543
  water media separations, 539
Physical vapor evaporation, 30
Physico-chemical sorting technique, 423—424, 424t
Physics-based systemic approach, 332—333, 344, 347, 353—354, 363—364, 367—375
Pigments, 141
Pigovian externality tax, 512
Plasterboard, as industrial by-product, 233t—238t
Plastering, 289, 291
Plastic Heroes, 506—507
Plastics, 293t, 469—470
  applications, 180f
  in packaging, 180—183, 300—301, 304, 413, 470, 506—507
  recycling, 182—184, 506—507
    impact, 186—189
    mechanical, 184—186
  use of, 180—182
  use in EU, 180f
Platinum, 33—37, 126—127
  substance flow analysis of, 34f
Platinum group metals (PGMs), 27—28, 33—36, 126—127
Platy particles, 44
Plumbing, 289
Policy instruments, 521—522
  comparing, 511—512
Policy Memorandum Prevention and Reuse of Waste Materials, 504—505
Polishing, 132
Polluter pays principle, 486, 506
Polycarbonate (PC), 181t—182t
Polyester fibers, 213—214
Polyethylene (PE), 179—180, 186—188, 426
Polyethylene terephthalate (PET), 184, 186—189, 187f, 300—301, 426
Polylactic acids (PLAs), 424—425
Polymer labeling processes, bibliographical research on, 431—433
Polymer recycling companies, 436
Polyolefins, 180—182
Polypropylene, 96, 98—99, 180, 426, 432—436
  concentration limits, 433—436
  experimental conditions, 433
Polystyrene (PS), 180
Polyurethane (PUR), 180
Polyvinyl chloride (PVC), 180—181, 186—188
Polyxene (Fe—Pt), 33—34
Poor-quality promotion, 522
Population models, 45—47
Porcelain stoneware tiles, 199
Positive externalities, generating, 484
Positive incentives for recycling, 486—487

Post-consumer waste, 212
  post-consumer textile waste (PCTW), 212, *see also* Old scrap
Postindustrial waste, 212
Potassium, 156–157
Poverty, 450–453
Powder-based catalysts, 130
Power generation, 470
Pozzolanic reaction, 204
Praseodymium, 132–136
Precast concrete frame, 286–287
Precious group metals (PGM), 557–558
Precious metals (PMs), 126–132
  production of, 127
  recycling, 126–127, 128f, 130f, 131f
  separation of, 109
Precipitation, 132–134
Pre-consumer scrap, 11
Prefabrication method, 287–288
Preparing for reuse, 531
Preprocessing, in informal sector, 442
Pressure electroslag remelting (PESR), 144, 144f
Price, market, 6
Primary and secondary refined production, 312f
Primary copper smelters, 87
Primary energy, 151, 152, 158–160, 188–189
  case study, 157–158, 157t
  implications
    building materials production, 160, 161t
    particleboards production, 158–160, 159t
Primary smelting, 129
  lead, 100–103, 102f
    Kivcet, 103
    QSL, 102
    SKS, 103
    TSL, 103
Printed circuit boards (PCBs), indium and, 30–31
Printed wire boards (PWBs), 332
  sorting of, 442–443
Printing inks, 177
  removal of, 170–171
Private and external costs, 484–485
Probability entropy method, 56
Process, recycling, 40
  carbon fibers, 270–274
  complex multimaterial consumer goods, 342–344
  copper, 87–91
    dust, 90–91
    slag, 90
    WEEE, 90
  dynamics, 50–51, 51f
  informal, recovery efficiencies in, 443
  lithium ion batteries, 145t
  spent petroleum catalysts, 142t
  zinc, 114f
Process efficiency, *see* Recycling efficiency
Process Simulation, 8, 93, 342, 343f, 344–353, 344f, 348f, 355f, 364–366
Producer Responsibility Obligations (Packaging Waste) Regulations, 196–197
Producer responsibility organization (PRO), 515
Product and recycling markets, imperfect competition in, 516
Product certification, 457
Product design, 22–25
Product manufacturing phase, 421–422
Product service systems, 453
Product-centric recycling, 12–15, 14f, 40, 313–329
  batteries, 321–328
    in electric vehicles, 327–328
    lead batteries, 327
    rechargeable batteries, 325–327
    zinc/alkaline batteries, 325
  catalyst compositions, 328–329
  complex multimaterial consumer goods recycling, 329–344
  EoL consumer products
    automotive applications, quantifying use and materials in, 316–318
    EoL developments and the urban mine, 313–315
    products, quantities and (critical) materials in, 313
    requirements for data collection of, 315–316
  lighting applications, 319–321
    fluorescent lamps, 319–321
    light emitting diodes, 321
    quantities of lamps and application of (critical) materials in lighting, 321
  quantifying products and materials in electronics, 318–319
Products, 12–13
Products Made from Mechanically Recycled Post-consumer and Postindustrial Waste, 214t
Promethium, 133–136
Prompt scrap, 18
Properties of particles, 45–47
  distribution, 46f
Property space, 45–47
  3D, 46f
Pulp rheology, 44–45
  apparent density, 44
  apparent viscosity, 44, 44t
  hindered settling, 44–45, 45t
Pulping, 171–172
  in cylindrical vessels, 172
  drum pulper, 172
Purification
  of steel scrap, 76–79
    dezincing of galvanized scrap, 76–78
    detinning of steel scrap, 78
  in zinc oxide production, 115
PV Cycle association, 33
PVC, 301
Pyrolysis, 271–272
Pyrometallurgy, 90, 92, 108, 119, 121, 134, 142–143, 264, 311, 338–339, 356–357
Pyrorefining, lead, 108–109
  decoppering, 108
  precious metals, separation of, 109
  removal of Sn, As, and Sb, 108

Q
Quality
  promotion, 522
  of scrap for steel production, 72–74
  steel, influence of tramp elements on, 75–76
  of zinc oxide, 116
Queneau-Schumann-Lurgi (QSL) furnace, 102
QuietLeig, 214t

R
Random sampling, 48
Raoultian activities and Raoultian solutions, 551
Rare earth containing magnets, 265–267

Rare earth containing magnets (*Continued*)
  chemical vapor transportation processing, 266–267
  hydrogen processing, 265–266
  liquid magnesium chloride, 266
  molten magnesium, 266
  processing with ammonium chloride, 267
Rare earth metals (REMs), 132–137, 491
Rare earth oxides (REOs), 133–135, 357
Rare metals, 5, 27, 125, 145–148

  examples, 31–36
  future perspectives, 36–37
  indium, 28–31
  electronic metals, 137–140
  precious metals, 126–127
  rare earth metals, 132–137, 491
  refractory metals, 140–145
Raw material, for copper recycling, 86–87
Raw material efficiency, paper, 168–169
Real Raoultian solution, 552
Rechargeable batteries, 325–327
Reclaimed asphalt, as industrial by-product, 233t–238t
Recovering plastics, practice of, 439–440
Recovery and disposal
  role of market, 533–535
  within the new waste hierarchy, 530–533
Recovery and recycling of scarce materials, 492
Recovery efficiencies in informal recycling processes, 443
Recovery operations, 530–531
Recovery rate
  defined, 12
  paper, 165–166, 166f
Rectangular-shaped hearth furnace, 122
Recyclability index (RI), 15
  DfR rules and guidelines, 364–367
  and eco-labeling of products, 362–367
Recycle recovery rate, 481
Recycled aggregate concrete
  applications, 385–386

Recycled carbon fibers (rCFs) *see* Carbon fibers recycling
Recycled content (RC), 12, 19–20
  for metals, 20–22, 21f, 259f
Recycled input ratio, *see* Recycled content (RC)
Recycling, defined, 10–12
Recycling centers, 412
Recycling certificates, 515
Recycling efficiency, 22–23
  defined, 12
  of metals, 19
Recycling of aluminum cans (RECAL), 522–523
Recycling rate (RR), 11, 256, 499
  constraints, 22–23
  defined, 12
  end-of-life (EOL-RR) for metals, 19–20, 20f, 22, 23f
Recycling statistics, 19–22
  current, perspectives on, 23–25
  relationship between recycled material value and material mixing, 24f
Recycling subsidy, 512–513
Recycling targets, setting, 533
Red mud, 245, 286
  as industrial by-product, 246
Redemption points, glass container, 193
Reduction, 13
  defined, 10
Redundant variable, 52
Refining, lead, 107–109
  electrolytic refining, 109
  pyrorefining, 108–109
Reforming catalysts, recycling of, 130–132
Refractories, 141
Refractory metals (RMs), 140–145
Refund or deposit fees, in packaging reuse system, 304
Refuse-derived fuel (RDF), 183, 186
Regular solution model, 552–553
Regulatory constraints, 431
Regulatory instruments, 476, 527–528
  EU waste framework, 529–535
    core definition of waste, 529–530
    goods with a positive economic value, 533
    recovery and disposal, 530–535

    resource efficiency and waste strategy, 528–529
Reinforced cement concrete slab and beam, 289
Renter–landlord problem, 521
Renewable materials, 9
Reprocessing, in plastic recycling, 185
Reserve base, 493
Reserve estimates, 493
Residues
  auto shredder residue, 77
  in blended cements, 222–225
  coal combustion residues, 246
  copper-containing, 116–119
  industrial residue types, 261t
  lead-containing, 116–119, 117t
  zinc-containing, 116–119, 117t
Resource efficiency, 3–8, 6f, 7f, 10, 15, 151, 155, 492, 498–499
  design for, 14f
  and waste strategy, 528–529
Resources, scarcity and geopolitics, 493
Response strategies, 492
Reuse, 13, 531
  defined, 11
  of materials, 459–464
    diverting scrap, 463–464
    reusing components, 462–463
    reusing material, 463
    reusing products, 460–462
  packaging, 303, 303t
Reverse vending machines, 193
Reynolds number (Re), 42–44
RFCS (Research Fund Coal and Steel), 77
Rhenium, 24, 27, 140–141
Rhodia, 134–136, 135f
Rhodium, 33–35, 126–127, 129
Roadmap on Resource Efficiency, 535
Roads, technical requirements for by-products used in, 242t–244t
Rönnskär, 129
Roofing, 290–291
Rosin–Rammler–Sperling–Bennett equation, 42
Rotary furnace, 105–106, 106f
Rotary melting furnace, 310f
Rotary wool forming, 192
Runkel's model, 516
Ruthenium, 33–35, 126–127

S

SafeLeig, 214t
SafeLeigh Premium, 214t
Sakamura Press Company, 458–459
Samarium, 132–136
Sampling, 47–48, 127–129
San Shing Fastech Company, 458–459
Sanitary landfilling, 498
Sanitation and public health, 497–498
Sauter mean diameter, 41
Scandium, 132, 136
Scarcity of materials, 491–492
School Can Program, 522–523
Scrap
  classification, 68
  copper, 86–87
    types, 87t
  diversion, 463–464
  rates, reducing, 457–459
  sizing, 40–41
  steel
    detinning of, 78
    dilution with ore-based iron units, 81
    European specifications, 691
    impurities in, 79
    processing, 68–70
    purification, 76–79
    quality, 72–74
    recycling, 68f
    smelting of, 70–72, 71f
    surface area effect on recovery, 309–310, 310f
Scrap Recycling Industry Inc., 86
Screen, 537–538
Screening
  in construction and demolition, 392, 395
  in paper recycling, 172–173
    low consistency screening, 174f
Screw grinding, concrete recycling method, 386
Screw press, 175, 176f
Seawater-neutralized red mud, 245
Second law of thermodynamics devil, 547
Secondary copper
  -containing material, 87t
  raw materials, 120t
Secondary resources, recovery of metals from, 255–268
  ferroalloys production from waste, 257–265

secondary raw materials, 259–262
spent oil refining catalysts, treatment and recycling activities, 263–265
rare earth containing magnets, 265–267
  chemical vapor transportation process, recycling by, 266–267
  liquid magnesium chloride, 266
  molten magnesium, 266
  processing with ammonium chloride, 267
  recycling by hydrogen processing, 265–266
secondary raw materials, prices of, 425
Secondary smelting, 129
  lead, 103–106
    shaft furnace, 104–105, 104f
    short rotary furnace, 105–106, 106f
Secondary zinc, sources and type of, 114t
Secondhand clothing export, 213
Secondhand sale, 461–462
Selective leaching, 129–130
Selectivity, in paper recycling, 176
Selenium, 27–28, 96
Semiconductor applications
  of europium, 31
  of gallium, 32
  of indium, 31
Separability curve, 56f, 57
  after breakage, 58f
Separate collection
  for recycling, 303–304, 303t
  and sorting, in MSW, 412–414
    mechanical separation of mixed waste, 413–414
    source separation of mixed dry recyclables, 413
Separation, 49f
  of large municipal solid waste, 379–384
  mechanical, 58–60, 59f, 413–414
  in paper recycling, 172–173
  of plastics, 184–185
  source separation, 413
  two-stream, 49f
Separation cut point, 60
Separation efficiency curve, 60, 60f
Sequence, recycling, 22f
Service performance, waste triangle, 414

Settling, particles, 43f, 44
  hindered settling, 44–45
S-glass, 192
Shaft furnace, 104–105, 104f, 121
Shaking table, 539–540
Shanks sorting lines, 381–382
Sharp Corporation, 30
Ship breaking, 463
Short rotary furnace, 105–106, 106f
  Na_2S-FeS phase diagram, 107f
Shredder, 68, 69f, 86
  sorting of material from, 70f, 80
Shredding, 49–50
  in plastic recycling, 185
Shui Kou Shan (SKS) oxygen bottom blowing, 103
Si-Al-Ca-Fe-oxides, 222t
Sieving, 40–41
Signal flow diagram, 51f
  for feedback control, 51f
Silicate products, 205, 205f
Silicon, 75–76
Silicon dioxide, 222t
Silver, 27, 33, 78–79, 95–96, 108–109, 126, 130
Silverware, 127
Simapro, 424–425
Simplicity, 450–453
Simulation, 8, 93, 340, 342, 348, 360, 363, 363f, 365–367, 370, 371, 373, 375, 515, 561, 562
Sink–float separation, 185
Sintering, 100
  additive, glass cullets as, 199
Size classification, 537
Slags, 262
  blast furnace slag, 222–224, 227, 241, 245
  copper, 87, 90, 92–93
  ferrochromium, 233t–238t
  ferrous, 246
  FGD (flue gas desulphurization), 233t–238t
  nonferrous, 233t–238t
  steel, 233t–238t
  zinc, 121–122, 122f, 122t
Slogan effectiveness, 524
Slushing, 171–172
Small household appliances (SHHAs)
  flowsheet and process simulation model, 349f
Smelting, 49–50
  copper, 86–88, 90–92, 91f

Smelting (*Continued*)
  lead, 100–106
    primary, 100–103
    secondary, 103–106
  steel, 70–72, 71f
Soda-iron sulfide, 78
Sodium carbonate, 99
Sodium hydrosulfite bleaching, 176
Sodium hydroxide, 99
  bleaching, 176
Sodium sulfate, 99–100
Sodium sulfide–iron sulfide, 78
Softening, 108
Soil structures in house building, technical requirements for by-products used in, 242t–244t
Solar energy, indium and, 31
Solders, indium, 30–31
Solid alternative fuels, 221t
Solid waste, 441
  management of, 152–154, *see also* Informal solid waste management, *see also* Large municipal solid waste (LMSW), *see also* Municipal solid waste (MSW)
Solid-state welding, 464
Solvay process, 195
Solvent extraction, 121, 133–134
Solving the e-Waste Problem (StEP) initiative, 443
Sortable materials, 423–424
Sorting, 436, 543
  batteries, 98–99, 98f
  in construction and demolition, 392–394
  color sorting of glass, 194–195
  electronic eye sorting, 194
  fast spectrometric sorting, 424t
  induction sorting, 184
  industrial sorting system, 434f
  of LMSW, 381
  of materials, 422, 426
    from shredder, 70f, 80
  in MSW management, 412–414
  physico-chemical sorting technique, 423–424, 424t
  of plastics, 184–185
  of PWBs, 442–443
  shanks sorting lines, 381–382
  of steel material from shredder, 70f, 80
  systems for paper grades, 167–168
Sorting analysis, 411

Special alloys, 140–141
Spectrometer
  fast spectrometric sorting, 424t
  Fluoro-max Jobin Yvon laboratory spectrometer, 433
Spectroscopy, 70
Spent oil refining catalysts, 263–265
  recycling of spent catalysts, 263–265
    combined processes, 264–265
    hydrometallurgical recycling, 263–264
    pyrometallurgical recycling, 264
    reduction of waste material amount, 263
Spent petroleum catalysts, 141
  recycling processes for, 142t
Sperrylite (PtAs2), 29–30, 32–34
Sputtering deposition techniques, 30
Stabilized earth brick technology, 288
Stainless steel, 67–68, 72–75
Starting-Lighting-Ignition (SLI)-type battery, 96
State-of-the-art integrated smelters, 442–443
State-of-the-art tracer technology, 432–433, 432t
Steel
  common elements in, 74f
  decarburization, behavior of impurity and tramp elements during, 75t
  distribution of use, 67f
  -framed commercial buildings, 462–463
  grades, types of, 79t
  and packaging waste, 298–302
  plant, improved processing at, 80
  product lifespan, 67t
  production of, 66f
  recycling, 40, 65, 466–468
    hindrances, 74–76
    rates, 66t
    scrap classification, 68
    scrap impurities, 79
    scrap processing, 68–70
    scrap purification, 76–79
    smelting of steel scrap, 70–72, 71f
    sustainable, measures to secure, 79–81
  scrap, 494–495
  slag, as industrial by-product, 233t–238t
  zinc-coated, 113

Steelmaking slag, 245
Stibiopalladinite (Pd3Sb), 33–34
Stock preparation for paper recycling, 169–178
  dewatering, 174–175
  dispersing and bleaching, 175–176
  floating and washing, 173–174
  graphic paper, 170–171, 171f
  limits and new trends in, 176–178
    climate change, 177–178
    energy demand, 177
    new printing inks, 177
    selectivity, 176
    water consumption, 177
  packaging paper, 169–170, 170f
  pulping/slushing, 171–172
  screening and cleaning, 172–173
Stocks, 494–495
Stokes diameter, 40–41
Stokes equation, 43
Stratified sampling, 48
Stripping, 201–202
Strontium, 192
Student t distribution, 47
Subsidies and public facilities, 486–487
Substance flow analysis of platinum, 34f
Sulfur, 81, 107–108, 141
Sulfuric acid, 96–98, 120–121
Super alloys, 140–141
Supercritical fluids (SCFs), 273–274
SuperLight Car project, 368–370
  electrical and electronic equipment, 370–373
Supplementary cementitious materials (SCMs), 222–224
Surface structures, technical requirements for by-products used in, 242t–244t
Sustainability, 1, 547
Sustainable materials management (SMM), 527
Swiss Ordinance on the return, the taking back, and the disposal of electrical and electronic appliances (ORDEA), 399, 401f
Synthetic fibers, 211
Systematic sampling, 48

T

Take-back schemes, WEEE, 398–401
Tantalum, 27, 140–141, 143, 143f
Target groups/audience, 522–523

Taxes, 486
Tax—subsidy combinations, 516
Technical-economic environment, 430
  constraints of, 431
Technosphere, 497—498
TeckCominco, 103
Tellurium, 27, 36, 107, 125
  reclamation and recycling routes for, 126f
  recycling, 139—140
Teniente slag-cleaning furnace, 90
Terbium, 132—136
Ternary particles, 55
Textile recycling, 211—218, 473
  conversion of carpet, 215
  conversion of mattresses, 214—215
  conversion to new products, 213—214
  diamonds, 216—217
  landfill and incineration, 216
  pyramid model, 213f
  recycling effort, 212—213
  secondhand clothing export, 213
  wipers, 215—216
Thematic Strategy on the Prevention and Recycling of Waste of January 2011, 529—530
Theory/tools of recycling, 39
  grade-recovery curves, 57—60, 57f
  liberation, 55—57
  mass balances and process dynamics, 48—51
  material balancing, 51—54
  particle size, 40—44
  properties and property spaces, 45—47
  pulp rheology, 44—45
  recycling process, 40
  sampling, 47—48
Thermal reduction process, 133—134
Thermal treatment
  of steel scrap, 76—77
  in zinc oxide production, 115
Thermochemical fiber reclamation, 270—274
  chemical recycling, 273—274
  fluidized bed process (FBP), 272—273
  pyrolysis, 271—272
Thermodynamics, 336—338
  carbon tragedy, 550
  chemical thermodynamics and reaction equilibrium, 548—549

of complex products, 336—338
equilibrium thermodynamics, 548
H_2 as reducer, 550
metals, consumption and availability of, 545
modeling and simulation, 93
recycling and extractive metallurgy, 546—547
for resource efficiency maximization, 338—340
second law of thermodynamics devil, 547
solutions and desired purity levels, 551
  free energy of mixing and entropy of mixing, 551
  ideal and real Raoultian activities and raoultian solutions, 551
  stability of oxides and metalcontaining minerals, 549—550
  very stable oxides, 550—551
Thermoplastics, 469—470
Thermoplasts, 180
Thermosets, 180
Thin strips casting, 80
Timber, 293t
Tin, 29—30, 33, 40, 73, 75—79, 79t, 85, 87, 93, 96, 107—109, 120—121, 129
  removal of, 108
  from steel scrap, 76—78
Titanium, 67—68, 140—141, 143—144
Titanium-aluminide scrap (γ-TiAl), 144—145, 144f
Top blown rotary converter (TBRC), 88, 89f, 90
  plus bath smelting, 88—90
Top submerged lance (TSL) smelting, 103
Topological redundancy, 52
Top-submerged lance (TSL), 312f
Tradable Recycling Certificates (TRCs), 515
Tramp elements
  defined, 74—75
  and steel recycling, 76, 80
  behavior during steel decarburization, 75t
  influence on steel quality, 75—76
Transaction costs, 514
Translational velocity of particles, 42—44
Treatment, defined, 11—12

TRITRACE, 432
Tungsten, 67—68, 140—141
200 mesh (200#) sieve, 41
Tyler series, 41

U
Ultrasound deinking, 177
Umicore, 129, 134, 135f, 139
  PM-integrated smelter-refinery facility, 130f
  precious metals recycling loop, 131f
Unalloyed scrap, copper, 86
Unbound aggregate, glass as, 201
  characteristics, 202t
Unemployment rates, 450—451
Unilateral commitments, 503
United Kingdom (UK)
  glass recycling in, 196—198, 197f
  routes, 198f
United Nations Environment Programme, 20
Unobservable variable, 52
Urban waste deposit, 395
Urban forest, 9, 494
Urban mine, 9, 494
Usage phase, 421—422
Used product, 430
Utilization
  of recovered paper, 165—166
  in different paper products, 168, 169f
  rate, 166, 166f
  of zinc oxide, 116, 116f

V
Vacuum arc remelting, 144—145
Vacuum induction melting (VIM), 144—145
VAL'EAS™ battery scrap recycling process, 134—135, 135f
Vanadium, 140—143, 258—259, 260t
Vanadium oxide, 262
Varta process, 104
Vehicles
  electric, batteries in, 327—328
    recycling routes for, 146f
  glass in, 192—193
  material composition of, 73f, see also End-of-life (EoL) vehicles
Very stable oxides, 550—551
Virgin carbon fibers (vCFs), mechanical properties of, 273f
Virgin material tax, 512—513

Virgin polymers, 422, 436
  tracing system, 431
Virgin pulps, 165
Virgin wood, 153–154
  particleboard production from, 158–160
Viscous behavior, 537
Volume distribution, particle, 41
Voluntary systems, in packaging recycling, 304
Voluntary and negotiated agreements (VA/NA), 503–504
  recycling policy, experiences in, 504–507
  lessons learned, 507–509

W

Waelz-kilns furnace, 73–74, 116–119, 117f
  products, analysis of, 118t
  reactions in, 118f
Wall construction, 289–290
Wallboards, as industrial by-product, 233t–238t
Wash-coating catalysts, 35
Washing
  in construction and demolition, 392, 395
  in paper recycling, 173–174
  in plastic recycling, 185
Waste
  composition, 408, 411
  core definition of, 529–530
  generation of, 480
  as opportunity for development, 441
  volume of, 480, see also Large municipal solid waste (LMSW), see also Municipal solid waste (MSW)
Waste Directive Guidelines, 530
Waste electric and electronic equipment (WEEE), 86, 129, 143, 148, 347–353
  application and results of process simulation/recycling, 347–353
    dynamic product and material recycling rate predictions, 347–348
    identification of the dispersion, occurrence and appearance of possible toxic/harmful elements, 353
    impact of operating modes of technology on total recycling of minor and commodity metals, 352–353
    mass flows of produced recyclates/recycling products and distribution/dispersion of (critical) materials over different recyclate streams, 348–349
    quality and composition calculation of recyclate streams and recycling products, 349–352
  average amounts of "critical" materials in, 320t
  long-term trends, 401–402
  management objectives, 398
  metals used for EEE, 399t
  recycling for copper, 90, 397–404, 439–440
  take-back schemes, 398–401
Waste Electrical and Electronic Equipment Directive (2003), 193
Waste Framework Directive, 498–499
Waste haven effect, 481
Waste ladder, 498–499
Waste management, 3–6, 10, 12–13, 15
  by-products in, 246
  chain, 10f, 19
  external costs of, 484–485
  glass, 195
  hierarchy, 10
  plastics, 179
  and reclamation, 239t–240t
  wood, 152–153
Waste management policy
  history of, 498–499
  policy design, integrating recycling in, 499–501
  recycling in, 497–498
Waste rock, as industrial by-product, 233t–238t
Waste triangle, 414
  cost performance, 414
  environmental performance, 414
  service performance, 414
Waste-to-energy facilities, 498
Wastewater treatment, by-products in, 246
Water consumption, in paper recycling, 177
Water media separations, 539
Water-borne preservatives, 156
Water-soluble inks, 177
Wertstofftonne (value-bin), 413
Weser-Metall GmbH, 103
Western Europe, paper consumption in, 166, 167f
Wet chemical leaching processes, 442–443
Wipers, 215–216
Wood, 151
  ash, 156–157, 245
  post-use management of, 152–157
    chemical preservatives, 156
    in integrated life-cycle flows, 154, 155f
    nutrient cycling, 156–157
    wood cascading, 155–156
Working hours, 450–451
World War II, recycling in, 497
Woven fabrics impregnation, 276–277

X

Xerox copying machines, 462
X-ray fluorescence, 432–433
  intensity, 432t
X-ray technology, 185

Y

Ye'elimite, 225
Yield, improving, 457–459
Ytterbium, 133–136
Yttrium, 132, 137–140

Z

Zabbaleen community, 441
Zero-liquid effluent paper mills, 169
Zero-waste goal, 211–212
Zinc, 27–30, 32, 36, 72–74, 85, 87, 95, 107, 109
  -containing residues, 116–119, 117t
  recycling, 113
    from copper industry dusts, 119–121
    EAF dust and other residues, 116–119
    processes, 114f
    slag fuming from lead metallurgy, 121–122, 122f, 122t

zinc oxide production from
    drosses, 114–116
removal from steel scrap, 76–78
secondary, sources and type of, 114t
slag, as industrial by-product,
    233t–238t
Zinc carbonate, 115, 121

Zinc hydroxide, 115
Zinc oxide
    production and utilized materials, 115f
    production from drosses, 114–116
    quality, market and utilization areas, 116, 116f

Zinc sulfide, 121
Zinc/alkaline batteries, 325
ZINCEX process, 121
Ziolites, 205
Zirconia, yttria-stabilized, 136–137
Zirconium, 140–141, 192